Satellite Aerosol Remote Sensing over Land

Alexander A. Kokhanovsky and Gerrit de Leeuw
(Editors)

Satellite Aerosol Remote Sensing over Land

 Springer

Published in association with
Praxis Publishing
Chichester, UK

Editors
Dr Alexander A. Kokhanovsky
Institute of Environmental Physics
University of Bremen
Germany

Professor Gerrit de Leeuw
Finnish Meteorological Institute & University of Helsinki, Finland
TNO, Utrecht, The Netherlands

SPRINGER–PRAXIS BOOKS IN ENVIRONMENTAL SCIENCES
SUBJECT *ADVISORY EDITOR*: John Mason M.B.E., B.Sc., M.Sc., Ph.D.
EDITORIAL *ADVISORY BOARD MEMBER*: Dr Alexander A. Kokhanovsky, Ph.D. Institute of Environmental
Physics, University of Bremen, Bremen, Germany

ISBN 978-3-540-69396-3 Springer Berlin Heidelberg New York

Springer is part of Springer-Science + Business Media (springer.com)

Library of Congress Control Number: 2008943659

Cover design: Jim Wilkie
Project copy editor: Mike Shardlow
Author-generated LaTex, processed by EDV-Beratung, Germany

Printed in Germany on acid-free paper

Table of contents

Group photo of the participants to the workshop on the Determination of Aerosol Properties Using Satellite Measurements (June 21-22, 2007, Bremen, Germany). Most of the papers discussed in this book have been presented and discussed at the Workshop.

1 I. Georgdzhayev	18 O. Hasekamp	35 T. Mielonen
2 W. Grey	19 W. Davies	36 A. Arola
3 D. Kaskaoutis	20 F.-M. Bréon	37 G. Mannarini
4 T. Holzer-Popp	21 L. Curier	38 B. Veihelmann
5 R. Kahn	22 K. Schmidt	39 G. Thomas
6 R. Levy	23 L. Istomina	40 P. Durkee
7 S. Kinne	24 Y. Bennouna	41 A. Sayer
8 K. Schepanski	25 C. Ritter	42 W. von Hoyningen-Huene
9 A. Devasthale	26 A. Richter	43 T. Wagner
10 M. Penning de Vries	27 A. Macke	44 G. Rohen
11 H.D. Kambezidis	28 J. Wauer	45 B. de Paepe
12 P. Glantz	29 R. Braak	46 P. Reichl
13 A. Natunen	30 C. Hsu	47 A. Kokhanovsky
14 S. Wagner	31 T. Rutz	48 L. Sogacheva
15 G. de Leeuw	32 Y. Govaerts	49 T. Dinter
16 R. Dlugi	33 R. Timmermans	
17 E. Bierwirth	34 R. Treffeisen	

List of contributors

Yasmine Bennouna
TNO Defense and Security
Oude Waalsdorpweg 63
2509 JG The Hague
The Netherlands
Yasmine.bennouna@tno.nl

Hanne Breitkreuz
German Aerospace Center (DLR)
German Remote Sensing Data Center
(DFD)
Oberpfaffenhofen
D-82234 Wessling
Germany
(now at Stadtwerke München GmbH,
München)
hanne.breitkreuz@gmx.de

Brian Cairns
NASA Goddard Institute for Space
Studies
2880 Broadway
New York
NY 10025
USA
Brian.Cairns-1@nasa.gov

Elisa Carboni
Atmospheric, Oceanic and Planetary
Physics
University of Oxford
Oxford OX1 3PU
UK
elisa@atm.ox.ac.uk

Jacek Chowdhary
Department of Applied Physics
and Applied Mathematics
Columbia University
2880 Broadway
New York
NY 10025
USA
jchowdhary@giss.nasa.gov

Lyana Curier
Finnish Meteorological Institute
Climate Change Unit
Erik Palmenin Aukio 1
PO Box 503
FI-00101 Helsinki
Finland
lyana.curier@gmail.com

Gerrit de Leeuw
Finnish Meteorological Institute
Climate Change Unit
Erik Palmenin Aukio 1
PO Box 503
FI-00101 Helsinki
Finland
gerrit.leeuw@fmi.fi

Jean-Luc Deuzé
Laboratoire d'Optique Atmosphérique
Cité Scientifique
Université de Lille 1
59655 Villeneuve d'Ascq
France
deuze@loa.univ-lille.fr

David J. Diner
Jet Propulsion Laboratory
California Institute of Technology
Pasadena, CA
USA
djd@jord.jpl.nasa.gov

Yves M. Govaerts
EUMETSAT
Am Kavalleriesand 31
D-64295 Darmstadt
Germany
yves.govaerts@eumetsat.int

Roy G. Grainger
Atmospheric, Oceanic and Planetary
Physics
University of Oxford
Oxford OX1 3PU
UK
r.grainger@physics.ox.ac.uk

William M. F. Grey
Meteorological Office
FitzRoy Road
Exeter EX1 3PB
UK
william.grey@metoffice.gov.uk

Thomas Holzer-Popp
German Aerospace Center (DLR)
German Remote Sensing Data Center
(DFD)
Oberpfaffenhofen
D-82234 Wessling
Germany
thomas.holzer-popp@dlr.de

Arkadii P. Ivanov
B. I. Stepanov Institute of Physics
National Academy of Sciences of Belarus
Pr. Nezavisimosti 68
220072 Minsk
Belarus
ivanovap@dragon.bas-net.by

Ralph A. Kahn
NASA Goddard Space Flight Center
Code 613.2
Greenbelt
MD 20771
USA
ralph.kahn@nasa.gov

Iosif L. Katsev
B. I. Stepanov Institute of Physics
National Academy of Sciences
of Belarus
Pr. Nezavisimosti 68
220072 Minsk
Belarus
katsev@light.basnet.by

Stefan Kinne
Max Planck Institute for Meteorology
Bundesstrasse 53
20146 Hamburg
Germany
Stefan.kinne@zmaw.de

Lars Klüser
German Aerospace Center (DLR)
German Remote Sensing Data Center
(DFD)
Oberpfaffenhofen
D-82234 Wessling
Germany
lars.klueser@dlr.de

Kirk Knobelspiesse
Department of Applied Physics
and Applied Mathematics
Columbia University
2880 Broadway
New York
NY 10025
USA
kdk2105@columbia.edu

Alexander A. Kokhanovsky
Institute of Environmental Physics
University of Bremen
O. Hahn Allee 1
D-28334 Bremen
Germany
alexk@iup.physik.uni-bremen.de

Pekka Kolmonen
Finnish Meteorological Institute
Climate Change Unit
Erik Palmenin Aukio 1
PO Box 503
FI-00101 Helsinki
Finland
pekka.kolmonen@fmi.fi

Alessio Lattanzio
MakaluMedia GmbH
Robert-Bosch Str. 7
D-64293 Darmstadt
Germany
alessio.lattanzio@eumetsat.int

Robert C. Levy
SSAI
10210 Greenbelt Road, Suite 600
Lanham
MD 20706
USA
levy@climate.gsfc.nasa.gov

Alexei Lyapustin
NASA Goddard Space Flight Center
Mail code 614.4
Greenbelt
MD 20771
USA
Alexei.I.Lyapustin@nasa.gov

John V. Martonchik
Jet Propulsion Laboratory
California Institute of Technology
Pasadena, CA
USA
jvm@jord.jpl.nasa.gov

Dmytro Martynenko
German Aerospace Center (DLR)
German Remote Sensing Data Center
(DFD)
Oberpfaffenhofen
D-82234 Wessling
Germany
dmytro.martynenko@dlr.de

Peter R. J. North
School of the Environment and Society
Swansea University
Singleton Park
Swansea
SA2 8PP
UK
P.R.J.North@swansea.ac.uk

Caroline A. Poulsen
Space Science and Technology
Department
Rutherford Appleton Laboratory
Chilton
Didcot OX11 0QX
UK
C.A.Poulsen@rl.ac.uk

Alexander S. Prikhach
B. I. Stepanov Institute of Physics
National Academy of Sciences
of Belarus
Pr. Nezavisimosti 68
220072 Minsk
Belarus
alexp@light.basnet.by

Andrew M. Sayer
Atmospheric, Oceanic and Planetary
Physics
University of Oxford
Oxford OX1 3PU
UK
sayer@atm.ox.ac.uk

Marion Schroedter-Homscheidt
German Aerospace Center (DLR)
German Remote Sensing Data Center
(DFD)
Oberpfaffenhofen
D-82234 Wessling
Germany
marion.schroedter-homscheidt@dlr.de

Richard Siddans
Space Science and Technology
Department
Rutherford Appleton Laboratory Chilton
Didcot OX11 0QX
UK
R.Siddans@rl.ac.uk

Larisa Sogacheva
Finnish Meteorological Institute
Climate Change Unit
Erik Palmenin Aukio 1
PO Box 503
FI-00101 Helsinki
Finland
larisa.sogacheva@fmi.fi

Anu-Maija Sundström
Division of Atmospheric Sciences and
Geophysics
Department of Physics
University of Helsinki
PO Box 68
FI-00014 Helsinki
Finland
anu-maija.sundstrom@helsinki.fi

Gareth E. Thomas
Atmospheric, Oceanic and Planetary
Physics
University of Oxford
Oxford OX1 3PU
UK
gthomas@atm.ox.ac.uk

Sébastien Wagner
Wagner Consulting
Soderstrasse 67
D-64287 Darmstadt
Germany
sebastien.wagner@eumetsat.int

Fabien Waquet
Laboratoire d'Optique Atmosphérique
Cité Scientifique
Université de Lille 1
59655 Villeneuve d'Ascq
France
waquet@loa.univ-lille.fr

Yujie Wang
NASA Goddard Space Flight Center
Mail code 614.4
Greenbelt
MD 20771
USA
Yujie.Wang@nasa.gov

Philip Watts
EUMETSAT
Am Kavalleriesand 31
D-64295 Darmstadt
Germany
philip.watts@eumetsat.int

Eleonora P. Zege
B. I. Stepanov Institute of Physics
National Academy of Sciences
of Belarus
Pr. Nezavisimosti 68
220072 Minsk
Belarus
eleonor@light.basnet.by

Foreword

Human activities greatly affect the composition of the atmosphere on all scales. The emission of potentially hazardous substances both in gaseous and condensed state contributes to increasing concentrations of many trace gases and particulate matter, with resulting impacts on climate change and air quality. Effects associated with gaseous emissions (e.g., carbon dioxide, methane, nitrogen oxides and sulfur dioxide) are well recognized and studied, and a range of satellite instruments is available for the determination of the concentrations of important trace gases. However, the scientific understanding of roles of aerosols in climate change is particularly poor and large uncertainties exist. Satellites are increasingly contributing to improve this situation and their use for air quality assessment is a topic of research worldwide.

This book is the first comprehensive overview of satellite techniques for the determination of aerosol optical thickness. The processes related to effects of light absorption and scattering by polydisperse aerosol media are smooth functions of wavelength. Therefore, the differential absorption methods, which often work well with gases, cannot be used and one needs to rely on the absolute values of the registered signals. These are influenced by different contaminating factors including calibration accuracy, presence of clouds and reflectance of the underlying surface. It is an extremely difficult task to infer the part of the signal related to aerosols from the total intensity detected by a satellite instrument, which is composed of contributions from both the atmosphere and underlying surface. The difficulties of retrieval are also due to the fact that, for the characterization of aerosols, it is not enough to specify their concentration (as it is for gases) – what is needed is information on particle size, along with concentration, chemical composition and particle shape as function of size. Often, not enough information is available to retrieve all of these properties. For the best retrieval results, comprehensive measurements of reflected light including its angular spectrum, polarization and spectral characteristics are required.

This book summarizes current knowledge on aerosol optical thickness retrieval from space, based on passive measurements of the reflected solar radiation. The book is written by experts in the field facing problems of retrievals of aerosol properties from space. Therefore it gives a flavor of current research on the frontier of satellite remote sensing. It shows advances and shortcomings of different techniques developed worldwide.

Mainz   Paul Crutzen
September, 2008

1 Introduction

Gerrit de Leeuw, Alexander A. Kokhanovsky

Atmospheric aerosol is a suspension of liquid and solid particles, with radii varying from a few nm to larger than 100 μm, in air. The particles can be directly emitted into the atmosphere (e.g., sea spray aerosol, dust, biomass or fossil fuel burning aerosol, volcanic ash, primary organic aerosol) or produced from precursor gases (e.g., sulfates, nitrates, ammonium salts, secondary organic aerosol). The total aerosol mass is dominated by aerosols produced from the surface due to natural processes such as the action of the wind (sea spray aerosol, desert dust). However, anthropogenic emissions of both primary particles and precursor gases contribute significantly to the total aerosol load [Andreae and Rosenfeld, 2008].

Aerosols have a large effect on air quality, in particular in densely populated areas where high concentrations of fine particulate matter (PM) are associated with premature death and decrease of life expectancy. Concentrations of PM are subject to regulation which in turn has economic effects. Aerosols also affect climate because the particles reflect solar radiation which cools the atmosphere, whereas absorbing particles (such as black carbon) warm the atmosphere; however, the net effect of aerosols is cooling. This is the direct radiative effect. Indirect effects occur due to the aerosol particles acting as cloud condensation nuclei and thus affecting cloud formation and the microphysical properties of clouds (cf. Andreae and Rosenfeld [2008]). Overall, aerosols partly offset warming due to greenhouse gases but the magnitude and scientific understanding of the aerosol effect on climate is low [IPCC, 2007]. Therefore, studies of aerosol properties are relevant to climate change, including investigations of anthropogenic influences on climate (forcing), and air quality.

Satellites are increasingly used to obtain information on aerosol properties (e.g., the aerosol optical depth (AOD), the columnar concentration of particles, their sizes), taking advantage of technical and scientific developments over the last decades. The purpose of this book is to provide an overview of current methods of retrieving aerosol properties from satellite observations. The application of satellite remote sensing for the determination of aerosol properties started some 30 years ago. A brief description of the history of aerosol observations from space is given by Lee et al. [2009]. One of the first retrievals of aerosol optical depth from spaceborne measurements of the spectral intensity of the reflected solar light was performed using observations from the Multi Spectral Scanner (MSS) onboard the Earth Resources Technology Satellite (ERTS-1) [Griggs, 1975; Mekler et al., 1977] and the first operational aerosol products were generated using data from the radiometer on board the TIROS-N satellite launched on 19 October 1978. The Advanced Very High Resolution Radiometer (AVHRR) [Stowe et al., 2002] onboard TIROS-N was originally intended for weather observations but its capability was expanded to the detection of aerosols. The Nimbus-7 was launched on 25 October 1978, carrying the Stratospheric Aerosol Measurement instrument (SAM) [McCormick et al., 1979] and the Total Ozone Mapping Spectrometer (TOMS) [Torres et al., 2002a]. Initially, retrievals were obtained only for measurements over water; aerosol retrieval results over land have started to become available on a regular basis only in the last decade.

The TOMS series was extended with the Ozone Monitoring Instrument (OMI) launched in 2004. The primary aerosol data delivered was Aerosol Index, a measure for absorbing aerosol, until an algorithm was developed to retrieve AOD as well [Torres et al., 1998, 2002a]. AVHRR flies on NOAA (National Oceanic and Atmospheric Administration) satellites and has provided continuous data since its first launch in 1978. The primary information delivered by AVHRR is the surface temperature, but the instrument has also provided AOD over the ocean [Geogdzhyev et al., 2002] and, more recently, over land [Hauser et al., 2005]. These long time series have been analyzed to show temporal trends in AOD [Mishchenko et al., 2007a]. The obtained results have shown the value of long-term global observations using the same instrument and the same data processing procedures.

Aerosol retrieval over ocean could be accomplished relatively easily due to the low surface reflectance in the near-infrared channels. In that case, the signal registered on a satellite is largely determined by light scattered in the atmosphere and the contribution of the surface (outside the glitter) is comparatively low. Over land the retrieval of aerosol properties is more complicated due to the relatively strong contribution of the land surface reflectance to the radiation measured at the top of the atmosphere. This contribution depends on land surface properties, varying from dark surfaces, such as forests in the visible, to very bright surfaces such as deserts, snow and ice fields. A further complication is the inhomogeneous distribution of surface types with a variety of spectral bidirectional reflectance distribution functions (BRDF) in a satellite ground scene. Over land, there are more intense, small-scale, and diverse sources than over ocean, contributing to the inhomogeneity of the scene. It becomes even more complicated when the horizontal photon fluxes from nearby pixels containing clouds (Nikolaeva et al., 2005; Marshak et al., 2008; Wen et al., 2008) or bright surfaces (so-called adjacency effects) need to be accounted for. This, together with uncertainty related to the choice of the aerosol model used in the retrieval process, leads to the severe problems of spectral AOD retrievals over land, as described, for example, by Liu and Mishchenko [2008].

Initially, satellite instruments were developed for other purposes than the retrieval of aerosol properties. Notably, they were designed for the retrieval of trace gas concentrations (e.g., TOMS, GOME (Global Ozone Monitoring Experiment), SCIAMACHY (SCanning Imaging Absorption SpectroMeter for Atmospheric CHartographY), OMI) or land/sea surface temperature and reflectance (e.g., AVHRR, SeaWiFS (Sea-viewing Wide Field-of-view Sensor), MERIS (MEdium Resolution Imaging Spectrometer), (A)ATSR ((Advanced) Along Track Scanning Radiometer). However, the instrument features (see Appendix) allowed for the retrieval of aerosol properties as well, albeit with variable success. Ideally, a sensor should have multiple wavelengths, from the ultraviolet (UV) to the thermal infrared (TIR), and multiple views, and should be sensitive to the polarization of reflected radiation. The combination of spectral polarization and multiple view measurements for a range of wavelengths is only available from the POLDER (POLarization and Directionality of the Earth Reflectance) [Deschamps et al., 1994] series of instruments, the latest of which is flying on PARASOL (Polarization and Anisotropy of Reflectances for Atmospheric Sciences coupled with Observations from a Lidar; see, for example, "http://smsc.cnes.fr/PARASOL" http://smsc.cnes.fr/PARASOL) as part of the A-Train. The GLORY mission [Mishchenko et al., 2007b] to be launched in 2009 will carry the Aerosol Polarimetry Sensor (APS). The APS will collect accurate multi-angle photo-polarimetric measurements of the Earth along the satellite ground track within a wide spectral range extending from the visible to the short-wave infrared. The data from this instrument are expected to provide aerosol retrievals with a higher accuracy then available from current instruments. However, the spatial resolution of APS is significantly lower than other instruments such as MISR and MODIS, so the scene heterogeneity may be a limiting factor for APS aerosol retrieval accuracy.

Results from the last decade show that it is possible to obtain a useful set of aerosol parameters even without using advanced multi-view instruments capable of detecting the polarization state of the reflected solar light. These include the AOD at various wavelengths and its wavelength dependence expressed as the Ångström coefficient. Principal component analysis shows what other aerosol parameters could be retrieved using a dedicated aerosol instrument. These could include, for example, for a bimodal lognormal size distribution, the effective radius and effective variance, and the refractive index (both real and imaginary parts) for both modes (Mishchenko et al., 2007b). Sensitivity studies and investigations of information content of satellite measurements with respect to the determination of aerosol properties were performed by, among others, Hasekamp and Landgraf [2005] and Veihelmann et al. [2007].

One of the first reliable retrievals of aerosol optical depth over land was made using the dual view of the Along-Track Scanning Radiometer (ATSR-2) [Veefkind et al., 1998], followed by retrievals using POLDER [Deuzé et al., 2001], MODIS (MODerate Resolution Imaging Spectroradiometer [Kaufman et al., 1997]) and MISR (Multiangle Imaging SpectroRadiometer [Martonchik et al., 1998]), i.e. instruments which were designed for aerosol retrievals. Also SeaWiFS and MERIS have been used for this purpose [von Hoyningen-Huene et al., 2003, 2006]. Retrieval techniques have improved to the extent that satellites are increasingly used for continuous observations of the aerosol distribution and composition on regional to global scales, complementary to ground-based observation networks and dedicated field campaigns, to assess the effects of aerosols on climate [Kaufman et al., 2002; IPCC, 2007] and, in particular, for calculating the aerosol radiative forcing [Charlson and Heintzenberg, 1995; Remer and Kaufman, 2006].

Satellite observations can also be used for air quality studies, in particular for the determination of the regional distribution of fine particulate matter (PM1, PM2.5, PM10), the concentrations of which are subject to regulation for health purposes. PM2.5 and PM10 can be directly derived from satellite observations of the AOD using empirical relations (e.g., Schaap et al. [2009] and references cited therein), through the assimilation of satellite-derived AOD values in chemical transport models (e.g., Collins et al. [2001]), and also directly from the measured spectral AODs using Mie theory [Kokhanovsky et al., 2006; von Hoyningen-Huene et al., 2008].

Land surface reflectance effects

The surface reflectance is particularly important for single-view instruments or algorithms using only one view from a multiple-view instrument. The most common assumption used with these instruments is the restriction to retrieval over dark surfaces. For wavelengths in the UV (and near-UV) this assumption works well over a wide variety of surfaces and forms the basis of, for example, the deep blue algorithm applied for MODIS [Hsu et al., 2004, 2006], the aerosol retrieval algorithms developed for SeaWiFS [von Hoyningen-Huene et al., 2003; Kokhanovsky et al., 2004], for MERIS [von Hoyningen-Huene et al., 2006; Vidot et al., 2008], and for the OMI multi-wavelengths algorithm [Torres et al., 2002b; Veihelmann et al., 2007; Curier et al., 2008].

However, for longer wavelengths the dark surface assumption generally does not apply for most land surfaces and other methods need to be used to account for the surface reflectance. Over land, the MODIS aerosol retrieval algorithm assumes a relationship (empirically derived) between the surface reflectances at wavelengths in the short wave infrared at 2.1 μm and in the visible channels (0.47 and 0.66 μm) [Kaufman et al., 1997; Levy et al., 2007a, 2007b]. In previous versions of the algorithm, the relationship was described with static ratios with aerosol assumed transparent at 2.1 μm. In dust conditions, aerosol transparency at 2.1 μm is not true. Therefore, in the Collection 5 version (described in Chapter 2), the observed (total) spectral reflectance is considered to be composed of surface and atmospheric contributions, such that the retrieval of the aerosols' spectral contribution is constrained by the surface reflectance relationships. Through the use of reflectance (directional) rather than albedo, the technique implicitly includes BRDF effects in the retrieval process. A major weakness of the algorithm, however, is that although the assumed surface reflectance relationships may be characteristic of some sort of averaged vegetated surfaces, deviations and variability of different surfaces can lead to systematic biases over certain regions.

With an accurate retrieval of aerosol optical depth (and aerosol type) over land, reported values can be used for the atmospheric correction needed to retrieve surface properties, such as the surface reflectance. Since successful aerosol retrieval relies on accurate characterization of the surface optical properties, surface and aerosol retrievals comprise an interdependent problem. Therefore, surface reflectance databases may be generated in conjunction with the construction of AOD databases (e.g., Grey et al. [2006]). In the past, when simultaneous AOD and surface reflectance retrievals were not possible, such databases were necessary. For example, one could begin by identifying cases where the measured top of atmosphere (TOA) reflectance was characteristic of the clearest and cleanest cases during a particular time period (say, a month [Koelemeijer et al., 2003]). By applying a radiative transfer model (RTM), and assuming some sort of background aerosol conditions, one could determine the appropriate surface reflectance to be used as a boundary condition for aerosol retrieval during that time period. Therefore, given accurate representation of surface reflectance, one could apply such a surface reflectance database and provide accurate aerosol retrievals even over very bright surfaces such as the Sahara desert [Kusmierczyk-Michulec and de Leeuw, 2005].

Only few instruments have more than one view over a particular pixel: (A)ATSR, POLDER/PARASOL and MISR. Hence these instruments provide more constraints on the path radiance and surface effects. Methods of deriving both surface and aerosol quantities were presented by Veefkind et al. [1998] for ATSR-2, by Deuzé et al. [2001] for POLDER and by Diner et al. [1998, 2005] for MISR. In addition to better decoupling of the surface effects, the multi-view instruments can also provide information about the aerosol phase function [Martonchik et al., 1998], and the height of aerosol plumes [Kahn et al., 2007, 2008].

Active sensors [Winker et al., 2003, 2007] and multi-angle imaging produce complementary information on the aerosol height. Multi-angle imaging provides maps of plume height and aerosol injection height near sources such as wildfires, volcanoes, and dust storms, where the aerosol is thick enough for features to be observed at multiple angles. Lidar has the sensitivity to see sub-visible layering globally, but the spatial coverage of CALIPSO (Cloud-Aerosol Lidar and Infrared Pathfinder Satellite Observation; http://www-calipso.larc.nasa.gov), for example, is 0.2% of the planet, once every 16 days, providing excellent coverage downwind of sources, but rarely if ever capturing actual source plumes during injection (Kahn et al., 2008).

Aerosol retrieval methods

Satellite observations are made using instruments in a certain orbit, with a certain view angle and for a solar zenith angle which varies with the season and the time of day. This geometry needs to be taken into account to properly retrieve atmospheric and surface properties. Aerosol particles scatter light in different directions with an angular distribution that depends on particle size, shape, and chemical composition as described by the scattering phase matrix.

The first step in a retrieval algorithm is cloud screening. Clouds have a very high reflectance that overwhelms the aerosol signal which renders cloud-contaminated pixels not suitable for aerosol retrieval. Several criteria may be applied for cloud detection [Ackerman et al., 1998]. Cloud detection is not the focus of this book although various chapters describe the cloud screening methods used for the algorithm discussed and the instrument characteristics available. These methods are based on a variety of principles including the analysis of spatial and temporal patterns, thresholds for the various channels from the UV to the thermal infrared, and also the synergetic use of other instruments including lidars.

The next problem, for clear sky measurements, is to account for the land surface contribution to the TOA reflectance. When this has been accomplished, the path radiance, which includes contributions from molecular scattering and absorption, remains. To properly account for molecular effects for the given sun-satellite geometry, a radiative transfer model is applied. Usually this is applied to a set of geometries to provide look-up tables (LUTs) which are used in the retrieval step to speed up the processing. The RTM is also referred to as forward modeling, the retrieval is referred to as inverse modeling.

The LUTs are usually prepared using vector radiative transfer calculations for a set of aerosol models which are representative for a certain area [Kaufman et al., 2001; Dubovik et al., 2002; Levy et al., 2007a, 2007b]. Ideally, the algorithm must have a way to select the most appropriate aerosol model or mixture of aerosol models. In many cases, however, aerosol type selection is based on a climatology (e.g., Levy et al. [2007a], Curier et al. [2008]). Such climatologies can be derived from observations (e.g., Dubovik et al. [2002]; Levy et al. [2007a]) or from results obtained using transport models for the area of interest [Curier et al., 2008]. In the future, it may be possible to use transport model forecasts [Verver et al., 2002] to constrain the retrieval.

A comparison of retrieval results over land was presented in Kokhanovsky et al. [2007, 2009] using results from most of the instruments and methods discussed here. The retrieval algorithms were applied to a single scene over Germany. This inter-comparison shows that the scene-averaged AOD as retrieved from different instruments is quite similar but significant differences occur as regards the spatial distribution. These observations were the basis for the discussion of the various algorithms. The next step will be further improvement and comparisons over other scenes which should reveal differences between algorithms (see, for example, Liu and Mishchenko [2008]).

Radiative transfer

Modern satellite retrievals of aerosol properties are based on the LUTs for several radiative transfer characteristics of cloud-free atmosphere including the atmospheric reflectance, R_a, the total transmittance from the TOA to the surface, T_1, (and from the surface to the satellite, T_2) and the spherical albedo of the atmosphere for illumination from below, r. LUTs are calculated using radiative transfer codes (see, for example, Rozanov and Kokhanovsky [2006]) for different illumination/observation geometries and aerosol phase functions, $p^a(\theta)$ (or matrices), dependent on the scattering angle θ, single scattering albedos, ω_0, and AODs, τ.

The TOA reflectance $R = \pi I/\mu_0 E_0$ (E_0 is the solar irradiance on the area perpendicular to the solar beam at the TOA, I is the intensity of reflected light, μ_0 is the cosine of observation angle) for an underlying Lambertian surface with albedo A at wavelength λ is presented as:

$$R = R_a + \frac{AT_1T_2}{1 - Ar}. \tag{1}$$

All parameters in Eq. (1) (except A) depend on the aerosol optical depth (or thickness) defined as

$$\tau(\lambda) = \int_0^H k_{ext}^a(z, \lambda) \, dz, \tag{2}$$

where H is the TOA height, $k_{ext}^a(z, \lambda)$ is the aerosol extinction coefficient at height z above the ground, for wavelength λ. The main task of aerosol retrievals is to determine the spectral dependence of τ from satellite measurements of the spectral reflectance R. The value of A (see Eq. (1)) must be either retrieved during the inversion process or eliminated (e.g., using multi-view observations of the same ground scene from different directions) in the retrieval process. Sometimes spectral surface reflectance databases (dependent on a given location) are used in the retrieval process. The applied LUTs for R_a, r, T_1, T_2 (dependent on $\tau, p^a(\theta), \omega_0$) differ mainly in the applied aerosol models which are either postulated or selected depending on the region of interest. Some of the retrievals incorporate the search for the best fit LUTs in the framework of the inversion routines (see, for example, Martonchik et al. [1998] and de Almeido Castanho et al. [2008]). The most accurate atmospheric aerosol model can be derived only if multi-view spectropolarimetry is used. This requires measurements of the angular and spectral distributions of the Stokes vector of reflected light [Mischenko et al., 2007b].

The value of r is smaller than 0.1 for most atmospheric situations and, therefore, for dark surfaces ($A \rightarrow 0$) one can neglect the product Ar in the dominator of Eq. (1), which leads to:

$$R = R_a + AT_1T_2. \tag{3}$$

Instead of albedo A one can use the surface BRDF for non-Lambertian surfaces. Simple parameterizations for transmittances T_1, T_2 and also the spherical albedo r, were derived by Kokhanovsky et al. [2005].

The value of R_a can be calculated using the RTM but for optically thin layers (e.g., $\tau \leq 0.01$), the following analytical result can be used [see, e.g., Kokhanovsky, 2008]:

$$R_a = \frac{\omega_0 p(\theta)}{4(\mu + \mu_0)}(1 - \exp(-M\tau)), \tag{4}$$

where $\mu = \cos\vartheta$, $\mu_0 = \cos\vartheta_0$, $M = \mu^{-1} + \mu_0^{-1}$, $\theta = \arccos(-\mu\mu_0 + ss_0\sin\varphi)$, $s = \sin\vartheta$, $s_0 = \sin\vartheta_0$, φ is the relative azimuth angle between the incident light and the direction of observation, ϑ_0 is the zenith incidence angle, ϑ is the zenith observation angle counted from the normal to the scattering layer. The optical thickness τ includes effects of both the aerosol and molecular scattering and absorption. The phase function $p(\theta)$ also contains contributions from scattering by both molecules and particles. The phase function can be calculated as follows:

$$p(\theta) = \frac{k_{sca}^a p_a(\theta) + k_{sca}^m p_m(\theta)}{k_{sca}^a + k_{sca}^m}. \tag{5}$$

Here k_{sca} is the scattering coefficient and symbols 'a' and 'm' denote scattering of light by aerosol particles and molecules, respectively. At wavelengths in the near-infrared spectral region (NIR), the ratio k_{sca}^m/k_{sca}^a is a small number and $p(\theta) \approx p^a(\theta)$.

Eq. (4) makes analytical determination of AOD possible:

$$\tau = \frac{1}{M}\ln\left[1 - \frac{R_a}{B}\right]^{-1}, \tag{6}$$

where

$$B = \frac{\omega_0 p(\theta)}{4(\mu + \mu_0)}. \tag{7}$$

For $\tau \rightarrow 0$ this results in

$$\tau = \frac{4\mu_0\mu R_a}{\omega_0 p(\theta)}. \tag{8}$$

Eq. (8) shows the importance of information on the phase function and single scattering albedo for aerosol retrievals.

Book outline

This book describes aerosol retrieval over land from the perspective of the instruments that are most commonly used for this purpose. It is based on a workshop conducted in Bremen (Germany) on June 21–22, 2007. The aim of the workshop was to bring together experts in the area of aerosol satellite remote sensing over land and discuss the results from current instruments.

In the workshop several retrieval techniques were discussed including:

- multispectral observations at a single viewing angle;
- multi-view spectral observations of the same target;
- multi-view spectral measurements of intensity and polarization of reflected light.

The state of the art on aerosol retrieval over land is discussed, based on a series of articles solicited to represent the instruments that are currently used for aerosol retrieval over land (MODIS, MISR, POLDER, AATSR, MERIS) in a sun-synchronous orbit and SEVIRI (in a geostationary orbit). For AATSR several retrieval algorithms have been developed which are based on different principles. The synergistic use of different instruments flying on the same platform, or on different platforms in a constellation such as the A-Train, is being explored. The synergy between AATSR and SCIAMACHY is used in the SYNAER (SYNergetic AERosol retrieval) algorithm. Alternative methods for aerosol retrieval are presented. In a final chapter the results are evaluated. The book is restricted to passive instruments, i.e. there is no chapter on lidar (light detection and ranging) which provides information on the vertical structure of aerosols and has recently become operational from space with CALIOP (Cloud-Aerosol LIdar with Orthogonal Polarization) flying on CALIPSO (Cloud-Aerosol Lidar and Infrared Pathfinder Satellite Observation) as part of the A-Train [Winker ct al., 2003, 2007]. Another restriction is that no results are presented from instruments with a relatively low spatial resolution such as GOME (http://www.temis.nl/intro.html) or OMI (http://disc.gsfc.nasa.gov/data/datapool/OMI/index.html), which nevertheless have contributed useful aerosol products. Yet another advanced satellite instrument not discussed in this book is the Global Imager (Nakajima et al., 1998, 1999 (see, e.g., http://suzaku.eorc.jaxa.jp/GLI/doc/index.html)).

It is noted that some algorithms are applied over an average of a larger number of pixels, others are applied to single pixels. The latter provide the highest possible spatial resolution, making possible the detection of small-scale features such as aerosol emission sources, at the expense of processing time and accuracy of the results.

In Chapter 2, Levy presents a comprehensive overview of the most recent algorithm developed for MODIS Collection 5 aerosol retrieval. This chapter starts out with a description of aerosol properties, in particular those of importance for retrieval: size distributions and optical properties. Also a general introduction is provided on aerosol ground-based measurements which are needed for validation and evaluation of retrieval products. After a general introduction to satellite remote sensing, the MODIS instrument and an overview of MODIS dark underlying surface algorithms are provided. The core of the chapter is the description of the algorithm applied over land, based on Levy et al. [2007a,b], including the aerosol climatology used, the treatment of the surface reflection (based on relations between the reflectances at 2.12 μm and 0.47 μm and those at 2.12 μm and 0.66 μm) and finally the inversion. Level 2 and level 3 satellite products are described and examples are presented, including their evaluation.

The dark field method discussed by Levy is only one of the many algorithms used for retrieving over-land aerosol properties using MODIS data. In particular we note that there is an entirely independent algorithm [Hsu et al., 2004, 2006] for deriving aerosol properties from MODIS data, using near-UV channels. This algorithm is optimized for visually bright surfaces (e.g. deserts), where the surface reflectance is very low in the near-UV and can be accounted for.

In Chapter 3, Lyapustin and Wang present an alternative method using a time series of up to 16 days in the MAIAC algorithm (Multi-Angle Implementation of Atmospheric Correction), which simultaneously derives the AOD and the surface BRF. During these 16 days, different view angles are provided which are used in the BRF retrieval. Assuming that the BRF changes little during this period and that the AOD is relatively invariant over short distances (25 km, see also Chapter 8 by Holzer-Popp et al. for a discussion of this subject), the AOD and BRF can be retrieved. The shape of the BRF is assumed similar for the 2.1 μm and blue MODIS band. The method is presented, including the radiative transfer model used, the aerosol and atmospheric correction algorithms and the cloud mask. Examples are given, including validation.

MERIS is an instrument designed to measure ocean color which requires very good radiometric performance. Other MERIS products are land surface properties and atmospheric properties such as information on clouds and aerosols. The official ESA (European Space Agency) aerosol product uses the algorithm developed by Santer et al. [1999, 2000]. The results have been evaluated by, for example, Höller et al. [2007]. Alternatively, non-operational scientific algorithms are available as well. In Chapter 4, Katsev et al. present the ART (Aerosol Retrieval Technique) for spectral AOD retrieval that uses radiative transfer computations in the retrieval process rather than pre-calculated LUTs that are used in most other algorithms. The core of the ART code is the RAY radiative transfer code [Tynes et al., 2001]. The aerosol retrieval technique is described, including the vector radiative transfer code, the atmospheric model, spectral models of underlying surfaces and the iteration process for the retrieval of spectral AOD. ART was applied to a scene over Europe using MERIS data and compared with retrievals over the same area using retrievals from other satellites (MODIS, MISR, AATSR) [Kokhanovsky et al., 2007]. The importance of the choice of the aerosol model is discussed in relation to the retrieval of the surface albedo. The ART uses several concepts of the Bremen AErosol Retrieval (BAER) algorithm introduced by von Hoyningen-Huene et al. [2003]. BAER was developed for single view instruments such as MERIS and SeaWiFS and makes possible the determination of both the surface albedo and AOD. One of the main features of BAER is the representation of the albedo of a given inhomogeneous surface as the weighted sum of the bare soil and the vegetated spectrum with the corresponding weight found from the analysis of the spectral top-of-atmosphere reflectance. The value of the NDVI (Normalized Differenced Vegetation Index) is used as a first guess for the weighting parameter in the iteration procedure.

The AATSR (see Chapter 6 for a description) is the third in a series of instruments (after ATSR and ATSR-2) with the main objective of obtaining information on the sea-surface temperature with high accuracy. However, the dual view (nadir and 55 ° forward) provided by the instrument renders it very useful in also obtaining information on aerosol properties over both land and ocean surfaces [Veefkind and de Leeuw, 1998; Veefkind et al., 1998]. The work of Veefkind was further pursued by several PhD students at TNO (Netherlands Organization for Applied Scientific Research). In 2007 the algorithm was transferred to

University of Helsinki and Finnish Meteorological Institute (Helsinki, Finland) for further development and both scientific and operational application as described in Chapter 5 by Curier et al. A detailed description of the cloud screening applied to AATSR is provided as well as the aerosol retrieval over land which relies on the dual view to eliminate land surface effects on the TOA radiation based on the assumption that the shape of the BRDF is similar for wavelengths in the vis-NIR part of the spectrum. The inversion is based on LUTs produced using the DAK (Double Adding KNMI) radiative transfer code developed by KNMI (Royal Netherlands Meteorological Institute) [de Haan et al., 1987]. The aerosol is modeled in terms of a multi-modal (usually only two modes are used) lognormal size distribution. Results are presented for a variety of different conditions and validated versus AERONET (AErosol RObotic NETwork; Holben et al., 1998) or MODIS data.

A different approach to retrieving information from AATSR dual view measurements is described in Chapter 6 by Grey and North. For the treatment of the surface the assumption on the BRDF wavelength dependence is replaced by a physical model for the surface reflection that accounts for the spectral change and the viewing angle [North et al., 1999]. This model results from consideration of effects of direct and diffuse atmospheric scattering. The forward model used for the calculation of LUTs is 6S [Kotchenova et al., 2008]. Chapter 6 starts with an overview of remote sensing of aerosols in the context of multi-view angle passive instruments and associated problems, followed by a description and practical implementation of their algorithm. Results are presented and validated versus AERONET data.

In Chapter 7, Thomas et al. describe the Oxford-RAL Aerosol and Cloud retrieval algorithm (ORAC) and its application to AATSR and SEVIRI. ORAC is an optimal estimation scheme that uses a bi-directional reflectance distribution function to describe the surface reflectance which allows for multiple views to be used in the retrieval. This is a significant improvement to the earlier version of ORAC which used a Lambertian approximation to describe the surface. Both algorithms are described in detail, as well as a third one that makes use of the thermal infrared channels to improve the detection of lofted dust above desert surfaces. The ORAC forward model uses the DIScrete Ordinates Radiative Transfer (DISORT) code [Stamnes et al., 1988] to predict atmospheric transmission and top-of-atmosphere reflectance based on aerosol phase functions, AOD, single scattering albedo, Rayleigh scattering and molecular absorption. In addition to aerosol optical depth and effective radius, the spherical albedo of the surface is also retrieved, using the MODIS land surface bi-directional reflectance product [Jin et al., 2003a,b] to provide a priori information. Examples of aerosol retrievals are presented for both single and dual view retrievals and evaluated versus AERONET data.

The retrieval of aerosol information from satellite sensors is an ill-posed problem. To obtain more information, the synergy between different sensors is exploited in the SYNAER algorithm described in Chapter 8 by Holzer Popp et al. SYNAER uses two sensors: AATSR and SCIAMACHY which both fly on ENVISAT. These instruments are briefly described and an analysis of the information content is presented. Here is assumed that the surface reflectance and the AOD are known from AATSR measurements and the extra information from using SCIAMACHY is analyzed. SYNAER consists of two steps. First the AOD is retrieved for AATSR dark pixels and, assuming spatial homogeneity, the obtained values are inter/extrapolated over brighter pixels in a scene to retrieve the surface reflectance of these pixels. These values in turn are used to determine a (weighted) average surface reflectance over the much larger (60×30 km^2) SCIAMACHY pixels. In the second step, based on knowing AOD and surface albedo and selected surface type (from NDVI, CVI (Calibrated Vegetation Index), brightness) from the first retrieval step, a choice between simulated aerosol spectra for 40 mixtures is made based on comparison with SCIAMACHY spectra. Dust over ocean is often misclassified as clouds and a specific scheme is implemented to account for such situations. Aerosol models used in SYNAER are discussed. Dark scene selection over ocean and over land and the selection of the most plausible aerosol type are discussed as well as validation and applications. Examples are presented.

Multi-view observations are available from instruments such as CHRIS-PROBA (Compact High Resolution Imaging Spectrometer – Project for On Board Autonomy), POLDER and MISR. In Chapter 9, Martonchik et al. present the retrieval of aerosol properties over land using MISR observations. MISR uses nine cameras each fixed at a particular view zenith angle in the along-track direction and having four spectral bands. Using the combination of viewing angles and spectral information, the retrieval of aerosol is based on several assumptions. The current MISR aerosol algorithm uses eight particle (six spherical and two non-spherical) types and 74 mixtures. Optical properties of spherical particle types are calculated using Mie theory, for non-spherical (dust) particles the discrete dipole approximation and the T-matrix technique are used. LUTs are created for use in the retrieval algorithm which is based on principal component analysis of the multispectral multi-angle MISR data. The retrieval over land is based on the assumption of spectral invariance of the ratio of the surface reflectance at two angles [Flowerdew and Haigh, 1995], similar to the method applied to AATSR [Veefkind et al., 1998] as described in Chapter 5. The MISR algorithm is extensively described in Chapter 9, including the applied cloud screening method. Results are presented, including a seasonal climatology for eight years (2000–2007) and comparisons with AERONET for different aerosol types, air quality and plume studies.

The use of polarimetric remote sensing is addressed in Chapter 10, where Cairns et al. describe polarization of light and review different instrumental approaches to Earth viewing measurements of polarization. Conclusions on the polarized reflectance of land surfaces are summarized and modeling for the land surface-atmosphere is discussed. Existing retrieval methods are discussed with a view on polarimetric remote sensing applied to POLDER and RSP (Research Scanning Polarimeter, an airborne instrument) [Cairns et al., 1999]. The RSP approach, an optimal estimation method, is planned to be used with the APS. The use of additional information contained in the measurements of the Stokes vector of the reflected light enables better determination of the aerosol model needed for the retrieval of AOD. However, the real crux of the method is in the fact that the polarized reflectance of underlying surfaces is usually grey. This implies that one can easily derive the polarized surface reflectance in the visible from measurements in the infrared, where the atmospheric contribution is low.

Aside from SEVIRI, all sensors discussed in Chapters 2 to 10 fly on polar orbiting satellites in a so-called low elevation orbit (LEO). This implies that they circle the Earth and observe a certain area depending on the swath width. When the swath is large enough, the whole Earth is observed daily (global coverage). Polar regions are observed several times per day and measurements at the equator are performed just once per day or even with larger temporal gaps depending on the swath of the instrument (see Appendix). For a higher observation rate, which is desirable for air-quality applications and monitoring of transport of pollution, other orbits are possible but obviously this will result in a smaller observation area. The highest temporal data rate is obtained using satellites in a geostationary orbit, i.e. at a fixed location above the equator (GEO). Examples of GEO instruments that are used for aerosol retrieval are GOES-ABI (Geostationary Operational Environmental Satellite - Advanced Baseline Imager) [Knapp, 2002] and SEVIRI (Spinning Enhanced Visible Infrared Radiometer) on MSG (Meteosat Second Generation). In Chapter 11 Govaerts et al. present an approach to retrieving AOD and surface BRDF using SEVIRI observations. SEVIRI is centered on the equator and Greenwich meridian and scans the Earth disk every 15 min in 11 spectral bands ranging from 0.6 to 13 μm with a sampling distance of the sub-satellite point of 3 km. Govaerts et al. apply the optimal estimation retrieval method in their Land Daily Aerosol (LDA) algorithm. The BRDF is delivered in three SEVIRI bands (0.635 μm (channel 1), 0.810 μm (channel 2), and 1.64 μm (channel 3)), and the AOD is delivered at 0.55 μm, determined using the wavelength dependence of aerosol models selected based on the measurements in channels 1 and 2. The forward model provides simulations for surface and atmospheric conditions for a range of geometries and observation conditions. Daily accumulated observations are used under different illumination geometries forming a virtual multiangular and multispectral database which is compared with forward model results. The optimal solution is obtained by minimizing the so-called cost function. Pre-defined aerosol classes are assumed to change little from one pixel to another but vary in time. Error and autocorrelation matrices are used as information sources. Error sources are discussed and results are compared with AERONET-retrieved AOD.

In the final Chapter 12, Kinne wraps up the current status of aerosol retrieval over both land and sea. Time averages of global multi-year satellite AOD data from MODIS (Collections 4 and 5) (2000–2005), MISR (2000–2005), TOMS (1979–2001), POLDER (1987, 2002), AVHRR NOAA (1981–1990) and AVHRR GACP (Global Aerosol Climatology Project) (1984–2001) are compared with a multi-year reference dataset provided by AERONET observations processed in a similar way. Kinne develops a scoring concept to objectively evaluate the various datasets considering bias as well as regional and seasonal differences. No single satellite dataset is better than any of the others, which all have merits for certain conditions. Kinne uses the satellite data to create global AOD maps and enhances this by adding AERONET data. Comparisons are presented with AOD results from global modeling exercises provided by AEROCOM (Aerosol Comparisons between Observations and Models, http://nansen.ipsl.jussieu.fr/AEROCOM/weboslo05.htm).

With these chapters, the book aims to provide an overview of the current state of the art of satellite aerosol retrieval over land. The book shows the significant progress in the development of aerosol remote sensing techniques since the review by King et al. [1999].

This book is dedicated to the memory of our friend and colleague Y. Kaufman (01.06.1948–31.05.2006), who made outstanding contributions in the field of aerosol remote sensing from space and ground.

Appendix:
Characteristics of optical instruments used in aerosol retrieval

Table 1.1. General characteristics of instruments currently used for aerosol retrieval (see Table 1.3 for links to websites providing detailed specifications of individual instruments).

Sensor	MERIS	AATSR	SeaWiFS	MODIS	PARA-SOL	MISR	AVHRR	TOMS	SEVIRI	OMI	GOME-2
Resolution at nadir [km]	0.3 (fine resolution) 1.2 (reduced resolution)	1.0	1.1	0.25 (bands 1-2) 0.50 (bands 3-7) 1.0 (bands 8-36)	6 × 7	0.275	1.1	39	1 (High res. VIS channel) 3 (IR and other VIS channels)	13 × 24 (UV-2 & VIS) 13 × 48 (UV-1)	80 × 40
Swath width [km]	1150	512	2801	2330	2400	380	2399	2800	Europe Africa S. America	2600	1920
Multi-view	No	2	No	No	Yes	9	No	No	No	No	No
Polarization	No	No	No	No	3	No	No	No	No	No	S and P in 312–790 nm channel
Platform	Envisat	Envisat	Seastar/ Orbview-2	Terra / Aqua	Myriade Series	Terra	POES	Nimbus-7 Earth Probe	MSG	AURA	METOP
Launch	March 2002	March 2002	August 1997	December 1999 / May, 2002	December 2004	December 1999	October 1978	November 1978	December 2005	July 2004	October 2006
Equator crossing time	ascending 10:00	ascending 10:00	descending 12:30	descending 10:30 / ascending 13:30	descending 13:30 / ascending 13:30	descending 10:30	descending 1:30–2:30 ascending 13:30–14:30	ascending noon	Image acquired every 15 min.	ascending 13:42	descending, 09:30
Heritage	–	ATSR-1 ATSR-2	–	–	Polder-1 Polder-2	–	AVHRR series	TOMS series	MVIRI[1]	TOMS	GOME

[1] Meteosat Visible and Infrared Imager

Table 1.2. Wavelength bands (in nm) of current instruments. Wavelength at the centre of each band is indicated (see Table 1.3 for links to websites providing detailed specifications of individual instruments).

MERIS	AATSR	SeaWiFS	MODIS	PARASOL	MISR	AVHRR	TOMS	MSG-SEVIRI
412.5		412	412.5				308, 312.5, 317.5, 322.3 331.2, 360	
442.5		443	443 469	443	446			
490		490	488	490				
510		510	531					
560	555	555	551 555	565	558			
620			645			630		635
665	659	670	667 678	670	672			
681.25								
705								
753.75			748					
760				763 765				
775		765						810
865	865	865	858 869.5	865	867			
890								
900			905 936 940	910		912		
			1240 1375	1020				
	1600		1640* 2130					1640
	3700		3750 3959 4050 4465 4516			3740		3920
								6250
			6715 7325 8550 9730					7350 8700 9660
	11000 12000		11030 12020 13335 13635 13935 14235			10800 12000		10800 12000 13400

* the 1640 nm band on MODIS-Aqua is non-functional;
+ GOME (240–790 nm), SCIAMACHY (240–2380 nm), OMI (264–504 nm) and GOME-2 (240–790 nm) are spectrometers which are not included in this table

Table 1.3. Websites providing specifications of instruments used for aerosol retrieval (Tables 1.1 and 1.2).

Instrument	website
MERIS	http://envisat.esa.int/handbooks/meris/CNTR3.htm#eph.meris.nstrumnt http://www.esa.int/esapub/bulletin/bullet103/besy103.pdf
AATSR	http://www.leos.le.ac.uk/aatsr/whatis/index.html
SeaWiFS	http://www.csc.noaa.gov/crs/rsªpps/sensors/seawifs.htm#specs
MODIS	http://modis.gsfc.nasa.gov/about/specifications.php
PARASOL	http://www-icare.univ-lille1.fr/parasol/?rubrique=mission_parasol
MISR	http://www-misr.jpl.nasa.gov/mission/mission.html
AVHRR	http://geo.arc.nasa.gov/sge/health/sensor/sensors/avhrr.html
TOMS	http://jwocky.gsfc.nasa.gov/n7toms/n7sat.html
SEVIRI	http://www.esa.int/esapub/bulletin/bullet111/chapter4_bul111.pdf
OMI	http://www.knmi.nl/omi/research/instrument/characteristics.php
GOME-2	http://www.eumetsat.int/HOME/Main/What_We_Do/Satellites/ EUMETSAT_Polar_System/Space_Segment/SP_1139327173571

References

Ackerman, S. A., K. I. Strabala, K. I. , Menzel, W. P. , Frey, R. A. , Moeller, C. C. , and Gumley L. E. , 1998, Discriminating clear sky from clouds with MODIS, *J. Geophys. Res.*, 103, 32,141–32,157.

Andreae, M.O., and Rosenfeld, D., 2008: Aerosol-cloud-precipitation interactions. Part 1. The nature and sources of cloud-active aerosols. *Earth-Science Reviews*, **89**, 13–41.

Collins, W.D., Rasch, P.J., Eaton, B.E., Khattatov, B.V., Lamarque, J.-F., and Zender, C.S., 2001: Simulating aerosols using a chemical transport model with assimilation of satellite aerosol retrievals: methodology for INDOEX. *J. Geophys. Res.*, **106**, 7313–7336.

Cairns, B., Russell, E.E., and Travis, L.D., 1999: The Research Scanning Polarimeter: calibration and ground-based measurements, *Proc. SPIE*, **3754**, 186–197.

Charlson, R.J., and Heintzenberg, J., 1995: *Aerosol Forcing of Climate*, Chichester: John Wiley.

Curier, R.L., Veefkind, J.P., Braak, R., Veihelmann, B., Torres, O., and de Leeuw, G., 2008: Retrieval of aerosol optical properties from OMI radiances using a multiwavelength algorithm: Application to western Europe. *J. Geophys. Res.*, **113** doi:10.1029/2007JD008738.

De Almeida Castanho, A.D., Vanderlei Martins, T., and Artaxo, P., 2008: MODIS aerosol optical depth retrievals with high spatial resolution over an urban area using the critical reflectance, *J. Geophys. Res.*, **113**, D02201, doi: 10.1029/2007JD008751.

De Haan, J.F. , Bosma, P.B., and Hovenier, J.W., 1987: The adding method for multiple scattering calculations of polarized light, *Astronomy Astrophysics*, **183**, 371–391.

Deschamps, P.Y., Breon, F.M., Leroy, M., Podaire, A., Bricaud, A., Buriez, J.C., and Seze, G., 1994: The POLDER Mission: Instrument characteristics and scientific objectives. *IEEE Trans. Geosci. Remote Sens.*, **32**, 598–615.

Deuzé, J.L., Bréon, F.-M., Devaux, C., Goloub, Herman, M., Lafrance, B., Maignan, F., Marchand, A., Nadal, F., Perry, G., and Tanré, D., 2001: Remote sensing of aerosols over land surfaces from POLDER-ADEOS-1 polarized measurements. *J. Geophys. Res.*, **106**(D5), 4913–4926.

Diner, D.J., Beckert, J.C., Reilly, T.H., Bruegge, C.J., Conel, J.E., Kahn, R., Martonchik, J.V., Ackerman, T.P., Davies, R., Gerstl, S.A.W., Gordon, H.R., Muller, J.-P., Myneni, R., Sellers, J., R. Pinty, B., and Verstraete, M.M., 1998: Multi-angle Imaging SpectroRadiometer (MISR) instrument description and experiment overview. *IEEE Trans. Geosci. Remote Sens.*, **36**, 1072–1087.

Diner, D.J., Martonchik, J.V., Kahn, R.A., Pinty, B., Gobron, N., Nelson, D.L., and Holben, B.N., 2005: Using angular and spectral shape similarity constraints to improve MISR aerosol and surface retrievals over land. *Remote Sens. Environ.*, **94**, 155–171.

Dubovik, O., Holben, B.N., Eck, T.F., Smirnov, A., Kaufman, Y.J., King, M.D., Tanré, D., Slutsker, I., 2002: Variability of absorption and optical properties of key aerosol types observed in worldwide locations. *J. Atmos. Sciences*, **59**, 590–608.

Flowerdew, R.J., and Haigh, J.D., 1995: An approximation to improve accuracy in the derivation of surface reflectances from multi-look satellite radiometers. *Geophys. Res. Lett.*, **22**, 1693–1696.

Geogdzhyev, I., Mishchenko, M., Rossow, W., Cairns, B., and Lacis, A., 2002: Global 2-channel AVHRR retrieval of aerosol properties over the ocean for the period of NOAA-9 observations and preliminary retrievals using NOAA-7 and NOAA-11 data, *J.Atmos.Sci.*, **59**, 262–278.

Grey, W.M.F. North, P.R.J. Los, S.O. Mitchell, R.M., 2006: Aerosol optical depth and land surface reflectance from multiangle AATSR measurements: global validation and intersensor comparisons. *IEEE Trans. Geosci. Remote Sens.*, **44**, 2184–2197.

Griggs, M., 1975: Measurements of atmospheric aerosol optical thickness over water using ERTS-1 data. *J. Air Pollut. Control. Assoc.*, **25**, 622-626.

Hasekamp, O.P., and Landgraf, J., 2005: Retrieval of aerosol properties over the ocean from multispectral single-viewing-angle measurements of intensity and polarization: retrieval approach, information content, and sensitivity study. *J. Geophys. Res.*, **110**, D20207, doi:10.1029/2005JD006212.

Hauser, A., Oesch, D., Foppa, N., and Wunderle, S., 2005: NOAA AVHRR derived aerosol optical depth over land. *J. Geophys. Res.*, **110**, D08204, doi:10.1029/2004JD005439.

Holben, B., Eck, T., Slutsker, I., Tanre, D., Buis, J.P., Setzer, A., Vermote, E., Reagan, J.A., Kaufman, Y.J., Nakajima, T., Lavenau, F., Jankowiak, I. and Smirnov, A.: AERONET, a federated instrument network and data- archive for aerosol characterization, *Rem.Sens.Environ.*, **66**, 1–66, 1998

Höller, R., Holzer-Popp, T., Garnesson, P., and Nagl, C., 2007: Using satellite aerosol products for monitoring national and regional air quality in Austria. *Proc. 'Envisat Symposium 2007', Montreux, Switzerland, 23–27 April 2007 (ESA SP-636)*.

Hsu, N.C., Tsay, S.C., King, M.D., and Herman, J.R., 2004: Aerosol properties over bright-reflecting source regions. *IEEE Trans. Geosci. Remote Sens.*, **42**, 557–569.

Hsu, N.C., Tsay, S.C., King, M.D., and Herman, J.R., 2006: Deep blue retrievals of Asian aerosol properties during ACE-Asia. *IEEE Trans. Geosci. Remote Sens.*, **44**, 3180–3195.

IPCC, 2007: *Climate Change 2007: The Physical Science Basis. Contribution of Working Group I to the Fourth Assessment Report of the Intergovernmental Panel on Climate Change* [Solomon, S., Qin, D., Manning, M., Marquis, M., Averyt, K.M.B., Tignor, M., Miller, H.L., and Chen, Z. (eds.)]. Cambridge, UK: Cambridge University Press.

Jin, Y., Schaaf, C.B., Woodcock, C.E., Gao, F., Li, X., Strahler, A.H., Lucht, W., and Liang, S., 2003a: Consistency of MODIS surface bidirectional reflectance distribution function and albedo retrievals: 1. Algorithm performance. *J. Geophys. Res.*, **108**(D5), doi:10.1029/2002JD002803.

Jin, Y., Schaaf, C.B., Woodstock, C.E., Gao, F., Li, X., Strahler, A.H., Lucht, W. and Liang, S., 2003b: Consistency of MODIS surface bidirectional reflectance distribution function and albedo retrievals: 2. Validation. *J. Geophys. Res.*, **108**(D5). doi:10.1029/2002JD002804.

Kahn, R.A., Li, W.-H., Moroney, D., Diner, D.J., Martonchik, J.V., and Fishbein, E., 2007: Aerosol source plume physical characteristics from space-based multiangle imaging. *J. Geophys. Res.*, **112**, D11205, doi:10.1029/2006JD007647.

Kahn R.A., Chen, Y., Nelson, D.L., Leung, F.-Y., Li, Q., Diner, D.J., Logan, J.A., 2008: Wildfire smoke injection heights: two perspectives from space, *Geophys. Res. Lett.*, **35**, L04809, doi:10.1029/2007GL032165.

Kaufman, Y.J., Tanré, D., Remer, L.A., Vermote, E.F., Chu, A., and Holben, B.N., 1997: Operational remote sensing of tropospheric aerosol over land from EOS moderate resolution imaging spectroradiometer. *J. Geophys. Res.*, **102**(D14), 17,051–17,067.

Kaufman, Y., Smirnov A., Holben B., and Dubovik O., 2001: Baseline maritime aerosol: methodology to derive the optical thickness and scattering properties, *Geophys. Res. Lett.*, **28**, 3251–3254.

Kaufman, Y.J, Tanré, D., and Boucher, O., 2002: A satellite view of aerosols in the climate system. *Nature* **419**, 215–223.

King, M.D., Kaufman Y.J., Tanré, D., and Nakajima, T., 1999: Remote sensing of tropospheric aerosols from space: past, present, and future. *Bull. Amer. Meteor. Soc.*, **80**, 2229–2260.

Knapp, K.R., 2002: Quantification of aerosol signal in GOES 8 visible imagery over the United States, *J. Geophys. Res.*, **107**, doi 10.1029/2001JD002001.

Koelemeijer, R.B.A., de Haan, J.F., and Stammes, P., 2003: A database of spectral surface reflectivity in the range 335–772 nm derived from 5.5 years of GOME observations, *J. Geophys. Res.*, **108**(D2), doi:10.1029/2002JD002429.

Kokhanovsky, A.A., 2008: *Aerosol Optics*, Berlin: Springer-Praxis.

Kokhanovsky, A.A., Hoyningen-Huene, W., Bovensmann, H., and Burrows J.P., 2004: The determination of the atmospheric optical thickness over Western Europe using SeaWiFS imagery, *IEEE Trans. Geosci. Remote Sens.*, **42**, 824–832.

Kokhanovsky, A.A., Mayer, B., and Rozanov, V.V., 2005: A parameterization of the diffuse transmittance and reflectance for aerosol remote sensing problems, *Atmos. Res.*, **73**, 37–43.

Kokhanovsky, A.A., von Hoyningen-Huene, W., and Burrows, J.P., 2006: Atmospheric aerosol load as derived from space, *Atmos. Res.*, **81**, 176–185.

Kokhanovsky, A.A., Breon, F.-M., Cacciari, A., Carboni, E., Diner, D., Di Nicolantonio, W., Grainger, R.G., Grey, W.M.F., Höller, R., Lee, K.-H., Li, Z., North, P.R.J., Sayer, A.M., Thomas, G.E., and von Hoyningen-Huene, W., 2007, Aerosol remote sensing over land: a comparison of satellite retrievals using different algorithms and instruments, *Atmospheric Research*, **85**, 372–394.

Kokhanovsky, A.A., Curier, R.L., de Leeuw, G., Grey, W.M.F., Lee, K.-H., Bennouna, Y., Schoemaker, R., and North, P.R.J., 2009: The inter-comparison of AATSR dual view aerosol optical thickness retrievals with results from various algorithms and instruments, *Int. J. Remote Sensing* (forthcoming).

Kotchenova, S.Y., Vermote, E.F., Levy, R., and Lyapustin, A., 2008: Radiative transfer codes for atmospheric correction and aerosol retrieval: intercomparison study, *Appl. Optics*, **47**, 2215-2226.

Kusmierczyk-Michulec, J., and de Leeuw, G., 2005: Aerosol optical thickness retrieval over land and water using Global Ozone Monitoring Experiment (GOME) data, *J. Geophys. Res.*, **110**, D10S05, doi:10.1029/2004JD004780.

Lee, K.H. , Li, Z., Kim, Y.J., and Kokhanovsky, A.A., 2009: Aerosol monitoring from satellite observations: a history of three decades, *Atmospheric and Biological Environmental Monitoring*, Y.J. Kim, U. Platt, M.B. Gu, and H. Iwahashi (eds.), Berlin: Springer (in press).

Levy, R.C., Remer, L.A., and Dubovik, O., 2007a: Global aerosol optical properties and application to Moderate Resolution Imaging Spectroradiometer aerosol retrieval over land. *J. Geophys. Res.*, **112**, D13210, doi:10.1029/2006JD007815.

Levy, R.C., Remer, L.A., Mattoo, S., Vermote, E.F., and Kaufman, Y.J., 2007b: Second-generation operational algorithm: Retrieval of aerosol properties over land from inversion of Moderate Resolution Imaging Spectroradiometer spectral reflectance. *J. Geophys. Res.*, **112**, D13211, doi:10.1029/2006JD007811.

Liu, L., and Mishchenko, M.I., 2008: Toward unified satellite climatology of aerosol properties: direct comparisons of advanced level 2 aerosol products, *J. Quant. Spectr. Rad. Transfer*, **109**, 2376–2385.

Marshak A., G. Wen, J. A. Coakley Jr., L. A. Remer, N. G. Loeb, R. F. Cahalan, 2008: A simple model for the cloud adjacency effect and the apparent bluing of aerosols near clouds, *J. Geophys. Res.*, **113**, D14S17, doi:10.1029/2007JD009196.

Martonchik, J.V., Diner, D.J., Kahn, R.A., Ackerman, T.P., Verstraete, M.M., Pinty, B., and Gordon, H.R., 1998: Techniques fot the retrieval of aerosol properties over land and ocean using multiangle imaging. *IEEE Trans. Geosci. Remote Sens.*, **36**, 1212–1227.

McCormick, M.P., Hamill, P., Pepin, P.J., Chu, W.P., Swissler, T.J., and McMaster, L.R., 1979: Satellite studies of the Stratospheric aerosol, *Bull. American Meteorol. Soc.*, **60**(9), 1038–1046.

Mekler, Y., Quenzel H., Ohring G., and Marcus I., 1977: Relative atmospheric aerosol content from ERS observations. *J. Geophys. Res.*, **82**, 967–972.

Mishchenko, M.I., Geogdzhayev, I.V., Rossow, W.B., Cairns, B., Carlson, B.E., Lacis, A.A., Liu, L., and Travis, L.D., 2007a: Long-term satellite record reveals likely recent aerosol trend. *Science*, **315**(5818), 1543. [DOI: 10.1126/science.1136709].

Mishchenko, M.I., Cairns, B., Kopp, G., Schueler, C.F., Fafaul, B.A., Hansen, J.E., Hooker, R.J., Itchkawich, T., Maring, H.B., and Travis, L.D., 2007b: Precise and accurate monitoring of terrestrial aerosols and total solar irradiance: introducing the Glory mission. *Bull. Amer. Meteorol. Soc.*, **88**, 677–691, doi:10.1175/BAMS-88-5-677.

Nakajima, T.Y., T. Nakajima, M. Nakajima, H. Fukushima, M. Kuji, A. Uchiyama, M. Kishino, 1998: Optimization of the Advanced Earth Observing Satellite II Global Imager Channels by use of Radiative Transfer Calculations, *Appl. Optics*, **37**, 3149–3163.

Nakajima, T., T. Y. Nakajima, M. Nakajima, and the GLI Algorithm Integration Team (GAIT), 1999: Development of ADEOS-II/GLI operational algorithm for earth observation, *SPIE Proceedings*, **3870**, 314–322.

Nikolaeva, O. V., L. P. Bass, T. A. Germogenova, A. A. Kokhanovsky, V. S. Kuznetsov, B. Mayer, 2005: The influence of neighboring clouds on the clear sky reflectance studied with the 3-D transport code RA-DUGA, *J. Quant. Spectr. Rad. Transfer*, **94**, 405–424.

North, P.R.J., Briggs, S.A., Plummer, S.E., and Settle, J.J., 1999: Retrieval of land surface bi-directional reflectance and aerosol opacity from ATSR-2 multi-angle imagery, *IEEE Trans. Geosci. Remote Sens.*, **37**(1), 526–537.

Remer, L.A., and Kaufman, Y.J., 2006: Aerosol direct radiative effect at the top of the atmosphere over cloud free ocean derived from four years of MODIS data, *Atmos. Chem. Phys.*, **6**, 237–253.

Rozanov, V.V., and Kokhanovsky, A.A., 2006: The solution of the vector radiative transfer equation using the discrete ordinates technique: selected applications, *Atmos. Res.*, **79**, 241–265.

Santer, R., Carrere, V., Dubuisson, P., and Roger, J.-C., 1999: Atmospheric corrections over land for MERIS, *Int. J. of Remote Sens.*, **20**, 1819–1840.

Santer, R., Carrere, V., Dessailly, D., Dubuisson, P., and Roger, J.-C., 2000: MERIS ATBD 2.15: Algorithm theoretical basis document, atmospheric correction over land. Technical report POTN-MEL-GS-0005, LISE.

Schaap, M., Apituley, A., Timmermans, R.M.A., Koelemeijer, R.B.A., and de Leeuw, G., 2009: Exploring the relation between aerosol optical depth and PM2.5 at Cabauw, The Netherlands. *Atmos. Chem. Phys.*, **9**, 909–925.

Stamnes, K., Tsay, S.-C., Wiscombe, W., and Jayaweera, K., 1988: Numerically stable algorithm for discrete-ordinate-method radiative transfer in multiple scattering and emitting layered media, *Appl. Optics*, **27**, 2502–2509.

Stowe, L.L., Jacobowitz, H., Ohring, G., Knapp, K.R., and Nalli, N.R., 2002: The Advanced Very High Resolution Radiometer (AVHRR) Pathfinder Atmosphere (PATMOS) climate dataset: initial analyses and evaluations, *J. Clim.*, **15**, 1243–1260.

Torres, O., Bhartia, P.K., Herman, J.R., Ahmad, Z., and Gleason, J., 1998: Derivation of aerosol properties from satellite measurements of backscattered ultraviolet radiation. Theoretical basis. *J. Geophys. Res.*, **103**, 17,099–17,110.

Torres, O., Bhartia, P.K., Herman, J.R., Sinyuk, A., Ginoux, P., and Holben, B., 2002a: A long term record of aerosol optical depth from TOMS observations and comparison to AERONET measurements, *J. Atm. Sci.*, **59**, 398–413.

Torres, O., Decae, R., Veefkind, J.P., and de Leeuw, G., 2002b: OMI aerosol retrieval algorithm. In: P. Stammes and R. Noordhoek (eds.), *OMI Algorithm Theoretical Basis Document*, Vol. III: *Clouds, Aerosols and Surface UV Irradiance*, ATBD-OMI-03, 46–71.

Tynes H.H., Kattawar, G.W., Zege, E.P., Katsev, I.L., Prikhach, A.S., and Chaikovskaya, L.I., 2001: Monte Carlo and multicomponent approximation methods for vector radiative transfer by use of effective Mueller matrix calculations, *Appl. Opt.*, **40**, 400–412.

Veefkind, J.P., and de Leeuw, G., 1998: A new algorithm to determine the spectral aerosol optical depth from satellite radiometer measurements. *J. Aerosol Sci.*, **29**, 1237–1248.

Veefkind, J.P., de Leeuw, G., and Durkee, P.A., 1998: Retrieval of aerosol optical depth over land using two-angle view satellite radiometry during TARFOX. *Geophys. Res. Lett.*, **25**(16), 3135–3138.

Veihelmann, B., Levelt, P.F., Stammes, P., and Veefkind, J.P., 2007: Simulation study of the aerosol information content in OMI spectral reflectance measurements. *Atmos. Chem. Phys.*, **7**, 3115–3127.

Verver, G.H.L., Henzing, J.S., de Leeuw, G., Robles Gonzalez, C., and van Velthoven, P.F.J., 2002: Aerosol retrieval and assimilation (ARIA). Final report Phase 1, NUSP-2, 02-09, KNMI-publication: 200.

Vidot J., Santer, R., and Aznay, O., 2008: Evaluation of the MERIS aerosol product over land with AERONET. *Atmos. Chem. Phys.*, **8**, 7603–7617.

von Hoyningen-Huene,W., Freitag, M., Burrows, J.B., 2003: Retrieval of aerosol optical thickness over land surfaces from top-of-atmosphere radiance. *J. Geophys. Res.*, **108**, 4260. doi:10.1029/ 2001JD002018.

von Hoyningen-Huene, W., Kokhanovsky, A.A., Burrows, J.B., Bruniquel-Pinel, V., Regner, P., Baret, F., 2006: Simultaneous determination of aerosol- and surface characteristics from top-of-atmosphere reflectance using MERIS on board ENVISAT, Adv. Space Res., **37**, 2172–2177.

von Hoyningen-Huene,W., Kokhanovsky, A.A., Burrows, T.P., 2008: Retrieval of particulate matter from MERIS observations, in Y.T. Kim, V. Platt (Eds), *Advanced Environmental Monitoring*, Berlin: Springer, pp. 190–202.

Wen G., A. Marshak, R. F. Cahalan, 2008: Importance of molecular Rayleigh scattering in the enhancement of clear sky reflectance in the vicinity of boundary layer cumulus clouds, *J. Geophys. Res.*, **113**, D24207, doi:10.1029/2008JD010592.

Winker, D.M., Pelon, J.R., and McCormick, M.P., 2003: The CALIPSO mission: spaceborne lidar for observation of aerosols and clouds, *Proc. of SPIE*, **4893**, 1–11.

Winker D.M., Hunt, W.H., and McGill, M.J., 2007: Initial performance assessment of CALIOP. *Geophys. Res. Lett.*, **34**, L19803, doi:10.1029/2007GL030135.

2 The dark-land MODIS collection 5 aerosol retrieval: algorithm development and product evaluation

Robert C. Levy

1. Introduction

Tropospheric aerosols significantly influence global climate, by changing the radiative energy balance as well as the hydrological cycle (e.g., IPCC [2007]). Also known as suspended airborne particles, or particulate matter (PM), aerosols are a component of smog and air pollution (e.g., USEPA [2003], Chen et al. [2002] and Dickerson et al. [1997]) and a regulated criteria air pollutant (e.g., http://www.epa.gov/oar/particlepollution/naaqs-rev2006.html). Aerosols are spatially and temporally inhomogeneous and, depending on aerosol type and meteorology, they may be found far from their sources. Spanning from nanometers (nm) to tens of micrometers (μm) in radius, aerosols are efficient at scattering solar radiation back to space, thus affecting photochemistry (e.g., Dickerson et al. [1997]) and changing the effective albedo of the Earth. Passive satellite sensors, from their vantage point above the atmosphere, observe the scattered solar radiation to measure the global aerosol distribution [Kaufman et al., 1997a]. The fundamental unit of measure for aerosol remote sensing is known as the aerosol optical depth (AOD), which is the integral of the aerosol light extinction over vertical path through the atmosphere, and is a function of wavelength λ. Typically, AOD (measured at $\lambda = 0.55\mu$m) ranges from near zero in pristine conditions to 1.0, 2.0 or even 5.0, during episodes of heavy pollution, smoke or dust.

The launch of the MODerate Imaging Spectroradiometer (MODIS, [Salomonson et al., 1989; King et al., 1992]) sensors aboard NASA's *Terra* (in 1999; [Kaufman et al., 1998]) and *Aqua* (in 2002; [Parkinson, 2003]) satellites, has led to 'quantitative' (accurate, precise and with known uncertainties) observation of global AOD. From MODIS-observed spectral reflectance, separate retrieval algorithms derive aerosol properties over ocean [Tanré et al., 1997; Remer et al., 2005], over dark land targets [Kaufman et al., 1997b; Levy et al., 2007b], and over bright land targets [Hsu et al., 2006], respectively. Each algorithm is customized for the surface type, taking advantage of known behavior of the surface optical properties. The combination of algorithms provides the basis for an *operational* algorithm (near real-time processing) for retrieving spatially continuous global AOD. The products are free and available to any investigator, and have been used for climate (e.g. IPCC [2001] and Yu et al. [2006], air quality (e.g., Chu et al. [2003], Al-Saadi et al. [2005], Engel-Cox et al. [2004, 2006] and Wang and Christopher, [2003]) as well as other applications. By 2004, enough MODIS data had been processed (known as 'collection 4', or c004) to enable extensive statistical evaluation of the aerosol products over dark land targets, both globally and regionally [Remer et al., 2005].

Remer et al. [2005] attempted to 'validate' the c004 products over dark-land targets, primarily by comparing them to collocated 'ground-truth' measurements made by the global AErosol RObotic sunphotometer NEtwork (AERONET). MODIS-derived AOD (τ_{MODIS}) regressed to sunphotometer-measured AOD (τ_{true}), with correlation coefficient $R = 0.80$. 68 % were within expected pre-launch uncertainty $\Delta\tau$ of

$$\Delta\tau = \pm 0.05 \pm 0.15\tau_{true}. \tag{1}$$

Although the good comparison was said to have *validated* the MODIS-derived AOD product, the same analysis showed that MODIS tended to be high for low AOD, and low for high AOD i.e.,

$$\tau_{MODIS} = 0.78\tau_{true} + 0.07 \tag{2}$$

at 0.55 μm [Remer et al., 2005]. The equation varied in different regions, including (somewhat embarrassingly) relatively poor comparison over the East Coast of the USA (the developmental home of the MODIS aerosol algorithm) (e.g., Levy et al. [2005]). The y-offset (MODIS biased high in low AOD conditions) implied errors induced by assuming inappropriate surface assumptions (boundary conditions). On the other hand, the less than one slope implied errors in the atmospheric assumptions. The combination of significant y-offset, and less than one slope implied that there was room to improve the MODIS algorithm, and was one focus of the MODIS Aerosol Science Team (MAST) since publication of Remer et al. [2005]. This paper describes development and evaluation of our 'second-generation' dark-land target aerosol retrieval.

2.   Properties of aerosols

Aerosols are a mixture of particles (in suspension of air) of different sizes, shapes, compositions, and chemical, physical, and thermodynamic properties. They range in size from nanometers to micrometers, spanning from molecular aggregates to cloud droplets. Aerosols between about 0.1 μm and 2.5 μm in radius are of main interest to climate precipitation, visibility, and human health studies. Most aerosols of interest are found in the troposphere and concentrated toward the Earth's surface (having a scale height about 2–3 km). Aerosol physical and chemical properties are determined by their sources and production processes. *Fine* aerosols (radius between 0.1 and 1.0 μm) are formed by coagulation of smaller nuclei (very fine aerosols), or produced directly during incomplete combustion (from biomass-burning or coal power plants). Aerosols larger than about 1.0 μm are known as *coarse* particles, and are primarily from mechanical erosion of the Earth's surface. Coarse particles include sea salt and soil dust lifted by winds.

Ambient aerosol distributions contain all size ranges. Coarse particles are usually small in number but can contain the largest portion of aerosol mass (or volume). Because of their larger size (and mass), coarse particles are usually quickly settled out of the atmosphere and are concentrated close to their sources. However, convection may lift them into prevailing winds, where they may be transported far from their source. Also known as the *accumulation* mode, the fine mode has the longest residence time (days to weeks) because it neither efficiently settles nor coagulates on its own. The fine mode contains the largest portion of aerosol surface area, and the greatest ability to scatter solar radiation. Here, we loosely define the *fine aerosol weighting* (*FW* or η) as the ratio of fine aerosol to the total X, but note that the exact definition for FW depends on the quantity represented by X (e.g., total mass, total AOD, etc.).

Many aerosols are hygroscopic}, meaning that they have the ability to absorb water vapor (e.g., Malm et al. [1994], Kotchenruther et al. [1999] and Gassó et al. [2003]) and thus become involved in cloud processes and the hydrologic cycle. For most accumulation-sized aerosols, the residence time is of the same order as water vapor in the atmosphere, usually about four to fourteen days (e.g., Hoppel et al. [2002]). Generally, more hygroscopic aerosols (known as hydrophilic, e.g., sulfate or sea salt) are spherical in shape, whereas those less hygroscopic (e.g., hydrophobic, e.g., soot or dust) tend to be non-spherical. Non-spherical aerosols may be clump-like (soot) or crystalline (certain dusts). Larger aerosols usually have shorter residence times (days to hours) due to dry deposition.

2.1   Properties of aerosol size distributions

For any size distribution of spherical particles, the number distribution as a function of radius, $N(r)$, is related to the volume V and area A distributions by:

$$\frac{\mathrm{d}N}{\mathrm{d}\ln r} = \frac{3}{4\pi r^3}\frac{\mathrm{d}V}{\mathrm{d}\ln r} = \frac{1}{\pi r^2}\frac{\mathrm{d}A}{\mathrm{d}\ln r}, \tag{3}$$

such that N_0, V_0, and A_0 are the totals of the corresponding distributions, i.e.,

$$N_0 = \int_0^\infty \frac{\mathrm{d}N}{\mathrm{d}\ln r}\,\mathrm{d}\ln r, \quad V_0 = \int_0^\infty \frac{\mathrm{d}V}{\mathrm{d}\ln r}\,\mathrm{d}\ln r, \quad A_0 = \int_0^\infty \frac{\mathrm{d}A}{\mathrm{d}\ln r}\,\mathrm{d}\ln r \tag{4}$$

and $\mathrm{d}N$; $\mathrm{d}V$; $\mathrm{d}A/\mathrm{d}\ln r$ are the number/volume/area size distributions with r denoting radius (in μm). Although the size distributions may be defined per any dimension, here we define them as 'per unit area'. For example, $\mathrm{d}V$ and V_0 have units of volume per area. The *in situ* aerosol measurement community commonly defines aerosol size in terms of diameter (for example, Seinfeld and Pandis [1998]), whereas the remote sensing community defines size by radius. Here, we define the means of the log-radii of the number (r_g) and volume (r_v) distributions, as

$$\ln r_g = \frac{\int\limits_{r_{\min}}^{r_{\max}} \ln r \frac{\mathrm{d}N(r)}{\mathrm{d}\ln r}\,\mathrm{d}\ln r}{\int\limits_{r_{\min}}^{r_{\max}} \frac{\mathrm{d}N(r)}{\mathrm{d}\ln r}\,\mathrm{d}\ln r}; \quad \ln r_v = \frac{\int\limits_{r_{\min}}^{r_{\max}} \ln r \frac{\mathrm{d}V(r)}{\mathrm{d}\ln r}\,\mathrm{d}\ln r}{\int\limits_{r_{\min}}^{r_{\max}} \frac{\mathrm{d}V(r)}{\mathrm{d}\ln r}\,\mathrm{d}\ln r} \tag{5}$$

and the standard deviations of the log-radii, σ_g and σ_v, as:

$$\sigma_g = \sqrt{\frac{\int\limits_{r_{\min}}^{r_{\max}} (\ln r - \ln r_g)^2 \frac{\mathrm{d}N(r)}{\mathrm{d}\ln r}\,\mathrm{d}\ln r}{\int\limits_{r_{\min}}^{r_{\max}} \frac{\mathrm{d}N(r)}{\mathrm{d}\ln r}\,\mathrm{d}\ln r}}; \quad \sigma_v = \sqrt{\frac{\int\limits_{r_{\min}}^{r_{\max}} (\ln r - \ln r_v)^2 \frac{\mathrm{d}V(r)}{\mathrm{d}\ln r}\,\mathrm{d}\ln r}{\int\limits_{r_{\min}}^{r_{\max}} \frac{\mathrm{d}V(r)}{\mathrm{d}\ln r}\,\mathrm{d}\ln r}}. \tag{6}$$

Note that σ is the log of the quantity defined within the *in situ* community.

Aerosol distributions are often approximately lognormal, so they are often assumed as such. For a lognormal, the moments of order k, M^k are

$$M^k = \int_0^\infty r^k \frac{dN}{d\ln r} \, d\ln r = (r_g)^k \exp(0.5k^2\sigma^2). \tag{7}$$

We can see that desirable properties of the lognormal distribution include

$$\sigma = \sigma_g = \sigma_v \quad \text{and} \quad r_v = r_g \exp(3\sigma^2). \tag{8}$$

For a single lognormal mode, the number size distribution is

$$\frac{dN}{d\ln r} = \frac{N_0}{\sigma\sqrt{2\pi}} \exp\left(-\frac{[\ln(r/r_g)]^2}{2\sigma^2}\right), \tag{9}$$

and the total volume and number are easily related by

$$N_0 = V_0 \frac{3}{4\pi r_g^3} \exp\left(-\frac{9}{2}\sigma^2\right). \tag{10}$$

This also leads to the definition of effective radius r_{eff} of a distribution, i.e.,

$$r_{eff} = \frac{M^3}{M^2} = \frac{\int_0^\infty r^3 \frac{dN}{d\ln r} \, d\ln r}{\int_0^\infty r^2 \frac{dN}{d\ln r} \, d\ln r} = \frac{3}{4}\frac{V_0}{A_0} = r_g \exp\left(\frac{5}{2}\sigma^2\right). \tag{11}$$

For aerosols composed of two or more modes, integration must be over both size bin and mode. For example, for a bimodal distribution,

$$r_{eff} = \frac{\int_0^\infty r^3 \frac{(dN_1 + dN_2)}{d\ln r} \, d\ln r}{\int_0^\infty r^2 \frac{(dN_1 + dN_2)}{d\ln r} \, d\ln r}. \tag{12}$$

2.2 Aerosol optical properties

Aerosols are important to Earth's climate and radiation because of their size. Particles most strongly affect the radiation field when their size is most similar to the wavelength of the radiation (e.g., Chandresekhar [1950]). Aerosols in the fine mode (0.1 to 1.0 μm) are similar in size to the wavelengths of solar radiation within the atmosphere, and are also the largest contributors to aerosol surface area. Radiation incident on aerosols may be absorbed, reflected or transmitted, depending on the chemical composition (complex refractive index, m) and orientation (if non-spherical) of the aerosol particles. Scattering and absorption quantities [Thomas and Stamnes, 1999] may be represented as functions of path distance (the scattering/absorption *coefficients*, β_{sca}/β_{abs}, each in units of [per length]), column number (the scattering/absorption *cross-sections*, $\sigma_{sca}/\sigma_{abs}$, each in units of [area]) or mass (the scattering/absorption *mass coefficients*, B_{sca}/B_{abs}, each in units of [area per mass]). The use of symbols is inconsistent within the literature, so symbols are defined for this work as those of Liou [2002]. *Extinction* (coefficient/cross-section/mass coefficient) is the sum of the appropriate absorption and scattering (coefficients/cross-sections/mass coefficients), e.g.,

$$\sigma_{ext}(\lambda) = \sigma_{sca}(\lambda) + \sigma_{abs}(\lambda) \tag{13}$$

for the cross-sections. These properties define the amount of radiation 'lost' from the radiation field, per unit of material loading, in the beam direction. Note that all of the parameters are dependent on the wavelength λ. The ratio of scattering to extinction (e.g., β_{sca}/β_{ext}) is known as the single scattering albedo (SSA or ω_0). As most aerosols are weakly absorbing in mid-visible wavelengths (except for those with large concentrations of organic/black carbon), extinction is primarily by scattering ($\omega_0 > 0.90$ at 0.55 μm). Black or elemental carbon (soot) can have $\omega_0 < 0.5$ [Bond and Bergstrom, 2006] especially near sources. Mineral dusts are unique in that they have a spectral dependence of absorption, such that they absorb more strongly in short visible and UV wavelengths ($\lambda < 0.47$ μm) than at longer wavelengths.

Properties of extinction (scattering and absorption) can be calculated, and are dependent on the wavelength of radiation, as well as characteristics of the aerosols' size distribution, chemical composition, and physical shape. For a single *spherical* aerosol particle, the combination of *complex refractive index* ($m + ki$), and *Mie size parameter*, X (relating the ratio of radius to wavelength, i.e., $X = 2\pi r/\lambda$) uniquely describe the scattering and extinction properties of the particle. The scattering/extinction *efficiency* (Q) for one particle is related to the cross-sections by

$$Q_{sca} = \sigma_{sca}/\pi r^2 \text{ and } Q_{ext} = \sigma_{ext}/\pi r^2. \tag{14}$$

The scattered photons have an angular pattern, known as the *scattering phase function* ($P_\lambda(\Theta)$), which is a function of the scattering angle (Θ) and wavelength. In other words, the Mie quantities describe the interaction between an incoming photon and aerosol particle, whether it is displaced, whether it is scattered, and toward which direction relative to the incoming path.

For a *distribution* of aerosol particles, one is concerned with the scattering by all particles within a space (e.g., per volume, per column). In general, since the average separation distance between particles is so much greater than particle radius, particles can be considered independent of each other. For a unit volume (or columnar surface area) containing N particles of varying r, the integrated extinction/scattering cross-sections (the extinction/scattering coefficients) are

$$\beta_{ext} = \int \sigma_{ext}(r)N(r) \, dr \text{ and } \beta_{sca} = \int \sigma_{sca}(r)N(r) \, dr. \tag{15}$$

Therefore, the scattering/extinction *efficiencies* for a representative single aerosol are

$$\bar{Q}_{ext} = \frac{\int \sigma_{ext}(r)N(r) \, dr}{\int \pi r^2 N(r) \, dr} \text{ and } \bar{Q}_{sca} = \frac{\int \sigma_{sca}(r)N(r) \, dr}{\int \pi r^2 N(r) \, dr}. \tag{16}$$

Light scattering by aerosols is a function of the wavelength, the aerosol size distribution, and the aerosol composition [Fraser, 1975]. Calculating the scattering properties at two or more wavelengths provides information about the aerosols' size. The Ångström exponent (a) relates the spectral dependence of the extinction (or scattering) at two wavelengths, $\lambda 1$ and $\lambda 2$:

$$a_{\lambda 1, \lambda 2} = \frac{-\log(\sigma_{p,\lambda 1}/\sigma_{p,\lambda 2})}{\log(\lambda 1/\lambda 2)}, \tag{17}$$

(e.g., Ångström [1929]; Eck et al. [1999]). Often the two wavelengths are defined in the visible or infrared (e.g., 0.47 and either 0.66 or 0.87 µm). Larger aerosol size is related to smaller values of a, such that aerosol distributions dominated by fine aerosols have $a \geq 1.6$, whereas those dominated by coarse aerosols have $a \leq 0.6$. Quadratic fits to more than two wavelengths, known as modified Ångström exponents (e.g., O'Neill et al. [2001]), can provide additional size information including the relative weighting of fine mode aerosol to the total (i.e., *FW*, or η).

The asymmetry parameter, g, represents the degree of asymmetry of the angular scattering (phase function), and is defined as:

$$g_\lambda = \frac{1}{2} \int_0^\pi P_\lambda(\Theta) \cos \Theta \sin \Theta \, d\Theta \tag{18}$$

Values of g range from -1 for entirely backscattered light to $+1$ for entirely forward scattering. For molecular (Rayleigh) scattering, $g = 0$. For aerosol, g typically ranges between 0.6 and 0.7 (mostly forward scattering), the lower values in dry (low relative humidity) conditions (e.g., Andrews et al. [2006]). g is strongly related to the aerosol size, and to the accumulation mode size, specifically.

The aerosol optical depth is the integral of the aerosol extinction coefficient over vertical path from the surface to the top of the atmosphere (TOA), i.e.

$$\tau(\lambda) = \int_0^{TOA} \beta_{ext,p}(\lambda, z) \, dz, \tag{19}$$

where the subscript p represents the contribution from the particles (to be separated from molecular or Rayleigh optical depth). Typically, AOD (at 0.55 µm) range from 0.05 over the remote ocean to 1.0, 2.0 or even 5.0, during episodes of heavy pollution, smoke or dust.

The *mass extinction coefficient*, B_{ext}, represents the area extinction for a unit mass of the aerosol (usually in units of [m²/g]). For a distribution of aerosols,

$$B_{ext} = \frac{3\bar{Q}_{ext}}{4\rho r_{eff}}, \tag{20}$$

where ρ is the average particle density (e.g., Chin et al. [2002]. From here, we define the *mass concentration coefficient (M_c)* as

$$M_c = \frac{1}{B_{ext}}, \tag{21}$$

which represents the aerosol mass for unit AOD, per unit surface area. If relating dry aerosol mass to ambient AOD, we also must take into account ambient relative humidity (RH) and aerosol hygroscopicity (e.g., Köpke et al. [1997]).

2.3   Aerosol measurement techniques

Numerous techniques are used to observe and quantify aerosol physical and chemical properties (e.g., Seinfeld and Pandis [1998]), either in situ or by remote sensing. Each of these techniques may be passive (operating under ambient conditions) or active (perturbed conditions). Combined surface and airborne measurements provide profiles of aerosol properties such as loading, size distribution, and chemistry. For example, the U.S. Environmental Protection Agency (US-EPA) monitors *in situ* aerosol mass concentrations at the surface by weighing dried (perturbed) aerosol mass collected on filters which provide PMn where $n = 1$, 2.5 or 10 µm depending on the largest particles measured. Other *in situ* techniques are based on optical measurements, and measure aerosol scattering within a cavity to determine extinction and backscatter coefficients (nephelometers). Aerosol absorption is commonly monitored by measuring the attenuation of light by aerosol collected on filter using absorption photometers (e.g., MAAP, aetholometer, PSAP). Yet, because nearly all *in situ* instruments collect the aerosols on a filter or within a cavity, they perturb the aerosols themselves.

By contrast, remote sensing techniques observe a radiation field as it interacts with the atmosphere and surface. *Active* remote sensing (e.g., radar or lidar) utilizes its own light source, whereas *passive* techniques use the ambient radiation (e.g., sunlight) as the source. Like *in situ* techniques, remote sensing includes ground-based (e.g. sunphotometer) and airborne radiometers. Radiometers can also be mounted on orbiting satellites (e.g., MODIS) for retrieving continuous information on regional and global scales [Kaufman et al., 1997a; King et al., 1999; Kaufman et al., 2002].

2.3.1 Passive remote sensing of direct beam extinction by sunphotometer (AERONET)

The simplest passive remote sensing technique is sunphotometry [Volz, 1959], where the solar disk is observed through a collimator. The sunphotometer measures the extinction of direct-beam radiation in distinct wavelength bands, and derives the aerosol contribution to the total extinction. The measurement assumes that the radiation has had little or no interaction with the surface or clouds, and that there is minimal (or known) gas absorption in the chosen wavelength, λ. In other words, sunphotometry is a basic application of the Beer–Bouguer–Lambert law, in the form of:

$$L_\lambda(\theta_0) = F_{0,\lambda}(\theta_0, d) \ \exp[-\tau_\lambda^t m^t(\theta_0)], \tag{22}$$

where L, F_0, d, θ_0, τ^t, and m are the measured solar radiance, extra-terrestial solar irradiance (irradiance outside the atmosphere), ratio of the actual and average Earth/sun distance, solar zenith angle, total atmospheric optical depth, and total relative optical air mass, respectively. The factor $\tau^t m^t$ is the only unknown (the other parameters can be calculated), and it can be further broken down as:

$$\tau_\lambda^t m^t = \tau_\lambda^R m^R + \tau_\lambda^a m^a + \tau_\lambda^g m^g \tag{23}$$

where the superscripts t, R, a and g refer to total, molecular (Rayleigh scattering), aerosol and gas absorption (variably distributed gases such as H_2O, O_3, NO_2, etc.), where the relative optical air masses of each component differ due to differing vertical distributions. The molecular portions of Eq. (23) are dependent only on the altitude of the surface target, and can be accurately calculated (e.g., Bodhaine et al. [1999]). The gas absorption portion, while varying in vertical profile by component, can be reasonably estimated. Therefore, since errors are well defined, estimation of AOD (τ^a, or hereby simplified as τ) is straightforward from a sunphotometer. When made at more than one wavelength, sunphotometers retrieve spectral (wavelength-dependent) τ, which in turn can be used to characterize the relative size of the ambient aerosol [Eck et al., 1999; O'Neill et al., 2003].

Although sunphotometers have been used for decades, the products provided by AErosol RObotic NEtwork (AERONET; Holben et al. [1998]) are considered state-of-the-art for consistent, calibrated and accessible spectral aerosol depth data. Operating at hundreds of sites globally, the AERONET sunphotometers (produced by Cimel Electronique in France) have been reporting at some sites since 1993 (e.g., http://climate.gsfc.nasa.gov). 'Sun' products are retrievals of spectral τ at several wavelengths (0.34, 0.38, 0.44, 0.67, 0.87 and 1.02 μm, and possibly others depending on the individual instrument), resulting from application of Eq. (23) to the observations of spectral extinction of the direct beam. In addition to spectral AOD, AERONET derives columnar water vapor (PW) from a water vapor absorbing channel (0.94 μm). Approximately every 15 minutes during the daytime, the sunphotometer points directly at the sun, taking spectral measurements in triplicate over 1.5 minutes. Cloud screening [Smirnov et al., 2000] is performed by limiting the variability within each triplet and compared to prior and subsequent triplets. Estimates of Ångström exponent, and separation into fine and coarse mode contributions, are easily computed via the spectral de-convolution algorithm of O'Neill et al. [2001].

Level 1 (raw data averages) and Level 1.5 (cloud screened data; e.g. Smirnov et al. [2000]) are provided in near real time to the user community. Level 2 data is considered calibrated, quality-assured data, meaning that the instrument has been corrected for optical drift and the products meet certain requirements. Since the upgrade to Level 2 requires the instrument to be taken from the field and re-calibrated, it may not be available for months or years after Level 1.5 is available. Recently, AERONET has gone through a re-processing of its direct sun products that is collectively known as 'Version 2'. AERONET derived estimates of spectral AOD are expected to be accurate within ± 0.02 (e.g., Holben et al. [1998]).

2.3.2 Passive remote sensing of sky scattering by sunphotometer (AERONET)

Collimated radiometers (e.g., AERONET) also can be pointed at discrete points in the sky to observe scattered sky radiance. Requiring additional assumptions as to the shape of the particles, interaction with the surface and multiple scattering processes, the technique in essence boils down to retrieval of the spectral, angular (Θ) aerosol scattering phase function, $P_\lambda(\Theta)$.

In addition to the direct 'sun' measurements, AERONET instruments are programmed to observe angular distribution of sky radiance, approximately every hour during the daytime. These 'sky' measurements are made in the almucantar (a circle made with constant zenith angle equal to solar zenith angle), and the principal plane (line of constant azimuth angle) in at least at four wavelengths (0.44, 0.67, 0.87 and 1.02 μm), in order to observe aerosol spectral scattering. These observations are controlled for quality, through rigorous cloud screening and requirements of angularly symmetric radiance. Sky radiance measurements are used to retrieve size distribution and scattering/extinction properties of the ambient aerosol field using spherical aerosol assumptions [Nakajima and King, 1990; Kaufman et al., 1994; Dubovik and King, 2000], and more recently, non-spherical assumptions [Dubovik et al. 2002b]. By assuming the ambient aerosol to be a homogeneous ensemble of polydisperse spheres and randomly oriented spheroids [Dubovik et al., 2006], the algorithm retrieves the volume distribution ($dV/d \ln r$) for 22 radius size bins and spectral complex refractive index (at wavelengths of sky radiance observations) that correspond to the best fit of both sun-measured AOD and almucantar sky radiances. The non-spherical fraction is modeled with distribution of aspect ratios retrieved [Dubovik et al., 2006)] that fit scattering matrices of mineral dust measured in the laboratory [Volten et al., 2001]. In either case, the modeling is performed using kernel look-up tables of quadrature coefficients employed in the numerical integration of spheroid optical properties over size and shape. These kernel look-up tables were generated using exact T-matrix code [Mishchenko and Travis, 1994] and the approximated geometric-optics-integral method of Yang and Liou [1996], that was used for size or shape parameters exceeding the convergence limits of T-matrix code. As a result the kernels cover a wide range of sizes ($\sim 0.12 \leq 2\pi r/\lambda \leq \sim 625$) and axis ratios ε ($0.3 \leq \varepsilon \leq 3$). The usage of kernel look-up tables allows quick and accurate simulations of optical properties of spheroids and therefore makes it possible to use a model of randomly oriented spheroids (introduced by Mishchenko et al. [1997] for desert dust) in AERONET operational retrievals.

The retrieved size distribution and complex refractive index uniquely determine the aerosol phase function (P) and single scattering albedo (ω_0), also provided as retrieved products. In addition, AERONET derives optical properties (τ, P and ω_0) and integral parameters of size distributions (total volume concentration C_v, volume median radius r_v, and standard deviation σ), separately for fine mode ($r \leq 0.6$ μm) and coarse mode ($r > 0.6$ μm) of the retrieved aerosol. Such a representation of AERONET retrievals is based on the convenient observation that the majority aerosol is bimodal. Although the parameters are simulated for each mode without assuming any particular shape or size distribution (see the formulation in Dubovik et al. [2002a]), they are analogous to corresponding parameters of lognormal size distributions described in Section 2.1 (V_0, r_v, σ). In fact, the assumption of lognormality allows accurate reproduction of aerosol optical properties in many cases (especially those dominated by fine mode), suggesting that these parameters represent lognormal properties of AERONET climatology [Dubovik et al., 2002a]. To ensure a large enough signal, retrievals of optical properties from sky radiance recommend the sun-observed ambient optical depth to be at least 0.4 at 0.44 μm.

Retrievals from both sun and sky AERONET measurements are controlled by rigorous calibration and cloud-screening processes. The results are also constrained by the criteria identified in sensitivity studies [Dubovik et al., 2000]. As discussed by Dubovik et al. [2002a], these selections yield accurate retrieval results that can be used as ground-truth estimates (for certain aerosol properties). These products are known as Level 2 AERONET products. It is noted the AERONET team has recently re-processed the combination of direct sun and sky products, known as 'Version 2' (http://aeronet.gsfc.nasa.gov), using improved characterization of the surface albedo around each site and modified spheroid axis ratio distributions (applied to the T-matrix code). This reprocessing has led to significant changes from Version 1 products at some sites, often lowering estimates of spectral ω_0.

2.3.3 Satellite passive remote sensing

While sunphotometers derive aerosol properties from measurements of extinction or skylight scattering in the downward direction, satellites derive aerosol properties from measurements of upscattered (reflected) radiation (e.g., Kaufman et al. [1997a]). The geometry of the satellite measurement is illustrated in Fig. 2.1, such that θ_0, θ and θ are the solar zenith, target view zenith and relative solar/target relative azimuth angles, respectively. The scattering angle, Θ is:

$$\Theta = \arccos(-\cos\theta_0 \cos\theta + \sin\theta_0 \sin\theta \cos\varphi), \qquad (24)$$

The normalized spectral radiance, or *reflectance*, ρ_λ is defined by

$$\rho_\lambda = L_\lambda \frac{\pi}{F_{0,\lambda} \cos(\theta_0)}. \qquad (25)$$

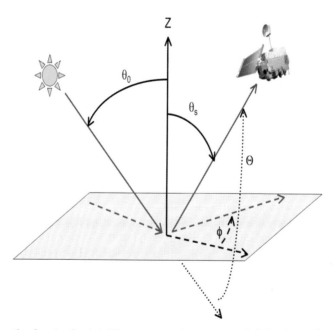

Fig. 2.1. Schematic of sun/surface/satellite remote sensing geometry, defining the angles as viewed from the surface target. The solid lines (and curves) represent solar zenith θ_0 and satellite view zenith θ angles (measured from the zenith, Z). The dashed lines (and curves) represent the relative azimuth angle φ (measured from the extension of the solar azimuth), whereas the dotted lines (and curves) represent the scattering angle Θ (measured from the extension of the direct beam). The Terra satellite icon is from the Earth Observatory (http://earthobservatory.nasa.gov).

Because the measured reflectance includes multiple contributions from the atmosphere and surface, reasonable assumptions must be made to separate them. We can assume that the reflectance observed at the TOA is a function of successive orders of radiation interaction within the coupled surface–atmosphere system. The observed spectral reflectance results from: scattering of radiation within the atmosphere without interaction with the surface (known as the 'atmospheric path reflectance'); the reflection of radiation off the surface that is transmitted to the TOA (the 'surface function'); and the reflection of radiation from outside the sensor's field of view (the 'environment function'). The environment function is often small so that to a good approximation, the angle-dependent TOA reflectance at a wavelength λ is described by:

$$\rho_\lambda^*(\theta_0, \theta, \varphi) = \rho_\lambda^a(\theta_0, \theta, \varphi) + \frac{T_\lambda(\theta_0) T_\lambda(\theta) \rho_\lambda^s}{1 - s_\lambda \rho_\lambda^s}, \tag{26}$$

where ρ_λ^a represents the atmospheric path reflectance, including aerosol and molecular contributions, T_λ are the 'upward (and downward) transmissions' (direct plus diffuse) defined for zero surface reflectance (reciprocity implied), s_λ is the *atmospheric backscattering ratio* (diffuse reflectance of the atmosphere for isotropic light leaving the surface), and ρ_λ^s is 'surface reflectance' [Kaufman et al., 1997a], which for now we assume to be Lambertian. Except for the surface reflectance, each term on the right-hand side of Eq. (26) is a function of the aerosol type (chemical composition, size distribution) and its columnar loading τ. Assuming dark and well-defined spectral surface reflectance, accurate measurements of TOA spectral reflectance can lead to retrievals of spectral τ and reasonable estimates of one or more aerosol size parameters [Tanré et al., 1996]. Note that in the context of satellite observations, the 'target' angles are referred to as 'sensor' angles or 'view' angles. In any case, the Earth's surface is considered the vantage point. As the surface becomes brighter, the term in the denominator becomes smaller, leading to larger possible errors in calculating atmosphere–surface interactions. The entire third term is zero if the surface is black ($\rho^s = 0$).

In essence, the goal of the satellite retrieval process is to first isolate the atmospheric path reflectance, then determine the portion that is contributed by aerosol (e.g. the aerosol path reflectance). Fig. 2.2 demonstrates the spectral response of aerosol scattering for a number of idealized aerosol types.

In order to reduce the computational cost of radiative transfer calculations at every satellite-observed pixel, most operational aerosol retrievals from a satellite make use of a look-up table (LUT). The LUT is a simulation of the atmospheric contribution to the TOA reflectance, namely the non-surface terms in Eq. (26). The LUT must be sufficiently representative of all reasonably likely atmospheric scenarios and satellite observations. Not only must the LUT span the real parameter space, it must have sufficient sensitivity to information contained in the measurements.

In addition to MODIS (this chapter), tropospheric aerosol properties have been operationally retrieved from passive (non-emitting), nadir-viewing, polar-orbiting satellite sensors, such as the Advanced Very High Resolution Radiometer (AVHRR) [Stowe et al., 1997; Husar et al., 1997; Higurashi et al., 1999], and the Total Ozone Mapping Sensor (TOMS) [Herman et al., 1997; Torres et al., 1998, 2002], both of which have been flown on a variety of satellites over the past two decades. The Multi-Angle Imaging Spectro-

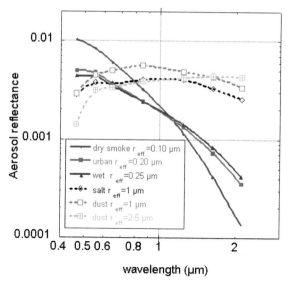

Fig. 2.2. Spectral dependence of aerosol reflectance for selected aerosol types viewed from the top of the atmosphere (for some arbitrary AOD). Aerosol reflectance depends on single scattering albedo, phase function and total aerosol loading. (Figure provided by Yoram Kaufman.)

Radiometer (MISR; e.g., Diner et al. [2007] and Kahn et al., [2005]), uses multiple angle views to retrieve detailed aerosol properties. Other passive sensors and their algorithms are described in this book, and have been compared by previous studies (e.g. Kokhanovsky et al. [2007, 2009]).

3. Aerosol remote sensing from MODIS

3.1 The MODIS instrument

The MODerate resolution Imaging Spectro-radiometer (MODIS) was designed with aerosol and cloud remote sensing in mind [King et al., 1992]. From polarorbit, approximately 700 km above the surface and a $\pm$ 55° view scan, MODIS views the Earth with a swath of about 2330 km, thereby observing nearly the entire globe on a daily basis (Fig. 3), and repeating orbits every 16 days. MODIS measures radiance in 36 wavelength bands, ranging from 0.41 to 14.235 μm [Salomonson et al., 1989], with nadir on-ground spatial resolutions between 250 meters and 1 km (off-nadir-angle pixels represent larger surface areas). Its measurements are organized into 5-minute sections, known as *granules*, each $\sim$ 2300 km long. MODIS actually flies on two NASA satellites, *Terra* and *Aqua*. *Terra* has a *descending* orbit (southward), passing over the equator about 10:30 local sun time, whereas *Aqua* is in *ascending* orbit (northward), so that it passes over the equator about 13:30 local sun time.

The combination of dark-target (over dark land and ocean) aerosol retrieval uses the seven so-called 'land' wavelength bands (listed in Table 2.1), which are all in atmospheric *windows* (little absorption by gases). Included in Table 2.1 are estimates of the central

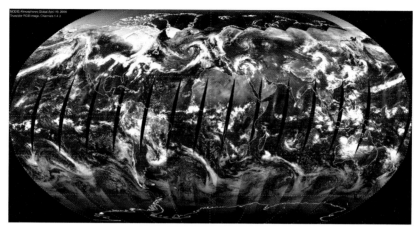

Fig. 2.3. Example of a composite MODIS 'RGB' image for a whole day, April 1, 2001 (Terra). 'RGB' or 'True-Color' images are a composite of reflectance in 0.47, 0.55 and 0.65 μm. Image created by Mark Gray.

wavelength in each band (obtained by integration of the channel-averaged response functions). The MODIS channels 1, 2, 3, 4, 5, 6 and 7 are known here as the 0.65, 0.86, 0.47, 0.55, 1.24, 1.64 and 2.12 μm channels, respectively. In addition, the aerosol algorithm makes use of radiance in other MODIS bands to help with cloud and surface screening. The 'Deep Blue' algorithm (e.g., Hsu et al. [2006]), uses a different set of wavelength bands, but is not discussed further here.

The MODIS instrument is designed to be spectrally stable and sufficiently sensitive to changes in aerosol properties (e.g., Guenther et al. [2002] and Xiong et al. [2003]). The spectral stability for each instrument is better than 2 nm (0.002 μm), with absolute radiance calibrated to within 2 % and is monitored by a solar diffuser screen onboard each satellite. The 'noise equivalent differential spectral reflectance' (Ne$\Delta\rho$) represents the sensitivity to changes in the signal, and is an inherent property of the instrument. The 'signal-to-noise ratio' (SNR) is defined as the ratio of the 'typical scene reflectance' (ρ^{ts}) and the Ne$\Delta\rho$. The Ne$\Delta\rho$ and the SNR specifications are given in Table 2.1. To be understood in the framework of aerosol remote sensing, the definition of SNR should be based on the

Table 2.1 Characteristics of MODIS channels used in aerosol retrieval

Band no.	Bandwidth (μm)	Weighted central wavelength (μm)	Resolution (m)	Ne$\Delta\rho$ ($\times 10^{-4}$)	Max ρ	Required SNR	ROD
1	0.620–0.670	0.646	250	3.39	1.38	128	0.0520
2	0.841–0.876	0.855	250	3.99	0.92	201	0.0165
3	0.459–0.479	0.466	500	2.35	0.96	243	0.1948
4	0.545–0.565	0.553	500	2.11	0.86	228	0.0957
5	1.230–1.250	1.243	500	3.12	0.47	74	0.0037
6	1.628–1.652	1.632	500	3.63	0.94	275	0.0012
7	2.105–2.155	2.119	500	3.06	0.75	110	0.0004

Notes: Ne$\Delta\rho$ corresponds to the sun at zenith ($\theta = 0°$), ROD is 'Rayleigh Optical Depth', SNR is 'Signal to Noise Ratio'

expected aerosol signal. Therefore, the 'noise equivalent differential optical depth' ($Ne\Delta\tau$) can be defined as

$$Ne\Delta\tau = Ne\Delta\rho \frac{4\cos(\theta_0)\,\cos(\theta_v)}{\omega_0 P(\Theta)} \qquad (27)$$

for small AODs. The least sensitivity to aerosol optical depth (largest noise) is expected when both sun and satellite are at nadir views ($\theta_0 = \theta_v = 0.0$), the sum of the Rayleigh and aerosol phase functions are near minimum ($\Theta \sim 90 - 120°$) and the channel used is the least sensitive (channel 7, at 2.12 μm). With a typical phase function value of 0.08 at 120°, a typical aerosol scene requires $Ne\Delta\tau \sim 1.5 \times 10^{-2}$. The 2.12 μm channel's 'typical scene optical depth' (τ^{ts}) is of the order of 0.01 or less, suggesting that the SNR defined by the ratio $\tau^{ts}/Ne\Delta\tau$ is SNR ~ 0.66. If one requires that the SNR is greater than 10 for sufficient sensitivity to aerosol variability, then a single 500-m pixel is insufficient.

However, if individual pixels are aggregated to larger areas, say to a grid of 10×10 km^2 (20×20 of 500-m pixels at nadir), then the noise is reduced by a factor of 400. Instead of 0.66, the SNR becomes 266. However, since clouds and surface inhomogeneities affect aerosol retrievals, not all pixels in the aggregate box may be suitable for aerosol retrieval. If only 5 % of the 500-m pixels are suitable for retrieval, the SNR is reduced to 13. Therefore, to require SNR > 10, 10×10 km^2 boxes can be safely used as the default retrieval size [Tanré et al., 1996]. Of course, if either the aerosol signal is larger or the noise is lower, then the retrieval may require fewer pixels.

3.2   Overview of MODIS dark-target algorithms

Since the launch of MODIS aboard *Terra* (in late 1999) and aboard *Aqua* (in early 2002), MODIS spectral reflectance observations have led to retrievals of spectral τ and a measure of aerosol size, known as the fine weighting (FW or η), each with 10 km resolution (at nadir). Separate dark-target algorithms derive aerosol properties over dark land and ocean [Remer et al., 2005], necessitated by different surface optical characteristics. The suite of MODIS dark-target algorithms were originally formulated by Kaufman et al. [1997a] over land and by Tanre et al. [1997] over water. By the time of MODIS launch on *Terra* (in December 1999), the algorithm had been already revised in order to align with actual MODIS specifications and operational needs. The operational algorithms and products have been continuously evaluated for self-consistency and comparability to other datasets, including AERONET [Remer et al., 2005]. MODIS algorithms are organized by 'versions' (e.g., vX.Y.Z, where X represents major 'science' update, Y represents minor updates, and Z represents bug fixes or otherwise presumably small updates; see http://modis-atmos.gsfc.nasa.gov/MOD04/history), whereas products are arranged as 'collections' (e.g., c00X, where X represents major science updates or reprocessing). After initial review by the MODIS science team, the products were released to the public as 'Collection 003'. Chu et al. [2002] and Remer et al. [2002] evaluated c003 products over land and ocean, respectively. Soon after *Aqua* was launched (in June 2002), the algorithm was applied to both MODIS instruments, beginning the product dataset known as 'collection 004' (c004). The v4.2.2 of the algorithm, described by Remer et al. [2005], was included in production of c004.

Whether the target is land or ocean, the algorithm must ensure that the target is free of clouds, snow, ice and extreme surface variability. A number of tests are performed to separate water bodies and land surfaces and to select appropriate pixels for retrieval (details in Remer et al. [2005] and MAST [2006]). Over either surface, some of the brightest and darkest pixels (within the 10-km box) are removed, in order to reduce residual cloud and surface contamination effects (such as cloud brightening, shadowing or adjacency effects). Both dark-target algorithms utilize look-up tables to simulate the radiative effects of a small set of aerosol types, loadings, and geometry that presumably span the range of global aerosol conditions [Kaufman et al., 1997b]. The goal of the algorithm is to select which of the LUT's simulated scenarios best matches the MODIS-observed spectral reflectance. To retrieve realistic aerosol properties, it is essential that the LUT represents realistic scenarios.

The first step in MODIS processing is to collect the raw data (known as Level 0), cut them into five minute chunks (known as granules), and present them as formatted data (Level 1A). Each granule is converted into calibrated radiance/reflectance and geo-location data (known as *Level 1B* or *L1B*). The aerosol retrieval uses calibrated reflectance data from the seven MODIS bands listed in Table 2.1. These reflectance data are first corrected (by about 1–2 %) for trace gas and water vapor columns, using 'ancillary' data from NCEP (National Centers for Environmental Prediction) analysis [MAST, 2006]. They are organized into 10×10 km boxes (e.g., 40×40 of 250 m data, 20×20 of 500 m data and 10×10 of 1 km data), and separated into land and ocean pixels. Depending on the relative dominance of either surface, the appropriate algorithm is assigned. Near coastlines, if any of the observed pixels are considered land, then the over-land algorithm is followed.

Primary products for dark-target algorithms include the total aerosol optical depth (τ) at 0.55 μm and an estimate of the fine aerosol weighting (η) to the total optical depth. During the course of either algorithm, certain criteria are evaluated and the final products are given Quality Assessment (QA) values to indicate subjective 'confidence', ranging from 3 (high) to 0 (none). The MODIS products include validated, not yet validated, and not able to be validated parameters; it is up to the user to determine how to utilize the QA information. It is also noted that whereas the definitions of τ are the same, the definitions of η are different for over dark land and ocean. It is instructive to briefly introduce the over (dark) ocean algorithm before describing the dark land algorithm in detail.

The main premise of the over-ocean algorithm (unchanged since inception through c005, e.g., Tanré et al. [1997], Levy et al. [2003], Remer et al. [2005], and MAST [2006]) is that the ocean reflectance is generally close to zero at red (0.66 μm) and longer wavelengths, providing a dark background to view aerosol. If all pixels in the 10×10 km box are identified as water pixels, the ocean algorithm is followed. First, obstructed pixels (cloudy, or otherwise unsuitable for retrieval) are removed, including: those within the glint mask (within 40° of the specular reflection angle), those flagged as cloudy [Platnick et al., 2003; Martins et al., 2002; Gao et al., 2002], and those that contain suspended river or other sediments [Li et al., 2003]. The remaining good pixels are sorted by their 0.86 μm brightness. Of these, the darkest and brightest 25 % are removed, theoretically eliminating residual cloud and or surface contamination. If at least 10 pixels remain in the 10×10 km box, then spectral reflectance statistics are calculated and used for the inversion. The over-ocean inversion attempts to minimize the difference between the observed spectral radiance in six MODIS channels and radiance pre-computed in a LUT (slightly modified for c005; MAST, 2006). The ocean LUT simulates the spectral reflectance arising from combined contributions of aerosol, molecular and ocean surface (chlorophyll, whitecaps and sunglint) radiation interactions. The LUT is computed by vector radiative transfer equations (to include polarization effects on the radiance) for 2304 sun/surface/satellite geometries and five total aerosol loadings, for four fine modes and five coarse modes [Remer et al., 2005]. The inversion first interpolates the LUT to match the sun/surface/satellite geometry of the observation. The major assumption is that the total aerosol contribution is composed of a single fine and single coarse mode. For each combination of fine and coarse modes (20 combinations) the inversion determines the total spectral τ and the *fine mode weighting* (η) to the total τ that minimizes the least squares difference error (ε) between the modeled and observed spectral reflectance. The fine/coarse mode combination providing the smallest ε is the final solution. A variety of other aerosol parameters are inferred, including the effective radius of the aerosol.

3.3 Overview of MODIS dark-target algorithm over land

Land surfaces do not provide the same uniform surface signal as the ocean. They are much more variable in their reflectance properties and therefore the algorithm must include additional steps to estimate the land surface contribution to the satellite-observed signal. If the surface is well behaved (i.e., it is either completely dark or its reflectance can be accurately modeled or assumed), the atmospheric signal may be sufficiently decoupled from the combined surface/atmosphere signal.

The aerosol retrieval over land uses spectral reflectance in five of the channels listed in Table 2.1, specifically the 0.66, 0.86, 0.47, 1.24 and 2.12 μm channels. Preliminary steps of the retrieval include testing the spectral observations to screen the 10 km box for clouds [Martins et al., 2002; Gao et al., 2002], snow and ice [Li et al., 2005], and sub-pixel water bodies such as ponds or swamps [Remer et al., 2005]. The pixels that remain are sorted by their relative reflectance (at 0.66 μm), such that the 20 % of the darkest pixels and 50 % of the brightest pixels are removed. The remaining pixels are expected to represent dark surface targets with the least amount of contamination from clouds (including cloud brightening and shadowing) as well as surface inhomogeneities. This means that, at most, 120 pixels remain from the original 400. The retrieval can proceed if at least 12 pixels (10 %; 3 % of the original 400 for sufficient SNR) remain. These remaining pixels are averaged, yielding one set of spectral reflectance values that are used to retrieve aerosol products representing 10 km.

The following subsections provide overview for each of the major updates implemented for MODIS's second-generation aerosol retrieval over land (e.g. Levy et al., 2007a, 2007b].

3.3.1 Aerosol model climatology

The tendency for MODIS c004 to under-predict AOD in conditions of high aerosol loading (regression slope less than one) suggested the assumed aerosol models were deficient. Levy et al. [2005] and Ichoku et al. [2002] demonstrated that updated aerosol model assumptions would lead to better retrievals over the East Coast of the USA and Southern Africa, respectively. The goal, therefore, is to derive a set of aerosol optical models that represent the range of global, realistic aerosol scenarios.

Levy et al. [2007a] describes a 'subjective' cluster analysis for AERONET data, using all AERONET Level 2 (Version 1) data that were processed as of February 2005, encompassing retrievals derived from spherical as well as spheroid models. Their analysis included only the data that met minimum quality parameters suggested by the AERONET team, including: τ at $0.44\,\mu$m greater than 0.4, solar zenith angle greater than $45°$, 21 symmetric left/right azimuth angles, and radiance retrieval error less than 4%. Even with these strict restrictions, the dataset comprised 13 496 retrievals made using spherical assumptions and 5128 that assumed mixtures of spheroids. Over a hundred land sites were represented. In order to differentiate between aerosol types, the data were separated into 10 discrete bins of AOD. Presumably, distinct aerosol types would stay distinct across bins, but changes as function of AOD would characterize the type of 'dynamic' (function of AOD) properties expected for some aerosol types (e.g., Remer and Kaufman [1998]). As spheroid-model solutions tend to be representative of coarse dust aerosol in the almucantar, spherical model solutions presumably are related with fine-dominated aerosol conditions.

Since neither total AOD nor particle shape were considered dependent variables (as were a similar clustering by Omar et al. [2005]), AERONET were separated into broad 'types', based on combination of single scattering albedo (ω_0) and asymmetry parameter g. Presumably, ω_0 differentiates between non-absorbing aerosols (such as urban/industrial pollution [Remer and Kaufman, 1998; Dubovik et al., 2002]) and more absorbing aerosols (such as savanna burning smoke [Ichoku et al., 2003; Dubovik et al., 2002a]), and g helps identify relative changes to the phase function that are related to changes in fine aerosol size. In the end, three distinct clusters were identified, separated primarily by their single scattering albedo in the visible. These fine-dominated aerosol 'types' included strongly

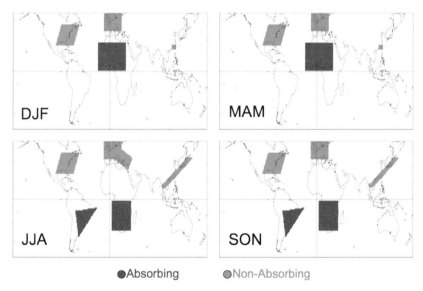

Fig. 2.4. Spherical fine-dominated aerosol type designated at $1° \times 1°$ per season. Red and green represent strongly absorbing ($\omega_0 \sim 0.85$) and weakly absorbing ($\omega_0 \sim 0.95$), respectively. Moderately absorbing ($\omega_0 \sim 0.91$) is assumed everywhere else (from Levy et al. [2007a]).

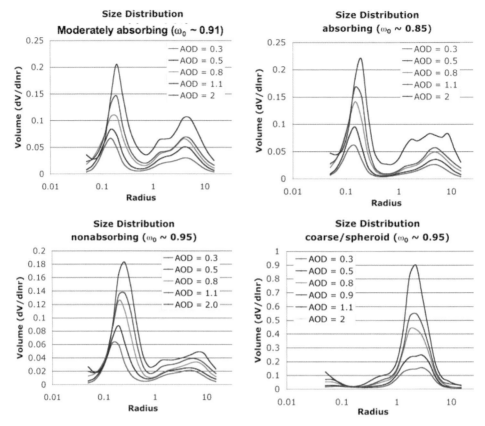

Fig. 2.5. Aerosol size distribution as a function of optical depth for the three spherical (moderately absorbing, strongly absorbing and weakly absorbing) and spheroidal (dust) models identified by clustering of AERONET (Levy et al. [2007a]). Please add: The units for each panel are μm for Radius and ($\mu m^3/\mu m^2$) for the volume concentration.

absorbing ($\omega_0 \sim 0.85$), *moderately absorbing* ($\omega_0 \sim 0.91$), and *weakly absorbing* ($\omega_0 \sim 0.95$).

As assumed for c004 [Remer et al., 2005], the theoretical over-land retrieval would require the fine-dominated aerosol type to be fixed, or prescribed for a given location and season. For each AERONET site, and for each season, Levy et al., [2007a] determined whether there was clear dominance of either the weakly absorbing or the strongly absorbing aerosol types. If neither extreme was dominant, then the site/season was assumed to be characterized by moderately absorbing aerosol. In most cases, the aerosol types assigned at each location were analogous to those assigned for c004, with exceptions where AERONET provided new information not known before 2003 (such as higher ω_0 over Southeast Asia; Eck et al. [2005]).

Fig. 2.4 assigns the fine-dominated aerosol type, as a function of season. While in most places the assignment directly follows from the logic described above, in some regions a subjective combination of literature and logic was needed to designate areas not otherwise characterized by AERONET. For example, even though there were few AERONET sites in

Table 2.2(a) Size properties of the aerosol models used for the c005 dark land LUT

Model	Mode	r_v (μm)	σ	V_0 ($\mu m^3/\mu m^2$)
Continental				
	1	0.176	1.09	0.305
	2	17.6	1.09	0.7364
	3	0.050	0.693	0.0105
Moderately absorbing (developing world)				
	1	$0.0203\tau + 0.145$	$0.1365\tau + 0.3738$	$0.1642\tau^{0.7747}$
	2	$0.3364\tau + 3.101$	$0.098\tau + 0.7292$	$0.1482\tau^{0.6846}$
Weakly absorbing (urban/industrial)				
	1	$0.0434\tau + 0.1604$	$0.1529\tau + 0.3642$	$0.1718\tau^{0.8213}$
	2	$0.1411\tau + 3.3252$	$0.1638\tau + 0.7595$	$0.0934\tau^{0.6394}$
Strongly absorbing (smoke)				
	1	$0.0096\tau + 0.1335$	$0.0794\tau + 0.3834$	$0.1748\tau^{0.8914}$
	2	$0.9489\tau + 3.4479$	$0.0409\tau + 0.7433$	$0.1043\tau^{0.6824}$
Spheroidal (dust)				
	1	$0.1416\tau^{-0.0519}$	$0.7561\tau^{0.148}$	$0.0871\tau^{1.026}$
	2	2.2	$0.554\tau^{-0.0519}$	$0.6786\tau^{1.0569}$

Listed for each model are two or more lognormal modes, where each mode is characterized by volume mean radius r_v, standard deviation σ, and total volume V_0. Radial properties (r_v and σ or the strongly absorbing and moderately absorbing model are defined for $\tau \leq 2.0$; for $\tau > 2.0$, $\tau = 2.0$ is assumed. Similarly, for the weakly absorbing and spheroidal/dust model, parameters are defined for $\tau \leq 1.0$. V_0 (for all models) is defined for all τ For the continental model, V_0 represents arbitrary τ.

Africa north of the equator, the known surface types and seasonal cycles suggest that heavy absorbing aerosol would be produced during the biomass burning season. Red designates regions where the strongly absorbing aerosol ($\omega_0 \sim 0.85$) is assigned, whereas green represents weakly absorbing ($\omega_0 \sim 0.95$) aerosol. The moderately absorbing ($\omega_0 \sim 0.91$) model is assumed everywhere else. These images were mapped onto a $1°$ longitude by $1°$ latitude grid, such that a fine aerosol type is assumed for each grid point, globally. This global map approach allows for easy alterations as new information becomes available.

Fig. 2.5 plots the averaged size distributions (as a function of τ for the four AERONET-derived models (three derived with spherical assumptions, one from spheroidal). Each aerosol 'model' can be approximated by the sum of two lognormal modes, either dominated by the fine mode (the three spherical models) or the coarse mode (the spheroidal model). Sea salt is not a separate aerosol type because, in general $\tau < 0.4$, and cannot be accurately derived via AERONET inversions. Therefore, sea salt properties can be thought of as a contribution to each of our aerosol models. Tables 2.2(a) and 2.2(b) list the size and optical properties representing the lognormal approximations to the AERONET-derived models, as well as the traditional continental model [Lenoble et al., 1984]. Note how the dynamic nature (function of τ) is contained in the formulas, and represents

Table 2.2(b) Optical properties of the aerosol models used for the c005 dark land LUT

Model	Mode	Refractive Index: $m = n - ki$	ω_0, g (defined at $\tau_{0.55} = 0.5$) (0.47/0.55/ 0.66/2.12 μm)
Continental			ω_0: 0.90/0.89/0.88/0.67 g: 0.64/0.63/0.63/0.79
	1	(1) $1.53 - 0.005i$ (2) $1.53 - 0.006i$ (3) $1.53 - 0.006i$ (4) $1.42 - 0.01i$	
	2	(1) $1.53 - 0.008i$ (2) $1.53 - 0.008i$ (3) $1.53 - 0.008i$ (4) $1.22 - 0.009i$	
	3	(1) $1.75 - 0.45i$ (2) $1.75 - 0.44i$ (3) $1.75 - 0.43i$ (4) $1.81 - 0.50i$	
Moderately absorbing (developing world)			ω_0: 0.93/0.92/0.91/0.87 g: 0.68/0.65/0.61/0.68
	1	$1.43 - (-0.002\tau + 0.008)i$	
	2	$1.43 - (-0.002\tau + 0.008)i$	
Weakly absorbing (urban/industrial)			ω_0: 0.95/0.95/0.94/0.90 g: 0.71/0.68/0.65/0.64
	1	$1.42 - (-0.0015\tau + 0.007)i$	
	2	$1.42 - (-0.0015\tau + 0.007)i$	
Strongly absorbing (Smoke)			ω_0: 0.88/0.87/0.85/0.70 g: 0.64/0.60/0.56/0.64
	1	$1.51 - 0.02i$	
	2	$1.51 - 0.02i$	
Spheroidal (dust)			ω_0: 0.94/0.95/0.96/0.98 g: 0.71/0.70/0.69/0.71
	1	(1) $1.48\tau^{-0.021}(0.0025\tau^{0.132})i$ (2) $1.48\tau^{-0.021}0.002i$ (3) $1.48\tau^{-0.021}(0.0018\tau^{-0.08})i$ (4) $1.46\tau^{-0.040}(0.0018\tau^{-0.30})i$	
	2	(1) $1.48\tau^{-0.021}(0.0025\tau^{0.132})i$ (2) $1.48\tau^{-0.021}0.002i$ (3) $1.48\tau^{-0.021}(0.0018\tau^{-0.08})i$ (4) $1.46\tau^{-0.040}(0.0018\tau^{-0.30})i$	

Each aerosol model is represented by two or more modes. Modal properties are determine as a function of τ, which for some aerosol types is related to increasing humidification. For each mode, the spectral refractive index ($m = n - ki$) is given, where 1–4 indicate wavelengths of 0.47, 0.55, 0.66 and 2.1 μm, respectively. If only one value is given, than it represents all wavelengths. Properties of the strongly absorbing and moderately absorbing model are defined for $\tau \leq 2.0$; for $\tau > 2.0$, $\tau = 2.0$ is assumed. Similarly, for the weakly absorbing and spheroidal/dust model, parameters are defined for $\tau = 1.0$. The last column lists the spectral single scattering albedo, ω_0 and asymmetry parameter, g, calculated for $\tau_{0.55} = 0.5$.

the properties of *ambient* aerosol. While optical properties of aerosol are an intrinsic property of the aerosol (and do not change with AOD), optical properties of the ambient aerosol may be changing with AOD, due to a correlative effect, such as increasing humidity.

3.3.2 Scattering/extinction properties of the aerosol models

Extinction/scattering coefficients and aerosol phase functions for the spherical aerosol models (Table 2.2) can be derived using a Mie code (spherical assumptions). For these calculations, the MIEV code [Wiscombe et al., 1980] was used. However, Mie theory is not appropriate for the coarse (spheroid) aerosol model. Therefore, in place of MIEV, Levy et al. [2007a] modeled the coarse aerosol properties with the same version of the T-matrix assumptions and kernels as used for the AERONET almucantar inversions [Dubovik et al., 2002b, 2006] prior to AERONET Version 2. The assumed spheroid axis ratio distribution is presented as Table 2.3.

Table 2.3 Axis ratio distribution used for spheroid optical property calculations

Axis ratio	Frequency
0.4019	0.14707
0.4823	0.10779
0.5787	0.10749
0.6944	0.06362
0.8333	0.
1.	0.
1.2	0.
1.44	0.09063
1.728	0.14186
2.0736	0.16846
2.48832	0.17308

Fig. 2.6 plots the final phase function at 0.55 μm for each model as well as spectral dependence of three parameters (τ, ω_0 and g), calculated for a reference $\tau_{0.55} = 0.5$. As an additional check, Levy et al. [2007a] compared the spectral dependence of the modeled aerosol types to some independent AERONET sun observations, and found that the aerosol models were reasonably representative of observed aerosol. The columnar mass concentration, M, for a given optical depth, is the product of the mass concentration coefficient (M_c; Eq. (21)) and the AOD, e.g.,

$$M = \tau M_c. \tag{28}$$

Table 2.4 lists the extinction, scattering and mass conversion factors for the four AERO-NET-derived aerosol models, along with the continental model for comparison. In each case, $\tau_{0.55} = 0.5$.

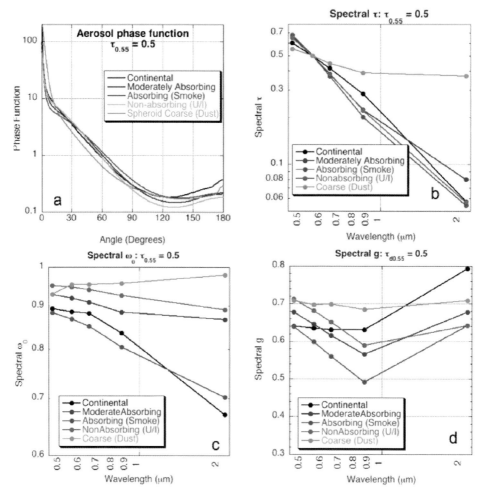

Fig. 2.6. Plot of optical properties for the five aerosol models of the dark-target LUT, for $\tau_{0.55=0.5}$. a) $P(\Theta)$ at 0.55 μm, b) $\tau(\lambda)$, c) $\omega_0(\lambda)$ and d) $g\lambda)$. [Levy et al., 2007a].

Table 2.4 Extinction/mass properties of the aerosol models

Model	ω_0	Q_{ext}	r_{eff} [μm]	B_{ext} [m^2/g]	M_c [μg/cm^2]
Continental	0.886	0.621	0.293	1.5910	62.8600
Moderately absorbing (developing)	0.910	1.018	0.261	2.9220	34.2230
Strongly absorbing (smoke)	0.869	0.977	0.256	3.5330	28.3070
Weakly absorbing (urban/industrial)	0.947	1.172	0.207	3.4310	29.1460
Spheroid (dust)	0.953	1.339	0.680	1.4770	67.6960

Listed for each model are the single scattering albedo, extinction efficiency, effective radius, mass extinction coefficient and mass concentration conversion factor. All optical parameters are defined at 0.55 μm, representing a loading of $\tau_{0.55} = 0.5$. For comparison, particle density is assumed 1 g/cm^3.

3.3.3 Radiative transfer and aerosol LUT calculation

The c004 MODIS LUT contained simulated aerosol reflectance in two channels (0.47 and 0.66 μm), calculated using a non-polarized (scalar) RT code [Dave, 1970]. Levy et al. [2004] found that neglecting polarization could lead to significant errors in top of atmosphere reflectance, resulting in significant errors in retrieving AOD. Therefore, vector RT code was preferred. Due to developer experience with the code, and satisfactory test performance compared to benchmarks, Levy et al., [2007a] selected the vector adding doubling code (RT3; Evans and Stephens [1991]) to simulate the atmospheric contribution to TOA reflectance in the LUT. For a given wavelength (Table 2.1), RT3 was run, assuming aerosol scattering/extinction properties (either derived from MIEV or T-matrix codes), Rayleigh optical depths (Table 2.1), realistic atmospheric profiles (US Standard Atmosphere; [US Government, 1976]), and surface reflectance (zero). Since the MODIS algorithm makes corrections for gas absorption (ozone, water vapor, etc.) to the TOA reflectance, gas absorption is not included in the RT runs.

For the c005 LUT, atmospheric path radiance (coupling of aerosol and Rayleigh) was calculated in four MODIS channels (0.47, 0.55, 0.66 and 2.12 μm, representing MODIS channels 3, 4, 1 and 7, respectively), to represent seven different loadings of each of the five aerosol types, and 2304 combinations of solar/surface/satellite scattering geometry. Loadings are indexed by AOD at 0.55 μm, having values of $\tau = 0.0$ (Rayleigh only), 0.25, 0.5, 1.0, 2.0, 3.0 and 5.0. The LUT includes information for nine solar zenith angles ($\theta_0 = 0.0$, 12.0, 24.0, 36.0, 48.0, 54.0, 60.0, 66.0 and 72.0), 16 sensor zenith angles ($\theta = 0.0$ to 65.0, approximate increments of 6.0), and 16 relative azimuth angles ($\varphi = 0.0$ to 180.0 increments of 12.0). The approximate increments of θ arise from the use of the Lobatto quadrature function in RT3, allowing for sensor zenith angles similar to those in the c004 LUT.

If there is nonzero surface reflectance, the second term in Eq. (26) is nonzero. RT3 code derives s and T, which are included in the LUT (as function of wavelength, aerosol model, aerosol loading and geometry). Additionally, the LUT includes values for the scattering and extinction efficiencies/coefficients/mass coefficients $Q/\beta/M_c$ of the aerosol models (also having similar functional dependencies).

3.3.4 Surface reflectance relationships

For c004, the surface reflectance in two visible channels (0.47 and 0.66 μm), are each assumed to be fixed ratios of that at 2.12 μm (e.g., Kaufman et al., 1997a; Remer et al., 2005], specifically 0.25 and 0.5, respectively. The empirical relationship was considered to be characteristic of the relationship of liquid water absorption and chlorophyll reflectance across different vegetation states [Kaufman and Remer, 1994; Kaufman et al., 2002]. However, studies such as Remer et al., [2001] and Gatebe et al., [2001] showed that the relationship of visible to 2.12 μm surface reflectance ('VISvs2.1') varies as a function of geometry and surface type.

Levy et al. [2007b] refined the VISvs2.1 relationships, to reduce aerosol retrieval errors due to insufficient surface information. They collected a set of over 10,000 global AERONET/MODIS co-locations, where the AOD was low (e.g., $\tau < 0.2$) as measured by AERONET. Presumably, in low AOD conditions, multiple scattering by aerosol is small, and thus the co-location is suitable for performing atmospheric correction (e.g., Kaufman and Sendra, 1988; Vermote et al., 1997b] to derive spectral surface reflectance properties. Given enough statistics with different scattering geometry, Levy et al., [2007b] presumed that surface BRDF effects could be approximated by the value of the surface reflectance for the relevant solar and satellite viewing geometry [Kaufman et al., 1997a] assuming a Lambertian surface.

Levy et al. [2007b] concluded that globally averaged ratios of the surface reflectance in the two visible channels (0.47 and 0.66 μm), to that in 2.12 μm, were closer to 0.28 and 0.55, respectively. However, they demonstrated better fits with lines that had both slope and offset terms. In addition, correlation of 0.47 and 0.66 μm surface reflectance was much higher than that at 0.47 and 2.12 μm. The presence of the y-offset is important, because even in the darkest, most water-laden vegetation, zero reflectance at 2.12 μm does not imply zero surface reflectance in the visible channels (e.g., Kaufman et al., [2002]).

Levy et al. [2007b] also found that the VISvs2.12 relationships were dependent on scattering angle (Θ), and influenced by the state of the surface vegetation. They separated the co-locations into over-urban and non-urban areas, by season (winter or summer) and by general location (mid-latitude or tropical). Generally, more vegetated surfaces (mid-latitude summer sites both urban and non-urban) displayed higher 0.66 to 2.12 μm surface reflectance ratios (ratio $>$ 0.55) than winter sites or tropical savannas and grasslands (ratio $>$ 0.55). Except for urban sites, the 0.47 μm to 0.66 μm ratio was relatively constant (ratio $\sim$ 0.52). The relationship of the surface reflectance ratios to known surface condition suggested a relationship to its vegetation amount/condition or 'greenness.' Since, the well-known Normalized Difference Vegetation Index (NDVI; e.g., Tucker and Sellers [1988]) is influenced by aerosol, Levy et al. [2007b] related the VISvs2.1 relationship to the $NDVI_{SWIR}$, defined as:

$$NDVI_{SWIR} = (\rho_{1.24}^m - \rho_{2.12}^m)/(\rho_{1.24}^m + \rho_{2.12}^m) \tag{29}$$

where $\rho_{1.24}$ and $\rho_{2.12}$ are the MODIS-measured reflectances of the 1.24 μm channel (MODIS channel 5) and the 2.12 μm channel (channel 7). These longer wavelengths are less influenced by aerosol (except for extreme aerosol loadings or dusts), and thus may be more useful for estimating surface condition. In aerosol-free conditions, $NDVI_{SWIR}$ is highly correlated with regular $NDVI$; values of $NDVI_{SWIR} > 0.6$ represent active vegetation, whereas $NDVI_{SWIR} < 0.2$ indicates dormant or sparse vegetation. The parameterization of the VISvs2.12 surface reflectance relationship, with functional dependence on $NDVI_{SWIR}$ and Θ, is written as:

$$\rho_{0.6}^s = f(\rho_{2.12}^s) = \rho_{2.12}^s * a_{0.66/2.12} + b_{0.66/2.12} \tag{30}$$

and

$$\rho_{0.47}^s = g(\rho_{0.66}^s) = \rho_{0.66}^s * a_{0.47/0.66} + b_{0.47/0.66},$$

where

$$a_{0.66/2.12} = a_{0.66/2.12}^{NDVI_{SWIR}} + 0.002\Theta - 0.27,$$

$$b_{0.66/2.12} = -0.00025\Theta + 0.033, \tag{31}$$

$$a_{0.47/0.66} = 0.49, \quad \text{and}$$

$$b_{0.47/0.66} = 0.005,$$

where in turn

$$a_{0.66/2.12}^{NDVI_{SWIR}} = 0.48; NDVI_{SWIR} < 0.25,$$

$$a_{0.66/2.12}^{NDVI_{SWIR}} = 0.58; NDVI_{SWIR} > 0.75, \qquad (32)$$

$$a_{0.66/2.12}^{NDVI_{SWIR}} = 0.48 + 0.2(NDVI_{SWIR} - 0.25); 0.25 \leq NDVI_{SWIR} \leq 0.75$$

It should be noted that while the large AERONET/MODIS co-location dataset was the broadest and most comprehensive data available for analyzing global surface reflectance relationships, it was still limited to AERONET site locations, which in turn are mostly concentrated in certain geographical regions. The parameterization leaves room for improvement, especially in non-sampled regions.

3.3.5 Inversion of spectral reflectance, including 2.12 μm

One major limitation of the c004 algorithms was that aerosol was assumed transparent in the 2.12 μm channel [Kaufman et al., 1997a; Remer et al., 2005]. Under a dust aerosol regime, aerosol transparency may be an extremely poor assumption. Even in a fine aerosol dominated regime, τ is nonzero. For the moderately absorbing aerosol model ($\omega_0 \sim 0.91$), $\tau_{0.55} = 0.5$ corresponds to $\tau_{2.12} \sim 0.05$ (Fig. 2.6). This is representative of 2.12 μm path reflectance of 0.005, rather than zero. Via 0.66 to 2.12 μm ratio of 0.55, this becomes a reflectance 'error' of 0.003 at 0.66 μm, in turn leading to error of ~ 0.03 in retrieved AOD. For low aerosol loadings, this can be a large relative error.

In the spirit of the MODIS aerosol over ocean algorithm [Tanré et al., 1997], the second generation of aerosol retrieval over dark land simultaneously inverts spectral reflectance to derive aerosol properties [Levy et al., 2007b]. Analogous to the ocean algorithm's attempt to combine fine and coarse aerosol *modes*, the new land algorithm combines the fine-dominated and coarse-dominated aerosol *models* (each bimodal) to match with the observed spectral reflectance. Thus, the inversion is free to assume that the 2.12 μm channel contains both surface and aerosol information, constrained by the parameterized VISvs2.12 surface reflectance relationships. Simultaneously inverting the aerosol and surface information in the three channels (0.47 μm, 0.66 μm and 2.12 μm) yields enough information, that with some assumptions (about spectral dependence), three parameters can be derived: $\tau_{0.55}$, $\eta_{0.55}$ and the surface reflectance ($\rho_{2.12}^s$). A cartoon representation of the inversion is given in Fig. 2.7.

We rewrite Eq. (26), considering that the calculated spectral total reflectance ρ_λ^* at the top of the atmosphere is approximately the weighted sum of the spectral reflectance from a combination of fine and coarse-dominated aerosol *models*, i.e.,

$$\rho_\lambda^* = \eta_\lambda^{*f} + (1-\eta)\rho_\lambda^{*c}, \qquad (33)$$

where ρ_λ^{*f} and ρ_λ^{*c} are each composites of surface reflectance ρ_λ^s and atmospheric path reflectance (aerosol + Rayleigh) of the separate aerosol models. That is, we have two equations

$$\rho_\lambda^{*f} = \rho_\lambda^{af} + \hat{T}_\lambda^f \rho_\lambda^s / (1 - s_\lambda^f \rho_\lambda^s) \quad \text{and} \quad \rho_\lambda^{*c} = \rho_\lambda^{ac} + \hat{T}_\lambda^c \rho_\lambda^s / (1 - s_\lambda^c \rho_\lambda^s), \qquad (34)$$

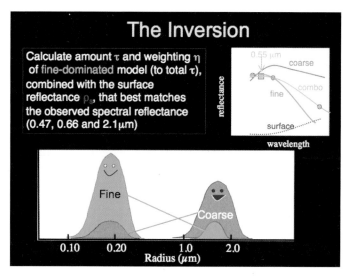

Fig. 2.7. Cartoon of inversion technique. Observed reflectance in three channels is represented by turquoise dots in the small sub-figure. Green represents fine-dominated model (and its induced path reflectance), whereas orange represents coarse-dominated model (and its induced path reflectance). Red denotes surface reflectance. The combination of fine, coarse and surface (combo) is the turquoise square and spectral dependence curve that most closely matches the observations (dots). The 'ghostlike' modal creatures were adapted from a presentation by Lorraine Remer.

where the symbols are defined as in Eq. (33), except for separation into fine and coarse aerosol models (denoted by the superscripts f and c, respectively), and $\hat{T}$ as the two-way transmission $T(\theta_0)T(\theta)$. Surface reflectance (ρ^s) is independent of aerosol type. The weighting parameter, η, is defined at $\lambda = 0.55\mu$m. This weighting parameter also represents the fraction of the total optical thickness at 0.55 μm contributed by fine (non-dust) aerosol [Remer et al., 2005]. Note that large surface reflectance and aerosol backscatter increase the surface/atmosphere couplings in Eq. (34). Nonetheless, for the conditions of the dark-target algorithm, the approximation of Eq. (33) is valid.

3.3.6 Mechanics of the algorithm

Fig. 2.8 is a flowchart of the second-generation land algorithm. The procedure collects Level 1B (L1B) spectral reflectance in eight wavelength bands (Table 2.1, plus 1.37 μm) at their finest spatial resolutions, as well as associated geo-location information. These L1B reflectance values are corrected for water vapor, ozone, and carbon dioxide obtained from ancillary NCEP analysis data files. Details of this gas correction and cloud masking are found online [MAST, 2006]. Basically, the high-resolution (20 × 20 at 500 m resolution) pixels in the 10 × 10 km box are evaluated pixel by pixel to identify whether the pixel is suitable for aerosol retrieval. Clouds [Martins et al., 2002], snow/ice [Li et al., 2003] and inland water bodies (via NDVI tests) are considered not suitable and are discarded (also explained in MAST [2006]).

The non-masked pixels are checked for their brightness. Pixels having measured 2.12 μm reflectance between 0.01 and 0.25 are grouped and sorted by their 0.66 μm reflectance. The brightest (at 0.66 μm) 50 % and darkest 20 % are discarded, in order to

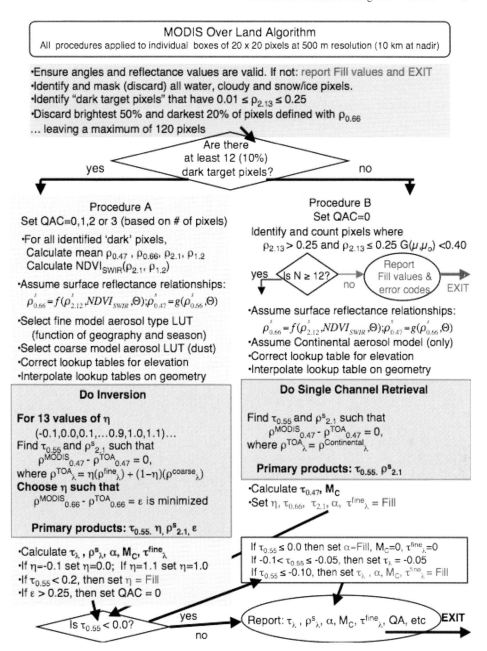

Fig. 2.8. Flowchart illustrating the derivation of aerosol over land for the new algorithm [Levy et al., 2007b].

reduce cloud and surface contamination and scale towards darker targets. If there are at least 12 pixels remaining (10 % of 30 % of the original 400), then the reflectance in each channel is averaged, yielding the 'MODIS-measured' spectral reflectance $\rho_{0.47}^m$, $\rho_{0.66}^m$, $\rho_{1.24}^m$ and $\rho_{2.12}^m$. These reflectance values are used for so-called 'Procedure A'. If less than 12 pixels remain, then 'Procedure B' (described later) is followed.

Along with any values of derived AOD and other aerosol products, the algorithm assigns a relative 'Confidence' for the derived products. This parameter, known as the 'Quality Assessment' (or QA) value, has values of 3, 2, 1, or 0, representing 'high', 'moderate', 'marginal', and 'low' confidence in the derived products. If following 'Procedure B', the QAC is automatically set to 0 (no confidence). The QAC will be discussed in later sections. Note that the logic for assigning QAC is completely new for c005 [MAST, 2006].

3.3.6.1 Correcting the LUT for elevation

A major improvement for the c005 is its treatment of elevated surface targets. This is described in detail by Levy et al., [2007b] and in MAST [2006]. Essentially, since Rayleigh optical depth (ROD or $\tau_{R,\lambda}$) is directly related to sea-level pressure, then an elevated target's 'effective' ROD is that scaled to its elevation (via the hydrostatic equation). In other words, a sea-level ROD for a longer wavelength is equivalent to an elevated ROD for a given wavelength. The algorithm makes use of the procedure described in Fraser et al. [1989], scaling the LUT to simulate different ROD by interpolating to an effective wavelength. For example, for a mountain site with elevation (Z) of 0.4 km, the effective wavelength (λ) increases by about 1.2 %. For the blue 0.47 μm channel, where $\lambda = 0.466$ μm and ROD 0.185. Assuming that gases and aerosols are optically well mixed in altitude, the parameter values of a 0.471 μm LUT can be acquired by interpolating (linearly as functions of log wavelength and log parameter) between the 0.47 μm (0.466 μm) and the 0.55 μm (0.553 μm) entries. Similar interpolations are performed for the other channels (for example, 0.55 μm would be adjusted to 0.559 μm). For the 0.4 km case, this means that lower values of TOA atmospheric path reflectance and higher values of transmission are chosen to represent a given aerosol model's optical contribution. However, also note that since the 0.55 μm channel has also been adjusted, the associated parameters of the look-up table (that are defined for specific τ) have been adjusted accordingly. In other words, the algorithm retrieves aerosol optical depth at the adjusted wavelength, which is equivalent to retrieving τ own to the surface elevation height.

Whereas most global land surfaces are at sea level or above, a few locations are below sea level ($Z = 0$). In these cases, the algorithm is allowed to extrapolate below 0.466 μm. Since the extrapolation is at most for a hundred meters or so, this is not expected to introduce large errors, and these cases are till retrieved. Note also that due to the extremely low ROD in the 2.12 μm channel, little is gained by adjusting this channel.

3.3.6.2 Procedure A: Inversion for darker surfaces

If following Procedure A (for dark surfaces), the QA 'confidence' is initially set to a value between 0 (none) and 3 (high), depending on the number of dark pixels remaining. In Procedure A, the algorithm assigns the fine aerosol model, based on the location and season. From the LUT, ρ^a, $\hat{T}$ and s (for the fine model and coarse model separately) are interpolated for angles (θ_0, θ and φ), and elevation, resulting in six values for each parameter, each one corresponding to a different aerosol loading (indexed by τ at 0.55 μm).

The 2.12 μm path reflectance is a non-negligible function of AOD, so that the surface reflectance becomes related to the AOD. For discrete values of η between -0.1 and 1.1 (intervals of 0.1), the algorithm attempts to find τ at 0.55 μm and the surface reflectance at 2.12 μm that exactly matches the MODIS measured reflectance at 0.47 μm. There will be some error, ε, at 0.66 μm. The solution is the one where the error at 0.66 μm is minimized. In other words,

$$\rho_{0.47}^m \rho_{0.47}^* = 0,$$
$$|\rho_{0.66}^m - \rho_{0.66}^*| = \varepsilon, \tag{35}$$
$$\rho_{2.12}^m - \rho_{2.12}^* 0,$$

where

$$\rho_{2.12}^* = \eta(\rho_{2.12}^{fa} + \hat{T}_{2.12}^f \rho_{2.12}^f/(1 - s_{2.12}^f \rho_{2.12}^s)) + (1-\eta)(\rho_{2.12}^{ca} + \hat{T}_{2.12}^c \rho_{2.12}^s/(1 - s_{2.12}^c \rho_{2.12}^s))$$
$$\rho_{0.66}^* = \eta(\rho_{0.47}^{fa} + \hat{T}_{0.47}^f g(\rho_{0.66}^s)/(1 - s_{0.47}^f g(\rho_{0.66}^s))) +$$
$$+ (1-\eta)(\rho_{0.47}^{ca} + \hat{T}_{0.47}^c g(\rho_{0.66}^s)/(1 - s_{0.47}^c g(\rho_{0.66}^s))),$$

and

$$\rho_{0.47}^* = \eta(\rho_{0.47}^{fa} + \hat{T}_{0.47}^f g(\rho_{0.66}^s)/(1 - s_{0.47}^f g(\rho_{0.66}^s))) +$$
$$+ (1-\eta)(\rho_{0.47}^{ca} + \hat{T}_{0.47}^c g(\rho_{0.66}^s)/(1 - s_{0.47} g(\rho_{0.66}^s))), \tag{36}$$

where in turn, $\rho^a = \rho^a(\tau)$, $\hat{T} = \hat{T}(\tau)$, $s = s(\tau)$ are each functions of τ (and geometry) indexed within the LUT, calculated for separate fine and coarse models. The surface reflectance relationships, $f(\rho_{2.12}^s)$ and $g(\rho_{0.66}^s)$ are described by Eqs (30) and (32). The algorithm includes choices for non-physical values of η (-0.1 and 1.1) to allow for the possibility of imperfect assumptions in either aerosol models or surface reflectance. Again, the primary products are $\tau_{0.55}$, $\eta_{0.55}$ and the surface reflectance ($\rho_{2.12}^s$). The fitting error (ε) is also noted.

Once the solution is found, a number of secondary products can also be calculated. These include the fine and coarse mode optical depths $\tau_{0.55}^f$ and $\tau_{0.55}^c$:

$$\tau_{0.55}^f = \tau_{0.55}\eta_{0.55} \quad \text{and} \quad \tau_{0.55}^c = \tau_{0.55}(1 - \eta_{0.55}), \tag{37}$$

the columnar mass concentration, M:

$$M = M_c^f \tau_{0.55}^f + M_c^c \tau_{0.55}^c, \tag{38}$$

the spectral total and model optical thicknesses τ_λ, τ_λ^f, and τ_λ^c:

$$\tau_\lambda = \tau_\lambda^f + \tau_\lambda^c, \tag{39}$$

where

$$\tau_\lambda^f = \tau_{0.55}^f(\bar{Q}_\lambda^f/\bar{Q}_{0.55}^f) \quad \text{and} \quad \tau_\lambda^c = \tau_{0.55}^c(\bar{Q}_\lambda^c/\bar{Q}_{0.55}^c), \tag{40}$$

the Ångström Exponent a:

$$a = - \ln(\tau_{0.47}/\tau_{0.65}) / \ln(0.4/0.65) \tag{41}$$

and the spectral surface reflectance ρ_λ^s, M_c^f and M_c^c are mass concentration coefficients for the fine and coarse model (e.g., Table 2.3), whereas Q_λ^f and Q_λ^c represent model extinction efficiencies at the wavelength, λ. If the resulting products are inconsistent, have large fitting error, or otherwise seem suspect, then QA confidence is lowered. Note that the subscripts in Eq. (41) refer to the MODIS channels, whereas the exact wavelengths are needed for the right-hand side.

3.3.6.3 Procedure B: alternative retrieval for brighter surfaces

The derivation of aerosol properties is still possible when the 2.12 μm reflectance is brighter than 0.25, but is expected to be less accurate [Remer et al., 2005], due to increasing errors from applying the VISvs2.12 relationship. However, if there are at least 12 cloud-screened, non-water pixels that satisfy

$$0.25 < \rho_{2.12}^m < 0.25G < 0.40, \tag{42}$$

where

$$G = 0.5 \left(\frac{1}{\mu} + \frac{1}{\sqrt{\mu_0}} \right), \tag{43}$$

then Procedure B is attempted. In this relationship μ_0 is the cosine of the solar zenith angle, and μ is cosine of the satellite view angle. Eq. (43) represents the combination of up and down slant paths of the radiation, taking into account the increase of atmospheric photon path for oblique angles. The contribution from the surface reflectance becomes less important, and the retrieval can tolerate higher surface reflectance [Remer et al., 2005].

Procedure B is analogous to 'Path B' of the c004 algorithm described in Remer et al. [2005], in that the continental aerosol model is assumed. Unlike c004, the continental aerosol properties are indexed to 0.55 μm, and we use the VISvs2.1 surface reflectance parameterization. Since η is undefined, the products for Procedure B include only $\tau_{0.55}$, the surface reflectance, and the fitting error. The QA is set to 0 (no confidence). Some of the secondary products are undefined.

3.3.7 Low and negative optical depth retrievals

A major philosophical change for the second-generation algorithm is that retrievals of negative AOD allowed. Given that there is both positive and negative noise in the MODIS observations, and that surface reflectance and aerosol properties may be under- or over-estimated depending on the retrieval conditions, it is statistically imperative to allow retrieval of negative AOD. In fact it is necessary for creating an unbiased dataset from any instrument. Without negative retrievals, AOD is biased by definition. However, a large negative retrieval indicates a situation outside the algorithm's solution space and should not be reported. The trick is to determine the cutoff between retrieved AOD that is essentially zero, and a value that is truly wrong. If within the expected error defined by Eq. (1), then values down to -0.05 should automatically be tolerated. Allowing for slightly higher uncertainty, the algorithm include τ retrievals down to -0.10 (twice the expected error in pristine aerosol conditions), but report these values as -0.05 and lower the QA value. Note that all retrievals with $-0.05 < \tau < 0$ are reported with high QA value $= 3$, unless identified as poor quality for some other reason. Some of the products that are retrieved or derived (such as η Ångström exponent) are set to zero or reported as not defined for negative retrievals. In cases of low $\tau(\tau < 0.2)$, η is too unstable to be retrieved with any accuracy. Therefore, η is reported as undefined even though other parameters (such as Ångström exponent and fine model τ) may be reported.

3.3.8 Sensitivity test

Because the second-generation aerosol algorithm utilizes MODIS channels with wide spectral range, the inversion (Procedure A) should be able to retrieve AOD with robustness, and have some sensitivity to the aerosol size (e.g., η). Levy et al. [2007b] described a sensitivity study that performed the following tests: (1) simulation of aerosol loadings and angles included within the LUT, with specific combinations of fine and coarse modes (distinct values of η that are not necessarily integral multiples of 0.1), (2) simulation of aerosol loadings not contained within the standard LUT (e.g., additional τ values) and (3) simulations for LUT conditions, but including one or more prescribed errors.

For most aerosol and geometrical conditions represented by the LUT, the algorithm successfully derived AOD within $\Delta\tau|0.01|$ for $\tau < 1.0$, and within $\Delta\tau|0.1|(10)$ for $\tau < 1.0$. Retrievals of surface reflectance were within 10% for most cases, and η was usually within 0.1 of the simulated value. When they simulated AOD conditions not within the standard LUT (e.g., $\tau = 0.35$, 1.5 and 6.0), the retrieval demonstrated 10 error in retrieved AOD, except for $\tau = 6.0$, which must be extrapolated for the standard LUT. Retrievals of η were usually qualitatively comparable with inputted values. Finally, products derived from perturbed input conditions (simulating errors in aerosol model, angular information, instrument calibration, assumed target elevation, etc.), demonstrated that retrieved AOD could generally be expected to lie within that described by Eq. (1).

3.4   L2 c005 products over dark land

Table 2.4 lists the 'dark-land target' L2 aerosol products (known as scientific datasets, or SDS). For each SDS, the table lists its name within the file, its dimension, and its 'type'. All products are derived spatially (135×203, representing 10×10 km resolution at nadir). For some products, a third dimension is listed, usually to indicate that it is reported at multiple wavelengths (also listed). A parameter's type may be Retrieved, Derived, Diagnostic, Experimental, or Joint. A *Retrieved* parameter is one that results directly from the inversion (e.g., 'Corrected_Optical_Depth_Land' SDS), whereas those *Derived* (such as the Ångström exponent) result from those retrieved directly. Products that are *Diagnostic* include the QA ('Quality_Assurance_Land' SDS), and parameters calculated during intermediate steps. These diagnostic parameters can be used to understand how the retrieval worked. Finally, *Joint* products are those that are composites of over-land and over-ocean aerosol retrievals ($0.55\,\mu m$ only), and are defined depending on the QA values assessed over the particular surface. Over land, the SDS 'Optical_Depth_Land_And_Ocean' is defined (equal to 'Corrected_Optical_Depth_Land') only when QA ≥ 1, and is the joint

Table 2.5 Land product contents of L2-MODIS aerosol file

Name of product (SDS): Symbol	Dimensions: Values of 3$^{\rm rd}$ dim.	Type of product
Corrected_Optical_Depth_Land:τ	X,Y,3: 0.47, 0.55, 0.66 μm	Retrieved Primary
Corrected_Optical_Depth_Land_wav2p1:τ	X,Y,1: 2.12 μm	Retrieved Primary
Optical_Depth_Ratio_Small_Land: η	X,Y: (for 0.55 μm)	Retrieved Primary
Surface_Reflectance_Land: ρ_s	X,Y,3: 0.47, 0.66, 2.12 μm	Retrieved Primary
Fitting_Error_Land: ε	X,Y: (at 0.66 μm)	Retrieved By-Product
Quality_Assurance_Land: QA	X,Y,5: 5 bytes	Diagnostic
Aerosol_Type_Land	X,Y:	Diagnostic
Angstrom_Exponent_Land: a	X,Y: (for 0.66/0.47 μm)	Derived
Mass_Concentration_Land: M	X,Y:	Derived
Optical_Depth_Small_Land: τ^f	X,Y,4: 0.47,0.55,0.66,2.12 μm	Derived
Mean_Reflectance_Land: ρ^m	X,Y,7: 0.47,0.55,0.66,0.86, 1.2,1.6,2.12 μm	Diagnostic
STD_Reflectance_Land	X,Y,7: 0.47,0.55,0.66,0.86, 1.2,1.6,2.12 μm	Diagnostic
Cloud_Fraction_Land	X,Y:	Diagnostic
Number_Pixels_Used_Land	X,Y:	Diagnostic
Optical_Depth_Land_And_Ocean	X,Y: 0.55 μm	Joint (QA ≥ 1)
Image_Optical_Depth_Land_And_Ocean	X,Y: 0.55 μm	Joint (QA ≥ 0)
Optical_Depth_Ratio_Small_Land_ And_Ocean	X,Y: 0.55 μm	Joint (QA ≥ 1)

$X = 135$; $Y = 203$. If there is a third dimension of the SDS, then the indices of it are given. The 'Retrieved' parameters are the solution to the inversion, whereas 'Derived' parameters follow from the choice of solution. 'Diagnostic' parameters aid in understanding of the directly Retrieved or Derived products. 'Joint' products are defined over land, depending on QA values.

product recommended for quantitative studies. İmage_Optical_Depth_Land_And_Ocean' is defined even when QA $= 0$, so that it is better suited for qualitative plume monitoring and plotting. More about each individual parameter is described in MAST [2006].

Fig. 2.9 shows c005 aerosol properties retrieved from the granule observed by *Terra* on May 4, 2003 (15:25 UTC) over the US east coast (same granule as presented in King et al. [2003]). Fig. 2.9(A) represents the 'RGB' or 'true-color' image, and is a composite of L1B reflectance in the 0.47, 0.55 and 0.66 μm channels. Figs 2.9(B) and (C) plot the primary products ($\tau_{0.55}$ and η retrieved from combination of dark-target (both land and ocean) algorithms (joint products with QA ≥ 1). Fig. 2.9(D) presents surface reflectance (ρ_s) over land only, as the third primary product retrieved from the dark-land algorithm. These plots indicate an aerosol plume, apparently transported from the Ohio Valley through Maryland and into the Atlantic. Within the plume, AOD is high ($\tau_{0.55} \sim 1.0$), and is dominated by fine particles $\eta \sim 1.0$ (plotted where $\tau_{0.55} > 0.2$). There is land/ocean continuity and we note that AERONET observations in Baltimore (MD_Science_Center) also indicate $\tau \sim 1.0$, dominated by fine aerosol.

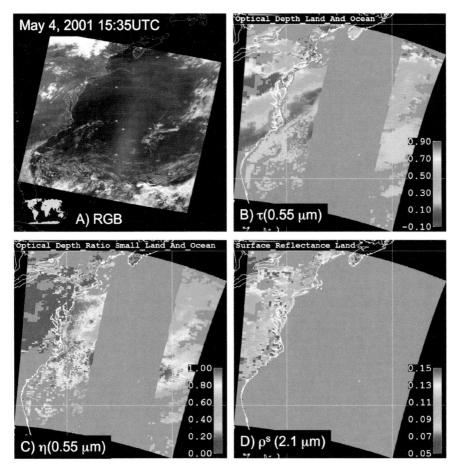

Fig. 2.9. Retrieved L2 aerosol and surface properties over the eastern USA on May 4, 2001 (same granule as King et al. [2003]). Panel (A) is a 'true-color' composite image of three visible channels, showing haze over the mid-Atlantic. Panels (B) and (C) display τ and η combined over land and ocean, showing an aerosol plume ($\tau \sim 1.0$) dominated by fine particles ($\eta \sim 1.0$). The transport of the aerosol into the Atlantic is well represented with good agreement between land and ocean. Note that over-land η is not reported when $\tau < 0.2$. Panel (D) shows the retrieved surface reflectance ρ_s (over land only).

3.5   Level 3 products

The L2 aerosol products are derived along the satellite track, only during the daylight, and where not obscured by clouds or over inappropriate surfaces. The MODIS orbit covers $> 70\%$ of the globe per day (during daylight hours), but it observes some places more than once, and others not at all. MODIS's 16-day repeat orbit ensures that all parts of the (daylight) globe are viewed, but that sampling and geometry (especially over the tropics) changes from day to day. Also, because of orbital geometry, L2 pixels near swath edges represent surface targets three times larger than the nadir 10×10 km resolution. Over certain regions and certain seasons, entire L2 pixels may be cloudy, resulting in no aerosol retrieval at all.

In many applications of satellite data (e.g., comparison with gridded model output), it is highly desirable that the satellite data be represented on regular grid. Therefore, the MODIS atmosphere data team (http://modis-atmos.gsfc.nasa.gov) has developed the concept of *Level 3* (*L3*; e.g., King et al. [2003]) products to represent the statistics of the L2 orbital products on a regular grid ($1° \times 1°$) and regular time intervals (daily, 8-day and monthly). The goal is to derive L3 products that preserve the integrity of the L2 data, but also have geophysical meaning. Because of both the variable sampling and variable confidence of the L2 data, it is not trivial to develop optimal L3 products.

3.5.1 L3-daily (D3)

Fig. 2.10 represents statistics for one day (May 4, 2003) of over-land L2 AOD on a $1° \times 1°$ grid. Fig. 2.10(A) presents the number of possible retrievals (pixel counts, or PC), while Fig. 2.10(B) represents the number of actual PC on this given day (May 4, 2003) over land. Possible PC ranges from > 120 at the poles in the summer hemisphere and at the equator in all seasons, to zero over areas in between orbits near the equator, and in polar darkness. Actual PC depends on MODIS distribution of cloud fields, inappropriate (bright, inhomogeneous) surface types, and other poor retrieval conditions. Fig. 2.10(C) presents the total confidence (sum of QA confidence values), which may range between zero and three times actual PC. Figs 2.10(D) and (E) map two representations of mean AOD. One is the simple average (known as the SDS's 'Mean') of the L2 values within the box (Fig. 2.10(D)). The other is known as the SDS's 'QA_Mean', and is derived from 'Confidence' weighted L2 (giving larger weighting to retrievals with larger QA). Values of $QA = 3$ are assigned a weighting $Q = 3$, whereas $QA = 0$ are weighted as $Q = 0$ (no weighting). Mathematically, the 'QA_Mean' AOD ($\bar{\tau}_{QA,j,l}$), is the QA-weighted average of the L2 AOD values i, in a grid box l and day j, i.e.,

$$\bar{\tau}_{QA,j,l} = \sum_i Q_{i,l}\tau_{i,l}/Q_{j,l}, \tag{44}$$

where

$$Q_{j,l} = \sum_i Q_{i,l}. \tag{45}$$

Fig. 2.10(E) plots the differences between the QA_Mean and Mean values for daily AOD over land (which are usually small). For each $1° \times 1°$ grid box, L3-daily products include PC, a QA histogram, and both estimates of the mean. Data files are known as 'MOD08_D3' for *Terra* and 'MYD08_D3' for *Aqua*, that we denote collectively as 'D3'.

3.5.2 L3-Monthly (M3)

Standard D3 products fully represent the statistics of the original L2 data (PC and QA, Mean and QA_Mean). Therefore, L3-monthly products (M3) are derived from D3 instead of from L2 (to save computational time and space).

One method for deriving gridded monthly AOD is to average the D3 data (equal weight per day, i.e., simple weighting) for every grid box. This means that days with more L2 data (higher PC) are treated the same as days with fewer. Instead we may choose to weight each

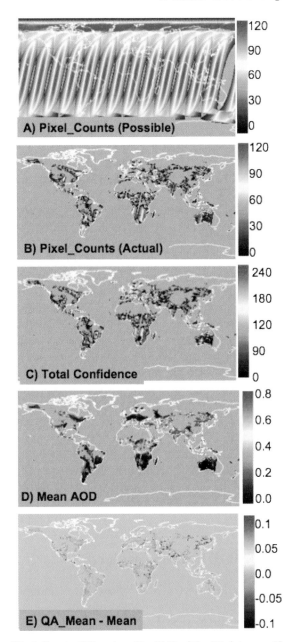

Fig. 2.10. Quantity (Pixel_Counts, PC) and quality (QA) of the L2 data on a $1° × 1°$ grid, for May 4, 2003. (A) Possible PC, (B) Actual PC, (C) Total Confidence (sum of QA), (D) computed daily Mean AOD, and (E) difference between quality weighted mean (QA_Mean) and Mean AOD. Note the different scales to the right of each panel.

day's contribution (to the mean) by its PC, so that a clear, well-sampled day gets more weight than a cloudy, poorly sampled day. We define the 'Pixel weighted' mean AOD ($\bar{\tau}_{k,l}$) for a given month, k, and grid box, l, as

$$\bar{\tau}_{K,l} = \sum_j P_{i,l}\bar{\tau}_{j,l} / \sum_j P_{i,l} = \sum_j \sum_i \tau_{i,j,l} / \sum_j P_{j,l}, \qquad 46$$

where $P_{j,l}$ and $\bar{\tau}_{j,l}$ are daily PC and daily AOD Mean. Thus, the gridded mean derived from basic Pixel weighting of gridded D3 data is equivalent to the mean of non-weighted L2 representing that grid box. Because Pixel weighting best represents the original L2 sampling, derived in 'clear sky' conditions, Pixel weighting characterizes global aerosol with a 'clear sky bias'.

We realize that aggregation (say from L2 to D3) over any spatial and/or temporal domain requires proper accumulation of weights over that domain. For any consequent aggregation (say from D3 to M3), we must decide whether (and how) to account for the first set of weights. Instead of deriving a monthly value from D3 that is equivalent to that from using L2, we might desire something else. For example, a particular day may be extremely cloudy, such that only a single (out of a possible 120) L2 value represents the entire grid box. Since we might expect this single L2 value to be cloud-contaminated in some way, we would not want to include it in derivation of a monthly mean. In fact, standard M3 processing requires that PC > 5 to include a day's contribution to the monthly mean. This 'threshold' Pixel weighting may mask heavily cloud-contaminated days that might otherwise influence a monthly mean. Across the entire globe, differences between basic Pixel weighting and threshold Pixel weighting are generally insignificant, but we note that threshold Pixel weighting is not equivalent to deriving M3 straight from L2. Derived from the D3 Mean, the standard M3 product (reported over land and ocean, separately) is known as the 'Mean_Mean'.

In fact, standard M3 processing also applies the same threshold Pixel weighting to the D3 QA_Mean to derive the M3's 'QA_Mean_Mean'. The QA_Mean_Mean is not a Confidence weighted product in the sense of the daily QA_Mean, because it may have been derived with higher weighting (larger PC) assigned to lower confidence data.

4.   Evaluation of MODIS c005 products

Evaluation of a satellite (or any) dataset refers to the exercise of understanding the integrity of the data under all measurement conditions. Validation implies quantitative assessment of the measurement uncertainty. The validation process asks questions about the precision, accuracy and consistency of the derived data products. We can focus on consistency ('Do the products represent physical quantities with no artificial boundaries?'), precision ('Do the products represent small enough increments of physical quantities?') or accuracy ('Can the products be matched with reference standards?'). In this chapter, we first discuss different evaluations of the L2 product, primarily via comparison with sunphotometry. Then we introduce some interesting new work assessing the meaning of a gridded L3 aerosol product, and how it may relate to estimates of 'global mean' AOD.

The first type of evaluation can be thought as a qualitative sanity check. Does the algorithm derive products that fit our expectations? We see from Fig. 2.9 that AOD seems continuous over land and ocean, and there are no obvious problems near mountains. Also, the aerosol gradients seem reasonable. As for the fine weighting (ηW) product, even with differing definitions over land and ocean, Fig. 2.9(C) shows relative continuity and few surprises.

4.1   Quantitative evaluation: comparison of L2 with AERONET

The spectral AOD is a physical quantity, resulting from the interaction of spectral radiation with a particular composition and amount of aerosol within the atmospheric column. If one assumes that sunphotometry provides the most simple and direct measurement of this quantity, then satellite-derived AOD should be directly comparable. Because the fine weighting (FW or η) is defined differently by AERONET and MODIS, we may compare them, but require additional assumptions. One way to get around direct FW comparison is by comparing 'Derived' products (e.g. α and τ^f) with those of AERONET.

We compare MODIS spectral AOD (L2) products to reference ground-based sunphotometers (AERONET) at over a hundred global sites. We use the spatial–temporal tech-

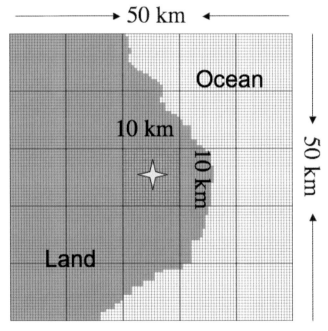

Fig. 2.11. Illustration of the Ichoku et al. [2002] spatial/temporal validation technique over a coastal AERONET site (star). The orange box is the 10×10 km MODIS retrieval containing the site. Since both MODIS over-land and over-ocean retrievals are performed in this case, both are averaged over the 50×50 km domain, and will be compared with the AERONET measurements. Any 10 km MODIS retrieval containing land is derived as land, whereas 100 % water is necessary for deriving as ocean. The tiny boxes represent the 20×20 original 500-m MODIS pixels within each 10 km. The time domain for AERONET is one hour ($\pm$ 30 minutes of overpass)

nique (Fig. 2.11) of Ichoku et al. [2002], such that the average of a 50×50 km area of MODIS products centered at the AERONET site (a 5×5 box $= 25$ retrievals at 10 km) is compared to the average of the AERONET direct-sun measurements within one hour of satellite overpass (normally four or five measurements). A valid collocation requires cloud-free conditions for both instruments (MODIS must retrieve at least five pixels out of 25 and AERONET must retrieve at least twice during the hour) and quality-assured data (MODIS with QA $= 3$, AERONET L2) for a valid match. Comparisons are performed for the MODIS pixels over ocean and over land separately.

4.1.1 Preliminary L2 validation using the test-bed

Levy et al. [2007b] provided a preliminary validation for c005 products, using a 6000-granule sample known as the 'test-bed'. The test-bed contained L1B reflectance data from a somewhat random combination of single granules, entire days, over multiple locations, time periods, and from both satellites. The same test-bed was used to derive products using both c005 and c004 algorithms. Histograms showed a dramatic reduction in global-derived AOD, especially in clean conditions. The test-bed's mean AOD dropped from 0.28 using c004 to 0.21 using c005. Comparison with AERONET AOD indicated significant improvement, such that the regression y-offset was reduced dramatically (from 0.097 to 0.029), and the slope was increased (from 0.901 to 1.009). In addition, the correlation (R) of the regression improved from 0.84 to 0.88. In general, the derived size parameter (η) had higher correlation using c005 ($R = 0.5$) instead of the c004 algorithm ($R = 0.26$), with marginal improvements in derived Ångström exponent and fine AOD [Levy et al., 2007b]. Fig. 2.12 (from Levy et al. [2007b]) compares the MODIS size parameters (η, a, τ_f) with the analogous parameters derived from AERONET sun observations [O'Neill et al., 2001].

While it is disappointing that MODIS-derived size parameters (η and a) compare so poorly to sunphotometer measurements, it should not be surprising given the results of the sensitivity study. The addition of even small errors in simulating the spectral dependence will lead to large errors in η, and therefore to a. In some regions, errors in surface reflectance parameterization dominate the spectral error, whereas, in others, the limited choice of aerosol optical models is the cause. Nonetheless, compared to c004 retrievals, where there was almost no spectrally dependent signal, we expected better skill in retrieving size parameters.

It should be noted that the input test-bed files used for each algorithm contained so-called 'c004 calibration' for the reflectance data. With the reprocessing to Collection 5, the MODIS Characterization Science Team (MCST) adjusted some channels' radiance/reflectance calibration coefficients by as much as 1 % from c004 values. 'True' validation exercises require data that have been produced using the consistent 'c005 calibration'.

4.1.2 Global c005 L2

Since official c005 processing began in early 2006, we have collected over 20 000 co-locations of MODIS L2 data and AERONET Level 2 (Version 2) data, using the algorithm of Ichoku et al. [2002]. This dataset covers the re-processed (and forward processed) entire MODIS (both *Terra* and *Aqua*) mission through early 2007. Fig. 2.13 compares MODIS

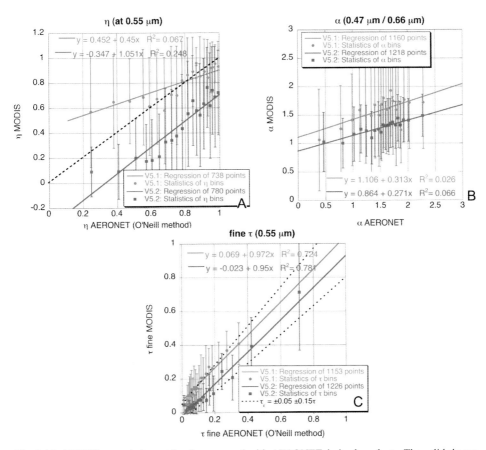

Fig. 2.12. MODIS aerosol size retrievals compared with AERONET-derived products. The solid shapes and error bars represent the mean and standard deviation of the MODIS retrievals, in 20 bins of AERONET-derived product. Both the retrievals from V5.1 (c004 algorithm; orange) and V5.2 (c005 algorithm; green) are shown. The regressions (solid lines) are for the cloud of all points (not shown). (A) η over land retrieved at 0.55 μm. Note that η is defined differently for MODIS and AERONET and that we only show results for $\tau 0.20$. (B) MODIS α (0.466/0.644 μm) over land; AERONET α interpolated to the same wavelengths. (C) MODIS fine τ over land retrieved at 0.55 μm, compared with AERONET fine τ interpolated to 0.55 μm by quadratic fitting and the O'Neill et al. [2003] method. The expected errors for MODIS ($\pm0.05 \pm0.15\tau$) are also shown (dashed lines). (Figure adapted from Levy et al. [2007b].)

dark target retrieved AOD to that from sunphotometer, for *Terra* (left) and *Aqua* (right). For both sensors, the quality of comparison is almost unimaginable; y-intercepts (absolute value) are less than 0.02, slopes are different than unity by only 1.6 %, and correlations (R are greater than 0.9.

At this point, the MODIS Aerosol Science Team (MAST) has not performed extensive evaluation of the c005 aerosol size products. Over regions heavily populated with AERONET sunphotometers (e.g., eastern North America, Western Europe), the fact that c005 development relied heavily on these data has resulted in qualitatively reasonable derivations of η and α. Over other regions (e.g., India, China), the MODIS estimated size parameters are marginally if at all improved from those derived via c004. These are regions

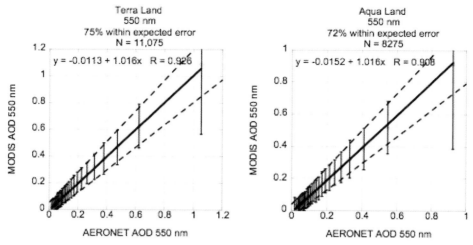

Fig. 2.13. MODIS aerosol optical depth (AOD at 0.55 mm) over dark land plotted against collocated AERONET observations for *Terra* (left) and *Aqua* (right). The data were sorted according to AERONET AOD, divided into 25 bins of equal observations. Points represent the means of each bin. Error bars represent the standard deviation of MODIS AOD within those bins. The highest AOD bin typically represents the mean of fewer observations than the other bins. AERONET AOD at 550 nm was interpolated from a log-log plot of standard wavelength observations. The regression line, regression equation and correlation were calculated from the full cloud of points before binning. Expected uncertainty is $\Delta\tau = \pm 0.05 \pm 0.15\tau$, and is shown in the plots by the dashed lines. (Figure from Remer et al. [2008a].)

where we have more variability of aerosol type, and therefore lower confidence in both aerosol model and atmospherically corrected surface reflectance parameterization.

4.1.3 Regional c005 evaluation

Global, long-term scatterplots are informative, but may hide systematic errors pertaining to certain regions. A few recent studies have focused on comparing c005 to c004 products over particular regions of the globe. It turns out that some of these locations were over AERONET sites that did not exist prior to c005 algorithm development, meaning they could be thought of as independent measures of the MODIS algorithm revision. Mi et al. [2007] and Jethva et al. [2007] showed significant improvement of AOD products over China (Xianghe and Taihu), and India (Kanpur), respectively. Yet over both countries, c005 size estimates showed no improvement. In contrast to that shown for c004 (e.g., Remer et al., 2005; Levy et al., 2005], the MODIS/AERONET comparisons over the East coast of the United States are much improved; now more than 70 % of the retrievals fall within expected error [Levy, 2007]. In this region, derived FW showed significant improvement from c004. Many other studies analyzing c005 MODIS data over various regions have recently been completed or in progress.

4.2 Impact of data weighting on L3 products

Recall that MODIS Level 3-daily (D3) data represent statistics of L2 (PC, QA, Mean and QA_Mean) on a $1° \times 1°$ grid. Standard gridded, monthly (M3) products are derived from threshold (PC > 5 per day) Pixel weighted D3 data, such that the 'Mean_Mean' product is similar to, but not the same as deriving from equal-weighted L2 data. The 'QA_Mean_- Mean' product, derived from Pixel weighted D3 QA_Mean, is completely different than derive from Confidence weighted L2 data. Levy et al. [2008] studied the impact of applying other weightings (instead of Pixel weighting) to derive monthly AOD from D3 data. They proposed a monthly product ('QA_Mean_QA_Mean') that is derived

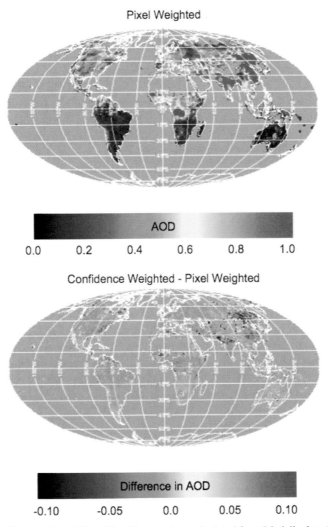

Fig. 2.14. May 2003 monthly AOD at $1° \times 1°$ resolution, calculated from L3-daily data. Top: Calculated via 'Pixel' weighting. Bottom: Difference between 'Confidence' weighting and 'Pixel' weighting. (Figure adapted from Levy et al. [2008].)

from Confidence weighted D3 QA_Mean, that represents Confidence weighted L2, i.e.,

$$\bar{\tau}_{QA,k,l} = \sum_j Q_{j,l}\bar{\tau}_{QA,j,l} / \sum_j Q_{j,l}. \tag{47}$$

Fig. 2.14 (bottom) displays a map of the difference when deriving monthly AOD derived from Confidence weighted D3, versus that computed from Pixel weighting (Fig. 2.14 (top)). Generally, differences between the two weightings are small. However, some aerosol hotspots (e.g., Asian dust/pollution plume) are marked by both large increases (> 0.1) and large decreases (< -0.1). Presumably, as the Asian plume is entrained within mid-latitude weather systems, optically thick aerosol and clouds may be confused. It is interesting that the African biomass-burning plume ($\tau \sim 1.0$) shows no significant differences between the two weightings.

4.3 Global mean AOD

Estimates of 'global mean' AOD are not standard MODIS products; however, they have been used as diagnostic of aerosol trends (e.g., Remer et al. [2008b]), and comparisons with other datasets (e.g., Kinne et al., 2006]. However, since data aggregation/weighting have such a huge impact on deriving mean values, users of the data should understand averaging logic. There are many issues that must be addressed when computing a 'global mean' value from irregularly sampled data such as MODIS. Since it is easy to show that standard gridded D3 data (PC, QA histogram, Mean and QA_Mean parameters) fully represent the sampling and confidence of L2 data (e.g., Eq. (42)), then we use the D3 data to illustrate the weighting/averaging issues (e.g., Levy et al. [2008]).

The order of averaging is important. One may choose to compute a monthly global mean by 'bucket' averaging (the order of temporal versus spatial averaging is irrelevant), or by 'ordered' averaging of D3 data (requiring first temporal averaging then spatial averaging, or vice versa). Analogous to Eq. (42), one can easily show that pixel weighting and 'bucket' averaging of a month's worth of D3 data is equivalent to deriving a monthly global mean directly from a bucket of a month's worth of equal weighted L2. Confidence weighted, bucket-averaged D3 are equivalent to Confidence weighted L2. 'Ordered' averaging implies that at least two steps are required in taking the average. For example, one may first derive gridded monthly data (a monthly map) by pixel weighting (Eq. (42)), but then derive the monthly mean by spatially averaging the map assuming a different weighting. The weighting for the spatial averaging may be equal weighting (i.e. simple average of the map), or may be something like latitude weighting (where the grid box is multiplied by the cosine of latitude, to represent diminishing surface area toward the poles). Latitude weighting favors information from tropical grid locations in the global mean. Of course, the weightings could be in the opposite order, spatial first (deriving a time series of daily global means) and then temporal (averaging the days). We can see how applying a pixel threshold (PC > 5 per day) will to lead to different value for monthly global mean. Fig. 2.15 illustrates (using artificial AOD and PC on a three-piece globe for a 2 day month) how the global monthly average is determined by the choices of ordering and weighting. For panel A, all paths to the global mean assume pixel weighting in the first step. The

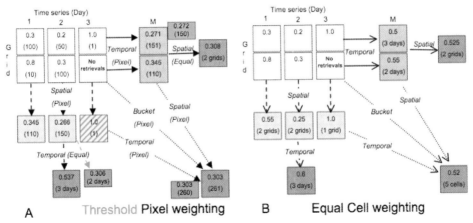

Fig. 2.15. Deriving $\bar{\bar{\tau}}$ from different spatial and temporal aggregations (order and weighting) of artificial D3 data (white boxes) for 3 days and two grids. Different color boxes and arrow types represent different paths toward $\bar{\bar{\tau}}$, (A) using pixel weighting in the first step, (B) using equal cell weighting at all steps. Mean AOD and corresponding pixel or cell count (in parentheses) are reported in each box along the path. Name of averaging and weighting (in parentheses) are reported along the arrows. Note that for day 3 there are no retrievals in the lower grid box, and only one representing the upper box. In Panel A, orange boxes represent application of 'threshold' pixel count weighting (PC $>$ 5) to exclude low sampled grid boxes/day, whereas black arrows and grey boxes point to the results derived without the threshold applied. Note how 'bucket' averaging represents consistent weighting along ordered averaging.

choice is whether to first derive a map then perform spatial averaging (dotted arrows), first derive a time series, then perform temporal averaging (dashed arrows), or perform the spatial and temporal average simultaneously. Each leads to a different estimate of the global mean.

The choices of the two weightings are also important. Levy et al. [2008] found that, depending on the ordering/aggregating/weighting assumptions, estimates of monthly global mean AOD over land vary by as much as 40 % for a given month (between 0.20 and 0.32 for May 2003). Pixel weighting represents the 'quantity' of L2 sampling, so that a monthly global mean will have a clear-sky bias. Confidence weighting represents the quality of L2 sampling, so that monthly global means are dominated regions where the aerosol retrieval has higher QA (darker targets, better constrained aerosol models and surface reflectance assumptions). During the course of a year (2003; Fig. 2.16), the Confidence weighting's dark target bias leads to estimates of monthly global mean that are lower (by $\Delta\tau \sim 0.01$) than that derived with pixel weighting's clear sky bias. Both weightings derive lower values ($\Delta\tau \sim 0.04$) than if we had applied equal weighting (simple averaging) instead. Presumably, due to tropical dominance of biomass burning and dust, Latitude weighting tends to derive higher estimates of global mean AOD.

We recall that global L2 validation shows that the expected uncertainty of an individual L2 value is $\Delta\tau = 0.05 + 0.15\tau$. Fig. 2.16 shows that the global mean for May 2003 (expected maximum from maximum of global dust transports), derived from equal weighted L2 is $\tau = 0.22$. This means that the uncertainty in estimating the global mean ($\Delta\tau = 0.12$) is larger than the expected uncertainty for retrieving that value ($\Delta\tau = 0.083$)!

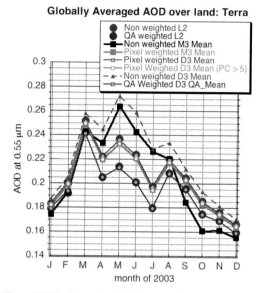

Fig. 2.16. Time series (*Terra*, 2003) of monthly, global mean AOD, computed from different aggregation/ weighting methods of L3-daily (D3) data (small squares/triangles, thin curves), L2 data (large dots) and L3-monthly (M3) data (large squares, thick curves). Both Mean and QA_Mean parameters are considered. Similar by colored curves/points indicate product equivalence (From Levy et al. [2008].)

5. Summary/Conclusion

As a result of obvious limitations to the MODIS c004 algorithm, Levy et al., [2007a, 2007b] developed a new algorithm for deriving aerosol properties over dark land targets. This 'second generation' algorithm required full re-evaluation of assumed aerosol models, surface reflectance parameterization, elevation correction, and retrieval philosophy. The resulting algorithm was implemented in early 2006, and so began the processing/re-processing known as 'collection 5' (c005). Initial algorithm evaluation (applied to a 6000-granule testbed), supported the conclusion of the sensitivity test, suggesting robust retrieval of total AOD, even in the presence of spectral and absolute reflectance errors. Global co-location with AERONET ($> 20\,000$ points) confirmed that MODIS does an excellent job at retrieving AOD [Remer et al., 2008a] over dark land surfaces ($\tau_{MODIS} = 1.016\tau + 0.01$; $R = 0.9$). Unfortunately, actual retrievals of aerosol size parameters (ηa) confirmed that in the presence of errors, this MODIS algorithm cannot accurately retrieve aerosol size. In the near future, as more aerosol data accumulates (from AERONET and other data sources), new or improved aerosol optical models may lead to better retrievals in some regions (e.g. Kanpur, India; H. Jethva, personal communication). Also, new understanding about spectral surface reflectance properties should also improve the resulting aerosol retrieval. Nonetheless, even with its limitations, the MODIS c005 data provides the most consistent, comprehensive, and easily available, global, long-term aerosol dataset.

Due to its relative success, MODIS data are (and have been) used for all sorts of applications, from monitoring and forecasting air quality, to characterizing aerosol effects on climate. Some of these applications (e.g., comparing to model output) suggest the use of Level 3 (gridded aggregations of Level 2 data) to represent aerosol data. However, due to issues of orbital geometry, the presence of clouds, inhomogeneous surfaces, algorithm logic, quality, etc, MODIS L2 sampling is highly variable. This means that derivations of L3 data are determined by choices of L2 weighting, and that estimates of global mean AOD are further complicated by choices of how to average the data (ordered averaging, or bucket?). Here, and in Levy et al. [2008], we showed that how one chooses to represent the MODIS data can lead to differences in global mean (May 2003) values by 40 % ($\Delta\tau \sim 0.12$), which has far more uncertainty than that of the L2 retrieval itself ($\Delta\tau = 0.05 + 0.15\tau = 0.083$).

In the spirit of Mishchenko et al. [2007], Remer et al. [2008b] used spatially averaged M3 (Pixel weighted D3) data, obtained from Giovanni (e.g., Acker and Leptoukh [2007] to identify trends in the MODIS mission time series. While not statistically significant at 95 %, they find a compelling trend ($\Delta\tau/\text{year} = -0.0026$) over land that might be attributable to efforts to clean up air pollution. Also, AOD trends from MODIS and AERONET are correlated over many sites. Yet, since the current Giovanni application does not consider the averaging inherent in M3, one cannot draw final conclusions. Clearly, any trend analysis requires further analysis of the uncertainties of averaging/aggregating the irregular sampled data, and should be taken up at regional levels.

References

Acker, J.G., and G. Leptoukh, 2007. Online analysis enhances use of NASA Earth science data, *EOS, Trans. of American Geophys. Union*, **88**, 14, doi: 10.1029/2007EO020003

Ahmad, Z., and R.S. Fraser, 1982: An iterative radiative transfer code for ocean–atmosphere systems. *J. Atmos. Sci.*, **39**, 656–665.

Al-Saadi, J., J. Szykman, R.B. Pierce, C. Kittaka, D. Neil, D.A. Chu, L. Remer, L. Gumley, E. Prins, L. Weinstock, C. MacDonald, R. Wayland, F. Dimmick, and J. Fishman, 2005: Improving national air quality forecasts with satellite aerosol observations. *Bull. Am. Met. Soc.*, **86**(9), 1249–1261.

Andrews, E., et al. (2006), Comparison of methods for deriving aerosol asymmetry parameter, *J. Geophys. Res.*, **111**, D05S04, doi:10.1029/2004JD005734.

Ångström, A., 1929: On the atmospheric transmission of sun radiation and on dust in the air. *Geograf. Ann. Deut.*, **11**, 156–166.

Bodhaine, B.A., N.B. Wood et al.,, 1999: On Rayleigh optical depth calculations. *J. Atmos. Ocean. Tech.*, **16**(11), 1854–1861.

Bond, T.C., and R.W. Bergstrom, Light absorption by carbonaceous particles: an investigative review, *Aerosol Science and Technology*, **40**(1), 27–67.

Chandresekhar, S., 1950: *Radiative Transfer*, Clarendon Press, London.

Chen, L.W.A., B.G. Doddridge, R.R. Dickerson, J.C. Chow, and R.C. Henry, 2002: Origins of fine aerosol mass in the Baltimore–Washington corridor: implications from observation, factor analysis, and ensemble air parcel back trajectories. *Atmos. Environ.*, **36**(28), 4541–4554.

Chin, M., P. Ginoux, et al., 2002: Tropospheric aerosol optical thickness from the GOCART model and comparisons with satellite and Sun photometer measurements. *J. Atmos. Sci.*, **59**(3), 461–483.

Chu, D.A., Y.J. Kaufman et al., 2002: Validation of MODIS aerosol optical depth retrieval over land, *Geophys, Res. Lett.*, **29**(12): art. no.-1617.

Chu, D.A., Y.J. Kaufman, G. Zibordi, J.D. Chern, J. Mao, C. Li, and B.N. Holben, 2003: Global monitoring of air pollution over land from EOS-Terra MODIS. *J. Geophys. Res.*, **108**(D21), 4661, doi:10.1029/2002JD003179.

Dave, J.V., 1970: Intensity and polarization of radiation emerging from a plane-parallel atmosphere containing monodispersed aerosols. *App. Optics*, **9**(12), 2673–2687.

Dickerson, R.R., S. Kondragunta, G. Stenchikov, K.L. Civerolo, B.G. Doddridge, and B.N. Holben, 1997: The impact of aerosols on solar ultraviolet radiation and photochemical smog. *Science*, **278**, 5339, 827–830.

Diner, D.J., L. Di Girolamo et al., 2007: Preface to the MISR Special Issue. *Rem. Sens. Environ.*, **107**(1–2): 1.

Dubovik, O., and M. D. King, 2000: A flexible inversion algorithm for retrieval of aerosol optical properties from Sun and sky radiance measurements, *J. Geophys. Res.*, **105**(D16), 20673–20696.

Dubovik, O., A. Smirnov, B.N. Holben et al., 2000: Accuracy assessments of aerosol optical properties retrieved from AERONET sun and sky radiance measurements. *J. Geophys. Res.*, **105**, 9791–9806.

Dubovik, O., B. N. Holben, et al., 2002a: Variability of absorption and optical properties of key aerosol types observed in worldwide locations. J. Atmos. Sci., 59(3), 590–608.

Dubovik, O., B.N. Holben et al., 2002b: Non-spherical aerosol retrieval method employing light scattering by spheroids. *Geophys. Res. Lett.*, **29**(10): art. no.-1415.

Dubovik, O., A. Sinyuk, T. Lapyonok, B.N. Holben, M. Mishchenko, P. Yang, T.F. Eck, H. Volten, O. Munoz, B. Veihelmann, W.T. van der Zander, M Sorokin, and I. Slutsker, 2006: Application of light scattering by spheroids for accounting for particle non-sphericity in remote sensing of desert dust. *J. Geophys. Res.*, **111**(D11208), doi:10.1029/2005JD006619.

Eck, T.F., B.N. Holben et al., 1999: Wavelength dependence of the optical depth of biomass burning, urban, and desert dust aerosols. *J. Geophys. Res.*, **104**(D24), 31333–31349.

Eck, T.F., B.N. Holben et al., 2005: Columnar aerosol optical properties at AERONET sites in central eastern Asia and aerosol transport to the tropical mid-Pacific. *J. Geophys. Res.*, **110**(D6).

Engel-Cox, J., C. Holloman, B. Coutant, and R. Hoff, 2004: Qualitative and quantitative evaluation of MODIS satellite sensor data for regional and urban scale air quality, *Atmos. Environ.*, **38**(16), 2495–2509.

Engel-Cox, J., R. Hoff, R. Rogers, F. Dimmick, A. Rush, J. Szykman, J. Al-Saadi, D. Chu, and E. Zell, 2006: Using particulate matter concentration and lidar and satellite optical depth to build a 3-dimensional integrated air quality monitoring capability, *Atmos. Environ.*, **40**, 8056–8067.

Evans, K.F., and G.L. Stephens, 1991: A new polarized atmospheric radiative-transfer model, *Quant. Spect. Rad. Trans.*, **46**(5), 413–423.

Fraser, R.S., 1975: Degree of inter-dependence among atmospheric optical thicknesses in spectral bands between 0.36–2.4 microns, *J. Applied Meteorology*, **14**(6), 1187–1196.

Fraser, R.S., R.A.Ferrare, Y.J. Kaufman, and S. Mattoo, 1989: Algorithm for atmospheric corrections of aircraft and satellite imagery. *NASA* Technical Memorandum No. 100751. NASA Goddard Space Flight Center, Greenbelt, MD, USA.

Gao, B.-C., Y.J. Kaufman, D. Tanr\'{e and R.-R. Li, 2002: Distinguishing tropospheric aerosols from thin cirrus clouds for improved aerosol retrievals using the ratio of 1.38-μm and 1.24-μm channels. *Geophys. Res. Lett.*, **29**, 1890, doi:10.1029/2002GL015475.

Gassó, S., and D.A. Hegg, 2003: On the retrieval of columnar aerosol mass and CCN concentration by MODIS. *J. Geophys. Res.*, **108** (D1), 4010, doi: 10.1029/2002JD002382.

Gatebe, C.K., M.D. King et al., 2001: Sensitivity of off-nadir zenith angles to correlation between visible and near-infrared reflectance for use in remote sensing of aerosol over land, *IEEE Trans. Geosci. Remote Sens.*, **39**(4), 805–819.

Guenther, B., X. Xiong, V.V. Salomonson, W.L. Barnes, and J. Young, 2002: On orbit performance of the EOS-MODIS: First year of data. *Rem. Sens. Environ.*, **83**, 16–30.

Hegg, D.A., and Y.J. Kaufman, 1998: Measurements of the relationship between submicron aerosol number and volume concentration, *J. Geophys. Res.*, **103**(D5), 5671–5678.

Herman, M., J. Deuzé, C. Devaux, P. Goloub, F. Bréon, and D. Tanré, (1997): Remote sensing of aerosols over land surfaces including polarization measurements and application to POLDER measurements, *J. Geophys. Res.*, **102**(D14), 17039–17049.

Higurashi, A., and T. Nakajima, 1999: Development of a two channel aerosol retrieval algorithm on global scale using NOAA/AVHRR. *J. Atmos. Sci.*, **56**, 924–941.

Holben, B.N., T.F. Eck, et al., 1998: AERONET – A federated instrument network and data archive for aerosol characterization, *Remote Sens. Environ.*, **66**(1), 1–16.

Hoppel, W.A., G.M. Frick, and J.W. Fitzgerald, 2002: Surface source function for sea-salt aerosol and dry deposition to the ocean surface, *J. Geophys. Res.*, **107**(D19), 4382, doi:10.1029/2001JD002014.

Hsu, N.C., S.C. Tsay, M.D. King, and J.R. Herman, 2006: Deep blue retrievals of Asian aerosol properties during ACE-Asia, *Ieee Trans. Geosci. Remote Sens.*, **44**, 3180–3195.

Hubanks, P.A., 2005: MODIS Atmosphere QA Plan for Collection 005. NASA Goddard Space Flight Center, Greenbelt, MD, USA.

Husar, R.J., Prospero, and L. Stowe, 1997: Characterization of tropospheric aerosols over the oceans with the NOAA advanced very high resolution radiometer optical thickness operational product, *J. Geophys. Res.*, **102**(D14), 16889–16909.

Ichoku, C., D.A. Chu et al., 2002: A spatio-temporal approach for global validation and analysis of MODIS aerosol products, *Geophys. Res. Lett.*, **29**(12): art. no.-1616.

Ichoku, C., L.A. Remer et al., 2003: MODIS observation of aerosols and estimation of aerosol radiative forcing over southern Africa during SAFARI 2000, *J. Geophys. Res.*, **108**(D13): art. no.-8499.

Intergovernmental Panel on Climate Change – IPCC, 2001: *Climate Change 2001: The Scientific Basis*, J.T. Houghton, Y. Ding, D.J. Griggs, M. Noguer, P.J. van der Linden, and D. Xiaosu (Eds), Cambridge University Press, Cambridge, UK.

Jethva, H., S.K. Satheesh, and J. Srinivasan, 2007: Assessment of second-generation MODIS aerosol retrieval (Collection 005) at Kanpur, India, *Geophys. Res. Lett.*, **34**, L19802, doi:10.1029/2007GL029647.

Kahn, R., B. Gaitley, J. Martonchik, D. Diner, K. Crean, and B. Holben, 2005: Multiangle Imaging SpectroRadiometer (MISR) global aerosol optical depth validation based on two years of coincident AERONET observations. *J. Geophys. Res.*, **110**(D10), doi: jd004706R.

Kaufman, Y.J., and L.A. Remer, 1994: Detection of forests using Mid-IR reflectance: an application for aerosol studies. *IEEE Trans. Geosci. Remote Sens.*, **32**, 672–683.

Kaufman, Y.J., and C. Sendra, 1988: Algorithm for atmospheric corrections of visible and Near IR satellite imagery, *Int. J. Rem. Sens.*, **9**, 1357–1381.

Kaufman, Y.J., A. Gitelson et al., [1994: Size distribution and scattering phase function of aerosol particles retrieved from sky brightness measurements. *J. Geophys. Res.*, **99**, 10341–10356.

Kaufman, Y.J., D. Tanré et al., 1997a: Operational remote sensing of tropospheric aerosol over land from EOS moderate resolution imaging spectroradiometer. *J. Geophys. Res.*, **102**(D14), 17051–17067.

Kaufman, Y.J., D. Tanré et al., 1997b: Passive remote sensing of tropospheric aerosol and atmospheric correction for the aerosol effect. *J. Geophys. Res.*, **102**(D14), 16815–16830.

Kaufman, Y.J., A.E. Wald, L.A. Remer, B.C. Gao, R.R. Li, and L. Flynn, 1997c: The MODIS 2.1-um channel – Correlation with visible reflectance for use in remote sensing of aerosol. *IEEE Trans. Geosci. Rem. Sens.*, **35**(5), 1286–1298.

Kaufman, Y.J., D.D. Herring, K.J. Ranson, and G.J. Collatz, 1998: Earth observing system AM1 mission to earth, *IEEE Trans. Geosci. Remote Sens.*, **36**(4), 1045–1055.

Kaufman, Y.J., N. Gobron, B. Pinty, J.L. Widlowski and M.M. Verstraete, 2002: Relationship between surface reflectance in the visible and mid-IR used in MODIS aerosol algorithm – theory, *Geophys. Res. Lett.*, **29**(18), art. no.-2116.

Kaufman, Y.J., O. Boucher et al., 2005: Aerosol anthropogenic component estimated from satellite data, *Geophys, Res. Lett.*, **32**(12).

King, M.D., Y.J. Kaufman, W.P. Menzel, and D. Tanre, 1992: Remote-sensing of cloud, aerosol, and water-vapor from the Moderate Resolution Imaging Spectrometer (MODIS). *IEEE Trans. Geosci. Rem. Sens.*, **30**(1), 2–27.

King, M.D., Y.J. Kaufman, D. Tanre, and T. Nakajima, 1999: Remote sensing of tropospheric aerosols from space: past, present, and future, *Bull. Am. Met. Soc.*, **80**(10), 2229–2259.

King, M.D., W.P. Menzel, Y.J. Kaufman, D. Tanre, B.-C. Gao, S. Platnick, S.A. Ackerman, L.A. Remer, R. Pincus, and P.A. Hubanks, 2003: Cloud and aerosol properties, precipitable water, and profiles of temperature and humidity from MODIS. *IEEE Trans. Geosci. Remote Sens.*, **41**, 442–458.

Kinne, S., M. Schulz, C. Textor, S. Guibert, Y. Balkanski, S.E. Bauer, T. Berntsen, T.F. Berglen, O. Boucher, M. Chin, W. Collins, F. Dentener, T. Diehl, R. Easter, J. Feichter, D. Fillmore, S. Ghan, P. Ginoux, S. Gong, A. Grini, J. Hendricks, M. Herzog, L. Horowitz, I. Isaksen, T. Iversen, A. Kirkevåg, S. Kloster, D. Koch, J.E. Kristjansson, M. Krol, A. Lauer, J.F. Lamarque, G. Lesins, X. Liu, U. Lohmann, V. Montanaro, G. Myhre, J. Penner, G. Pitari, S. Reddy, O. Seland, P. Stier, T. Takemura, and X. Tie, 2006: An AeroCom initial assessment optical properties in aerosol component modules of global models, *Atmos. Chem. Phys.*, **6**, 1815–1834.

Kokhanovsky, A.A., F.-M. Breon et al., 2007: Aerosol remote sensing over land: a comparison of satellite retrievals using different algorithms and instruments, *Atmos. Res.*, **85**, 372–394, doi:10.1016.atmosres.2007.02.008.

Kokhanovsky, A.A., R.L. Curier, G. de Leeuw et al., 2009, The inter-comparison of AATSR dual view aerosol optical thickness retrievals with results from various algorithms and instruments, accepted for publication in *Int. J. Remote Sensing*, in press.

Köpke, P., M. Hess, I. Schult, and E.P. Shettle, 1997: Global aerosol data set, Report No. 243, Max-Planck-Institut für Meteorologie, Hamburg, ISSN 0937–1060.

Kotchenruther, R.A., P.V. Hobbs, and D.A. Hegg, 1999): Humidification factors for aerosols off the mid-Atlantic coast of the United States, *J. Geophys. Res.*, **104**(D2).

Lenoble, J., and C. Brogniez, 1984: A comparative review of radiation aerosol models, *Beitr. Phys. Atmos.*, **57**(1), 1–20.

Levy, R.C., 2007: Retrieval of tropospheric aerosol properties over land from inversion of visible and near-infrared spectral reflectance: application over Maryland. PhD dissertation, Atmospheric and Oceanic Sciences, University of Maryland, http://hdl.handle.net/1903/6822.

Levy, R.C., L.A. Remer, D. Tanré, Y.J. Kaufman, C. Ichoku, B.N. Holben, J.M. Livingston, P.B. Russell, and H. Maring, 2003: Evaluation of the MODIS retrievals of dust aerosol over the ocean during PRIDE. *J. Geophys. Res.*, **108**(D14), 10.1029/2002JD002460.

Levy, R.C., L.A. Remer et al., 2004: Effects of neglecting polarization on the MODIS aerosol retrieval over land, *IEEE Trans. Geosci. Remote Sens*, **42**(10), 2576–2583.

Levy, R.C., L.A. Remer et al., 2005: Evaluation of the MODIS aerosol retrievals over ocean and land during CLAMS, *J. Atmos. Sci.*, **62**(4), 974–992.

Levy, R.C., L.A. Remer, and O. Dubovik, 2007a: Global aerosol optical properties and application to MODIS aerosol retrieval over land. *J. Geophys. Res.*, **112**, D13210, doi:10.1029/2006JD007815.

Levy, R.C., L. Remer, S. Mattoo, E. Vermote, and Y.J. Kaufman, 2007b: Second-generation algorithm for retrieving aerosol properties over land from MODIS spectral reflectance. *J. Geophys. Res.*, **112**, D13211, doi:10.1029/2006JD007811.

Levy, R.C., V. Zubko, G. Leptoukh, A. Gopalan, L.A. Remer, 2008: A critical look at deriving monthly aerosol optical depth from MODIS orbital data, *Geophys. Res. Letts.*, in press.

Li, R.R., Y.J. Kaufman, B.C. Gao, and C.O. Davis, 2003: Remote sensing of suspended sediments and shallow coastal waters. *IEEE Trans. Geosci. Remote Sens.*, **41**(3), 559–566.

Li, R.R., L. Remer et al., 2005: Snow and ice mask for the MODIS aerosol products. *IEEE Geo. and Rem. Sens. Lett.*, **2**(3), 306–310.

Liou, K.-N., 2002: *An Introduction to Atmospheric Radiation*, 2nd edition, Academic Press, San Diego, CA.

Malm, W.C., J.F. Sisler, D. Huffman, R.A. Eldred, and T.A. Cahill, 1994: Spatial and seasonal trends in particle concentration and optical extinction in the United States, *J. Geophys. Res.*, **99**(D1), 1347–1370.

Martins, J.V., D. Tanré, L.A. Remer, Y.J. Kaufman, S. Mattoo, and R. Levy, 2002: MODIS Cloud screening for remote sensing of aerosol over oceans using spatial variability. *Geophys. Res. Lett.*, **29**(12), 10.1029/2001GL013252.

MAST-MODIS Aerosol Science Team, 2006: Algorithm Theoretical Basis Document (ATBD): algorithm for remote sensing of tropospheric aerosol from MODIS, available from http://modis-atmos.gsfc.nasa.gov/MOD04_L2.

Mi, W., Z. Li, X. Xia, B. Holben, R. Levy, H. Zhao, H. Chen, and M. Cribb, 2007: Evaluation of the Moderate Resolution Imaging Spectroradiometer aerosol products at two Aerosol Robotic Network stations in China. *J. Geophys. Res.*, **112**, D22S08, doi:10.1029/2007JD008474.

Mie, G., 1908: A contribution to the optics of turbid media: special colloidal metal solutions (in German: Beitrage zur Optik trüber Medien speziell kolloidaler Metallösungen). *Ann. Phys.* **25**: 377–445.

Mishchenko, M.I., and L.D. Travis, 1994: Light scattering by polydisperse rotationally symmetric non-spherical particles: linear polarization. *J. Quant. Spectrosc. Radiat. Transfer*, **51**, 759–778, doi:10.1016/0022-4073(94)90130-9.

Mishchenko, M.I., and L.D. Travis, 1997: Satellite retrieval of aerosol properties over the ocean using polarization as well as intensity of reflected sunlight, *J. Geophys. Res.*, **102**, 16989–17013.

Mishchenko, M.I., I.V. Geogdzhayev et al., [2007: Long-term satellite record reveals likely recent aerosol trend. *Science*, **315**, 1543, doi:10.1126/science.1136709.

Nakajima, T., and M.D. King, 1990: Determination of the optical thickness and effective particle radius of clouds from reflected solar radiation measurements. Part I: Theory, *J. Atmos. Sci.*, **47**, 1878–1893.

Nakajima, T., G. Tonna, R.Z. Rao, P. Boi, Y. Kaufman, and B. Holben, 1996: Use of sky brightness measurements from ground for remote sensing of particulate polydispersions. *Appl. Optics*, **35**,(15), 2672–2686.

Omar, A.H., J.G. Won, D.M. Winker, S.-C. Yoon, O. Dubovik, and S. Fan, 2005: Development of global aerosol models using cluster analysis of Aerosol Robotic Network (AERONET) measurements, *J. Geophys. Res.*, **110**(D10S14).

O'Neill, N.T., A. Ignatov, B.N. Holben, and T.F. Eck, 2000: The lognormal distribution as a reference for reporting aerosol optical depth statistics; empirical tests using multi-year, multi-site AERONET sunphotometer data, *Geophys Res. Lett.*, **27**(20), 3333–3336.

O'Neill, N.T., O. Dubovik, and T.F. Eck, 2001: Modified angstrom exponent for the characterization of submicrometer aerosols, *Appl. Optics*, **40**(15), 2368–2375.

O'Neill, N.T., Eck, T.F., Smirnov, A., Holben, B.N. and Thulasiraman, S., 2003: Spectral discrimination of coarse and fine mode optical depth, *J. Geophys. Res.*, **108**, doi:10.1029/2002JD002975.

Parkinson, C.L., 2003: Aqua: an earth-observing satellite mission to examine water and other climate variables. *IEEE Trans. Geosci. Remote Sens.*, **41**(2), 173–183.

Platnick, S., M.D. King, S.A. Ackerman, W.P. Menzel, B.A. Baum, J.C. Riedi, and R.A. Frey, 2003: The MODIS cloud products: algorithms and examples from Terra, *IEEE Trans. Geosc. Remote Sens.*, **41**(2), 459–473.

Remer, L.A., and Y.J. Kaufman, 1998: Dynamic aerosol model: Urban/industrial aerosol, *J. Geophys. Res.*, **103**(D12), 13859–13871.

Remer, L.A., A.E. Wald, et al., 2001: Angular and seasonal variation of spectral surface reflectance ratios: implications for the remote sensing of aerosol over land, *IEEE Trans. Geosci. Remote Sens.*, **39**(2), 275–283.

Remer, L.A., D. Tanre, Y.J. Kaufman, C. Ichoku, S. Mattoo, R. Levy, D.A. Chu, B.N. Holben, O. Dubovik, A. Smirnov, J.V. Martins, R.-R. Li, and Z. Ahmad, 2002: Validation of MODIS aerosol retrieval over ocean. *Geophys Res. Lett.*, **29**, doi: 10.1029/2001GL013204.

Remer, L.A., Y.J. Kaufman et al., 2005: The MODIS aerosol algorithm, products, and validation, *J. Atmos. Sci.*, **62**(4), 947–973.

Remer, L.A., R.G. Kleidman, R.C. Levy, Y.J. Kaufman, D. Tanré, S. Mattoo, J.V. Marins, C. Ichoku, I. Koren, H. Yu, and B.N. Holben, 2008a: An emerging global aerosol climatology from the MODIS satellite sensors, *J. Geophys. Res.*, **113**, D14S07, doi:10.1029/2007JD009661.

Remer, L.A., I. Koren, R.G. Kleidman, R.C. Levy, J. V. Martins, K-M Kim, D. Tanré, S. Mattoo and H. Yu, 2008b: Recent short term global aerosol trends over land and ocean from MODIS Collection 5 retrievals, submitted 2008 to *Geophys. Res. Letts.*, in press.

Salomonson, V.V., W.L. Barnes, P.W. Maymon, H.E. Montgomery, and H. Ostrow, 1989: MODIS, advanced facility instrument for studies of the Earth as a system, *IEEE Trans. Goesci. Remote Sens.*, **27**, 145–153.

Seinfeld, J.H., and S.N. Pandis, 1998. *Atmospheric Chemistry and Physics: From Air Pollution to Climate change*, p. xxvii, John Wiley, New York.

Shettle, E.P., and R.W. Fenn, 1979: Models for the aerosols of the lower atmosphere and the effects of humidity variations on their optical properties, AFCRL-TR-79-0214, *Air Force Research Papers*, L.G. Hanscom Field, Bedford, MA.

Smirnov, A., B.N. Holben, T.F. Eck, O. Dubovik, and I. Slutsker, 2000: Cloud-screening and quality control algorithms for the AERONET database. *Remote Sens. Environ.*, **73**(3), 337–349.

Stowe, L.A., Ignatov, and R. Singh, 1997: Development, validation, and potential enhancements to the second-generation operational aerosol product at the National Environmental Satellite, Data, and Information Service of the National Oceanic and Atmospheric Administration, *J. Geophys. Res.*, **102**(D14), 16923–16934.

Tanré, D., M. Hermon, and Y.J. Kaufman, 1996: Information on aerosol size distribution contained in solar reflectance spectral radiances, *J. Geophys. Res.*, **101**(D14), 19043–19060.

Tanré, D., Y.J. Kaufman et al., 1997: Remote sensing of aerosol properties over oceans using the MODIS/EOS spectral radiances, *J. Geophys. Res.*, **102**(D14), 16971–16988.

Thomas, G.E., and K. Stamnes, 1999: *Radiative transfer in the atmosphere and ocean*, Cambridge University Press, New York.

Torres, O., P.K. Bhartia, J.R. Herman, Z. Ahmad, and J. Gleason, 1998: Derivation of aerosol properties from satellite measurements of backscattered ultraviolet radiation: theoretical bases. *J. Geophys. Res.*, **103**, 17009–17110.

Torres, O., P.K. Bhartia, J.R. Herman, A. Sinyuk, P. Ginoux, and B. Holben, 2002: A long-term record of aerosol optical depth from TOMS observations and comparison to AERONET measurements. *J. Atmos. Sci.*, **59**, 398–413.

Tucker, C.J., 1979: Red and photographic infrared linear combinations monitoring vegetation, *Remote Sens. Environ.*, **8**, 127–150.

Tucker, C.J., and P.J. Sellers, Satellite remote sensing of primary production. *Int. J. Remote Sensing*, **7**, 1395–1416.

Twomey, S., 1977: *Atmospheric Aerosols*. Elsevier, Amsterdam.

U.S. Environmental Protection Agency (USEPA), 2003: National Air Quality and Emissions Trends Report, EPA 454/R-03-005, Research Triangle Park, NC, (http://www.epa.gov/airtrends)

U.S. Government Printing Office, 1976: *U.S. Standard Atmosphere*, Washington, DC.

Van de Hulst, H.C., 1984: *Light Scattering by Small Particles*, Dover Publications, New York.

Vermote, E.F., D. Tanré et al., 1997a: Second Simulation of the Satellite Signal in the Solar Spectrum, 6S: an overview, *IEEE Trans. Geosci. Remote Sens.*, **35**(3), 675–686.

Vermote, E.F., N. ElSaleous, C.O. Justice, Y.J. Kaufman, J.L. Privette, L. Remer, J.C. Roger, and D. Tanre, 1997b: Atmospheric correction of visible to middle-infrared EOS-MODIS data over land surfaces: background, operational algorithm and validation. *J. Geophys. Res. – Atmos.*, **102**(D14), 17131–17141.

Volten, H., O. Munoz, E. Rol, J.F. de Haan., W. Vassen., J.W. Hovenier., K. Muinonen, and T. Nousiainen, 2001: Scattering matrices of mineral aerosol particles at 441.6 nm and 632.8 nm, *J. Geophys. Res.*, **106**, 17375–17401.

Volz, F.E., 1959: Photometer mit Selen-photoelement zurspektralen Messung der Sonnenstrahlung und zer Bestimmung der Wallenlangenabhangigkeit der Dunsttrubung. *Arch. Meteor. Geophys Bioklim.* **B10**, 100–131.

Wiscombe, W., 1980: Improved Mie scattering algorithms. *Appl. Opt.*, **19**, 1505–1509.

Wang, J., and S.A. Christopher, Intercomparison between satellite-derived aerosol optical thickness and PM2.5 mass: implication for air quality studies, *Geophys. Res. Lett*, **30**, 2095, doi:10.1029/2003GL018174.

Yu, H., Y.J. Kaufman, M. Chin, G. Feingold, L. Remer, T. Anderson, Y. Balkanski, N. Bellouin, O. Boucher, S. Christopher, P. DeCola, R. Kahn, D. Koch, N. Loeb, M.S. Reddy, M. Schulz, T. Takemura, and M. Zhou, 2006: A review of measurement-based assessments of aerosol direct radiative effect and forcing, *Atmos. Chem. Phys.*, **6**, 613–666.

Xiong, X., J. Sun, J. Esposito, B. Guenther, and W.L. Barnes 2003: MODIS reflective solar bands calibration algorithm and on-orbit performance, *Proceedings of SPIE Optical Remote Sensing of the Atmosphere and Clouds*, **III**, 4891, 95–104.

3 The time series technique for aerosol retrievals over land from MODIS

Alexei Lyapustin, Yujie Wang

1. Introduction

Atmospheric aerosols interact with sunlight by scattering and absorbing radiation. By changing the irradiance at the Earth surface, modifying cloud fractional cover and microphysical properties and by a number of other mechanisms, they affect the energy balance, hydrological cycle, and planetary climate [IPCC, 2007]. In many world regions there is a growing impact of aerosols on air quality and human health.

The Earth Observing System [NASA, 1999] initiated high-quality global Earth observations and operational aerosol retrievals over land. With the wide swath (2,300 km) of MODIS instrument, the MODIS *Dark Target* algorithm [Kaufman et al., 1997; Remer et al., 2005; Levy et al., 2007] currently complemented by the *Deep Blue* method [Hsu et al., 2004] provides a daily global view of planetary atmospheric aerosol. The MISR algorithm [Martonchik et al., 1998; Diner et al., 2005] makes high-quality aerosol retrievals in 300-km swaths covering the globe in 8 days.

With the MODIS aerosol program being very successful, there are still several unresolved issues in the retrieval algorithms. The current processing is pixel-based and relies on a single-orbit data. Such an approach produces a single measurement for every pixel characterized by two main unknowns, aerosol optical thickness (AOT) and surface reflectance (SR). This lack of information constitutes a fundamental problem of remote sensing which cannot be resolved without a priori information. For example, the MODIS *Dark Target* algorithm makes spectral assumptions about surface reflectance, whereas the *Deep Blue* method uses an ancillary global database of surface reflectance composed from minimal monthly measurements with Rayleigh correction. Both algorithms assume a Lambertian surface model.

The surface-related assumptions in the aerosol retrievals may affect subsequent atmospheric correction in an unintended way. For example, the *Dark Target* algorithm uses an empirical relationship to predict SR in the Blue (B3) and Red (B1) bands from the 2.1 μm channel (B7) for the purpose of aerosol retrieval. Obviously, the subsequent atmospheric correction will produce the same SR in the red and blue bands as predicted, i.e. an empirical function of $\rho_{2.1}$. In other words, the spectral, spatial and temporal variability of surface reflectance in the Blue and Red bands appears "borrowed" from band B7. This may have certain implications for the vegetation and global carbon analysis because the chlorophyll-sensing bands, B1 and B3, are effectively substituted in terms of variability by band B7, which is sensitive to plant liquid water.

This chapter describes a new recently developed generic aerosol-surface retrieval algorithm for MODIS. The Multi-Angle Implementation of Atmospheric Correction (MAIAC) algorithm simultaneously retrieves AOT and surface bi-directional reflection factor (BRF) using the time series of MODIS measurements.

MAIAC starts with accumulating 3 to 16 days of calibrated and geo-located level 1B (L1B) MODIS data. The multi-day data provide different view angles, which are required for the surface BRF retrieval. The MODIS data are first gridded to 1 km resolution in order to represent the same surface footprint at different view angles. Then, the algorithm takes advantage of the following invariants of the atmosphere–surface system: (1) the surface reflectance changes little during accumulation period, and (2) AOT changes little at short distances ($\sim$ 25 km), because aerosols have a mesoscale range of global variability of $\sim$ 50–60 km [Anderson et al., 2003]. Under these generic assumptions, the system of equations becomes over-defined and formally can be resolved. Indeed, we define the elementary processing area as a block with the size of $N \sim 25$ pixels (25 km). With K days in the processing queue, the number of measurements exceeds the number of unknowns

$$KN^2 > K + 3N^2 \quad \text{if} \quad K > 3, \tag{1}$$

where K is the number of AOT values for different days, and 3 is the number of free parameters of the Li-Sparse Ross-Thick (LSRT) [Lucht et al., 2000] BRF model for a pixel.

To simplify the inversion problem, the algorithm uses BRF, initially retrieved in B7, along with an assumption that the shape of BRF is similar between the 2.1 μm and the Blue spectral band:

$$\rho_{ij}^{\lambda}(\mu_0,\mu;\phi) = b_{ij}^{\lambda}\rho_{ij}^{B7}(\mu_0,\mu;\phi). \tag{2}$$

The scaling factor b is pixel-, wavelength-, and time-dependent. This physically well-based approach reduces the total number of unknown parameters to $K + N^2$. Below, factor b is called spectral regression coefficient (SRC).

The assumption (2) of similarity of the BRF shape is robust for most land cover types because the surface absorption coefficient, or inversely, surface brightness, is similar in the visible and shortwave infrared (SWIR) spectral regions, and because the scale of macroscopic surface roughness, which defines shadowing, is much larger than the wavelength [Flowerdew and Haigh, 1995]. One obvious exception is snow, which is very bright in the visible wavelengths and dark in the SWIR. The principle of spectral similarity of the BRF shape was extensively tested and implemented in ATSR-2 [Veefkind et al., 1998] and MISR [Diner et al., 2005] operational aerosol retrievals.

The MAIAC algorithm is based on minimization of an objective function, so it can directly control the assumptions used. For example, the objective function is high if surface changed rapidly or if aerosol variability was high on one of the days. Such days are filtered and excluded from the processing. The algorithm combines the block-level and the pixel-level processing, and produces the full set of parameters at 1 km resolution.

From a historical prospective, the new algorithm inherits multiple concepts developed by the MISR science team, from using the rigorous radiative transfer model with non-Lambertian surface in aerosol/surface retrievals [Diner et al., 1999, 2001] to the concept of image-based rather than pixel-based aerosol retrievals [Martonchik et al., 1998]. The latter idea, in a different implementation, was proposed in the *Contrast Reduction* method by Tanre et al. [1988], who showed that consecutive images of the same surface area, acquired on different days, can be used to evaluate the AOT difference between these days.

MAIAC is a complex algorithm which includes water vapor retrievals, cloud masking, aerosol retrievals and atmospheric correction. The separate processing blocks are inter-dependent: they share the data through the common algorithm memory and may update each other's output. For example, the cloud mask is updated during both aerosol retrievals and atmospheric correction. Section 2 of this chapter provides an overview of MAIAC processing. Section 3 presents the radiative transfer basis for the aerosol retrievals and atmospheric correction algorithm, which are described in sections 4 and 5, respectively. Section 6 describes the MAIAC cloud mask algorithm. Finally, section 7 presents examples of MAIAC performance and results of AERONET validation. The chapter is concluded with a summary (Section 8).

2. MAIAC overview

The block-diagram of MAIAC algorithm is shown in Fig. 3.1.

1. The received L1B data are gridded, split in 600 km Tiles, and placed in a Queue with the previous data. The size of the Tile is selected to fit the operational memory of our workstation. As a reminder, MODIS uses 1000 km Tiles in operational processing. In order to limit variation of the footprint with changing view zenith angle (VZA), the resolution is coarsened by a factor of 2. For example, the grid cell size is 1 km for MODIS 500 m channels B1–B7. We use the MODIS land gridding algorithm [Wolfe et al., 1998] with minor modifications that allow us to better preserve the anisotropy of signal in the gridded data when measured reflectance is high, for example over snow, thick clouds or water with glint.

2. The column water vapor is retrieved for the last Tile using MODIS near-IR channels B17–B19 located in the water vapor absorption band 0.94 μm. This algorithm is a modified version of Gao and Kaufman [2003]. It is fast and has the average accuracy of $\pm 5 - 10\%$ over the land surface [Lyapustin and Wang, 2007]. The water vapor retrievals are implemented internally to exclude dependence on other MODIS processing streams and unnecessary data transfers.

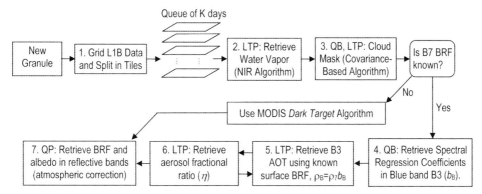

Fig. 3.1. Block-diagram of MAIAC algorithm. The initial capital letters indicate spatial and temporal domains of operations, for example at pixel- (P) or/and block- (B) level, and using the data of the last Tile only (LT) or using the full time series of the Queue (Q).

3. The time series of measurements helps to develop a high-quality cloud mask (CM). It is based on the notion that the surface spatial pattern is stable and reproducible in the short time frame in cloud-free conditions, whereas clouds randomly disturb this pattern. The algorithm uses covariance analysis to identify cloud-free blocks. On this basis, it builds a reference clear-skies image of the surface, which is further used in the pixel-level cloud masking. The MAIAC CM algorithm has an internal land–water–snow dynamic classification, which guides the algorithm flow.

4. The *main algorithm* simultaneously retrieves the block-level AOT for K-days and N^2 values of the spectral regression coefficient b_{ij} for the Blue (B3) band. This algorithm turns on when the B7 BRF is known. Otherwise, MAIAC implements a simplified version of the MODIS *Dark Target* algorithm.

5. The AOT computed in the previous step has a low resolution of 25 km. On the other hand, knowledge of SRC provides the Blue band BRF from Eq. (2) at a grid resolution. With the boundary condition known, the Blue band AOT in this step is retrieved at high 1 km resolution.

6. The ratio of volumetric concentrations of coarse-to-fine aerosol fractions (η) is calculated for the last Tile at the grid resolution. This parameter selects the relevant aerosol model and provides spectral dependence of AOT for the atmospheric correction. The AOT and parameter η retrievals are done simultaneously, which is indicated by two arrows between processing blocks 5 and 6.

7. Finally, surface BRF and albedo are retrieved at grid resolution from the K-day Queue for the reflective MODIS bands.

2.1 Implementation of time series processing

The MAIAC processing uses both individual grid cells, also called pixels below, and fixed-size (25×25 km^2) areas, or blocks, required by the cloud mask algorithm and SRC retrievals. In order to organize such processing, we developed a framework of C ++ classes and structures (algorithm-specific Containers). The class functions are designed to handle processing in the various time–space scales, for example at the pixel- versus block-level, and for a single (last) day of measurements versus all available days in the Queue, or for a subset of days which satisfy certain requirements (filters). The data storage in the Queue is efficiently organized using pointers, which avoids physically moving the previous data in memory when the new data arrive.

The structure of the Queue is shown schematically in Fig. 3.2. For every day of observations, MODIS measurements are stored as Layers for reflective bands 1–7 and thermal band 31, all of which are required by the CM algorithm. Besides storing gridded MODIS data (*Tiles*), the Queue has a dedicated memory (Q-memory) which accumulates ancillary information about every block and pixel of the surface for the cloud mask algorithm (*Refcm* data structure). It also keeps information related to the history of previous retrievals, for example spectral surface BRF parameters and albedo. Given the daily rate of MODIS observations, the land surface is a relatively static background. Therefore, knowledge of the previous surface state significantly enhances both the accuracy of the cloud detection, and the quality of atmospheric correction, for example, by imposing a requirement of consistency of the time series of BRF and albedo.

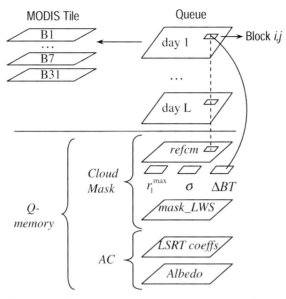

Fig. 3.2. Structure of Queue for ASRVN processing. The Queue, designed for the sliding window algorithm, stores up to 16 days of gridded MODIS observations at 1-km resolution. The CM algorithm uses MODIS bands 1–7 and band 31, which are stored as Layers (double-indexed arrays) shown in the upper-left corner. A dedicated Q-memory is allocated to store the ancillary information for the CM algorithm, such as a reference clear-skies image (*refcm*), block-level statistical parameters $\{r1_{max};\ \sigma_1;\ \Delta BT\}$, and results of dynamic land–water–snow classification (*mask_LWS*). This information is updated with latest measurements (day L) once a given block is found cloud-free, thus adapting to changing surface conditions. The Q-memory also stores results of previous reliable BRF retrievals for MODIS bands 1–7.

3. Radiative transfer basis

MAIAC resulted from an effort to develop an operational algorithm with explicit minimization where parameters of the surface BRF model can be calculated analytically from measurements. A similar approach developed by Martonchik et al. [1998] for MISR features a relatively small size of the look-up table (LUT) and a high efficacy, which is critically important for operational algorithm. We will be using a high-accuracy semi-analytical formula for the top-of-atmosphere (TOA) radiance derived with Green's function method [Lyapustin and Knyazikhin, 2001; Lyapustin and Wang, 2005]. Below, τ is atmospheric optical thickness, πS_λ is spectral extraterrestrial solar irradiance, and $s = (\mu = \cos\theta,\ \phi)$ is a vector of direction defined by zenith (θ) and azimuthal (ϕ) angles. The z-axis is pointed downwards, so $\mu_0 > 0$ for the solar beam and $\mu < 0$ for the reflected beam. The TOA radiance $L(s_0, s)$ is expressed as a sum of the atmospheric path radiance (D), and surface-reflected radiance (L_s), directly and diffusely transmitted through the atmosphere:

$$L(s_0, s) = D(s_0, s) + L_s(s_0, s)\mathrm{e}^{-\tau/|\mu|} + L_s^d(s_0, s). \tag{3}$$

The surface-reflected radiance is written as:

$$L_s(s_0, s) \cong S_\lambda \mu_0 \mathrm{e}^{-\tau/\mu_0}\{\rho(s_0, s) + ac_0\rho_1(\mu)\rho_2(\mu_0)\} + \frac{a}{\pi}\int_{\Omega^+} D_s(s_0, s')\rho(s', s)\mu'\mathrm{d}s' \tag{4}$$

where D_s is path radiance incident on the surface, c_0 is spherical albedo of the atmosphere, and

$$\rho_1(\mu) = \frac{1}{2\pi} \int\limits_{\Omega^+} \rho(s',s) \, ds', \quad \rho_2(\mu_0) = \frac{1}{2\pi} \int\limits_{\Omega^-} \rho(s_0,s) \, ds. \qquad (5)$$

a is a multiple reflection factor, $a = (1 - q(\mu_0)c_0)^{-1}$, where q is surface albedo. The diffusely transmitted surface-reflected radiance at the TOA is calculated from L_s with the help of 1D diffuse Green's function of the atmosphere:

$$L_s^d(s_0,s) = \int\limits_{\Omega^-} G^d(s_1,s)L_s(s_0,s_1) \, ds_1. \qquad (6)$$

The function πG^d is often called bi-directional upward diffuse transmittance of the atmosphere. The method of its calculation was discussed in detail by Lyapustin and Knyazikhin [2001]. The surface albedo is defined as a ratio of reflected and incident radiative fluxes at the surface:

$$q(\mu_0) = F^{Up}(\mu_0)/F^{Down}(\mu_0), \qquad (7a)$$

$$F^{Down}(\mu_0) = \pi S_\lambda \mu_0 \, e^{-\tau/\mu_0} + \int\limits_{\Omega^+} D_s(s_0,s')\mu' \, ds' = F_s^{Dir}(\mu_0) + F_s^{Dif}(\mu_0), \qquad (7b)$$

$$F^{Up}(\mu_0) = \pi S_\lambda \mu_0 \, e^{-\tau/\mu_0} q_2(\mu_0) + \int\limits_{\Omega^+} \mu' q_2(\mu')D_s(s_0,s') \, ds', \quad q_2(\mu_0) = \frac{1}{\pi} \int\limits_{\Omega^-} \rho(s_0,s)\mu \, ds.$$

$$(7c)$$

These formulas give an explicit expression for the TOA radiance as a function of surface BRF. As shown in Lyapustin and Knyazikhin [2001], the expression for TOA radiance is reduced to a classical equation of Chandrasekhar [1960] when surface reflectance is Lambertian. The accuracy of the above formulas is high, usually within a few tenths of a percent. Below we will use the TOA reflectance, which is defined as

$$R_\lambda = L_\lambda/(\mu_0 S_\lambda). \qquad (8)$$

3.1 Expression for the TOA reflectance using LSRT BRF model

Based on the described semi-analytical solution, TOA reflectance can be expressed as an explicit function of parameters of the BRF model. We are using a semi-empirical Li Sparse – Ross Thick (LSRT) BRF model [Lucht et al., 2000]. This is a linear model, represented as a sum of Lambertian, geometric-optical, and volume scattering components:

$$\rho(\mu_0,\mu,\phi) = k^L + k^G f_G(\mu_0,\mu,\phi) + k^V f_V(\mu_0,\mu,\phi). \qquad (9)$$

It uses predefined geometric functions (kernels) f_G, f_V to describe different angular shapes. The kernels are independent of the land conditions. The BRF of a pixel is characterized by a combination of three kernel weights, $\vec{K} = \{k^L, k^G, k^V\}^T$. The LSRT model is used in the operational MODIS BRF/albedo algorithm [Schaaf et al., 2002].

The substitution of Eq. (9) into Eqs. (3)–(7) and normalization to the reflectance units gives the following expressions for the surface-reflected signal (the last two terms of Eq. (3)):

$$R_s(\mu_0, \mu, \phi) = e^{-\tau/\mu_0}\{k^L + k^G f_G(\mu_0, \mu, \phi) + k^V f_V(\mu_0, \mu, \phi) +$$

$$+ ac_0 \rho_1(\mu) \rho_2(\mu_0)\} + a\mu_0^{-1}\{k^L E_0^d(\mu_0) + k^G D_G^1(\mu_0, \mu, \phi) + k^V D_V^1(\mu_0, \mu, \phi)\}, \quad (10)$$

$$R_s^d(\mu_0, \mu, \phi) = e^{-\tau/\mu_0} \times \{[k^L G^{av}(\mu) + k^G G_G^1(\mu_0, \mu, \phi) +$$

$$+ k^V G_V^1(\mu_0, \mu, \phi)] + ac_0[k^L G^{av}(\mu) + k^G G_G^{11}(\mu) + k^V G_V^{11}(\mu)]\rho_2(\mu_0)\} +$$

$$+ a\mu_0^{-1}\{k^L E_0^d(\mu_0)G^{av}(\mu) + k^G H_G^1(\mu_0, \mu, \phi) + k^V H_V^1(\mu_0, \mu, \phi)\}. \quad (11)$$

The surface albedo is written as:

$$q(\mu_0) = E_0^{-1}(\mu_0)\{\mu_0 e^{-\tau/\mu_0} q_2(\mu_0) + k^L E_0^d(\mu_0) + k^G D_G^3(\mu_0) + k^V D_V^3(\mu_0)\}. \quad (12)$$

Different functions of these equations represent different integrals of the incident path radiance (D_s) and atmospheric Green's function (G) with the BRF kernels. They were described by Lyapustin and Wang [2005] along with the method of numerical calculation. Below, we give only the integral expressions for these functions:

$$\rho_1(\mu) = k^L + k^G f_G^1(\mu) + k^V f_V^1(\mu), \quad (13)$$

$$\rho_2(\mu_0) = k^L + k^G f_G^2(\mu_0) + k^V f_V^2(\mu_0), \quad (14)$$

$$q_2(\mu_0) = k^L + k^G f_G^3(\mu_0) + k^V f_V^3(\mu_0), \quad (15)$$

$$D_k^1(\mu_0, \mu, \phi - \phi_0) = \frac{1}{\pi}\int_0^1 \mu' \, d\mu' \int_0^{2\pi} d\phi' D_s(\mu_0, \mu', \phi' - \phi_0) f_k(\mu', \mu, \phi - \phi'), \quad (16)$$

$$D_k^3(\mu_0) = \frac{1}{\pi}\int_0^{2\pi} d\phi' \int_0^1 \mu' f_k^3(\mu') D_s(\mu_0, \mu'; \phi') \, d\mu', \quad (17)$$

$$G^{av}(\mu) = \int_{-1}^0 d\mu_1 \int_0^{2\pi} G^d(\mu_1, \mu, \phi - \phi_1) \, d\phi_1, \quad (18)$$

$$G_k^{11}(\mu) = \int_{-1}^0 f_k^1(\mu_1) \, d\mu_1 \int_0^{2\pi} G^d(\mu_1, \mu, \phi - \phi_1) \, d\phi_1, \quad (19)$$

$$G_k^1(\mu_0, \mu, \phi - \phi_0) = \int_{-1}^0 d\mu_1 \int_0^{2\pi} G^d(\mu_1, \mu, \phi - \phi_1) f_k(\mu_0, \mu_1, \phi_1 - \phi_0) \, d\phi_1, \quad (20)$$

$$H_k^1(\mu_0,\mu,\phi-\phi_0) = \int\limits_{-1}^{0} d\mu_1 \int\limits_{0}^{2\pi} G^d(\mu_1,\mu,\phi-\phi_1) D_k^1(\mu_0,\mu_1,\phi_1-\phi_0)\, d\phi_1. \quad (21)$$

The subscript k in the above expressions refers to either geometric-optical (G) or volumetric (V) kernels, and the supplementary functions of the BRF kernels are given by:

$$f_k^1(\mu) = \frac{1}{2\pi} \int\limits_{0}^{1} d\mu' \int\limits_{0}^{2\pi} f_k(\mu',\mu,\phi'-\phi)\, d\phi', \quad (22a)$$

$$f_k^2(\mu_0) = \frac{1}{2\pi} \int\limits_{-1}^{0} d\mu_1 \int\limits_{0}^{2\pi} f_k(\mu_0,\mu_1,\phi_1-\phi_0)\, d\phi_1, \quad (22b)$$

$$f_k^3(\mu') = \frac{1}{\pi} \int\limits_{-1}^{0} \mu\, d\mu \int\limits_{0}^{2\pi} f_k(\mu',\mu,\phi-\phi')\, d\phi. \quad (22c)$$

The diffuse and total spectral surface irradiance are calculated from Eq. (7b) as:

$$E_0^d(\mu_0) = F^{Dif}(\mu_0)/(\pi S_\lambda),\; E_0(\mu_0) = F^{Down}(\mu_0)/(\pi S_\lambda). \quad (23)$$

Let us rewrite Eqs. (10)–(11) separating the kernel weights. First, separate the small terms proportional to the product $c_0 \rho_2(\mu_0)$ into the nonlinear term:

$$R^{nl}(\mu_0,\mu) = a c_0 \rho_2(\mu_0) e^{-\tau/\mu_0} \{ e^{-\tau/|\mu|} \rho_1(\mu) + k^L G^{av}(\mu) + k^G G_G^{11}(\mu) + k^V G_V^{11}(\mu) \}. \quad (24)$$

Second, collect the remaining multiplicative factors for the kernel weights:

$$F^L(\mu_0,\mu) = (e^{-\tau/\mu_0} + a\mu_0^{-1} E_0^d(\mu_0))(e^{-\tau/|\mu|} + G^{av}(\mu)), \quad (25)$$

$$F^k(\mu_0,\mu;\phi) = \{ e^{-\tau/\mu_0} f_k(\mu_0,\mu,\phi) + a\mu_0^{-1} D_k^1(\mu_0,\mu,\phi) \} e^{-\tau/|\mu|} +$$
$$+ e^{-\tau/\mu_0} G_k^1(\mu_0,\mu,\phi) + a\mu_0^{-1} H_k^1(\mu_0,\mu,\phi), k = V,\, G. \quad (26)$$

With these notations, the TOA reflectance becomes:

$$R(\mu_0,\mu,\phi) = R^D(\mu_0,\mu,\phi) + k^L F^L(\mu_0,\mu) + k^G F^G(\mu_0,\mu,\phi) +$$
$$+ k^V F^V(\mu_0,\mu,\phi) + R^{nl}(\mu_0,\mu). \quad (27)$$

This equation, representing TOA reflectance as an explicit function of the BRF model parameters, provides the means for an efficient atmospheric correction.

Let us derive a modified form of this equation which is used in the aerosol retrievals. The last nonlinear term of Eq. (27), which describes multiple reflections of the direct-beam sunlight between the surface and the atmosphere, is small ($R^{nl} \propto q c_0$), and can be neglected for simplicity of further consideration. The functions F^k are still weakly nonlinear via parameter a, which describes multiple reflections of the diffuse incident sunlight. By setting $a = 1$, we omit this nonlinearity and Eq. (27) becomes a linear function of the BRF parameters. With an additional assumption of spectral invariance of the BRF shape (Eq. (2)), Eq. (27) can be rewritten for the pixel (i, j) and observation day k as:

$$R_{ij}^k(\lambda) \cong R^D(\lambda, \tau^k) + b_{ij}(\lambda) Y_{ij}(\lambda, \tau^k), \tag{28}$$

where $b_{ij}(\lambda)$ is spectral regression coefficient for a given spectral band, and function

$$Y_{ij}(\lambda, \tau^k) = k_{ij}^{L,B7} F^L(\lambda, \tau^k) + k_{ij}^{G,B7} F^G(\lambda, \tau^k) + k_{ij}^{V,B7} F^V(\lambda, \tau^k) \tag{29}$$

can be calculated from the look-up table (LUT) for a given geometry, AOT and wavelength, once the BRF parameters in band B7 for the pixel (i, j) are known. The LUT stores functions f_k^1, f_k^2, f_k^3, which depend on geometry of observations, and functions $D_k^1, D_k^3, G^{av}, G_k^1,$ $G_k^{11}, H_k^1, E_0^d, E_0, R^D$, which depend on geometry, selected aerosol model and AOT. Index k refers to either volumetric (V) or geometric-optical (G) BRF kernel function. The pressure correction and water vapor correction of the LUT functions are performed with the algorithm described in Lyapustin and Wang, [2007]. The absorption by stratospheric ozone is corrected by dividing the measured reflectance by ozone transmittance [e.g., see Diner et al., 1999].

4. Aerosol algorithm

The aerosol algorithm consists of two steps: deriving spectral regression coefficients (SRCs), and retrieving AOT and aerosol fractional ratio. The SRC retrievals use parametric Eq. (28).

4.1 SRC retrievals

Let us assume that the ancillary information for the aerosol retrievals, including water vapor, cloud mask, and surface BRF in band B7, is available. Let us also assume that gridded TOA MODIS reflectance data is available for $3 \leq K \leq 16$ cloud-free days, which form the processing Queue. Our goal is to derive the set of K AOT values for different days (orbits), and N^2 SRC values for the Blue band (B3) for a given 25-km block of the surface. The SRC algorithm is implemented in three steps:
(1) select the clearest day from the Queue;
(2) calculate the AOT difference for every day with respect to the clearest day, $\Delta \tau^k = \tau^k - \tau_0$;
(3) find AOT on the clearest day, τ_0. At this step, the algorithm simultaneously generates the full set of spectral regression coefficients.
The first task is solved as follows. Initially, the SRCs are calculated for every day and every pixel separately using Eq. (28) for AOT = 0. For a given pixel, the coefficient b_{ij}^k is lowest on the clearest day because its value is increased by the path reflectance on hazier days. Therefore, the clearest day is selected as a day with the lowest on average set of coefficients b_{ij}^k in the block.

In the next step (2), the AOT difference between the day k and the clearest day is calculated independently for every day of the Queue by minimizing the difference

$$F_1^k = \frac{1}{N^2} \sum_{i,j} \{b_{ij}^{Clear} - b_{ij}^k(\Delta\tau^k)\}^2 = \min\{\Delta\tau^k\}. \tag{30}$$

The SRCs for the clearest day (b_{ij}^{Clear}) have been calculated for $\tau_0 = 0$ in step (1). When solving Eq. (30), SRCs for the day k are re-calculated for the increasing values of AOT from the LUT τ^k ($\Delta\tau^k = \tau^k - \tau_0 = \tau^k$) until the minimum is reached. This operation is equivalent to simultaneous removal of bias and 'stretching' the contrast for a given block that minimizes the overall difference.

In step (3), AOT on the clearest day is found by minimization of *rmse* between the theoretical reflectance and the full set of measurements for K days and N^2 pixels:

$$F_2 = \sum_K \sum_{i,j} \{R_{ij}^{Meas,k} - R_{ij}^{Th,k}(\tau_0 + \Delta\tau^k)\}^2 = \min\{\tau_0\}. \tag{31}$$

To calculate theoretical reflectance with Eq. (28), one needs to know the coefficients b_{ij}. These are calculated using the first assumption described in Introduction, namely that the surface reflectance changes little during K days. Therefore, for a given pixel and given value τ_0, the SRC can be found by minimizing the *rmse* over all days of the Queue:

$$F_{ij} = \sum_K \{R_{ij}^{Meas,k} - R_{ij}^{Th,k}(\tau^k)\}^2 = \min\{b_{ij}\}, \tau^k = \tau_0 + \Delta\tau^k, \tag{32}$$

which is solved by the least-squares method ($\partial F_{ij}/\partial b_{ij} = 0$) with the analytical solution:

$$b_{ij} = \sum_K [R_{ij}^{Meas,k} - R^D(\tau^k)]Y_{ij}(\tau^k) / \sum_K \{Y_{ij}(\tau^k)\}^2. \tag{33}$$

Thus, given the aerosol model, Eq. (31) becomes parameterized in terms of the only parameter τ_0. Eqs (31) and (32) are positively defined quadratic forms which have unique solutions. To solve these equations numerically, the MAIAC algorithm incrementally increases the AOT (e.g., τ_0 in Eq. (31)) using the LUT entries, until the minimum is found. Because the discretization of LUT in AOT is relatively coarse, the algorithm finds the 'bend' point, where function F_2 starts increasing, approximates the last three points, encompassing the minimum, with quadratic function, and finds the minimum analytically. The set of SRCs is calculated with the final value τ_0 from Eq. (33).

This algorithm was developed and optimized through a long series of trial and error. It requires at least three clear or partially clear days in the Queue for the inversion, with at least 50 % of the pixels of the block being clear for three or more days. The algorithm has a self-consistency check, verifying whether the main assumptions hold. This is done during step (2) processing. If the surface had undergone a rapid change during the accumulation period (e.g., a snowfall, or a large-scale fire, flooding or rapid land-cover conversion, with the size of disturbance comparable to the block size), or if the AOT changes significantly inside a given block on day k, then the value of *rmse* $\sqrt{F_1^k}$ remains high. Currently, the algorithm excludes such days from the processing Queue based on a simple empirically established threshold $\sqrt{F_1^k} \geq 0.03$. In regular conditions, the value $\sqrt{F_1^k}$ is usually lower than 0.01–0.015.

Retrieving SRCs is a well-optimized and a relatively fast process. Nevertheless, in order to reduce the total processing time, MAIAC makes these retrievals with the period of 2–3 days, which is sufficient to track a relatively slow seasonal variability of the land surface. For every block, the retrieved spectral regression coefficients are stored in the Q-memory, along with the band B7 LSRT coefficients. They are used as ancillary information for the aerosol retrievals at 1 km grid resolution, which are described next.

4.2 Aerosol retrievals

With spectral regression coefficients retrieved, the surface BRF in every grid cell in the Blue band becomes known (Eq. (2)). Further, the AOT and aerosol fractional ratio are retrieved at 1-km resolution from the last Tile of MODIS measurements.

This algorithm requires a set of aerosol models with increasing particle size and asymmetry parameter of scattering. The aerosols are modeled as a superposition of the fine and coarse fractions, each described by a log-normal size distribution. For example, for the continental USA we are currently using the weak absorption model with the following parameters for the fine (F) and coarse (C) fractions: median radius $R_v^F = 0.14$ μm, $R_v^C = 2.9$ μm; standard deviation $\sigma^F = 0.38$ μm $\sigma^C = 0.75$ μm; spectrally independent real part of refractive index $n_r = 1.41$, and imaginary part of refractive index $n_i = \{0.0044, 0.0044, 0.0044, 0.002, 0.001\}$ at wavelengths $\{0.4, 0.55, 0.8, 1.02, 2.13\}$ μm, respectively. Parameter n_i is linearly interpolated between five grid wavelengths. By varying the ratio of volumetric concentrations of coarse and fine fractions, $\eta = C_v^{Coarse}/C_v^{Fine}$, a wide range of asymmetry (size) parameter is simulated. The LUT is originally computed for the fine and coarse fractions separately. When MAIAC reads the LUT, it generates a series of mixed aerosol LUTs for different values $\eta = \{0.2; 0.5; 1; 2; 5; 10\}$, which are stored in the operational memory. In this sequence, value $\eta = 0.5$ gives a model that is close to the urban continental weak absorption (GSFC) model from AERONET classification [Dubovik et al., 2002], whereas the values $\eta = 2–5$ are more representative of the mineral dust. Following the MISR aerosol algorithm [Diner et al., 1999, 2001; Martonchik et al., this volume Ch. 9], a modified linear mixing algorithm [Abdou et al., 1997] is used to mix the LUT radiative transfer (RT) functions for the fine and coarse fractions. This algorithm retains high accuracy with increasing AOT and aerosol absorption.

For each pixel, the retrieval algorithm goes through a loop of increasing values of fractional ratio η, and using known surface BRF $\rho_{ij}^{B3}(\mu_0, \mu; \phi) = b_{ij}^{B3} \rho_{ij}^{B7}(\mu_0, \mu; \phi)$ it computes AOT (τ_{ij}) in the Blue band by fitting theoretical TOA reflectance to the measurement

$$R^{Theor,B3}(\eta; \tau_{ij}) = R_{ij}^{Meas,B3}. \tag{34}$$

In the next step, a spectral residual is evaluated using the Blue (B3), Red (B1), and SWIR (B7) bands:

$$\chi_{ij} = \sum_{\lambda} \{R_{ij}^{Meas,\lambda} - R_{ij}^{Theor,k}(\tau^\lambda(\eta))\}^2 = \min\{\eta\}. \tag{35}$$

The procedure is repeated with the next value η until the minimum is found. Theoretical reflectance in Eq. (35) is computed with the LSRT BRF parameters from the previous cycle of atmospheric correction, which are stored in the Q-memory.

Because MODIS measurements provide only a spectral slice of information, MAIAC does not attempt MISR-like retrievals for multiple aerosol models with different absorption and sphericity of particles. Instead, it follows the MODIS *Dark Target* approach [Levy, 2007] where the aerosol fractions and their specific absorption properties are fixed regionally.

The spectral sensitivity of measurements to variations of the aerosol model in clear atmospheric conditions, especially at longer wavelengths, is limited. Currently, growth of the MODIS footprint with the scan angle is the main source of uncertainty in MAIAC's knowledge of the surface spectral BRF. These errors, although small, can be costly if a very asymmetric aerosol model with large AOT values is selected when the atmosphere is actually very clean. For these reasons, the full minimization procedure Eqs (34) and (35) is performed only when the retrieved optical thickness for the standard continental model ($\eta = 0.5$) exceeds 0.3. Otherwise, a single value of $\eta = 0.5$ is used and AOT is reported for these background conditions.

The retrieval examples given below were generated using the single low absorption aerosol model described above. In future, the aerosol model will be geographically prescribed based on the AERONET climatology [Holben et al., 2001]. The following work is underway: we are studying the AERONET-based classification used in the MODIS aerosol algorithm [Levy, 2007] and plan to investigate a MISR level 3 aerosol product, which provides an independent global aerosol climatology over land.

5. Atmospheric correction

Determination of spectral surface BRF is an integral part of MAIAC's aerosol retrieval process. Below, we describe the atmospheric correction algorithm implemented in MAIAC.

Once the cloud mask is created and aerosol retrievals performed, the MAIAC algorithm filters the time series of MODIS measurements for every pixel and places the remaining clear-skies data in a 'container'. The filter excludes pixels with clouds and cloud shadows, as well as snow-covered pixels as detected by the CM algorithm during land–water–snow classification. We also filter out pixels with high AOT (> 0.9) where sensitivity of measurements to surface reflectance decreases. The container stores measurements along with the LUT-based RT functions for the cloud-free days of the Queue. If the number of available measurements exceeds three for a given pixel, then the coefficients of LSRT BRF model are computed. The retrieval diagram is shown in Fig. 3.3.

5.1 Inversion for LSRT coefficients

In operational MODIS land processing, BRF is determined in two steps: first, the atmospheric correction algorithm derives surface reflectance for a given observation geometry using Lambertian approximation [Vermote et al., 2002], and next, three LSRT coefficients are retrieved from the time series of surface reflectance accumulated for a 16-day period [Schaaf et al., 2002]. The Lambertian assumption simplifies the atmospheric correction but creates biases in the surface reflectance which depend on the observation geometry and atmospheric opacity. It is known that Lambertian assumption creates a flatter BRF pattern while the true BRF is more anisotropic [Lyapustin, 1999].

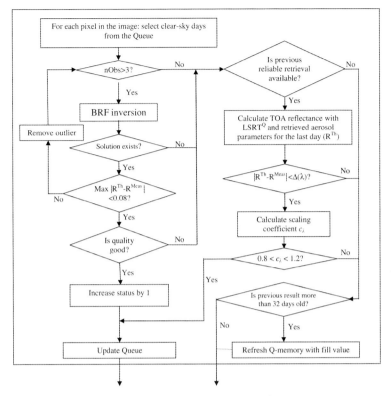

Fig. 3.3. Block-diagram of MAIAC atmospheric correction algorithm.

The MAIAC algorithm derives LSRT coefficients directly by fitting the measured TOA reflectance accumulated for a period of 4–16 days. The inversion is based on Eq. (27) derived earlier. This equation provides an explicit parameterization of TOA reflectance in terms of the BRF model parameters $\vec{K} = \{k^L,\ k^G,\ k^V\}^T$.

The quasi-linear form of equation (27) leads to a very efficient iterative minimization algorithm:

$$RMSE = \sum_j (r_j^{(n)} - F_j^L k^{L(n)} - F_j^V k^{V(n)} - F_j^G k^{G(n)})^2 = \min_{\{K\}},\ r^{(n)} = R - R^D - R^{nl(n-1)},$$
(36)

where index j lists measurements for different days, and n is the iteration number. Eq. (36) provides an explicit least-squares solution for the kernel weights. In a matrix form, the solution is written as:

$$\vec{K}^{(n)} = A^{-1} \vec{b}^{(n)},$$
(37)

where

$$
A = \begin{bmatrix} \sum_j (F_j^L)^2 & \sum_j F_j^G F_j^L & \sum_j F_j^V F_j^L \\ \sum_j F_j^G F_j^L & \sum_j (F_j^G)^2 & \sum_j F_j^V F_j^G \\ \sum_j F_j^V F_j^L & \sum_j F_j^V F_j^G & \sum_j (F_j^V)^2 \end{bmatrix}, \vec{b}^{(n)} = \begin{bmatrix} \sum_j r_j^{(n)} F_j^L \\ \sum_j r_j^{(n)} F_j^G \\ \sum_j r_j^{(n)} F_j^V \end{bmatrix}.
$$

In the first iteration, the small nonlinear term is set to zero, $R_j^{nl(0)} = 0$, and the multiple reflection factor a (see Section 3) is set to one, $a^{(0)} = 1$. These parameters are updated once, after the BRF coefficients are calculated in the first iteration. Except for snow-covered surfaces, the problem converges with high accuracy in two iterations in all conditions because the nonlinear terms are small. The described algorithm is very efficient computationally.

5.2 Solution selection and update

Although the LSRT model leads to an efficient BRF retrieval algorithm, there are several caveats associated with this model. The LSRT kernels are not orthogonal, they are not positive-only functions, and they are normalized in somewhat arbitrary fashion not linked to the radiative transfer. These factors reduce the stability of solution upon small perturbation of measurements and may lead to non-uniqueness of solution. The high goodness-of-fit at the measurement angles does not always guarantee the correct shape of the retrieved BRF, and may result in negative BRF values at other angles. The albedo, being an integral function of BRF, is especially sensitive to the peculiarities of a particular BRF shape. For these reasons, we developed several tests to remove unrealistic solutions.

The initial validation of the solution (see Fig. 3.3) checks that the maximal difference over all days of the Queue between measured and computed TOA reflectance does not exceed a specified threshold ($|R^{Meas} - R^{LSRT}| < 0.08$). The day (measurement) with higher deviation is excluded from the Queue and the retrieval is repeated.

If the solution provides a good agreement with measurements for all days, the algorithm verifies that values of the direct-beam albedo (q; also function $q_2(\mu_0)$ in Eq. (7c)) at solar zenith angle SZA $= 15°, 45°, 60°$ are positive. Finally, the new solution must be consistent with the previous solution: $|q(45°) - q^{Pr\ ev}(45°)| < \Delta(\lambda)$. Δ is the band-dependent threshold currently equal to 0.04 (blue), 0.05 (green and red), 0.1 (NIR and shortwave infrared bands, B4–B6), and 0.06 for band B7. The thresholds are relatively loose to allow variations in the solution for surface reflectance. The consistency of the time series of BRF and albedo is characterized by a parameter *status*. Initially, the confidence in the solution is low (*status* $= 0$). Each time, when the new retrieval agrees with the previous retrieval, the value of *status* increases by 1. When the *status* reaches the value of 4, the retrieval is considered reliable.

When the new solution is validated, the coefficients of BRF model and direct-beam albedo $q(45°)$, stored in the Q-memory are updated. The update is performed with relaxation designed to mitigate random noise of retrievals:

$$
\vec{K}_\lambda^{New} = (\vec{K}_\lambda^{New} + \vec{K}_\lambda^{Prev})/2. \tag{38}
$$

This updating method increases the quality of BRF and albedo product when the surface is relatively stable, but it delays response of solution to the surface changes.

Often, the solution for some pixels or for the full area cannot be found because of a lack of clear-sky measurements. In these cases, MAIAC assumes that the surface does not change and fills in the gaps with the previous solution for up to a 32-day period. This is the most natural way of gap-filling with specific solution for a given pixel under the assumption of stable surface. The gap-filled pixel is marked as 'Extended' in the quality assurance (QA) value with parameter QA. *nDelay* giving the number of days since the last reliable solution.

5.3 MAIAC surface reflectance products

MAIAC computes two main products at 1-km resolution for seven 500-m MODIS bands, i.e. a set of BRF coefficients and the surface albedo. The albedo is defined by Eq. (7a) as a ratio of surface-reflected to incident radiative fluxes. Thus, it represents a true albedo at a given solar zenith angle in ambient atmospheric conditions, the value of which can be directly compared to the ground-based measurements.

MAIAC also computes several derivative products useful for science data analysis and validation:

(1) *NBRF*: a Normalized BRF, which is computed from LSRT parameters for the common geometry of nadir view and SZA $= 45°$. This product is similar to MODIS NBAR (nadir BRF-adjusted reflectance) product (part of MOD43 suite). With the geometry variations removed, the time series of *NBRF* is useful for studying vegetation phenology, performing surface classification, etc.

2) *IBRF*: an instantaneous (or one-angle) BRF for specific viewing geometry of the last day of observations. This product is calculated from the latest MODIS measurement assuming that the shape of BRF, known from previous retrievals, has not changed. To illustrate the computation of *IBRF*, let us rewrite Eq. (27) for the measured TOA reflectance as follows:

$$R(\mu_0,\mu,\phi) = R^D(\mu_0,\mu,\phi) + cR^{Surf}(\mu_0,\mu,\phi), \tag{39}$$

where R^{Surf} combines all surface related terms and can be calculated using previous solution for BRF (BRF_λ) and retrieved aerosol information. c is spectrally dependent scaling factor. Then,

$$IBRF_\lambda(\mu_0,\mu,\phi) = c_\lambda BRF_\lambda(\mu_0,\mu,\phi). \tag{40}$$

Below, this algorithm will be referred to as scaling. This description was given as an illustration. In reality, R^{Surf} is a nonlinear function so that parameter c_λ and *IBRF* are computed accurately using the formulas of Section 3.

The algorithm computing scaling coefficient (and *IBRF*) is shown in Fig. 3.3 on the right. First, the algorithm filters out measurements which differ from theoretically predicted TOA reflectance based on previous solution (R_Q^{LSRT}) by more than factor of $\Delta(\lambda)$. Then, scaling coefficients are computed, and the consistency requirement is verified as follows: $0.8 < c_\lambda < 1.2$. If all conditions are satisfied and the *status* of pixel is high (*status* ≥ 4), then the Q-memory is updated with the scaled solution:

$$\vec{K}_{\lambda}^{New} = \frac{c_{\lambda} + 1}{2} \vec{K}_{\lambda}^{Prev}. \tag{41}$$

The time series processing is intrinsically controversial when surface changes rapidly. On the one hand, one needs all available cloud-free measurements and maximal time window in order to reduce the *rmse*. This approach, which mitigates the noise of measurements, including that of gridding and residual clouds, and which ensures robust BRF shape, is best for stable periods, for example for natural ecosystems in summer time in mid-latitudes. On the other hand, detecting and tracking surface changes like spring green-up or fall senescence requires the least possible number of days in the Queue. Such retrievals tend to have more spatial and spectral noise. Moreover, it is difficult to assess reliability of such solutions when the surface reflectance changes daily with possible data gaps due to clouds. Based on our experience, the combination of one-day solution (*IBRF*) and 16-day solution (*NBRF*) with an update from the last day of measurements (Eq. (41)) combine both the accuracy and an ability to track surface changes.

Besides faster response to the surface change, our update strategy assures fast removal of the retrieval artifacts, mainly residual clouds, which turn out to be the most common problem.

While the response of the 16-day solution may still be delayed, the *IBRF* tracks spectral changes immediately. The update of Q-memory with latest measurements with Eq. (41) was found to significantly accelerate response of the LSRT coefficients ($\vec{K}_{\lambda}$), and hence of NBRF, to changing surface conditions. Overall, the *IBRF* is better suited for analysis of the fast surface processes and detection of the rapid surface changes.

6.   MAIAC cloud mask

The MAIAC cloud mask is a new algorithm making use of the time series of MODIS measurements and combining an image and pixel-based processing. With a high frequency of MODIS observations, the land surface can be considered as a static or slowly changing background as opposed to ephemeral clouds. This offers a reliable way of developing the 'comparison clear-skies target' for the CM algorithm. An early example of such an approach is the ISCCP CM algorithm [Rossow and Garder, 1993] developed for geostationary platforms. It builds the clear-skies composite map from the previous measurements and infers CM for every pixel by comparing a current measurement with the clear-skies reference value. The uncertainty of the reference value, caused by the natural variability and sensor noise, is directly calculated from the measurements.

The MAIAC cloud mask is a next step in evolution of this idea. It uses covariance analysis to build reference clear-skies images (*refcm*) and to accumulate a certain level of knowledge about every pixel of the surface and its variability, thus constructing fairly comprehensive comparison targets for cloud masking. The reference image contains a clear-skies reflectance in MODIS band 1 (0.645 μm). In order to account for the effects related to scan angle variation, e.g., pixel size growth, surface BRF effect or reduction of contrast at higher view zenith angles (VZA), two reference clear-skies images are maintained by the algorithm, *refcm1* for VZA $0 < 45°$ and *refcm2* for VZA $= 45 - 60°$. In addition to *refcm*, the Q-memory also stores the maximal value ($r1_{max}$) and the variance (σ_1) of reflectance in band 1 as well as the brightness temperature contrast ($\Delta BT = BT_{max} - BT_{min}$) for each 25×25 km^2 block. Analysis of MODIS data shows that thermal contrast (ΔBT) is a rather stable metric of a given land area in clear conditions. In partially cloudy conditions, the contrast increases because BT_{min} is usually lower over clouds.

The new CM algorithm has an internal surface classifier, producing a dynamic land–water–snow (LWS) mask, and a surface change mask. These are an integral part of MAIAC guiding both cloud masking and further aerosol–surface reflectance retrievals when the surface changes rapidly as a result of fires, floods or snow fall/ablation. The cloud mask generated by the CM algorithm is updated during aerosol retrievals and atmospheric correction, which makes it a synergistic component of MAIAC. This complex approach increases the overall quality of cloud mask.

Below, we briefly describe the algorithm constructing the reference clear-skies image (*refcm*), and an overall decision logic in cloud masking. Further details of the algorithm are given elsewhere [Lyapustin et al., 2008].

6.1 Building reference clear skies image

The clear-sky images of a particular surface area have a common textural pattern, defined by the surface topography, boundaries of rivers and lakes, distribution of soils and vegetation, etc. This pattern changes slowly compared with the daily rate of global Earth observations. Clouds randomly change this pattern, which can be detected by covariance analysis. The covariance is a metric showing how well the two images X and Y correlate over an area of $N \times N$ pixels,

$$cov = \frac{1}{N^2} \sum_{i,j=1}^{N} \frac{(x_{ij} - \bar{x})(y_{ij} - \bar{y})}{\sigma_x \sigma_y}, \quad \sigma_x^2 = \frac{1}{N^2} \sum_{i,j=1}^{N} (x_{ij} - \bar{x})^2. \tag{42}$$

A high covariance of two images usually implies cloud-free conditions in both images, whereas low covariance usually indicates presence of clouds in at least one of the images. Because covariance removes the average component of the signals, this metric is equally successful over the dark and bright surfaces and in both clear and hazy conditions if the surface spatial variability is still detectable from space.

The core of the MAIAC CM algorithm is initialization and regular update of the reference clear-skies image for every block. The *refcm* is initially built from a pair of images for which covariance is high, and caution is exercised to exclude correlated cloudy fields. The algorithm calculates a block-level covariance between the *new Tile* and the previous *Tiles*, moving backwards in the Queue until either the 'head' of Queue is reached, in which case initialization fails and the algorithm would wait for the new data to arrive, or until clear conditions are found. The latter corresponds to high covariance ($cov \geq 0.68$) and low brightness temperature contrast in the block for both days, $\Delta BT = BT_{max} - BT_{min} < \Delta_1$. The initial value of threshold Δ_1 is currently defined as $\Delta_1 = 25 + dT(h)$ K. Factor $dT(h)$ accounts for the surface height variations in the block and is defined for an average lapse rate, $dT(h) = 0.0045(h_{max} - h_{min})$, where h (km) is surface height over the sea level. Once the image *refcm* is initialized, the algorithm begins to use the block-specific value of the brightness temperature contrast $Q.\Delta BT$, which is stored in the Q-memory.

After initialization, the algorithm uses the *refcm* to compute covariance with the latest measurements. Once clear conditions are found, *refcm* and block-parameters $\{rl_{max}; \sigma_1; \Delta BT\}$ are updated. With this dynamic update, the *refcm* adapts to the gradual land-cover changes related to the seasonal cycle of vegetation. The rapid surface change events (e.g., snowfall/ablation) are handled through repetitive re-initialization which is performed each time when covariance of the latest Tile with *refcm* is found to be low.

Following the covariance calculation, the algorithm looks for clouds at the pixel level. For regular surfaces, not covered by snow, cloud detection is based on a simple postulate that clouds are usually colder and brighter than the surface:

$$\text{IF}(BT_{ij} \leq BT_G - 4)\text{AND}(rl_{ij} > refcm.rl_{ij} + 0.05) \Rightarrow CM\_PCLOUD, \qquad (43)$$

where BT_{ij} is measured brightness temperature and rl_{ij} is measured B1 reflectance. The reference surface reflectance for every pixel is provided by the *refcm* clear-skies image, whereas an estimate of the ground brightness temperature BT_G comes either from the clear land pixels detected by a *Dense Vegetation* (or high NDVI) spectral test for a given block, or from the cloud-free neighbor blocks, identified by high covariance.

The final values of the MAIAC CM are clear (CM_CLEAR, CM_CLEAR_WATER, CM_CLEAR_SNOW), indicating surface type as well, possibly cloudy (CM_PCLOUD), and confidently cloudy (CM_CLOUD). The value of CM_SHADOW is used for pixels defined as cloud shadows. Shadows are detected with a simple threshold algorithm which compares the latest MODIS measurement (ρ^{meas}) with predicted reflectance (ρ^{pred}) based on the LSRT coefficients from the previous retrievals:

$$\text{IF}\rho^{meas} < \rho^{pred} - 0.12 \Rightarrow CM\_SHADOW. \qquad (44)$$

The shadow algorithm uses MODIS band 5 (1.24 μm), which has little atmospheric distortion and is bright over land so that the change of reflectance due to cloud shadow is easy to detect above the noise level.

The covariance component of MAIAC algorithm, which offers a direct way to identify clear conditions, renders another commonly used value of cloud mask – 'possibly clear' redundant.

6.2. Performance of MAIAC CM algorithm

The algorithm performance has been tested at scales of 600–1800 km using the 2004–2005 MODIS Terra data for northeastern USA, southern Africa (Zambia), the Amazon region (Brazil), the Arabian peninsula, and Greenland. The testing was done for at least half a year of continuous data in each case, using visual analysis and comparison with MODIS collection 5 operational cloud mask (MOD35).

Fig. 3.4 shows a case of cloud detection over receding snow for three winter days (36–37, 42) of 2005 for northeastern USA. The area of the image is 600 $\times$ 600 km^2. The two RGB images have a different normalization, helping visual distinction between snow and clouds. The MAIAC cloud mask is shown on the right, and the MODIS collection 5 (MOD35) reprojected and gridded cloud mask is shown at the bottom. The conditions represent different degrees of cloudiness over the land. Day 35 is entirely cloud-free over land. MAIAC CM algorithm gives an accurate overall classification. Thin ice on

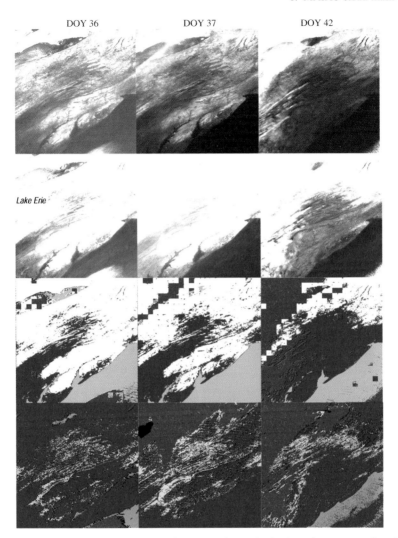

Fig. 3.4. Example of MAIAC (third row) and MOD35 (bottom) cloud mask over snow from MODIS Terra for days 36, 37 and 42 of 2005 (DOY is day of year). The image shows 1 Tile (600 × 600 km²) for northeastern USA. The top two RGB images have a different normalization helping visual distinction between snow and clouds. Legend for MOD35 CM: Blue – clear, Green – possibly clear, Yellow – possibly cloudy, Red – cloudy, Black – undefined. Legend for MAIAC CM: Blue, Light Blue, and White – clear (land, water and snow, respectively), Yellow – possibly cloudy, Red – cloudy.

Lake Erie is partly misclassified as clouds. It is not as bright as snow in the visible bands, and has a higher than snow reflectance in the shortwave infrared (2.1 μm). The same holds true for the block of land and some pixels in the transitional zone from snow to land, which are masked as clouds. As explained earlier, the error is expected in these cases. On day 36, MAIAC accurately detects a cloud stretching across Lake Erie. There are two large cloud systems on day 42, in the left upper and left bottom parts of the image, captured well by the algorithm. These images also show the strong retreat of the snow line by day 42, and a high

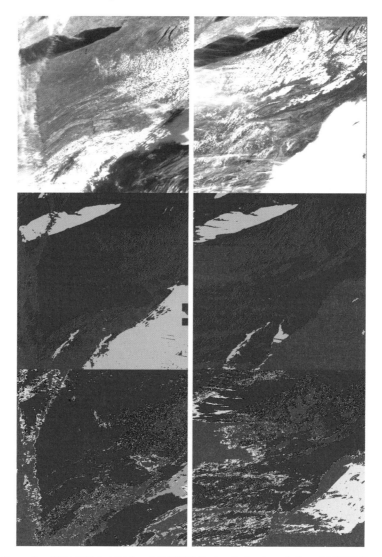

Fig. 3.5. Example of MAIAC (middle) and MOD35 (bottom) cloud mask from MODIS Terra data for days 138 (left) and 152 (right) of 2005. The image shows the same Tile (northeastern USA) as in Fig. 3.4.

quality of snow mapping by MAIAC. The MOD35 product accurately detects clouds, but it also overestimates cloudiness over snow on all three days, with the highest error on day 37.

Fig. 3.5 compares the cloud mask of the two algorithms for the late spring of 2005 for the same region. Over land, the accuracy is similar. Some difference exists with regards to thin cirrus, or otherwise semitransparent clouds. MAIAC CM does not explicitly try to mask these clouds. Created for the purpose of aerosol retrievals and atmospheric correction, the algorithm maximizes the volume of data available for the atmospheric correction. Our study shows that achievable accuracy of surface reflectance retrievals through thin

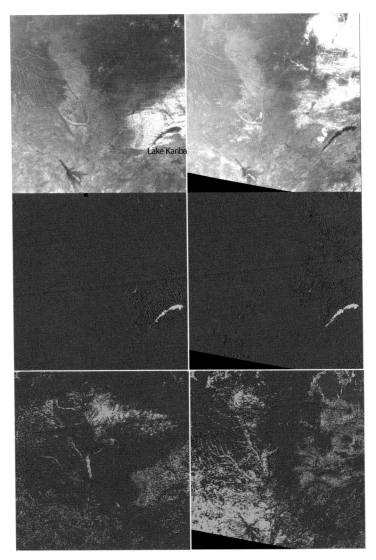

Fig. 3.6. Example of MAIAC (middle) and MOD35 (bottom) cloud mask at the beginning of dry season for Zambia, Africa, from MODIS Terra data for days 130 (left) and 141 (right) of 2005. The image shows four Tiles (1,200 × 1,200 km²).

cirrus is sufficiently high [Lyapustin and Wang, 2007] but more investigation is necessary. Another notable difference is cloud detection over the water. The current version of the algorithm had been developed for the land applications and cloud detection over water at this stage is rudimentary.

A large-scale comparison of cloud mask products is shown in Fig. 3.6 for a 1200 × 1200 km² region of the African Savanna (Zambia). This is a region of intense biomass burning in the dry season. The MAIAC and MOD35 cloud masks are generally comparable. MAIAC is a little more sensitive, detecting more clouds. One large difference

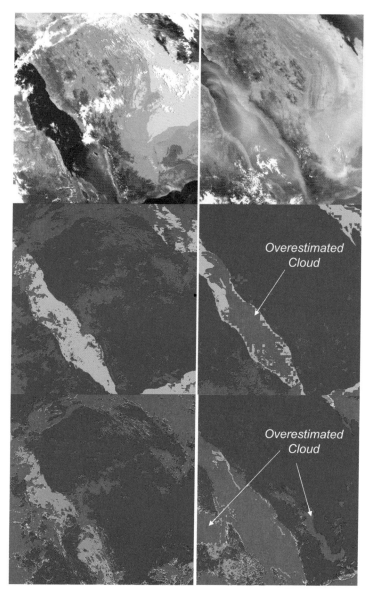

Fig. 3.7. Example of MAIAC (middle) and MOD35 (bottom) cloud mask for the Arabian Peninsula from MODIS Terra data for days 145 (left) and 207 (right) of 2005. The image shows nine Tiles (1800×1800 km^2).

is the large number of 'possibly clear' pixels in MOD35 when the algorithm cannot declare clear conditions with confidence. This category is not used in MAIAC, which has a covariance criterion and ancillary *refcm* data to identify clear conditions. This feature is particularly appealing to land applications, sometimes significantly increasing the volume of measurements, which may be confidently used in the atmospheric correction and in further applied analysis.

A final example of the cloud mask comparison for the large (1800 × 1800 km²) bright desert area of the Arabian Peninsula is shown in Fig. 3.7 for days 145 and 207 of 2005. Here, the MAIAC cloud mask is shown in the middle of the image and the MOD35 product is shown at the bottom. Except for a few small differences, the products agree quite well for day 145. On day 207, MOD35 overestimates cloudiness masking the dust-storm areas as clouds.

These examples show that the MAIAC CM algorithm demonstrates a high accuracy of cloud discrimination over land. It offers potential improvements to the operational MODIS cloud mask when land surface changes rapidly, and over bright snow and ice.

7. MAIAC examples and validation

The MAIAC performance has been tested extensively for different world regions using 50 × 50 km² subsets of MODIS TERRA data centered on AERONET sites. Because the algorithm is synergistic and the quality of aerosol retrievals and atmospheric correction are mutually dependent, the testing includes analysis of all main components of the algorithm. We are using both visual analysis, which remains unsurpassed in complex quality assessments of imagery products, and direct validation of aerosol retrievals by AERONET described in Section 7.1. One example of such testing is shown in Fig. 3.8 for the Goddard Space Flight Center (GSFC), USA site. The image shows 15 successive MODIS Terra observations for the end of June to early July (Fig. 3.8(a)), and for December (Fig. 3.8(b)) of 2000. The left two columns show MODIS top of atmosphere RGB reflectance. The images are normalized differently to help visual separations of clouds and aerosols from the surface signal. With the view geometry varying, the images collected at nadir have a better spatial resolution and contrast than those observed at the edge of scan (viewing zenith angle (VZA) ~ 55°) despite aggregation to 1 km. This source of noise notwithstanding, the images display a well-reproducible spatial pattern in cloud-free conditions, which is the basis of the cloud mask algorithm. Because of the re-projection, the top-left and bottom-right corners appear not covered by measurements.

Columns 3–7 and 9 show products of MAIAC processing: RGB NBRF (normalized BRF for the standard viewing geometry of VZA = 0°, SZA = 45°; cloud mask; RGB IBRF (instantaneous, or one-angle, BRF for the viewing geometry of latest observation); Blue band AOT (scale 0–1); spectral regression coefficients for the Blue and Red bands (scale 0.1–0.8); NBRF and MODIS L1B TOA reflectance for band B7 (2.1 μm) (scale 0–0.4).

The last column shows the ratio of volumetric concentrations of the coarse and fine aerosol fractions (C_v^C/C_v^F). In these retrievals, values of ratio $\eta = \{0.5; 1; 2; 4; 10\}$ were used, which are represented by colors magenta, blue, light blue, green and yellow, respectively. One more model used in the retrievals was a liquid water cloud model with the median droplet diameter of 5 μm. The cloud model was used in aerosol retrievals together with aerosol models to detect residual clouds. Cases when this model was selected are shown in red. The aerosol model with $\eta = 10$, which is a usually unrealistic combination of the fine and coarse fractions, was also used for cloud detection. Although this model may provide a spectrally neutral extinction similar to a cloud, it absorbs in the shortwave spectral region whereas the cloud does not. Overall, the aerosol retrievals provide a valu-

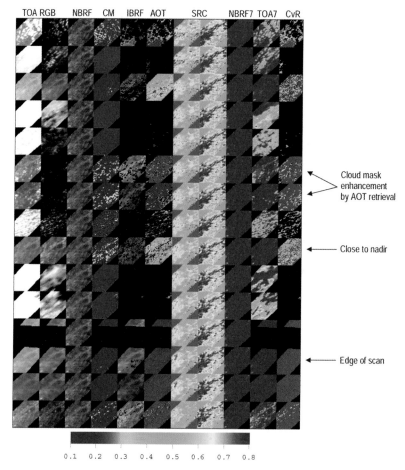

Fig. 3.8(a). Example of MAIAC processing for 50-km MODIS Terra subsets for the GSFC site. Shown are 15 consecutive observations for days 175–189 of 2000. The left two columns show differently normalized TOA RGB MODIS gridded reflectance. Next shown are the following MAIAC products: RGB NBRF; cloud mask; RGB IBRF; AOT at 0.47 μm (scale 0–1); spectral regression coefficients for the blue and red bands (scale 0.1–0.8); NBRF and MODIS L1B TOA reflectance for band B7 (2.1 μm) (scale 0–0.4); ratio of volumetric concentrations (C_v^C / C_v^F). The color bar is shown for SRCs.

able enhancement to the cloud mask (yellow and red colors in the last column). Fig. 3.8 shows detection of small popcorn clouds in summer (Fig. 3.8(a)) and semitransparent clouds in winter (Fig. 3.8(b)), as well as extensions on the cloud boundaries, which are difficult to detect by any specialized cloud mask algorithm.

The NBRF and IBRF images are shown as true color RGB composites. The RGB images are produced using the Red (B1), Green (B4) and Blue (B3) channels with equal weights. One can see that the quality of atmospheric correction is generally good although there are still some artifacts related to aerosol retrievals, such as color distortions in the IBRF image (Fig. 3.8(a), third row). The work is ongoing to resolve this and some other remaining issues of MAIAC algorithm.

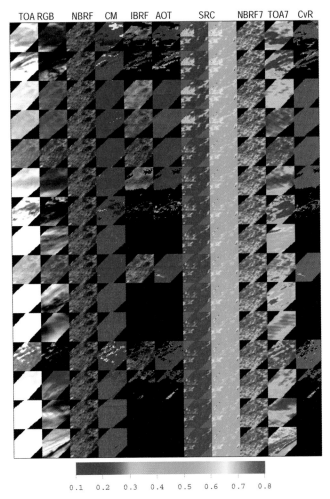

TOA RGB   NBRF   CM   IBRF   AOT   SRC   NBRF7 TOA7 CvR

0.1   0.2   0.3   0.4   0.5   0.6   0.7   0.8

Fig. 3.8(b). Example of MAIAC processing for the GSFC site, days 337–349 of 2000. The dark red color in CM image shows detected cloud shadows.

Columns 7–8 in Fig. 3.8 show the derived spectral regression coefficients in the Blue and Red bands. These retrievals are temporally consistent during the short time interval. A comparison of the summer and winter seasons shows an obvious seasonal trend of SRCs. These retrievals can be validated indirectly by comparing aerosol results with the AERO-NET measurements.

7.1 Validation

Fig. 3.9 and 3.10 show scatterplots of MAIAC AOT versus AERONET AOT in the Blue and Red bands. Following MODIS validation strategy [Remer et al., 2005], AERONET measurements are averaged over ±30 min interval of Terra satellite overpass. MAIAC retrievals are averaged over 20 km^2 area.

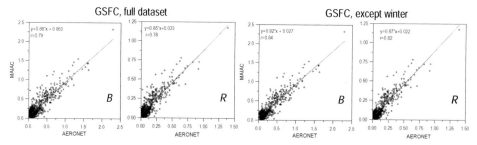

Fig. 3.9. Comparison of MAIAC AOT with AERONET data for GSFC (USA) using MODIS Terra data for 2000–2007. The left plots for both Blue (*B*, 0.47 μm) and Red (*R*, 0.64 μm) bands show the full dataset (740 points). The right plot shows the reduced dataset where winter days with snow on ground were filtered (587 points).

Fig. 3.9 gives a comparison for the GSFC site. The overall agreement is good with relatively high correlation coefficient ($r \sim 0.78$) and slopes of regression, which are close between the Blue and Red bands (0.88 and 0.85). The offset is positive in both bands (0.053 and 0.033, respectively). It can be explained by a limited sensitivity of the method at low AOT values, by residual cloud contamination, and by snow contamination in wintertime. The identification of pixels partially covered by snow is a very complex problem, and a small fraction of undetected snow may notably increase the retrieved AOT. In fact, snow is a significant source of bias in the MAIAC retrievals, as shown in the right plots, where winter days with snow on the ground were manually filtered. Snow filtering reduces the bias by a factor of 2 in the Blue band; it also increases correlation coefficient and slope of regression.

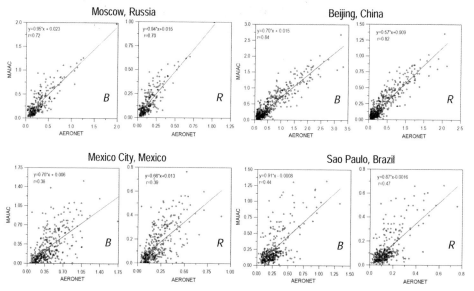

Fig. 3.10. Comparison of MAIAC AOT with AERONET data for Moscow (2001–2007, 238 points), Beijing (2001–2007, 504 points), Mexico City (2000–2007, 342 points) and Sao Paulo (2000–2007, 327 points), using MODIS Terra data.

Fig. 3.10 compares MAIAC AOT with AERONET data for several large cities of the world with medium-to-high levels of pollution, including Moscow, Beijing, Mexico City and Sao Paulo. In addition to pollution, there are several dust storms per year over Beijing with dust blown from the nearest Gobi desert located to the north of the city. MAIAC results compare well with the AERONET data for Moscow and Beijing with high correlation. There is considerably more scattering in the cases of Mexico City and Sao Paulo. One can see that the slopes of regression are high ($\sim 0.9 - 0.95$) for Moscow and Sao Paulo. In these cases, the MAIAC aerosol model with relatively low absorption works reasonably well. However, the slope of regression drops down to $\sim 0.6 - 0.7$ for Beijing and Mexico City, indicating that aerosol is significantly more absorbing for these cities.

There is some correlation of noise in the retrievals over Mexico City and Sao Paulo with the viewing geometry, specifically in the forward versus backward scattering directions. MAIAC tends to underestimate AOT in the forward scattering directions and overestimate it for the backscattering view geometry. This is the reason for much higher scattering in the scatterplots for these cities. One possible explanation of this behavior is an uncompensated surface BRF effect. Over bright surfaces, such as Mexico City and Sao Paulo, the shape of BRF seems to be more anisotropic in the visible bands than in the SWIR, which is also brighter. We plan to address this issue in further research.

7.2   Examples of MAIAC aerosol retrievals

At present, we have evaluated performance of MAIAC over the different world regions for an extended period of time. Typically, we order MODIS data for large areas of several thousand square kilometers for at least one year, and process the full set of data. Two examples of the large-scale AOT retrievals from MODIS Terra are shown in Figs 3.11 and 3.12. Fig. 3.11 shows smoke from biomass-burning during the dry season over an area of 1200×1200 km^2 in Zambia, Africa. The TOA image for the day 205 shows dozens of small-to-large fires. The fine 1-km resolution allows MAIAC to resolve and trace plumes of the individual fires. The fire plumes disappear at the coarse 10-km resolution of operational MODIS aerosol product MOD04 shown on the inset. The comparison shows that the magnitude of MOD04 and MAIAC retrieved AOT and its spatial distribution is rather similar, although there are certain differences depending on the surface type and geometry of observations. This particular example shows that, through significantly higher spatial resolution, MAIAC offers quantitatively new information about aerosols and their sources unavailable before. The gradient of AOT at 1-km resolution is high enough to implement an automatic delineation algorithm for the smoke plume detection, with the data that could be used in different applications, such as air quality.

Another example of MAIAC aerosol retrievals over a large portion of bright Arabian Peninsula (area 1800×1800 km^2 for day 207 of 2005 is shown in Fig. 3.12. The conditions are rather complex on this day. On one hand, the dust is transported across the Red Sea from Sudan (Africa). The wind does not penetrate the mountains along the peninsular's western shore. It is clear on the top of the mountain ridge, and the dust is concentrated along the shore, as can be seen both from the MODIS RGB image and from the AOT image. On the other hand, a separate internal dust storm has developed in the southern part of peninsula, with winds carrying dust in the northwest direction. For comparison,

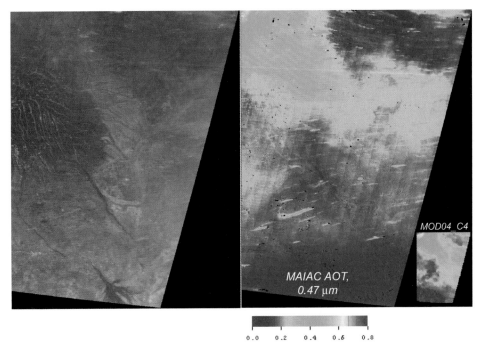

Fig. 3.11. Fires during dry biomass-burning season in Zambia, Africa, for day 205 of 2005 (area $1,200 \times 1,200$ km^2). The 1-km gridded MODIS Terra TOA RGB image is shown on the left and MAIAC-retrieved AOT at 0.47 μm is on the right. The AOT scale is the same for MOD04 and MAIAC. The high resolution (1-km) of AOT product allows detecting and tracing individual fire plumes. The inset shows result of the MODIS dark target algorithm MOD04_C4.

Fig. 3.13 shows the true color RGB image of the surface NBRF for this area. The bright surface feature, corresponding to the epicenter of the dust storms, is absent on the NBRF image, which confirms that this event is indeed a local dust storm.

8.   Concluding remarks

MAIAC is a new algorithm which uses time series processing and combines image- and pixel-level processing. It includes a cloud mask and generic aerosol–surface retrieval algorithm. The suite of MAIAC products includes column water vapor, cloud mask, dynamic mask of standing water and snow, AOT at 0.47 μm and the ratio of volumetric concentrations of the coarse and fine fractions, and spectral surface reflectance metrics, which include LSRT coefficients, albedo, NBRF and IBRF. The suite of products is generated in a systematic and mutually consistent way to satisfy the energy conservation principle. In other words, the radiative transfer calculation with the given set of parameters closely corresponds to measurements. All products are produced in gridded format at a resolution of 1 km.

A high spatial resolution of MAIAC (1 km vs 10 km for operational MODIS aerosol product) allows a new type of analysis and applications. One demonstrated example is a

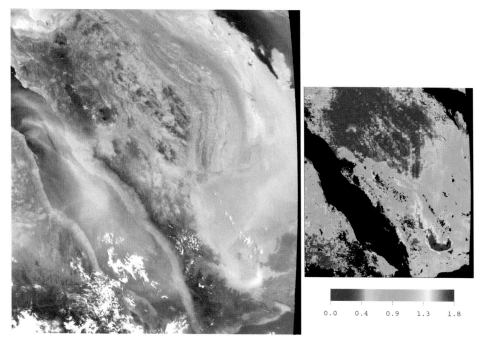

Fig. 3.12. MODIS Terra RGB TOA image and MAIAC AOT at 0.47 μm over Arabian Peninsula (area 1800 × 1800 km^2) for day 207 of 2005.

possibility of detection and tracing fire plumes from biomass burning. A high resolution of 1 km makes this application possible, whereas most of the information disappears at coarse, 10-km, resolution. We plan to apply MAIAC to study aerosol and their sources over large urban centers to complement the air quality analysis.

The current performance of the algorithm is not yet fully optimized. Nevertheless, MAIAC is already sufficiently fast for operational processing: it takes ≈ 50 seconds of one single-core AMD Opteron-64 processor to process one Tile (600 × 600 km^2) of MODIS data. The operational testing of MAIAC is planned to begin in 2009 in collaboration with the University of Wisconsin and the GSFC-based MODIS land processing team.

Acknowledgments The algorithm MAIAC was developed by Dr Lyapustin and Dr Wang with support by the NASA EOS Science (Dr D. Wickland) grant.

Fig. 3.13. RGB image of surface NBRF (BRF for a fixed geometry, VZA=0°, SZA = 45° for the Arabian Peninsula for day 184 of 2005. The image is built with equal weights for RGB bands.

References

Abdou, W., J.V. Martonchik, R.A. Kahn, R.A. West, and D.J. Diner, 1997: A modified linear-mixing method for calculating atmospheric path radiances of aerosol mixtures. *J. Geophys. Res.*, **102**, 16883–16888.

Anderson, T.L., R.J. Charlson, D.M. Winker, J.A. Ogren, and K. Holmen, 2003: Mesoscale variations of tropospheric aerosols, *J. Atm. Sci.*, **60**, 119–136.

Chandrasekhar, S. Radiative Transfer (Dover, New York, 1960).

Diner, D.J., J. Martonchik et al., 1999: MISR level 2 surface retrieval algorithm theoretical basis. JPL D-11401, Rev. D.

Diner, D.J., W. Abdou, T. Ackerman et al., 2001: MISR level 2 aerosol retrieval algorithm theoretical basis, JPL D-11400, Rev. E.

Diner, D.J., J.V. Martonchik, R.A. Kahn, B. Pinty, N. Gobron, D.L. Nelson, and B.N. Holben, 2005: Using angular and spectral shape similarity constraints to improve MISR aerosol and surface retrievals over land, *Rem. Sens. Environ.*, **94**, 155–171.

Dubovik, O., B. Holben, T.F. Eck, A. Smirnov, Y.J. Kaufman, M.D. King, D. Tanre, and I. Slutzke, 2002: Variability of absorption and optical properties of key aerosol types observed in worldwide locations. *J. Atmos. Sci.*, **59**, 590–608.

Flowerdew, R.J., and J.D. Haigh, 1995: An approximation to improve accuracy in the derivation of surface reflectances from multi-look satellite radiometers. *Geophys. Res. Lett.*, **22**, 1693–1696.

Gao, B.C., and Y.J. Kaufman, 2003: Water vapor retrievals using Moderate Resolution Imaging Spectroradiometer (MODIS) near-infrared channels. *J. Geophys. Res.*, **108**(D13), 4389–4399, doi:10.1029/2002JD003023.

Holben, B.N., T.F. Eck, I. Slutsker, D. Tanré, J.P. Buis, A. Setzer, E. Vermote, J.A. Reagan, Y.J. Kaufman, T. Nakajima, F. Lavenu, I. Jankowiak, and A. Smirnov, AERONET – a federated instrument network and data archive for aerosol aharacterization. *Rem. Sens. Environ.*, **66**, 1–16, 1998.

Holben, B., et al., 2001: An emerging ground-based aerosol climatology: aerosol optical depth from AERONET. *J. Geophys. Res.*, **106**, 9807–9826.

Hsu, N.C., S.C. Tsay, M.D. King et al., 2004: Aerosol properties over bright-reflecting source regions. *IEEE Trans. Geosci. Rem. Sens.*, **42**, 557–569.

Intergovernmental Panel on Climate Change: Climate Change 2007: Synthesis Report. Available at: http://www.ipcc.ch/ipccreports/ar4-syr.html.

Kaufman, Y.J., D. Tanre et al., 1997: Operational remote sensing of tropospheric aerosol over land from EOS moderate resolution imaging spectroradiometer. *J. Geophys. Res.*, **102**, 17051–17067.

Levy, R.C., L. Remer, S. Mattoo, E. Vermote, and Y.J. Kaufman, 2007: Second-generation algorithm for retrieving aerosol properties over land from MODIS spectral reflectance. *J. Geophys. Res.*, **112**, D13211, doi: 10.1029/2006JD007811.

Lucht, W., C.B. Schaaf, and A.H. Strahler, 2000: An algorithm for the retrieval of albedo from space using semiempirical BRDF models. *IEEE Trans. Geosci. Remote Sens.*, **38**, 977–998.

Lyapustin, A.I., 1999: Atmospheric and geometrical effects on land surface albedo, *J. Geophys. Res.*, **104**, 4123–4143.

Lyapustin, A., and Yu. Knyazikhin, 2001: Green's function method in the radiative transfer problem. I: Homogeneous non-Lambertian surface, *Appl. Optics*, **40**, 3495–3501.

Lyapustin, A., and Wang, Y., 2005: Parameterized code Sharm-3D for radiative transfer over inhomogeneous surfaces, *Appl. Optics*, **44**, 7602–7610.

Lyapustin, A., and Y. Wang, 2007: MAIAC – Multi-Angle Implementation of Atmospheric Correction for MODIS. Algorithm Theoretical Basis Document. Available at: http://neptune.gsfc.nasa.gov/bsb/subpages/index.php?section=Projects&content=SHARM, section MAIAC ATBD.

Lyapustin, A., Y. Wang, and R. Frey, 2008: An automatic cloud mask algorithm based on time series of MODIS Measurements, *J. Geophys. Res.*, **113**, D16207, doi:10.1029/2007JD009641.

Martonchik J.V., D.J. Diner, R.A. Kahn, et al., 1998: Techniques for the retrieval of aerosol properties over land and ocean using multiangle imaging. *IEEE Trans. Geosci. Remote Sens.*, **36**, 1212–1227.

NASA, 1999: *EOS reference handbook: a guide to Earth Science Enterprise and the Earth Observation System* (p. 355), M. King and R. Greenstone (Eds), EOS Project Science Office, NASA/Goddard Space Flight Center, Greenbelt, MD.

Remer, L., Y. Kaufman, D. Tanre et al., 2005: The MODIS aerosol algorithm, products, and validation, *J. Atm. Sci.*, **62**, 947–973.

Rossow, W.B., and L.C. Garder, 1993: Cloud detection using satellite measurements of infrared and visible radiances for ISCCP. *J. Climate*, **6**, 2341–2369.

Schaaf C.B., F. Gao, A.H. Strahler, et al., 2002: First operational BRDF, albedo nadir reflectance products from MODIS, *Rem. Sens. Environ.*, **83**, 135–148.

Tanre, D., P.Y. Deschamps, C. Devaux, and M. Herman, 1988: Estimation of Saharan aerosol optical thickness from blurring effect in Thematic Mapper data. *J. Geophys. Res.*, **93**, 15955–15964.

Veefkind, J.P., G. de Leeuw, and P. Durkee, 1998: Retrieval of aerosol optical depth over land using two-angle view satellite radiometry during TARFOX. *Geophys. Res. Lett.*, **25**, 3135–3138.

Vermote, E.F, N.Z. El Saleous, and C.O. Justice, 2002: Atmospheric correction of MODIS data in the visible to middle infrared: first results. *Rem. Sens. Environ.*, **83**, 97–111.

Wolfe, R.E., Roy, D.P., and E. Vermote, 1998: MODIS land data storage, gridding, and compositing methodology: Level 2 grid. *IEEE Trans. Geosci. Remote Sens.*, **36**, 1324–1338.

4 Iterative procedure for retrieval of spectral aerosol optical thickness and surface reflectance from satellite data using fast radiative transfer code and its application to MERIS measurements

Iosif L. Katsev, Alexander S. Prikhach, Eleonora P. Zege, Arkadii P. Ivanov, Alexander A. Kokhanovsky

1. Introduction

The retrieval of aerosol characteristics over land from satellite data has been a challenge up to now. Currently, several well known techniques for retrieving aerosol optical thickness (AOT) have been developed for satellite instruments including MODIS (Kaufman et al., 1997; Remer et al., 2005), MERIS (Santer et al., 1999, 2000), POLDER (Deuze et al, 2001), and MISR (Martonchik et al., 2002). While each technique has its own merits, the accuracy of AOT retrieval still needs further clarification and improvement. From this point of view the inter-comparison and verification of these techniques undertaken recently (Kokhanovsky et al., 2007) is of a great importance.

   To provide the required accuracy of the retrieval procedures, realistic atmospheric models and accurate radiative transfer computations should be used. These procedures employ look-up-table (LUT) techniques to ensure fast processing of satellite data. The LUT techniques employ the preliminary computation of the radiative characteristics for the chosen atmospheric parameters set by a verified and reliable radiative code as well as fitting procedures in the retrieval process. Hence, LUTs save time for radiative transfer calculation in the processing of satellite data. However, there is a price to be paid for this advantage as LUT techniques have some inherent restrictions. First, LUTs are based on a huge bulk of previously computed data, and this bulk has to be recomputed every time the atmospheric model is changed. The restriction of the LUT volume, the necessity of having particular LUTs for a given instrument, and difficulties in the application of some standard techniques for data processing (e.g., the least-mean-squares method to determine the AOT and Ångström exponent) may be considered as additional limitations of the LUT technique.

   In this paper we describe the newly developed aerosol retrieval technique (ART) for spectral AOT retrieval that uses radiative transfer computations in the process of retrieval rather than a LUT. This approach can be applied operationally only if the accurate and extremely fast radiative transfer code is used. Previously, we have developed the RAY code (Tynes et al., 2001) for simulation of the radiative transfer in the atmosphere–underlying surface system with regard to polarization that meets these requirements. This code is based on the results derived by Zege et al. (1993) and Zege and Chaikovskaya (1996). It has been applied with success for the solution of many problems. RAY is a core of the ART code for the retrieval of AOT, Ångström exponent, and surface spectral albedo. RAY's high processing speed allows the use of iterative radiation transfer computations in the processing of satellite data for AOT retrieval, eliminating the need for LUT techniques. This fast AOT retrieval technique has some additional merits, namely any set of wavelengths from any satellite optical instruments can be processed. In the ART algorithm the computations of the derivatives of the reflectance at top of atmosphere (TOA) with respect to the sought-for values are performed. It provides the application of the least-mean-squares technique to process the AOT and Ångström exponent retrieval using spectral satellite data. The ART approach allows one to change atmospheric models in the retrieval process. It uses minimal simplifications in the computation of the radiative transfer through atmosphere, accounts for the effect of polarization, and includes the radiative interaction between atmospheric layers.

We have employed ideas and features from earlier algorithms (Kaufman et al., 1997; Santer et al., 2000; von Hoyningen-Huene et al., 2003) in the ART algorithm. Particularly, the pixel sorting technique is very close to that of the MODIS algorithm (Kaufman et al., 1997). Additionally, the spectral model of the underlying surface is taken from Bremen AErosol Retrieval (BAER) code (von Hoyningen-Huene et al., 2003).

This chapter presents the first step in developing the AOT retrieval algorithm using a fast RT code. It is still not a mature algorithm assigned for routine satellite data processing and retrieving the global AOT distribution. We demonstrate here the ART version for the middle latitudes. Below we give a brief description of the RAY code (Section 2.1 and Appendix), the discussion of the radiative characteristics (Section 2.2), the structure of the atmospheric model used (Section 2.3), brief information on the spectral models of underlying surfaces (land, water) (Section 2.4) and the description of the ART code with preliminary data sorting and iteration procedures to retrieve the AOT, the Ångström exponent and the spectral signature of an underlying surface (Section 2.5). The results of retrievals, comparisons with AERONET data, and with results of other codes are presented in Section 3. A short conclusion summarizes the results of the ART deployment.

2.   The aerosol retrieval technique

2.1   The vector radiative transfer code

The RAY code (Tynes et al., 2001) is assigned to calculate radiance and polarization of radiation at wavelengths in the UV, visible and IR parts of the electromagnetic spectrum for atmosphere–ocean and atmosphere–land systems including realistic aerosol and trace gas models and bi-directional reflectivity of underlying surfaces. Accurate and extremely fast computations with RAY along with a possibility of detailed and realistic modeling of light scattering media are due to the original numerical algorithm used, which couples the newly developed two-component approach of the vector radiative transfer theory with the traditional adding–doubling technique. The two-component technique to solve the vector radiative transfer equation (VRTE) is a cornerstone and defines the architecture of the whole algorithm. It combines two approaches recently developed by the authors, namely, the multi-component approach (Zege et al., 1993) and the new approximate theory of polarized light transfer (Zege and Chaikovskaya, 1996). Because thus far this code has only

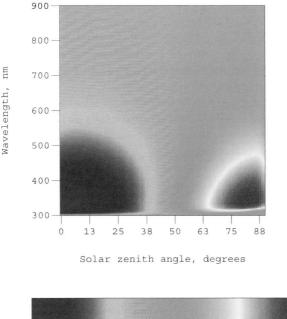

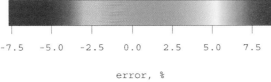

error, %

Fig. 4.1. The error of the scalar approximation (ε, in percent) for calculations of the TOA radiances at different wavelengths and solar zenith angles ϑ_0 (LOWTRAN maritime atmosphere, the nadir observation. AOT is equal to 0.086 at 550 nm). For the considered geometry, the scattering angle is equal to $\pi - \vartheta_0$. As one might expect, the largest deviations occur at scattering angles of $90°$ and $180°$, where the effects of the polarization are at maximum for single ($90°$) and double ($180°$) scattering regimes in the case of purely molecular scattering atmosphere.

been briefly outlined (Tynes et al., 2001; Chaikovskaya et al., 1999), the theoretical approach, architecture, specific features and accuracy estimation are given in the Appendix.

Validation of the RAY code showed (Tynes et al., 2001) that it provides highly accurate data in a fraction of the time required by the Monte Carlo and other methods. In addition, to run the RAY code a powerful computer is not required; the RAY code can be run quickly on an ordinary personal computer. These properties make this approach a practical technique for nearly real-time simulations and processing of spectral satellite data.

The importance of accounting for the polarization effects in the radiative transfer processes is demonstrated in Fig. 4.1, where the error $\varepsilon = (I_{\uparrow sca} - I_{\uparrow vect})/I_{\uparrow vect}$ of the scalar approximation for the LOWTRAN maritime aerosol model with an aerosol optical thickness of 0.086 at 550 nm is shown. Here $I_{\uparrow sca}$ and $I_{\uparrow vect}$ are the TOA radiances computed with the scalar RTE (without including polarization) and with VRTE (including polarization), respectively. As might be expected, the polarization effect on radiances depends strongly on the scattering angle and the wavelength. This figure demonstrates these features. Computed intensities of the reflected light at wavelengths above 550 nm are less affected by the effect of light polarization than intensities at shorter wavelengths. This is related to the decrease of the Rayleigh scattering contribution as compared to scattering by aerosol particles for larger wavelengths. Clearly, polarization effects are more important for scattering angles about $90°$, where the scalar radiative transfer theory overestimates the top-of-atmosphere reflectance by almost 10 %. Underestimation of reflectance by about 8 % is possible at a scattering angle of $180°$ as demonstrated in Fig. 4.1 (the solar zenith angle is zero degrees, the nadir observation). The radiances for solar angles of about $45°$ at the nadir observations are least affected by polarization as shown in Fig. 4.1.

2.2 Radiative characteristics

Various radiative characteristics of a light scattering layer are used in the AOT retrieval procedures described below. For the reader's convenience we will introduce them now.

First, let us introduce the concept of TOA reflectance that is the reflection function at the top of the atmosphere defined as

$$R_{TOA}(\lambda,\mu,\mu_0,\phi) = \frac{\pi I_\uparrow(\lambda,\mu,\mu_0,\phi)}{\mu_0 E_0(\lambda)}, \tag{1}$$

where $E_0(\lambda)$ is the extraterrestrial irradiance incident normally on a given unit area at the top-of-atmosphere, $I_\uparrow(\lambda,\mu,\mu_0,\phi)$ is the measured TOA radiance, μ_0 and μ are cosines of the incidence and observation zenith angles, and ϕ is a difference between the azimuth angles of the incidence and observation directions. Hereafter we will use also other characteristics of light reflection and transmission phenomena. Table 4.1 summarizes the definitions and notations of these values.

In Table 4.1, $I_\downarrow(\lambda,\mu,\mu_0,\phi)$ is the radiance transmitted by a layer. For brevity the angular arguments in the reflection and transmission functions will be omitted, i.e. $R_{TOA}(\lambda) \equiv R_{TOA}(\lambda,\mu,\mu_0,\phi)$ and $T(\lambda) \equiv T(\lambda,\mu,\mu_0,\phi)$. But we will keep the argument μ_0 in the notations of the reflection $r(\lambda,\mu_0)$ and transmission $t(\lambda,\mu_0)$ coefficients for di-

Table 4.1. Characteristics of light reflection and transmission by a plane layer

Radiative characteristic	Notation and definition
Reflection function	$R(\lambda,\mu,\mu_0,\phi) = \dfrac{\pi I_\uparrow(\lambda,\mu,\mu_0,\phi)}{\mu_0 E_0(\lambda)}$
Reflection coefficient for directional illumination (plane albedo)	$r(\lambda,\mu_0) = \dfrac{1}{\pi} \int\limits_0^{2\pi} d\phi \int\limits_0^1 R(\lambda,\mu,\mu_0,\phi)\mu d\mu$
Reflection coefficient for diffuse illumination (spherical albedo)	$r_s(\lambda) = 2\int\limits_0^1 r(\lambda,\mu_0)\mu_0 d\mu_0$
Transmission function	$T(\lambda,\mu,\mu_0,\phi) = \dfrac{\pi I_\downarrow(\lambda,\mu,\mu_0,\phi)}{\mu_0 E_0(\lambda)}$
Transmission coefficient for directional illumination	$t(\lambda,\mu_0) = \dfrac{1}{\pi} \int\limits_0^{2\pi} d\phi \int\limits_0^1 T(\lambda,\mu,\mu_0,\phi)\mu d\mu$
Transmission coefficient for diffuse illumination	$t_d(\lambda) = 2\int\limits_0^1 t(\lambda,\mu_0)\mu_0 d\mu_0$

rectional illumination particularly to distinguish them from the reflection $r_s(\lambda)$ and transmission $t_d(\lambda)$ coefficients under diffuse illumination.

2.3   The atmospheric model

The atmospheric model input includes temperature and pressure profiles, profiles of trace gas concentrations and stratification of aerosol microstructure and concentration. The optical properties of the atmosphere (Mueller matrix elements, extinction and absorption coefficients) are computed with this input as the superposition of the molecular scattering, gas absorption, and aerosol absorption and scattering. Note, that tropospheric aerosol properties over land may have very large spatial and temporal variations. Upper layers are much more stable in space and time in comparison with labile over-surface layers, and hence the parameters of the upper layers could be taken the same for all pixels of the processing image area. The radiative characteristics of the upper atmosphere can be computed one time for all pixels of the processing area. The specific feature of the ART approach is the use of the direct radiative transfer computation in the satellite data processing. With the goal to speed up these computations the stratified atmosphere is divided into two parts:

(1) The layer '1' (lower layer) of the atmosphere is a layer of the lower troposphere up to the height H. The value of H is the same for all pixels of the processing image area and can be defined for each processing area. This layer includes aerosol scattering and absorption, molecular scattering, and gas absorption (see Fig. 4.2). The aerosol in this layer may have very large spatial and temporal variations. The AOT of this layer is supposed to vary over the processing image area and this value is retrieved for each pixel independently. Note that the surface altitude, and hence the molecular scattering and gas absorption, can differ between pixels (see below how the pixel altitude is regarded). This feature may be particularly important for mountainous regions. To speed up the ART calculations, the radiative characteristics of the layer '1' for every pixel are computed as for a homogeneous layer with the characteristics of scattering and absorption averaged over this layer. The computations performed showed that ignoring the stratification inside the layer '1' leads to a relative error in the reflectance at the top of atmosphere of less than 0.2 % at any wavelength in the visible.

(2) Layer '2' (upper layer, i.e. above the altitude H) includes the stratosphere and upper and middle troposphere. Naturally, layer '2' is characterized by a vertical stratification of aerosol and gaseous concentrations, pressure and temperature profiles. For computation this layer is presented as N homogeneous sub-layers with optical characteristics averaged over this sub-layer. As was explained above, the radiation characteristics of the stratified layer '2' can be computed once for all pixels of the processing area. This feature reduces the computation volume considerably and is a main reason for the described partitioning of the atmosphere in two specified layers.

Let us emphasize that the division of atmosphere into two layers in ART does not imply any additional assumptions. The radiative interaction between these layers is included in the computation process. The optical model of the atmosphere is flexible and can be changed easily if it is needed. The developed databases of aerosol optical properties allow for adjustment of an aerosol model using any a priori available information (for example, on the presence of turbid sub-layers in the layer '2' due to a volcano eruption or Saharan dust

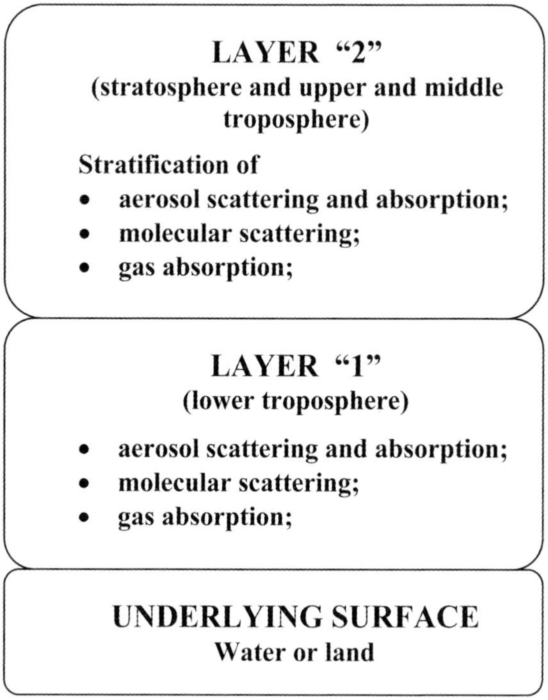

LAYER "2"
(stratosphere and upper and middle troposphere)

Stratification of
- aerosol scattering and absorption;
- molecular scattering;
- gas absorption;

LAYER "1"
(lower troposphere)

- aerosol scattering and absorption;
- molecular scattering;
- gas absorption;

UNDERLYING SURFACE
Water or land

Fig. 4.2. The atmospheric model.

transport). If such information is absent and the chosen model of aerosol in the layer '2' is not representative, the effect of this error in the aerosol model of layer '2' on the retrieved value of AOT will be partially corrected through the retrieved value of the AOT of the layer '1'.

The profiles of optical characteristics, i.e. the extinction and scattering coefficients and Mueller (single scattering) matrices, are calculated as superposition of the contribution of all absorbing and scattering components (gases and aerosol). Radiative characteristics are calculated accounting for polarization using the VRTE. The importance of the VRTE used for the calculation of radiative characteristics was discussed in Section 2.1.

In the particular version of ART used for retrievals presented in this paper the radiative characteristics of layer '1' are calculated without accounting for polarization. This speeds up calculations. Because the Rayleigh optical thickness of layer '1' is small, and as a rule the aerosol scattering in this layer prevails, the disregard of polarization in the calculation for the lower layer results in small errors in the calculation of the TOA reflectance. The relative error of computation of the TOA reflectance if one neglects the polarization effects for the layer '1' is depicted in Fig. 4.3. It is seen that the maximal error occurs at small values of the AOT in the lower layer '1' and for scattering angles of about $90°$. At AOT $= 0.1$ the relative error due to disregarding polarization achieves 1.8%. But in the most typical situations (when the scattering angle is more than $110°$) the errors are smaller and do not exceed 1% (see Fig. 4.3). Note that if necessary, the ART allows for the full account of the polarization effects in all atmospheric layers.

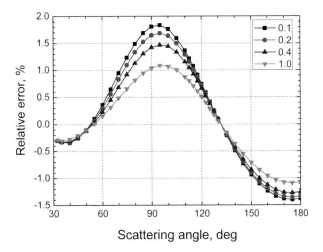

Fig. 4.3. The relative error of computation of the TOA reflectance due to neglecting the polarization effect in layer '1', $\lambda = 412.5$ nm. Solar zenith angle is equal to $60°$, azimuth angle is 0 or $180°$. Numbers in the legend denote the values of AOT of layer '1'.

The optical characteristics (optical thickness τ, single scattering albedo ω, and phase function $p(\theta)$) of this uniform layer '1' are calculated including contributions of molecular scattering, gas absorption and absorption and scattering by aerosol, namely

$$\tau = \tau_R + \tau_{aer} + \tau_g, \tag{2}$$

$$\omega = \frac{\omega_R \tau_R + \omega_{aer} \tau_{aer}}{\tau_R + \tau_{aer} + \tau_g}, \tag{3}$$

where τ_R, τ_{aer}, and τ_g are the optical thicknesses corresponding to molecular scattering, light extinction by aerosol particles, and gaseous absorption contributions, respectively. Eq. (3) also includes the single scattering albedo of aerosol particles (ω_{aer}) and that of Rayleigh scattering processes ($\omega_R = 1$).

As for the phase function of layer '1', it is enough to calculate the coefficients x_n of the expansion of the phase function $p(\theta)$ into a series of Legendre polynomials:

$$x_n = \frac{\tau_R x_{n,R} + \omega_{aer} \tau_{aer} x_{n,aer}}{\tau_R + \omega_{aer} \tau_{aer}}. \tag{4}$$

Here $x_{n,R}$ and $x_{n,aer}$ are the coefficients of the expansion of the Rayleigh and aerosol phase functions, respectively, into a series of Legendre polynomials. These coefficients x_n are used in the adding method procedure to calculate the radiative interaction between layers '1' and '2'. Because the Rayleigh scattering phase function comprises only the zeroth and second Legendre polynomials, we have:

$$x_n = \frac{0.5\delta_{n2}\tau_R + \omega_{aer}\tau_{aer} x_{n,aer}}{\tau_R + \omega_{aer}\tau_{aer}} \tag{5}$$

at $n \geq 1$ and also $x_0 = 1$ by definition. Here δ_{n2} is the Kronecker delta symbol equal to one at $n = 2$ and zero, otherwise.

Adjusting characteristics of layer '1' allows one to account for the variations of the optical thickness of the Rayleigh atmosphere as a function of altitude over sea level and to the deviation of the atmospheric pressure from the normal pressure at sea level. These variations are included as a correction of the Rayleigh optical thickness of a layer '1', i.e. as the additional Rayleigh optical thickness

$$\Delta\tau_R = \tau_R - \tau_R^0, \tag{6}$$

where τ_R^0 is the Rayleigh optical thickness at normal conditions with 'land' pixels at sea level ($h = 0$), and τ_R is the Rayleigh optical thickness under real conditions. The value of this correction of the Rayleigh optical thickness of a layer '1' can be presented as a sum of two terms:

$$\Delta\tau_R = \Delta\tau_R^P + \Delta\tau_R^h. \tag{7}$$

The first term in Eq. (7) that accounts for the deviation $\Delta P_0 = P - P_0$ of the atmospheric pressure P at sea level from the normal value P_0 is given as

$$\Delta\tau_R^P = \frac{\Delta P_0}{P_0} \tau_R^0. \tag{8}$$

The second term in Eq. (7) is equal to

$$\Delta\tau_R^h = \frac{\Delta P_h}{P_0} \tau_R^0 = \tau_R^0 \left(e^{-h/h_0} - 1 \right), \tag{9}$$

where ΔP_h is the deviation of the atmospheric pressure from its normal value P_0 at the sea level, when the area corresponding to the considered image pixel is located at the altitude h. We assume that $h_0 = 8000$ m in this work (see also Santer et al., 2000).

Note that changes of water vapor and ozone optical thicknesses in the layer '1' over elevated targets are neglected for the following reasons. The total optical thicknesses of water vapor and ozone in the atmosphere layer 0–2 km in the spectral range of interest is small. For instance, the ozone optical thickness in this layer is no more than 0.0006 even inside the absorption band 510–665 nm.

2.4 Spectral models of underlying surfaces

As was already mentioned in Section 1, the spectral models of the underlying surfaces are taken from the code described by von Hoyningen-Huene et al. (2003). Similar to von Hoyningen-Huene et al. (2003), it is taken that a linear combination of the basic vegetation spectra $r_{veg}(\lambda)$ and soil $r_{soil}(\lambda)$ describes the land spectral albedo:

$$r_s(\lambda) = c r_{veg}(\lambda) + (1 - c) r_{soil}(\lambda). \tag{10}$$

Similarly, a linear combination of the basic spectra of clear ocean water $r_{clear}(\lambda)$ and coastal water $r_{coastal}(\lambda)$ represents the water spectral albedo:

$$r_s(\lambda) = cr_{clear}(\lambda) + (1 - c)r_{coastal}(\lambda). \tag{11}$$

All basic spectra are given by von Hoyningen-Huene et al. (2003). Thus, the spectral albedo of the surface (land or water) is characterized by the only parameter c. Just this surface parameter is determined in the retrieval process. The normalized differential indices serve as a zeroth approximation for the parameter c. This normalized differential vegetation index for the 'land' pixels is defined as

$$\mathrm{NDVI} = \frac{R(865 \text{ nm}) - R(665 \text{ nm})}{R(865 \text{ nm}) + R(665 \text{ nm})}. \tag{12}$$

For 'water' pixels the normalized differential pigment index (NDPI) is used. It is defined as:

$$\mathrm{NDPI} = \frac{R(443 \text{ nm}) - R(560 \text{ nm})}{R(490 \text{ nm})}. \tag{13}$$

2.5 Brief description of the aerosol retrieval technique

The ART algorithm is designed to retrieve the AOT, Ångström exponent and underlying spectral surface albedo. The spectral radiance at the top of the atmosphere measured by any satellite optical instrument in N spectral channels is the input to the ART algorithm. As has been already underlined, because ART does not use the LUT technique, the choice of the spectral channels is very flexible. Their number and particular wavelengths depend on the spectral characteristics of the satellite optical instrument used. In the examples below satellite data in nine MERIS spectral channels specified by the wavelengths

$$\lambda = 412.5, \ 442.5, \ 490, \ 510, \ 560, \ 620, \ 665, \ 865 \text{ and } 885 \text{ nm} \tag{14}$$

are used. These channels are hardly affected by gaseous absorption, except for the ozone absorption at wavelengths of 510–665 nm. For example, according to the LOWTRAN Mid-Latitude Summer molecular-gas model optical thicknesses of the water vapor at 620 nm, 665 nm, and 885 nm are equal to 10^{-29}, $2.8 \cdot 10^{-12}$ and 0.075, respectively; ozone optical thicknesses at 560 nm and 620 nm are equal to 0.033 and 0.035, respectively.

Fig. 4.4 presents the flowchart of ART. The atmospheric model was described in Section 2.3; the underlying surface model was discussed in Section 2.4. The preparation of the input data includes operations described in steps 1–3 below. Steps 4–8 describe the sequence of the retrieval operations.

1. *Discarding pixels containing sub-pixel clouds.* The pixels with $R_{TOA}(560) \geq 0.4$ or $\xi = R_{TOA}(412)/R_{TOA}(443) \leq 1.16$ are considered as containing a cloud. The first inequality immediately discards pixels containing comparatively thick clouds. As was shown by Kokhanovsky (2004), when clouds completely cover a pixel, as a rule $R_{TOA}(560) \geq 0.2$. But the choice of the threshold criterion as $R_{TOA}(560) \geq 0.2$ (instead of $R_{TOA}(560) \geq 0.4$) might lead to discarding not only cloud pixels, but pixels with highly reflective surfaces (for instance, sands, deserts). However, lowering of the cloud threshold criterion to the level $R_{TOA}(560) \geq 0.2$ does not guarantee discarding pixels containing sub-pixel clouds.

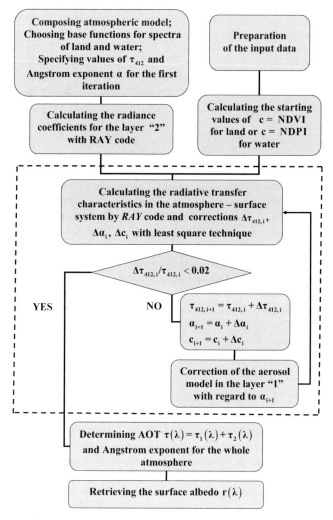

Fig. 4.4. Flowchart of the ART algorithm.

The second criterion $\xi \leq 1.16$ introduced above serves to discard pixels containing sub-pixel clouds. Figs 4.5 and 4.6 help to explain this criterion. Fig. 4.5 shows the spectral dependence of the reflectance measured by the MERIS instrument at TOA for 25 pixels in an area near the city of Hamburg. As seen in the spectral region $\lambda = 400 \div 500$ nm, where the underlying surface contributes very little, the reflectance for all pixels is similar. Only few pixels with different reflectance make exclusion. The natural supposition is that these pixels contain sub-pixel clouds or thin clouds. In Fig. 4.6 one can see the ratio ξ for all 25 pixels of this box. This ratio differs considerably from the mean value $\xi \approx 1.2$ for five pixels that apparently contain sub-pixel clouds.

Variations of the ratio ξ for pixels, which are free from sub-pixel clouds, arise from changes of the surface albedo. These variations are small because the surface albedo poorly contributes to the TOA reflectance as compared to the contribution of the Ray-

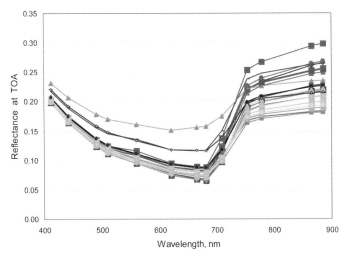

Fig. 4.5. Spectral dependence of the TOA reflectance measured by the MERIS instrument for 25 pixels in a box.

leigh scattering in the wavelength range 412–443 nm. Just this feature along with strong spectral dependence of the Rayleigh scattering makes the recommended choice of the wavelengths $\lambda = 412.5$ nm and $\lambda = 443$ nm to be expedient. For the same reasons the criterion $\xi \leq 1.16$ should be applied in the retrieval algorithm before subtraction of the Rayleigh scattering. As one can see from Fig. 4.4, the criterion $\xi \leq 1.16$ looks promising for discarding pixels containing sub-pixel clouds. The estimations show that the use of the criterion $\xi \leq 1.16$ leads to discarding pixels where clouds with the reflectance more than 0.2 cover 20 % and more of the pixel area.

2. *Separating 'land' ($R_{TOA}(885$ nm$) \geq 0.1$) and 'water' ($R_{TOA}(0.885$ nm$) < 0.1$) pixels (von Hoyningen-Huene et al., 2003) and discarding 'bad' pixels in the group of 'land' pixels using the criterion NDVI (Eq. (12)).* Following Remer et al. (2005), pixels classified as 'land' at step 2 with NDVI < 0.1 are considered as pixels containing sub-pixels 'water' and are discarded.

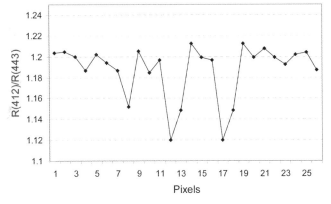

Fig. 4.6. The ratio R_{TOA} (412 nm)$/R_{TOA}$ (443 nm) for 25 pixels of a box.

3. *Compiling the box of neighbors 'good' pixels of the same type ('land' or 'water').* The box size depends on the spatial resolution of the deployed satellite instrument, number of pixels in the processing image, the scenario, and the problem under consideration. Particularly separate pixels can be processed independently, for instance while detecting sub-pixel fires. In the examples below a box of 5×5 pixels is used. For this box the average value of $\bar{R}_{TOA}(\lambda)$ is calculated. In doing so, 20 % of the pixels with minimal values of $R_{TOA}(665\ nm)$ and 30 % of the pixels with maximal values of $R_{TOA}(665\ nm)$ are discarded from the data to be averaged. The determined average value of $\bar{R}_{TOA}(\lambda)$ is ascribed to the central pixel in the box.

   This averaging procedure implies that the optical characteristics of the aerosol layers are practically the same for all pixels in the compiled pixel box, i.e. the scale of the spatial changes of the aerosol parameters is much larger than the pixel size. It decreases the effects of random errors of measurements and of small-scale variations of the surface albedo. Displacement of the pixel box by one column or one row in the array of the image frame provides retrieval of the moving-average value of the AOT in the retrieval procedure. Discarding pixels with maximal and minimal values of $R_{TOA}(665\ nm)$ in the compiled pixel box from the averaging procedure allows one to eliminate or at least to decrease the effect of pixels which either partially contain clouds (maximal values of $R_{TOA}(665\ nm)$) or include cloud shadows (minimal values of $R_{TOA}(665\ nm)$).

4. *For layer '2', the reflectances at the illumination from the top $R_2(\mu,\mu_0,\phi)$ and from the bottom $R_2{}^*(\mu,\mu_0,\phi)$ as well as the transmittance at the illumination from the bottom $T_2{}^*(\mu,\mu_0,\phi)$ are computed with the RAY code.* These radiative transfer characteristics are computed at the set of Gaussian quadratures taking into account the stratification of atmosphere and light polarization effects as a solution of the vector radiative transfer equation. To be more specific, the azimuth Fourier harmonics of the reflectance $R_2^{(k)}(\mu,\mu_0)$, $R_2{}^{*(k)}(\mu,\mu_0)$, and the transmittance $T_2{}^{*(k)}(\mu,\mu_0)$ (k is a number of the azimuth harmonics) are computed in the doubling scheme (Lenoble, 1985) and used in the retrieval procedure to include the radiative interaction between layers '1' and '2'.

5. *For each central pixel (with average reflectance for a box), a correspondent normalized differential index is estimated: NDVI (Eq.(12)) for 'land' pixels or NDPI (Eq.(13)) for 'water' pixels.*

6. *The iteration process (see Section 2.6) with the least-mean-squares algorithm using the average values (over the box) $\bar{R}_{TOA}(\lambda)$ in seven spectral channels ($\lambda = 412.5,\ 442.5,\ 490,\ 510,\ 560,\ 620$ and 665 nm) is carried out.* In this process the aerosol optical thickness $\tau(\lambda)$ and Ångström exponent a of layer '1' are retrieved with the assumption

$$\tau(\lambda) = \beta\lambda^{-a}. \tag{15}$$

The land spectral albedo is described by Eq. (10); the water spectral albedo is given by Eq. (11). The value of the parameter c in Eq. (10) or Eq. (11), and values of $\tau(412)$ and a are determined in the iteration process with the least-mean-squares method. The value of $c_0 = $ NDVI for land pixels ($c_0 = $ NDPI for water pixels) is taken for the zeroth iteration.

The key question is the choice of the tropospheric aerosol model in layer '1' that determines the spectral dependence of the single scattering albedo and, what is most important, the aerosol phase function. The ART algorithm includes the procedure of correction and adapting the aerosol model using the value of Ångström exponent a obtained at the previous step of iteration in the iteration process.

7. *The spectral optical thickness of the total atmosphere*

$$\tau_{total}(\lambda) = \tau(\lambda) + \tau_{strat}(\lambda), \tag{16}$$

$\tau_{strat}(\lambda)$ being the optical thickness of aerosol in the stratified upper layer '2', and the Ångström exponent for the total atmosphere is calculated.

8. *The spectral albedo of land $r_s(\lambda)$ is calculated with the new corrected model of tropospheric aerosol and retrieved dependence $\tau(\lambda)$.* When the underlying surface is land, the surface albedo r_s is defined from the well-known equation

$$R_{TOA}(\mu,\mu_0,\phi) = R_a(\mu,\mu_0,\phi) + \frac{t_a(\mu)t_a(\mu_0)r_s}{1 - r_s r_{sa}{}^*}, \tag{17}$$

where $R_a(\mu,\mu_0,\phi)$ is the TOA reflectance of the whole atmosphere above a black underlying surface, $t_a(\mu_0)$ is the integral transmittance (direct + diffuse) of the whole atmosphere, $r_{sa}{}^*$ is the spherical albedo of the atmosphere at illumination of atmosphere bottom upwards. The definitions of all these values are given in Table 4.1. Eq. (17) accounts for the reciprocity theorem (Zege et al., 1991) and is obtained assuming a Lambertian underlying surface.

When the underlying surface is water it is necessary to separate the reflection by the water body (inner component) and reflection by the atmosphere–water interface. The inner component can be considered as diffuse with spherical albedo r_{sw}. The surface reflection depends on the incidence and observation angles and on the ratio of diffuse to direct components of the solar light after propagating through the atmosphere. This chapter is directed mainly to development and testing the AOT retrieval over land. Therefore, we will consider the reflection from sea only in a first approximation. Outside the glint region this is approximately:

$$R_{TOA}(\mu,\mu_0,\phi) = R_a(\mu,\mu_0,\phi) + \frac{t_a(\mu)t_a(\mu_0)[r_{sw} + R_{surf}^{ef}(\mu,\mu_0,\phi)]}{1 - r_{sa}{}^*(r_{sw} + r_{sf})}, \tag{18}$$

where $r_{sf} \approx 0.06$ (Ivanov, 1975) is the integral Fresnel coefficient of water surface reflection under diffuse illumination (the spherical albedo of the sea surface). In this relation we have not included the reflection by whitecaps.

As was already mentioned, the value of the effective reflectance of the water surface $R_{surf}^{ef}(\mu,\mu_0,\phi)$ depends on the ratio of diffuse to direct fluxes of the sunlight that illuminates a water surface after propagation through the atmosphere. The value of $t_a(\mu)t_a(\mu_0)R_{surf}^{ef}(\mu,\mu_0,\phi)$ can be approximately defined as a sum of three components (outside the glitter):

$$t_a(\mu)t_a(\mu_0)R_{surf}^{ef}(\mu,\mu_0,\phi) = t_a^0(\mu)r_f(\mu)t_a^s(\mu_0) + t_a^s(\mu)r_f(\mu_0)t_a^0(\mu_0) + t_a^s(\mu)r_{sf}t_a^s(\mu_0), \tag{19}$$

where $t_a^0(\mu_0)$ and $t_a^s(\mu_0)$ are direct and diffuse components of the atmospheric transmittance, respectively, $t_a(\mu_0) = t_a^0(\mu_0) + t_a^s(\mu_0)$, $r_f(\mu_0)$ is the integral reflectance for a water surface under directional illumination. For not very oblique incidence and observation angles (μ, $\mu_0 > 0.5$), the Fresnel coefficient $r_f(\mu_0)$ could be taken as $r_f(\mu_0) \approx 0.02$ (Ivanov, 1975). As seen, the first term in Eq. (19) describes the contribution of sunlight scattered into atmosphere on its way to the water surface, reflected by this surface and propagated back to the receiver without scattering. The non-scattered Sun radiation reflected by the sea surface and scattered on its way to the receiver is described by the second term. The third term in Eq. (19) includes the contribution of the radiation scattered both on its way to and from the sea surface.

The listed steps and operations make the skeleton of the ART algorithm. The iteration procedure used will be described in Section 2.6.

2.6   Iteration process for the retrieval of the spectral aerosol optical thickness

Let us consider the spectral dependence $\bar{R}_{TOA}(\lambda)$ for a given observation geometry as a function of the surface spectral albedo $r_s(\lambda)$ and aerosol optical thickness of the layer '1' $\tau_1(\lambda)$ assuming that in accordance to (15)

$$\tau(\lambda) = \tau_{412}(\lambda/\lambda_0)^{-a}, \tag{20}$$

where $\lambda_0 = 412.5$ nm, τ_{412} is AOT of the layer '1' at the wavelength $\lambda_0 = 412.5$ nm. Hence, the function $\bar{R}_{TOA}(\lambda)$ depends on two parameters (τ_{412}, a) and also the spectral ground albedo $r_s(\lambda)$.

The surface albedo $r_s(\lambda)$ is described by Eq. (10) for land and by Eq. (11) if water is an underlying surface. If spectra $r_{veg}(\lambda)$ and $r_{soil}(\lambda)$ (or $r_{clear}(\lambda)$ and $r_{coastal}(\lambda)$) are specified, the spectrum $r_s(\lambda)$ depends on the only parameter that is the weight coefficient c. Thus, the spectrum $\bar{R}_{TOA}(\lambda)$ depends on the following three parameters: τ_{412}, a and c in the framework of the retrieval scheme described here.

The parameters τ_{412}, a, and c do not depend on the wavelength by definition and can be derived using the least squares method by iteration process:

$$\begin{aligned} \tau_{412,\,i+1} &= \tau_{412,\,i} + \Delta\tau_{412,\,i} \\ a_{i+1} &= a_i + \Delta a_i. \\ c_{i+1} &= c_i + \Delta c_i \end{aligned} \tag{21}$$

The corrections $\Delta\tau_{412,i}$, Δa_i, and Δc_i are determined from the following set of equation

$$\bar{R}_{TOA}(\lambda_j) = \bar{R}_{TOA,i}(\lambda_j) + \frac{d\bar{R}_{TOA,i}(\lambda_j)}{d\tau_{412,i}}\Delta\tau_{412,i} + \frac{d\bar{R}_{TOA,i}(\lambda_j)}{da_i}\Delta a_i + \frac{d\bar{R}_{TOA,i}(\lambda_j)}{dc_i}\Delta c_i, \tag{22}$$

$$j = 0, 1, \dots N-1$$

where $\bar{R}_{TOA}(\lambda_j)$ are satellite processed data, and values $\bar{R}_{TOA,i}(\lambda_j)$ are computed characteristics.

The derivatives $d\bar{R}_{TOA,i}(\lambda_j)/d\tau_{412,i}$, $d\bar{R}_{TOA,i}(\lambda_j)/da_i$, $d\bar{R}_{TOA,i}(\lambda_j)/dc_i$ at the point $(\tau_{412,i},\ a_i,\ c_i)$ with regard to relations (20) and (10) (or (11)) are calculated using the following equations:

$$\frac{d\bar{R}_{TOA,i}(\lambda_j)}{d\tau_{412,i}} = \frac{d\bar{R}_{TOA,i}(\lambda_j)}{d\tau_i(\lambda_j)}\frac{d\tau_i(\lambda_j)}{d\tau_{412,i}} = \frac{d\bar{R}_{TOA,i}(\lambda_j)}{d\tau_i(\lambda_j)}\left(\frac{\lambda_j}{\lambda_0}\right)^{-a}, \tag{23}$$

$$\frac{d\bar{R}_{TOA,i}(\lambda_j)}{da_i} = \frac{d\bar{R}_{TOA,i}(\lambda_j)}{d\tau_i(\lambda_j)}\frac{d\tau_i(\lambda_j)}{da_i} = -\frac{d\bar{R}_{TOA,i}(\lambda_j)}{d\tau_i(\lambda_j)}\tau_{412,i}\left(\frac{\lambda_j}{\lambda_0}\right)^{-a_i}\ln\left(\frac{\lambda_j}{\lambda_0}\right), \tag{24}$$

$$\frac{d\bar{R}_{TOA,i}(\lambda_j)}{dc_i} = \frac{d\bar{R}_{TOA,i}(\lambda_j)}{dr_{s,i}(\lambda_j)}\frac{dr_{s,i}(\lambda_j)}{dc_i} = \frac{d\bar{R}_{TOA,i}(\lambda_j)}{dr_{s,i}(\lambda_j)}\left[r_{veg}(\lambda_j) - r_{soil}(\lambda_j)\right] \tag{25}$$

for land pixels and

$$\frac{d\bar{R}_{TOA,i}(\lambda_j)}{dc_i} = \frac{d\bar{R}_{TOA,i}(\lambda_j)}{dr_{s,i}(\lambda_j)}\frac{dr_{s,i}(\lambda_j)}{dc_i} = \frac{d\bar{R}_{TOA,i}(\lambda_j)}{dr_{s,i}(\lambda_j)}\left[r_{clear}(\lambda_j) - r_{coastal}(\lambda_j)\right] \tag{26}$$

for water pixels.

The derivatives $d\bar{R}_{TOA,i}(\lambda_j)/d\tau_i(\lambda_j)$ and $d\bar{R}_{TOA,i}(\lambda_j)/dr_{s,i}(\lambda_j)$ are calculated with the RAY code using the finite difference technique (small variations of parameters $\tau_i(\lambda_j)$ and $r_{s,i}(\lambda_j)$).

The iterations (21) stop if

$$\left|(\tau_{412,i+1} - \tau_{412,i})/\tau_{412,i}\right| \leq \delta, \quad \delta = 0.02. \tag{27}$$

The zeroth approximation for the iteration process is chosen as: $a_0 = 1.3$, $\tau_{412} = 0.3$ and $c_0 = $ NDVI (or $c_0 = $ NDPI in the case of water). It was concluded from multiple calculations that the iteration process convergence is hardly sensitive to the choice of the zeroth approximation for the parameters a and τ_{412}. Fig. 4.7 confirms this statement. It shows the correlation between the AOT values retrieved with different zeroth approximations for the parameter τ_{412} (to be more specific, the starting values of τ_{412} were taken equal to 0.1 and 0.4).

Although the starting AOT values are drastically different, the retrieved values are pretty close. To conclude this section, one more note is necessary. As known, the Ångström exponent for the majority of atmospheric aerosol types is in the range $-0.5 < a < 2$ (Holben et al., 2001). The average a value for climatological model (WMO, 1986) is about $a = 1.3$. In the described iteration process the values of a sometimes appear to be beyond these limits. This leads to the divergence of the iteration process. We found that this occurs when there is a large difference between unknown real surface spectral albedo $r_s(\lambda)$ and the spectral model used. To limit the possible range of the Ångström exponent values, we use the following equation for the determination of the value a_{i+1} in the iterative process instead of Eq. (21):

$$a_{i+1} = a^* + (a_i + \Delta a_i - a^*)\exp\left[-(a_i + \Delta a_i - a^*)^2/2\sigma_a^2\right]. \tag{28}$$

The values of a^* and σ_a can be defined by a user. In our case $a^* = 1.3$, $\sigma_a \approx 0.23$. At $|a_i + \Delta a_i - a^*| < 3\sigma_a$ we have $a_{i+1} = a_i + \Delta a_i$. Therefore when the value of $a_i + \Delta a_i$ is close to a^*, Eq. (28) practically coincides with Eq. (21). At $|a_i + \Delta a_i - a^*| \geq 3\sigma_a$ we have $a_{i+1} \approx a^*$.

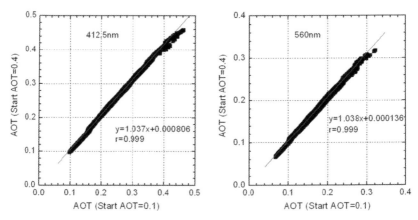

Fig. 4.7. Correlation between the AOT values retrieved with starting values $\tau_{412} = 0.1$ and $\tau_{412} = 0.4$ at $\lambda = 412.5$ nm (left) and 560 nm (right), r is the correlation coefficient.

2.7 Notes about the resemblance and difference between ART and other AOT retrieval algorithms

As seen from the previous discussion, the ART has benefited a lot from other AOT retrieval algorithms. For instance, from the BAER algorithm (von Hoyningen-Huene et al., 2003) we took the base spectral functions of surface albedo. From the MODIS algorithm (Kaufman et al., 1997) we took the idea of compiling pixels in boxes. This compilation implies averaging of data and diminishes the effect of random errors. The process of the 'good'/'bad' pixels sorting and 'bad' pixels discarding has a lot in common with the MODIS algorithm. The procedure of sorting 'water' and 'land' pixels is the combination of methods used in both algorithms, namely the value of the spectral reflectance at TOA and the value of the NDVI are jointly used. The procedures of the iteration adjustment of an aerosol model exists in both MODIS and ART algorithms.

One of the most important characteristics of any satellite retrieval algorithm is the computation time because of the need to process a large amount of pixels. Although the ART code does not use LUT, but rigorous computations of the radiative transfer in the iteration retrieval process, the retrieval procedure with ART code is fast. The retrieval of the AOT, Ångström parameter, and land reflectance for a pixel takes about 0.02 s; an area with $2 \cdot 10^5$ pixels is processed in about 1 hour on an ordinary PC (Pentium 2.6 GHz, 1 GB RAM). We have been developing a much faster semi-analytical technique that uses the combination of analytical solutions and numerical computations.

In the present-day AOT retrieval algorithms the Rayleigh optical thickness is considered as a well known parameter and its contribution to the registered radiance can be easily taken into account. Sometimes the Rayleigh reflectance $R_R(\mu, \mu_0, \phi)$ is simply subtracted from the apparent reflectance $R_{TOA}(\mu, \mu_0, \phi)$ at the top of atmosphere after correction for gaseous absorption, no coupling of Rayleigh and aerosol scattering being included. As a rule, the coupling of layers '1' and '2' is not regarded properly as well. In the MERIS algorithm (Santer et al., 2000), as in some other algorithms, the interaction between layers is considered using the following equation.

$$R_{TOA}(\mu,\mu_0,\phi) = R_R(\mu,\mu_0,\phi) + \frac{t_R(\mu)t_R(\mu_0)R_G(\mu,\mu_0,\phi)}{1 - r_{sG}r_{sR}}, \tag{29}$$

where $R_G(\mu,\mu_0,\phi)$ is the aerosol–ground system reflectance, $t_R(\mu_0)$ and $t_R(\mu)$ are the downward and upward Rayleigh transmittances, respectively, and r_{sR} and r_{sG} are the spherical albedos relating to the Rayleigh and aerosol–ground systems, respectively.

Eq. (29) is correct only in the assumption of the Lambertian distribution of the reflectance $R_G(\mu,\mu_0,\phi)$ at the top of the aerosol layer. Besides, this relation is more accurate for the narrower angular distributions of light transmittance $T_R(\mu,\mu_0,\phi)$. It was noted in (Santer et al., 2000) that assumption (29) has to be done because of the lack of a priori information about the aerosol scattering and about the land surface characteristics.

Without above assumptions the following more accurate equation should be used instead of Eq. (29):

$$R_{TOA}(\mu,\mu_0,\phi) = R_R(\mu,\mu_0,\phi) + \frac{T_R * R_G * T_R}{1 - r_{sG}r_{sR}}, \tag{30}$$

where the notation $Z = X*Y$ means (Lenoble, 1985)

$$Z(\mu,\mu_0,\phi - \phi_0) = \frac{1}{\pi} \int\limits_0^{2\pi} d\phi' \int\limits_0^1 X(\mu,\mu',\phi - \phi')Y(\mu',\mu_0,\phi' - \phi_0)\, \mu'\, d\mu'. \tag{31}$$

It means that

$$T_R * R_G * T_R = \frac{1}{\pi^2} \int\limits_0^{2\pi} d\phi' \int\limits_0^1 T_R(\mu,\mu',\phi - \phi')\, \mu'\, d\mu'$$

$$\times \int\limits_0^{2\pi} d\phi'' \int\limits_0^1 R_G(\mu',\mu'',\phi' - \phi'')T_R(\mu'',\mu_0,\phi - \phi_0)\, \mu''\, d\mu'' \tag{32}$$

Our computations have shown that the assumption (29) can lead to noticeable errors in the computations of the aerosol–ground system reflectance $R_G(\mu,\mu_s,\phi)$, overestimations and underestimations of this value are both possible. The error depends on the sun position, observation angle, surface albedo, and wavelength. To study the accuracy of Eq. (29), let us compare the value

$$\tilde{R}_a(\mu,\mu_0,\phi) = \frac{[R_{TOA}(\mu,\mu_0,\phi) - R_R(\mu,\mu_0,\phi)](1 - r_{sG}r_{sR})}{t_R(\mu)t_R(\mu_0)} \tag{33}$$

following from approximation given by Eq. (29) in the case of the black surface ($r_s = 0$) with the accurate value of $R_a(\mu,\mu_0,\phi)$. Here it follows: $r_a = r_{aG}$ at $r_s = 0$. This comparison is shown in Fig. 4.8. The relative error

$$\varepsilon = \frac{\tilde{R}_a(\mu,\mu_0,\phi) - R_a(\mu,\mu_0,\phi)}{R_a(\mu,\mu_0,\phi)} \tag{34}$$

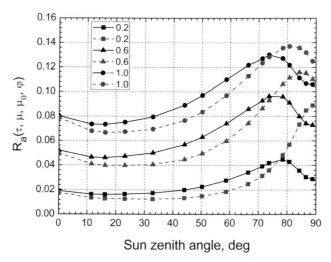

Fig. 4.8 Comparison of exact values of $R_a(\mu,\mu_o,\phi)$ (dashed lines) with $\tilde{R}_a(\mu,\mu_0,\phi)$ computed in approximation (29), (33) (solid lines) at $\lambda = 412.5$ nm and nadir observation. Numbers in the legend denote the values of AOT.

of computation of the value $\tilde{R}_a(\mu,\mu_0,\phi)$ is depicted in Fig. 4.9. All functions in Eq. (33) and the accurate values of the reflectance $R_a(\mu,\mu_0,\phi)$ of the aerosol layer were computed with the RAY code. As seen, this error may be very pronounced and can even exceed 50 % at small values of AOT. Note that the increase of the surface albedo reduces this error.

In contrast, in the ART algorithm the radiative interaction between the aerosol–ground system and layer '2' is computed accurately with regard to the multiple re-reflections and real angular distribution of the radiation transmitted and reflected by the layers.

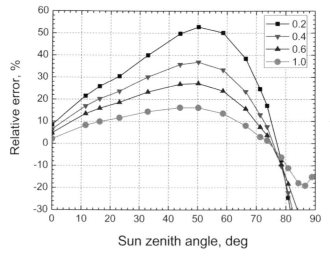

Fig. 4.9. Relative error of $\tilde{R}_a(\mu,\mu_0,\phi)$ defined by Eq. (34) for different AOTs (numbers in the legend) at $\lambda = 412.5$ nm and nadir observation.

3. Results

3.1 Inter-comparisons of the retrieved AOT with results of other algorithms and AERONET data

The values of the AOT at 550 nm retrieved with the ART code were compared with AOT values retrieved with other algorithms.

In the examples shown below we used the well-known Standard Atmosphere model (WMO, 1986). The top boundary of the layer '1' was chosen as $H = 2$ km (the altitude of the highest point in the processing area was about 700 m). It was supposed that this layer contained aerosol of Continental model. Within this model the aerosol is considered as a mixture of three fractions with log-normal particle size distributions. The first fraction contains water soluble (WS) particles, the second one includes aerosol of the soil origin (dust), and the third fraction includes soot particles. Variations of the relative concentrations of water-soluble and dust aerosols alter the Ångström exponent. The Continental model with volume concentrations equal to 29 % WS, 70 % dust and 1 % soot is recommended in (WMO, 1986) as typical for continental areas.

In accordance with the Standard Atmosphere model (WMO, 1986), the aerosol in the stratosphere and upper troposphere were supposed to consist of H_2SO_4 droplets, and an aerosol optical thickness of the atmospheric layers in the altitude range 5.9–50 km at $\lambda = 550$ nm is taken equal to 0.019. The aerosol optical thickness of the middle troposphere in the altitude interval 2–5.9 km is taken to be equal to 0.02. It is supposed to contain Continental aerosol. The atmospheric model used allows the altitude stratification of all constituents. Particularly, vertical profiles of pressure, temperature, and gas concentration can be specified. For example, in our case we use the LOWTRAN Mid-Latitude Summer molecular-gas model (see Table 4.2) which consisted of 48 layers. As a result the stratified atmosphere model can consist of N layers, where $N \leq 52$.

For ART retrieval testing we used the data of a recently performed inter-comparison (Kokhanovsky et al., 2007). In this work the performance of different algorithms was tested for a site in Europe, where multiple and near-simultaneous satellite data were available. As many as 10 different algorithms for the AOT retrieval that used data taken by six satellite optical instruments currently operated in space were inter-compared.

The explored site included the cloudless ground scene in central Europe (mainly, Germany) on October 13, 2005 (10:00 UTC). The latitude range was 49–53 N and the longitude range was 7–12 E. More detailed results are given by Kokhanovsky et al. (2007), for a smaller area (9–11.5 E, 52–52.5 N) as well. Several AERONET instruments operated at the time of satellite measurements.

The average values of AOT and standard deviations for different instruments and algorithms, taken from paper by Kokhanovsky et al. (2007), are presented in Tables 4.3 and 4.4 as well as the retrieval data obtained with ART. Note that the time difference between the measurements of ENVISAT and Terra is 30 min. The data in the tables present the total atmosphere aerosol optical thickness, including both layers '1' and '2'. As seen from the tables, the ART data are in a good agreement with results of MISR JPl, MODIS NASA, MERIS BAER, and MERIS ESA retrieval algorithms.

Inter-comparisons between different satellite products and data from ALRONET stations located in central Europe were performed by Kokhanovsky et al. (2007). AERONET

Table 4.2. The LOWTRAN Mid-Latitude Summer molecular atmosphere model (Kneizys, 1996). Here N refers to the total number of molecules at a given altitude.

Altitude (km)	P (mb)	T (K)	N (cm^{-3})	H_2O (cm^{-3})	O_3(cm^{-3})
120	2.27E − 05	380	4.33E + 11	8.65E + 04	2.16E + 02
115	3.56E − 05	316.8	8.14E + 11	1.95E + 05	4.07E + 03
110	6.11E − 05	262.4	1.69E + 12	4.72E + 05	8.43E + 04
105	1.17E − 04	222.2	3.81E + 12	1.30E + 06	7.63E + 05
100	2.58E − 04	190.5	9.81E + 12	3.92E + 06	3.92E + 06
95	6.25E − 04	178.3	2.54E + 13	1.37E + 07	1.78E + 07
90	1.64E − 03	165	7.20E + 13	6.12E + 07	5.40E + 07
85	4.48E − 03	165.1	1.97E + 14	2.61E + 08	1.12E + 08
80	1.20E − 02	174.1	4.99E + 14	1.05E + 09	9.98E + 07
75	3.00E − 02	196.1	1.11E + 15	3.27E + 09	2.11E + 08
70	6.70E − 02	218.1	2.23E + 15	8.23E + 09	8.90E + 08
65	1.39E − 01	240.1	4.19E + 15	1.85E + 10	3.35E + 09
60	2.72E − 01	257.1	7.66E + 15	3.83E + 10	9.96E + 09
55	5.15E − 01	269.3	1.39E + 16	7.41E + 10	2.49E + 10
50	9.51E − 01	275.7	2.50E + 16	1.37E + 11	7.00E + 10
47.5	1.29E + 00	275.2	3.40E + 16	1.87E + 11	1.19E + 11
45	1.76E + 00	269.9	4.72E + 16	2.57E + 11	2.13E + 11
42.5	2.41E + 00	263.7	6.62E + 16	3.51E + 11	3.91E + 11
40	3.33E + 00	257.5	9.37E + 16	4.78E + 11	7.07E + 11
37.5	4.64E + 00	251.3	1.34E + 17	6.69E + 11	1.16E + 12
35	6.52E + 00	245.2	1.93E + 17	9.53E + 11	1.71E + 12
32.5	9.30E + 00	239	2.82E + 17	1.37E + 12	2.28E + 12
30	1.32E + 01	233.7	4.09E + 17	1.92E + 12	2.86E + 12
27.5	1.91E + 01	228.45	6.05E + 17	2.69E + 12	3.63E + 12
25	2.77E + 01	225.1	8.91E + 17	3.74E + 12	4.28E + 12
24	3.22E + 01	223.9	1.04E + 18	4.17E + 12	4.17E + 12
23	3.76E + 01	222.8	1.22E + 18	4.71E + 12	4.16E + 12
22	4.37E + 01	221.6	1.43E + 18	5.14E + 12	4.14E + 12
21	5.10E + 01	220.4	1.68E + 18	5.78E + 12	4.02E + 12
20	5.95E + 01	219.2	1.97E + 18	6.49E + 12	3.93E + 12
19	6.95E + 01	217.9	2.31E + 18	7.39E + 12	3.47E + 12
18	8.12E + 01	216.8	2.71E + 18	8.55E + 12	2.71E + 12
17	9.50E + 01	215.7	3.19E + 18	1.02E + 13	2.23E + 12
16	1.11E + 02	215.7	3.73E + 18	1.23E + 13	2.24E + 12
15	1.30E + 02	215.7	4.37E + 18	1.48E + 13	2.18E + 12
14	1.53E + 02	215.7	5.14E + 18	2.57E + 13	2.26E + 12
13	1.79E + 02	215.8	6.01E + 18	4.81E + 13	1.80E + 12
12	2.09E + 02	222.3	6.81E + 18	2.01E + 14	1.52E + 12
11	2.43E + 02	228.8	7.69E + 18	7.35E + 14	1.38E + 12
10	2.81E + 02	235.3	8.65E + 18	2.14E + 15	1.13E + 12
9	3.24E + 02	241.7	9.71E + 18	4.01E + 15	1.08E + 12
8	3.72E + 02	248.2	1.09E + 19	7.02E + 15	9.91E + 11
7	4.26E + 02	254.7	1.21E + 19	1.24E + 16	9.41E + 11
6	4.87E + 02	261.2	1.35E + 19	2.04E + 16	8.65E + 11
5.9	4.93E + 02	261.8	1.37E + 19	2.14E + 16	8.62E + 11
5	5.54E + 02	267.2	1.50E + 19	3.34E + 16	8.28E + 11
4	6.28E + 02	273.2	1.67E + 19	6.35E + 16	8.03E + 11
3	7.10E + 02	279.2	1.84E + 19	1.10E + 17	7.78E + 11
2	8.02E + 02	285.2	2.04E + 19	1.97E + 17	7.52E + 11
1	9.02E + 02	289.7	2.26E + 19	3.11E + 17	7.53E + 11
0	1.01E + 03	294.2	2.49E + 19	4.68E + 17	7.52E + 11

Table 4.3. Average AOTs and standard deviations at $\lambda = 550$ nm for different instruments and algorithms in the selected area (7–12E, 49–53N)

Instrument	Algorithm	Average AOT	Standard deviation	Spatial resolution of reported AOT, km²	Platform
MERIS	ART	0.16	0.035	5 × 5	ENVISAT
MERIS	ESA	0.13	0.05	1 × 1	ENVISAT
MERIS	BAER	0.18	0.03	1 × 1	ENVISAT
MISR	JPL	0.14	0.03	17.6 × 17.6	TERRA
MODIS	NASA	0.14	0.04	10 × 10	TERRA
AATSR	AATSR-2	0.23	0.05	3 × 3	ENVISAT
AATSR	AATSR-1	0.26	0.1	3 × 3	ENVISAT

Table 4.4. The same as in Table 4.3 except for a smaller area (9–11.5E, 52–52.5N)

Instrument	Algorithm	Average AOT	Standard deviation	Spatial resolution of reported AOT,km²	Platform
MERIS	ART	0.18	0.025	5 × 5	ENVISAT
MERIS	ESA	0.21	0.05	1 × 1	ENVISAT
MERIS	BAER	0.20	0.02	1 × 1	ENVISAT
MISR	JPL	0.16	0.02	17.6 × 17.6	TERRA
MODIS	NASA	0.15	0.03	10 × 10	TERRA
AATSR	AATSR-2	0.22	0.06	3 × 3	ENVISAT
AATSR	AATSR-1	0.30	0.06	3 × 3	ENVISAT

is a network of identical sunphotometers designed specifically for the validation of satellite aerosol retrievals. We also compared ART retrieval with AERONET data. Fig. 4.10 (left), where the ART retrieval data are plotted versus AERONET data for $\lambda = 550$ nm, demonstrates a pretty good agreement. AERONET AOT data at 550 nm are obtained by interpolation between $\lambda = 440$ nm and $\lambda = 670$ nm.

Here the correlation coefficient is equal to 0.84. For comparison, other satellite products versus AERONET, taken from Kokhanovsky et al. (2007), are demonstrated in Fig. 4.10 right.

With ART the values of the AOT at $\lambda = 412.5$ nm and the Ångström exponent are directly retrieved, the values of AOTs at other wavelengths, including $\lambda = 550$ nm, are calculated through these values. It is interesting to compare data of ART and AERONET at different wavelengths. Such a comparison at $\lambda = 440$ nm and $\lambda = 670$ nm is presented in Fig. 4.11 and shows satisfactory agreement.

Let us compare the values of the AOT retrieved with the ART and other algorithms. In Fig. 4.12 the correlation of the AOT at 550 nm obtained for the whole test area (7–12 E, 49–53 N) using MERIS-ART and MISR-JPL algorithms is shown. As seen, the AOT values retrieved with both algorithms are pretty close. The difference is maximal at small AOT values. The most probable cause of this discrepancy is the difference between the tropospheric aerosol models used and, particularly, the phase functions used. Fig. 4.13, where the correlation of AOT at 550 nm obtained using MERIS-ART and MODIS-NASA algorithms is shown, leads to the similar propositions.

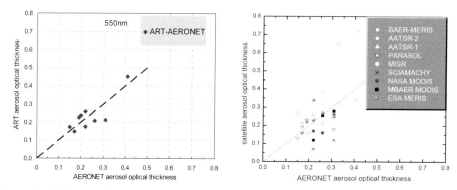

Fig. 4.10. AOT at $\lambda = 550$ nm retrieved with ART versus AERONET (left) and different satellite products versus AERONET (Kokhanovsky et al., 2007) (right).

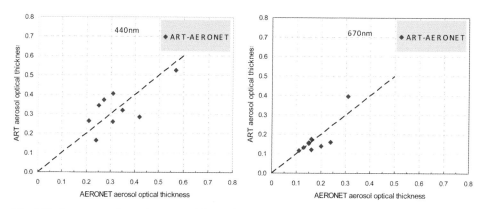

Fig. 4.11. Inter-comparisons between ART and AERONET products at $\lambda = 440$ nm (left) and $\lambda = 670$ nm (right).

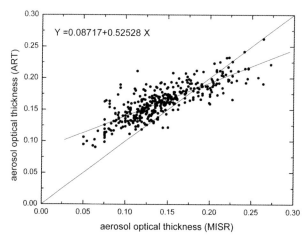

Fig. 4.12. Correlation between the AOT (550 nm) values retrieved with ART and MISR algorithms for the large test area (7–12 E, 49–53 N).

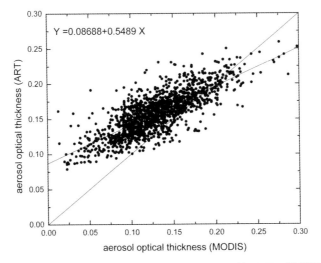

Fig. 4.13. Correlation between the AOT (550 nm) values retrieved with ART and MODIS algorithms for the large test area (7–12 E, 49–53 N).

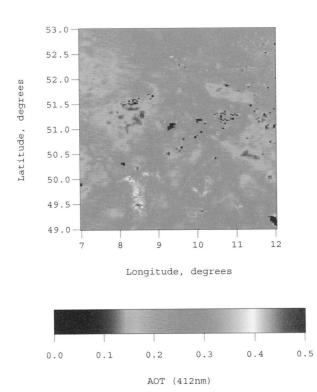

Fig. 4.14. AOT at $\lambda = 412.5$ nm for the large test area (7–12 E, 49–53 N) studied by Kokhanovsky et al. (2007).

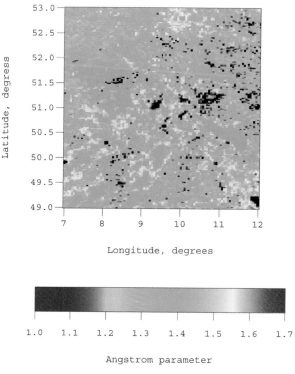

Longitude, degrees

1.0   1.1   1.2   1.3   1.4   1.5   1.6   1.7

Angstrom parameter

Fig. 4.15. Ångström parameter for the large test area (7–12 E, 49–53 N).

Figs 4.14 and 4.15 show the spatial distribution of the retrieved AOT at 412 nm and the Ångström exponent. It follows from Fig. 4.14 that the AOT spatial distribution is very inhomogeneous with higher AOTs at the southern part of the study area, where AOT reaches 0.4. The background aerosol (blue color) has an AOT around 0.15. The background Ångström exponent is around 1.2 (see Fig. 4.15) with larger values in areas, where the AOT is also larger, which points at industrial pollution sources dominated by fine mode aerosol particles.

3.2 Importance of the choice of the aerosol model

As mentioned above, a very important issue is the choice of the aerosol model for layer '1'. This model assumes not only the value of the single scattering albedo, but, what is the most important, also the phase function. In our runs we used the Standard Atmosphere model (WMO, 1986) with the Continental aerosol model in the lower layer '1'.

Let us estimate the influence of the chosen phase function on the retrieved AOT value. Fig. 4.16 demonstrates the phase function of the Continental model at $\lambda = 550$ nm and a phase function $p(\theta)$ (θ is the scattering angle) that was measured earlier in Germany (von Hoyningen-Huene et al., 2003) (we will refer to it as the 'experimental' phase function). Here the phase function is normalized as $(1/2) \int_0^\pi p(\theta) \, \sin \theta d\theta = 1$. As seen, in the angle

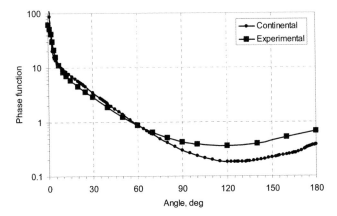

Fig. 4.16. Continental and experimental phase functions at $\lambda = 550$ nm.

range $\sim 100 \div 150$ degrees mainly used for AOT retrieval, the values of the experimental phase function are about two times larger than the values of the phase function of the Continental model.

Values of the AOT retrieved with the ART algorithm but with two different aerosol models (Continental and experimental) are compared in Fig. 4.17. The experimental model is rather simple: the single scattering albedo is equal to 0.9 (as for the Continental model) and the phase function is considered as independent of the wavelength. The correlation between these data is rather high, but AOT values retrieved with the experimental phase function are 1.5 times smaller.

This difference is easy to understand. Let us consider the simplest case of a small AOT, when the single scattering approximation can be used. In this case the reflectance of the aerosol layer '1' is proportional to the product of the single scattering albedo ω, phase function $p(\theta)$, and AOT τ, i.e.

$$R_2(\mu, \mu_0, \phi) \sim \omega p(\theta) \tau. \tag{35}$$

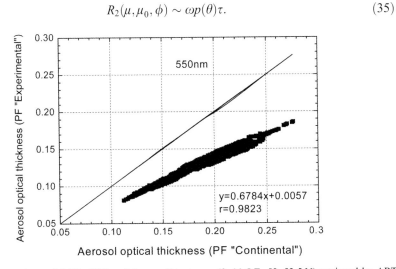

Fig. 4.17. Comparison of AOTs (550 nm) for small test area (9–11.5 E, 52–52.5 N) retrieved by ART algorithm using Continental and experimental phase functions, r is the correlation coefficient.

It means that, for example, a twofold increase of the value of the phase function $p(\theta)$ in the considered direction leads to a twofold decrease of the retrieved values of the AOT.

3.3 Retrieval of the surface albedo

One of the most important problems of satellite remote sensing is the retrieval of the true spectral albedo $r_s(\lambda)$ of the underlying surfaces (atmospheric correction). After retrieving $\tau(\lambda)$ the spectral albedo $r_s(\lambda)$ of the underlying Lambertian surface can be simply calculated from Eq. (17) if the surface is land:

$$r_s = \frac{R_{TOA}(\mu,\mu_0,\phi) - R_a(\mu,\mu_0,\phi)}{t_a(\mu)t_a(\mu_0) + r_{sa}*[R_{TOA}(\mu,\mu_0,\phi) - R_a(\mu,\mu_0,\phi)]}. \tag{36}$$

In the case of water the value of r_w is retrieved from Eq. (18) as follows:

$$r_{sw} = \frac{[R_{TOA}(\mu,\mu_0,\phi) - R_a(\mu,\mu_0,\phi)](1 - r_{sa}*r_{sf}) - t_a(\mu)t_a(\mu_0)R_{surf}^{ef}(\mu,\mu_0,\phi)}{t_a(\mu)t_a(\mu_0) + r_{sa}*[R_{TOA}(\mu,\mu_0,\phi) - R_a(\mu,\mu_0,\phi)]}. \tag{37}$$

Let us study how the choice of the aerosol model, particularly, the aerosol phase function, influences the spectral surface albedo retrieval. The comparison of the spectral surface albedo values for two wavelengths retrieved with Continental and experimental phase functions is shown in Fig. 4.18. The correlation of these data is very high.

Hence, the following important conclusion results from this study: the retrieved spectra of the surface albedo are comparatively stable whereas the retrieved values of AOT are sensitive to the choice of the aerosol model. This insensitivity of the retrieved spectra of the surface albedo to variations of the aerosol phase function $p(\theta)$ follows immediately from the fact that the reflectance of layer '1' is proportional to the product of the phase function $p(\theta)$ and AOT (see Eq. (35)). This statement is more accurate as the the optical

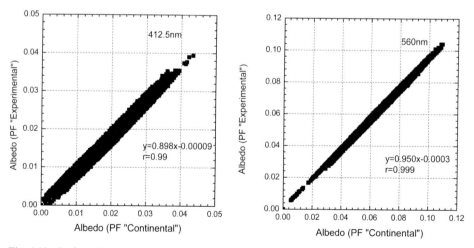

Fig. 4.18. Surface albedo retrieved with the experimental phase function versus the same value retrieved with the Continental phase function at wavelengths of 412.5 nm (left) and 560 nm (right) for the small test area (9–11.5 E, 52–52.5 N) used by Kokhanovsky et al. (2007). r is the correlation coefficient.

thickness of the layer '1' is smaller. Indeed, for a thin layer a twofold increase of the value of the phase function $p(\theta)$ in the considered direction leads to a twofold decrease of the retrieved values of the AOT. The product $p(\theta)\tau$ and the value of $R_a(\mu, \mu_0, \phi)$ in Eqs. (36) and (37) change only slightly. At small AOT such changes of τ lead to modest variations of the transmittance $t_a(\mu)$ as well.

4. Conclusion

A new aerosol retrieval technique for the spectral AOT retrieval from data of satellite optical instruments has been presented. This technique incorporates some ideas and features from the most elaborated retrieval techniques (e.g., those of MODIS and MERIS). Particularly, pixel sorting is close to those of the MODIS algorithm; the starting spectral model of the underlying surface is taken from the BAER code. The main distinction of this new technique is that it does not use LUT (as practically all AOT retrieval codes do) but includes radiative transfer computations in the process of retrieval. The base of the developed approach is the accurate and extremely fast radiative transfer code RAY for the simulation of the radiative transfer in the atmosphere–underlying surface system taking into account light polarization developed by authors earlier (see Appendix). RAY's high processing speed allowed us to use iterative radiation transfer computations in the processing of satellite data for AOT retrieval. The AOT retrieval with ART being comparable in the processing time with other techniques has some additional merits, particularly

- ART uses minimal simplifications in the computation of the radiative transfer through the atmosphere, for instance it accurately accounts for the interaction between atmospheric layers; and also ART includes the effect of polarization on radiance.
- In the AOT and Ångström exponent retrieval the least-mean-squares technique is applied (RAY provides fast computation of the derivatives with respect to retrieved values).
- Any set of wavelengths and also data of many satellite optical instruments can be processed without any changes in the ART code.
- The atmospheric model can be easily changed in the retrieval process.

The evaluation of the ART retrieval technique by comparison with AERONET data and results of AOT and land spectral albedo retrieval with other codes showed that its accuracy is at least as good as that in other widely used approaches.

It is worth noting that the developed technique can be easily generalized to process the spectral multi-angle and polarization satellite data.

Appendix: The vector radiative transfer code

A.1 Theory

One of the main keys to the success of our retrieval algorithm is the speed and accuracy of the RAY code (Tynes et al., 2001; Chaikovskaya et al., 1999) for the computation of the polarized radiative transfer in atmosphere–land and atmosphere–ocean systems. Below we give a brief description of this code: outline of the underlying theory, the code architecture, and the means used to speed up computations.

Let us consider the scattering of a polarized light beam in a plane-parallel turbid medium. The z-axis of the Cartesian coordinate system XYZ is set normal and downwards to the scattering layer upper boundary $z = 0$. This upper boundary is uniformly illuminated by an infinitely wide beam. A Stokes vector $\mathbf{I}$ of the scattered light depends on the optical depth $\tau = \int_0^z \sigma_e(z')\,dz'$, where $\sigma_e(z)$ is the extinction coefficient, and on the beam direction $\mathbf{n}(\mu, \phi)$. The Stokes vector of scattered light $\mathbf{I}(\tau, \mathbf{n})$ at the depth τ in the direction $\mathbf{n}(\mu, \phi)$ is related to the Stokes vector of the incident radiation $\mathbf{I}_0(\mathbf{n_0})$, where $\mathbf{n_0}(\mu_0, \phi_0)$ is the incident light direction, as follows:

$$\mathbf{I}(\tau, \mathbf{n}) = \frac{1}{\pi} \int \int \mathbf{G}(\tau, \mathbf{n}, \mathbf{n}_0) \mu_0 \mathbf{I}_0(\mathbf{n}_0)\,d\mathbf{n}_0. \tag{38}$$

The Green's matrix $\mathbf{G}(\tau, \mathbf{n}, \mathbf{n_0})$ obeys the vector radiative transfer equation (VRTE) (Zege and Chaikovskaya, 1996)

$$B\{\mathbf{G}(\tau, \mathbf{n}, \mathbf{n_0})\} = \frac{\omega(\tau)}{4\pi} \int \int \mathbf{Z}(\tau, \mathbf{n}, \mathbf{n}') \mathbf{G}(\tau, \mathbf{n}', \mathbf{n}_0)\,d\mathbf{n}' \tag{39}$$

and the boundary conditions

$$\mathbf{G}(\tau = 0, \mu, \mu_0, \phi - \phi_0) = \pi \mathbf{E} \delta(\tau) \delta(\mathbf{n} - \mathbf{n_0}), \quad \mu \in [0, 1]$$

$$\mathbf{G}(\tau = \tau_0, \mu, \mu_0, \phi - \phi_0) = 0, \qquad \qquad \mu \in [0, -1] \tag{40}$$

Here

$$B = \mu \frac{\partial}{\partial \tau} + 1 \tag{41}$$

is the differential operator, $\omega(\tau)$ and $\mathbf{Z}(\tau, \mathbf{n}, \mathbf{n}')$, the single scattering albedo and phase matrix, respectively (their argument τ hereafter will be omitted), $\mathbf{E}$ is the 4×4 unit matrix, $\delta(\tau)$ and $\delta(\mathbf{n} - \mathbf{n_0})$ are the delta functions, and τ_0 is the optical thickness of a scattering layer. Eq. (39) presents four coupled integro-differential radiative transfer equations, each one being for the Green's matrix column.

The phase matrix $\mathbf{Z}(\mathbf{n}, \mathbf{n}')$ is related to the Mueller (single-scattering) matrix $\mathbf{F}(x)$ as follows (Chandrasekhar, 1960):

$$\mathbf{Z}(\mathbf{n}, \mathbf{n}') = \mathbf{L}(\pi - \chi_2')\mathbf{F}(x)\mathbf{L}(-\chi_1'), \tag{42}$$

where $x = \cos\theta$, θ is the scattering angle,

$$x = \cos\theta = \mu\mu' + \sqrt{1 - \mu^2}\sqrt{1 - \mu'^2}\cos(\phi - \phi'), \quad -1 \le \mu, \mu' \le 1 \tag{43}$$

and $\mathbf{L}(\chi)$ is the well-known rotation matrix (Chandrasekhar, 1960).

When a medium phase function is strongly forward peaked, then our way to solve Eq. (39) consists of the following steps.

1. The phase function $F_{11}(x)$ (note that $F_{11}(x) = Z_{11}(x)$) is represented as a sum of two components, the first one differs from zero only at small scattering angles $\theta \le \theta_{\lim} \sim 10^0$, i.e.,

$$Z_{11}(x) = A_1 Z_{11}^f(x) + (1 - A_1)Z_{11}^d(x), \tag{44}$$

A_1 is the normalization constant. The first term, for example, can be determined as follows

$$A_1 Z_{11}^f(x) = \begin{cases} Z_{11}(x) - Z_{11}(X), & \text{if } x \geq X, \\ 0, & \text{if } x < X, \end{cases} \tag{45}$$

$X = \cos\theta_{\lim}$. The total phase and Green's matrices are also presented by the sums of two components:

$$\hat{Z}(\mathbf{n}, \mathbf{n}') = A_1 \hat{Z}^f(\mathbf{n}, \mathbf{n}') + (1 - A_1)\hat{Z}^d(\mathbf{n}\mathbf{n}'), \tag{46}$$

$$\hat{G}(b, \mathbf{n}, \mathbf{n}_0) = \hat{G}^f(b, \mathbf{n}, \mathbf{n}_0) + \hat{G}^d(b, \mathbf{n}, \mathbf{n}_0), \tag{47}$$

and $\hat{Z}^f$ is given by $\hat{Z}_{11}^f \tilde{Z}$, as well as $\hat{Z}^d$ is given by $Z_{11}^d \tilde{Z}$, where $\tilde{Z}_{ik} = Z_{ik}/Z_{11}$.

2. Eq. (39) is then transformed, using Eqs (42) and (43) and taking into account the feature of matrices $\hat{Z}^f$ and $\hat{G}^f$ to have forward-peaked diagonal functions. Two different radiative transfer problems are formulated instead of the original one. The first component $\mathbf{G}^f(\tau, \mathbf{n}, \mathbf{n}_0)$ of the solution satisfies the independent transfer equation with the small-angle phase matrix

$$B\{\mathbf{G}^f(\tau, \mathbf{n}, \mathbf{n}_0)\} = \frac{\omega A_1}{4\pi} \int\int \mathbf{Z}^f(\mathbf{n}, \mathbf{n}')\mathbf{G}^f(\tau, \mathbf{n}', \mathbf{n}_0)\, d\mathbf{n}', \tag{48}$$

and the boundary condition

$$\mathbf{G}^f(\tau = 0, \mu, \mu_0, \phi - \phi_0) = \pi\mathbf{E}\delta(\tau)\delta(\mathbf{n} - \mathbf{n}_0), \quad \mu \in [0, 1]. \tag{49}$$

Let us agree to use the term 'small-angle' VRTE in referring to Eq. (48). For the second component of the Green's matrix we have:

$$B\{\mathbf{G}^d(\tau, \mathbf{n}, \mathbf{n}_0)\} = \frac{\omega}{4\pi} \int\int \mathbf{Z}(\mathbf{n}, \mathbf{n}')\mathbf{G}^d(\tau, \mathbf{n}', \mathbf{n}_0)\, d\mathbf{n}' +$$

$$+ \frac{\omega(1 - A_1)}{4\pi} \int\int \mathbf{Z}^d(\mathbf{n}, \mathbf{n}')\mathbf{G}^f(\tau, \mathbf{n}', \mathbf{n}_0)\, d\mathbf{n}' \tag{50}$$

Thus, the problem of strongly anisotropic multiple light scattering described by Eq. (39) is decomposed into two simpler problems based on the solution of Eqs (48) and (50). Up to this point, no simplifying assumptions have been made. The set of Eqs (48) and (50) is completely equivalent to the original VRTE (39). But these new transfer equations are still complicated. The described algorithm is based on the transformation of Eqs (48) and (50), which makes their forms suitable for solving by known semi-analytical methods.

3. Using theory developed by Zege and Chaikovskaya (1996), with the main idea of separating the equations for diagonal and non-diagonal elements of the matrix $\mathbf{G}^f(\tau, \mathbf{n}', \mathbf{n}_0)$, the small-angle matrix problem is reduced to
   - scalar small-angle RTEs for leading diagonal elements of matrix $\mathbf{G}^f(\tau, \mathbf{n}', \mathbf{n}_0)$; each equation can be solved by a small-angle scalar technique;
   - simple quasi-single scattering solution for the small non-diagonal elements.

4. Taking into account the properties of the matrix $\mathbf{G}^f(\tau, \mathbf{n}', \mathbf{n}_0)$, the following equation is used in the code to derive the Green's matrix component $\mathbf{G}^d(\tau, \mathbf{n}, \mathbf{n}_0)$ (see Eq. (50)):

$$B_d\{\mathbf{G}^d(\tau, \mathbf{n}, \mathbf{n}_0)\} = \frac{\omega_d}{4\pi} \int \int \mathbf{Z}^d(\mathbf{n}, \mathbf{n}') \mathbf{G}^d(\tau, \mathbf{n}', \mathbf{n}_0) \, d\mathbf{n}' + \frac{\omega_d}{4} \mathbf{Z}^d(\mathbf{n}, \mathbf{n}_0) \, \exp\left(-\frac{\tau_d}{\mu_0}\right)$$

(51)

with

$$B_d = \mu \frac{\partial}{\partial \tau_d} + 1, \quad \omega_d = \frac{\omega(1 - A_1)}{1 - \omega A_1}, \quad \tau_d = (1 - \omega A_1)\tau.$$

(52)

Eq. (51) is the conventional VRTE for the case of an effective scattering medium, which is characterized by the extinction coefficient $\sigma_e(1 - \omega A_1)$, by the single scattering albedo ω_d and also by the truncated phase matrix $\mathbf{Z}^d(\mathbf{n}, \mathbf{n}')$.

The advantage of this approach, named the two-component one, arises from a different character of a new pair of equations. The first equation is defined only at small scattering angles and has the effective single scattering albedo smaller than 0.5, therefore it can be solved within the semi-analytical small-angle approach (Zege et al., 1991; Zege and Chaikovskaya, 1996). The second equation, with the truncated phase function, can be easily solved using the Fourier expansion with adding–doubling methods. Eq. (51) is much simpler as compared to Eq. (39) (forward scattering peaks are removed). The chart of the approach used in the framework of RAY is given in Fig. 4.19.

This two-component approach makes a basis of the RAY algorithm. The time saving is achieved owing to the economical arrangement of the code at the every step and mainly due to following newly developed features and findings:

1. The basic method used, namely, the two-component approach, that is the principal source of the saving computation time practically without a loss of the accuracy.

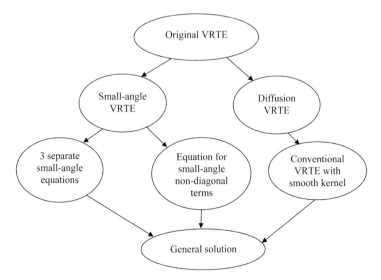

Fig. 4.19. Chart of the RAY code.

The accurate semi-analytical solution is obtained for the small angle part of the Green's matrix, using theory developed Zege and Chaikovskaya (1996). The diffuse part of the Green's matrix is the solution of the traditional VRTE with a truncated phase matrix $\hat{Z}^d(x)$, which is very smooth in comparison with initial phase matrix $\hat{Z}(x)$. Unlike $\hat{Z}(x)$, the truncated phase matrix $\hat{Z}^d(x)$ requires a reasonable number M (up to 30) of the generalized spherical functions in its expansion with sufficient accuracy. The value of M is the crucial number, which mainly determines the computation time (the computation time changes at least as M^3).

2. The fundamental symmetry and reciprocity relationships to decrease the volume of calculations, particularly, while computing the Fourier harmonics in the adding–doubling technique, are used.

3. Analytical approximations of a sum of high number of the multi-dimensional integrals describing the re-reflections between atmospheric sub-layers in the doubling procedure are used. Also asymptotic solutions of the radiative transfer equation are used for thick layers. Therefore, not only aerosols but also thick cloud layers can be studied with the developed technique.

4. The analytical description of the behavior of the Green's matrix elements in the vicinity of points $\mu = 1$ and $\mu_0 = 1$ is used.

5. Separate exact computations of the Green's matrix component for singly scattered radiation are performed.

A.2 The accuracy of the code

Several approximations are used to speed up radiative transfer calculations and in this sense the developed code provides approximate results. However, the error of the corresponding approximations is much smaller than that due to a priori assumptions, which are inherent to the retrieval process (e.g., the assumption on the surface reflectance and the atmospheric state model). Clearly, theoretical tools for the solution of any physical problem must be adequate, e.g., with respect to the information content for a given set of measurements. Nevertheless, the accuracy of the code is of a primary importance.

The first step to check the code accuracy was to compare the computed reflection and transmission matrix for the Rayleigh layer with well known data of Coulson et al. (1960). Our data coincide within five digits with the tabular data presented by Coulson et al. (1960).

As numerous comparisons with other codes have shown, RAY achieves a similar accuracy as compared to other codes and provides highly accurate data in a fraction of the time required by the Monte Carlo and other methods (Tynes et al., 2001). For instance, the difference between RAY and SCIATRAN computations (Rozanov et al., 2005) for molecular and aerosol atmospheres is smaller than 0.5 % for all Stokes vector components (see Fig. 4.20). In addition, the RAY code does not require the use of very powerful computers; it runs quickly on an ordinary personal computer. It is important that the RAY code provides the tradeoff between time and accuracy of computations and allows one to achieve practically any accuracy at the expense of computation time. These properties make this code a practical technique for fast simulations of polarized light propagation through light scattering systems, particularly through the terrestrial atmosphere.

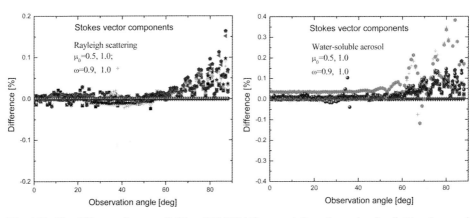

Fig. 4.20. The difference between RAY an SCIATRAN computations for molecular (left) and aerosol (right) atmospheres for all Stokes vector components.

Acknowledgments. This work benefited through numerous discussions and collaboration with many scientists. In particular, we are grateful to W. von Hoyningen-Huene, J. P. Burrows, V. Rozanov, Y. Kaufman, R. Levy, T. Nakajima, I. N. Polonsky, and L. I. Chaikovskaya.

This work was partially done in the framework of European Integrated Project DAMOCLES that is financed by the European Union in the 6th Framework Programme for Research and Development. Research of A. A. Kokhanovsky was supported by University of Bremen, European Space Agency (ESA) and German Space Agency (DLR). Authors are thankful to ESA for providing MERIS data used in this research.

References

Chaikovskaya L.I., I.L. Katsev, A.S. Prikhach, and E.P. Zege, 1999: Fast code to compute polarized radiation transfer in the atmosphere–ocean and atmosphere–earth systems, IGARSS'99, IEEE, International Geoscience and Remote Sensing Symposium, Hamburg, Germany, Congress Centrum Hamburg Remote Sensing of the System Earth a Challenge for the 21th Century Proceedings, CD-ROM.

Chandrasekhar S., 1960: *Radiative Transfer*, Oxford University Press, London.

Coulson K.L., J.V. Dave, and Z. Sekera, 1960: *Tables Related to Radiation Emerging from a Planetary Atmosphere with Rayleigh Scattering*, University of California Press, Berkeley CA.

Deuze, J.L., F.M. Breon, and C. Devaux et al., 2001: Remote sensing of aerosols over land surfaces from POLDER-ADEOS-1 polarized measurements. *Journal of Geophysical Research*, 106, 4913–4926.

Holben, B.N. et al., 2001: An emerging ground-based aerosol climatology: aerosol optical depth from AERONET, *Journal of Geophysical Research*, 106, 12067–12097.

Ivanov A.P., 1975: *Physical Principles of Hydrooptics*, Nauka i Tekhnica, Minsk.

Kaufman Y.J., D. Tanre , H.R. Gordon, T. Nakajima, J. Lenoble, R. Frouin, H. Grassl, B.M. Herman, M.D.King, and P.M. Teillet, 1997: Passive remote sensing of tropospheric aerosol and atmospheric correction for the aerosol effect, *Journal of Geophysical Research*, 102, 16815–16830.

Kneizys F.X., L.W. Abreu, G.P. Anderson, J.H. Chetwynd, E.P. Shettle, A. Berk, L.S. Bernstein, D.C. Robertson, P. Acharya, L.S. Rothman, J.E.A. Selby,W.O. Gallery, and S.A. Clough, 1996: 'The MODTRAN 2_3 report and LOWTRAN 7 model,' (Ontar Corporation, North Andover, M).

Kokhanovsky A.A., 2004: Reflection of light from nonabsorbing semi-infinite cloudy media. A simple approximation, *Journal of Quantitative Spectroscopy and Radiative Transfer*, 85, 25–33.

Kokhanovsky, A.A. et al., 2007: Aerosol remote sensing over land: a comparison of satellite retrievals using different algorithms and instruments, *Atmos. Res.*, 85, 372–294.

Kokhanovsky, A.A., W. von Hoyningen-Huene, H. Bovensmann, and J.P. Burrows, 2004: The determination of the atmospheric optical thickness overWestern Europe using SeaWiFS imagery, *IEEE Transac. Geosci. Rem. Sens.*, 42, 824–832.

Lenoble J., 1985: *Radiative Transfer in Scattering and Absorbing Atmospheres: Standard Computational Procedures*, A. Deepak Publishing, Hampton, VA.

WMO International Association for Meteorology and Atmospheric Physics Radiation Commission, „A preliminary cloudless standard atmosphere for radiation computation," World Climate Program, WCP-112 WMOyTD-#24 ∼World Meteorological Organisation, Geneva, 1986.

Martonchik, J.V., D.J. Diner, K.A. Crean, and M.A. Bull, 2002: Regional aerosol retrieval results from MISR. *IEEE Trans. Geosci. Remote Sens.*, 40, 1520–1531.

Remer L.A., Y.J. Kaufman, D. Tanré, S. Mattoo, D.A. Chu, J.V. Martins, R.R. Li, C. Ichoku, R.C. Levy, R.G. Kleidman, T.F. Eck, E. Vermote, and B.N. Holben, 2005: The MODIS Aerosol Algorithm, Products, and Validation, *J. Atm. Sci.*, 62, 947–973.

Rozanov, A., V. Rozanov, M. Buchwitz, A. Kokhanovsky, and J.P. Burrows, 2005: 'SCIATRAN 2.0 – A new radiative transfer model for geo-physical applications in the 175 –2400 nm spectral region', *Adv. Space Res.*, 36, 1015–1019.

Santer, R. et al., 2000: Atmospheric product over land for MERIS level 2. MERIS Algorithm Theoretical Basis Document, ATBD 2.15, ESA.

Santer, R., V. Carrere, P. Dubuisson, and J.C. Roger, 1999: Atmospheric corrections over land for MERIS. *Int. J. Remote Sens.*, 20, 1819–1840.

Tynes H., G.W. Kattawar, E.P. Zege, I.L. Katsev, A.S. Prikhach, and L.I. Chaikovskaya, 2001: Monte Carlo and multicomponent approximation methods for vector radiative transfer by use of effective Mueller matrix calculations, *Appl. Opt.*, 40, 400–412.

von Hoyningen-HueneW., M. Freitag, and J.B. Burrows, 2003: Retrieval of aerosol optical thickness over land surfaces from top-of-atmosphere radiance, *Journal of Geophysical Research*, 108(D9), 4260, doi:10.1029/2001JD002018.

Zege E.P. and L.I. Chaikovskaya, 1996: Newapproach to the polarized radiative transfer problem, *Journal of Quantitative Spectroscopy and Radiative Transfer*, 55, 19–31.

Zege E.P., A.P. Ivanov, and I.L. Katsev, 1991: *Image Transfer through a Scattering Medium*, Springer-Verlag, Heidelberg.

Zege E.P., I.L. Katsev, and I.N. Polonsky, 1993: Multicomponent approach to light propagation in clouds and mists, *Appl. Opt.*, 32, 2803–2812.

5   Aerosol retrieval over land using the (A)ATSR dual-view algorithm

Lyana Curier, Gerrit de Leeuw, Pekka Kolmonen, Anu-Maija Sundström, Larisa Sogacheva, Yasmine Bennouna

1.   Introduction

Aerosols play an important role in climate and air quality. They have a direct effect on climate by scattering and/or absorbing the incoming solar radiation [Haywood and Boucher, 2000]. Reflection of solar radiation increases the atmospheric albedo, causing a negative radiative effect and therefore cooling of the atmosphere. On local scales absorbing aerosols can cause net positive radiative forcing resulting in warming of the atmospheric layer. Aerosols have an indirect effect on climate through their influence on cloud microphysical properties and, as a consequence, on cloud albedo and precipitation. The aerosol net effect on the Earth's radiative balance depends on the aerosol chemical and physical properties, the surface albedo and the altitude of the aerosol layer [Torres et al., 1998]. The uncertainty in the effect of aerosols on climate stems from the large variability of aerosol sources, i.e., their concentrations and physical, chemical and optical properties, in combination with their short atmospheric residence time of a few days. In the IPCC (2007) [Forster et al., 2007] assessment report the total direct aerosol radiative forcing as derived from models and observations is estimated to be -0.5 [± 0.4] Wm^{-2}, with a medium-low level of scientific understanding. The radiative forcing due to the effect on cloud albedo is estimated as -0.7 [$-1.1,+0.4$] Wm^{-2} with a low level of understanding. Long-lived greenhouse gases are estimated to contribute $+2.63$ [± 0.26] Wm^{-2}. Improved satellite measurements have contributed to the increase of the level of scientific understanding since the third IPCC assessment report in 2001. The continued improvement of aerosol retrieval from satellite-based instruments, to provide consistent information with a known level of accuracy, is important to further our understanding of climate and climate change.

Experimental data on aerosol properties are available from ground based *in situ* and remote sensing observations. However, often they are only representative for local situations and cannot be used for estimates of effects on regional to global scales. Furthermore, they are part of several networks which often are disconnected and the data are available in different formats, on different timescales and measured using different procedures and correction factors. Although efforts have been made to harmonize some of these datasets (e.g., GAW and EMEP), the data availability on a global or regional scale is still a concern. In addition, the area coverage is very sparse for most continents.

An alternative to provide aerosol properties on regional to global scales with a temporal resolution of one per day, is offered by sun-synchronous satellites [Forster et al., 2007]. Satellite data provide a consistent set of aerosol data on regional to global scales, determined with the same instrument and the same procedure but at the expense of accuracy and detail. Therefore, the use of satellites is complementary to ground-based networks but cannot replace them. In addition, satellite data are not always available, e.g., in the presence of clouds when retrieval of aerosol information is not possible, or over bright surfaces when most retrievals are currently not reliable.

Instruments used for aerosol measurements from space are often designed for other tasks such as the measurement of trace gases, surface temperature or ocean color. However, their characteristics make them also suitable for aerosol retrieval, e.g., AVHRR, TOMS, OMI, ATSR-2, AATSR, MERIS. In recent years, instruments dedicated to the remote sensing of aerosols and clouds have been developed such as MODIS, MISR and the POLDER series. Algorithms have been developed for over 25 years for application over sea and for a decade for application over land. However, the quality of the results is different for different instruments, although they are converging as shown by comparison exercises [e.g., Myhre et al., 2004, 2005; Kokhanovsky et al., 2007]. Kokhanovsky et al. [2009] evaluated results from the Advanced Along-Track Scanning Radiometer (AATSR) for the retrieval of aerosol properties over land. The major feature of the AATSR instruments is the dual-view. A scene is first observed at a forward angle of 55° and then, approximately 150 s later, at nadir. This dual-view capability is at the basis of the aerosol retrieval algorithms used with this instrument because it allows us to eliminate land surface effects from the total measured reflectance. More detail on this instrument and its use for aerosol retrieval is presented in this contribution.

2.   AATSR instrument

The Advanced Along-Track Scanning Radiometer (AATSR) instrument onboard the European ENVISAT satellite flies at an altitude of approximately 800 km in a sun-synchronous polar orbit. Like its predecessor ATSR-2 on ERS-2, the AATSR has seven wavelength bands in the visible and infrared parts of the spectrum (0.55, 0.67, 0.87, 1.6, 3.7, 11 and 12 μm). The instrument has a conical scanning mechanism providing two views of the same location with a resolution of 1×1 km^2 at nadir view. The radiometer views the surface at a forward angle of 55° and 150 s later at nadir. The swath width of 512 km results in an overpass over a given location at mid-latitudes once every three days at mid-latitudes.

3.   Aerosol retrieval

In the absence of clouds, the upwelling radiance at the top of the atmosphere can be decomposed into three main contributions:
- reflection by the surface
- molecular (Rayleigh) scattering
- aerosol scattering

All three components are affected by transmission and absorption between the height of scattering and the TOA. Absorption is due to molecular and aerosol effects. For aerosol retrieval, wavelengths are chosen where molecular absorption is minimal. In the visible and near-infrared domains, the Rayleigh contribution to the total radiance is relatively small as compared to wavelengths in the UV, but not negligible.

The surface contribution depends on the surface type. Over dark surfaces, such as vegetation in the UV–visible or the sea surface in the visible and near-infrared, the surface reflectance is very small. Over brighter surfaces the contribution is significant and of similar magnitude or larger than the aerosol contribution, thus the accurate retrieval of aerosol information requires either a precise characterization of the surface, or a method of eliminating the surface effect on the TOA reflectance.

The retrieval of aerosol information from satellite data is done in several steps. The first step is cloud screening. For clear areas the surface contribution is eliminated, to retain the atmospheric contributions to the TOA radiance. Contributions due to gaseous species can be well-estimated by the use of radiative transfer models, which leaves the aerosol contribution to be determined. These steps cannot usually be separated because of cross-terms. Therefore a radiative transfer model, including all processes, is used in combination with a retrieval code. Because radiative transfer models are computationally heavy, calculations are made for discrete situations (solar zenith angle, viewing zenith angle, relative azimuth angle, see Table 5.3) and results are stored in look-up tables (LUTs) which are used in the actual retrieval.

4.   Cloud screening

In the presence of clouds, the aerosol contribution to the TOA reflectance is completely masked by the cloud contribution. Therefore, aerosol properties can be retrieved accurately only over cloud-free areas. To this end, the semi-automatic algorithm to discriminate between cloudy and clear pixels from ATSR-2 data developed by Koelemeijer et al. [2001], based on the work of Saunders and Kriebel [1988], was implemented. It was fully automated to allow for the processing of large amounts of data, over large areas [Robles-Gonzalez, 2003]. The cloud detection procedure consists of four standard cloud detection tests which are applied to individual pixels and are classified as clear if and only if all tests indicate that no cloud is detected. These tests are based on the evaluation of histograms of either brightness temperature or reflectance. To determine whether a pixel is cloud-contaminated, thresholds are set which depend on the surface properties and the sun–satellite geometry. The thresholds are determined for each individual frame of 512×512 pixels, in two steps. First, a fast-fourier transform low-pass filter is applied to smooth the data. Then the extrema (maximum, minimum and inflexion points) of the smooth histogram are computed by applying a modified Lagrange interpolation, and from this set of extrema the thresholds for each test are determined.

The first test (T1) is the gross cloud test. It uses the 12 μm channel and makes use of the fact that the temperature of the atmosphere decreases with altitude. As a result, the temperature of an optically thick cloud at high altitude will be significantly lower than the surface temperature. If the measured brightness temperature is below the set threshold, the pixel is marked as cloudy.

The second test (T3) exploits the fact that clouds are brighter than the underlying surface and checks the reflectance at 0.67 μm for each pixel. Cloudy pixels will present a higher reflectance than non-cloudy pixels. For each frame, the reflectances are histogrammed and a threshold is selected to discriminate between cloudy and clear pixels.

The third test (T4) is based on the reflectance ratio at 0.87 μm and 0.67 μm. It is based on the difference between the spectral behavior of the surface and the cloud at these wavelengths. Cloud reflectances are similar in these channels whereas surface reflectances are generally different. Over land the surface reflectance at 0.85 μm tends to be higher than the one at 0.67 μm while over water the reflectance at 0.67 μm tends to be higher than the one at 0.87 μm. Therefore, three cases can be observed: a reflectance ratio of 1 which denotes the presence of clouds, a reflectance ratio lower than 1 characterizes water surfaces whereas land surfaces feature a ratio higher than one.

The fourth test (T5) is made to detect semi-transparent clouds such as cirrus for which the difference between the brightness temperatures at 12 μm and at 11 μm is large. The temperature difference is computed for each pixel and compared to a tabulated threshold value. When the difference exceeds the threshold, the pixel is flagged as cloudy. This last test must be used carefully and is not used for the analysis of large amounts of data.

Finally, after application of the retrieval algorithm, a spatial variability test can be applied as a post-processing step to remove extremes in a 10 × 10 pixel frame. This effectively takes care of enhanced aerosol optical depth (AOD) values in the vicinity of clouds [de Leeuw et al., 2007].

4.1 Cloud screening: example

The effectiveness of the cloud screening procedure outlined above was tested versus the semi-automatic procedures by Robles-Gonzalez [2003]. In this section we provide an example of a comparison with the MODIS cloud screening results.

Fig. 5.1 shows the cloud mask derived for August 10, 2004. The pixels flagged as clear by all tests are colored in blue, whereas all other pixels have been detected as cloudy by one or more tests. Table 5.1 provides a description of the different values of the cloud quality

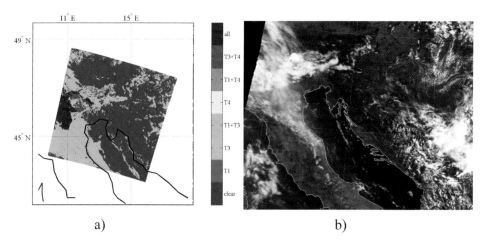

a)   b)

Fig. 5.1. (a) Composite map of the cloud mask derived by the ATSR cloud screening protocol applied to AATSR for a scene over Italy on August 10, 2004; the color code indicates clear sky conditions (dark blue) and different cloud tests (Tn, n – number of the cloud test) for each pixel. Pixels for which all tests indicate the presence of a cloud are dark red. Clear pixels, i.e. no test indicated the presence of clouds, are dark blue. (b) RGB picture derived from channels 1, 3 and 4 of MODIS for August 10, 2004.

Table 5.1. Cloud flag coding used for the AATSR cloud tests as described in the text and used to produce Fig. 5.1.

Flag number	Interpretation
clear	Clear
T1	Flag by test 1
T3	Flag by test 2
T1 + T3	Flag by tests 1 and 2
T4	Flag by test 3
T1 + T4	Flag by tests 1 and 3
T3 + T4	Flag by tests 2 and 3
all	Flag by all tests

flag. For comparison, an RGB picture from MODIS is shown in Fig. 5.1(b). Cloud patterns are similar in both images.

A quantitative comparison for the same scene is presented in Fig. 5.2, where the different color codes designate complete agreement (dark blue) and complete disagreement between the AATSR and MODIS cloud masks, and values in between. It is noted that the ATSR cloud screening protocol designates pixels as either cloudy or clear, whereas the MODIS cloud screening protocol flags pixels in four different ways, cloudy, probably cloudy, probably clear or clear. Thus, pixel information from AATSR and MODIS have been combined into an unsigned integer, such as the two first bits code for the MODIS cloud mask and the third one for the AATSR cloud mask. This is summarized in Table 5.2 which also includes the interpretation for each observed value in the combined cloudmask. The composite map in Fig. 5.2 was derived by meshing the AATSR-derived cloud-mask into the MODIS cloud-mask grid.

The histogram of the available pixels for this comparison in Fig. 5.2(b) shows that the cloud screening of the MODIS and AATSR algorithms is in total agreement for 83 % of the pixels. The cloud masks were in complete disagreement for 12 % of the pixels. For 5 % of the pixels the comparison is not conclusive as the pixels were flagged as probably clear or probably cloudy by MODIS.

It is noted that disagreements mainly occur near cloud edges and areas with low cloud cover. This reflects that the cloud macrostructure is often not well defined, due to humidity gradients where cloud condensation nuclei may be activated (cloud formation) or hygroscopic aerosol particles and cloud droplets adjust to the local relative humidity by taking up or releasing water vapor which in turn affects their size and thus optical properties. Because the atmosphere is turbulent, local fluctuations of temperature and dew-point temperature occur, i.e. relative humidity fluctuations resulting in local fluctuations in the aerosol size distribution, extinction and AOD. (Note that the response time of accumulation mode aerosol particles and small cloud droplets is fast enough to follow these fluctuations [Andreas, 1992].) As a result, on a pixel-to-pixel base the AOD may strongly fluctuate in the vicinity of clouds. By averaging over a larger area (e.g., 10 × 10 pixels) pixels) and removing pixels deviating from the average by more than a preset fraction, the cloud influence can be strongly reduced [de Leeuw et al., 2007].

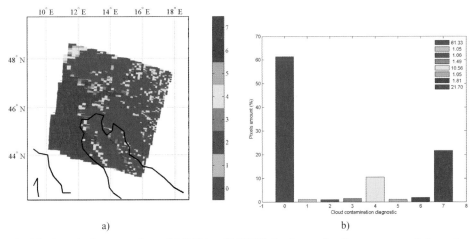

a)   b)

Fig. 5.2. Quantitative comparison of AATSR and MODIS cloud screening for a scene over Italy on August 10, 2004. (a) Composite map of the combined cloud-mask. (b) Histogram of the number of pixels versus the cloud-mask flag. Dark blue (0) and red (7) represent pixels when AATSR and MODIS diagnostics agree. Orange and yellow represent pixels when AATSR and MODIS disagree. For more detailed explanation of the colors, see column 'Combined' in Table 5.2.

Table 5.2. Flag description of the combined cloud mask, Fig. 5.2.

MODIS	AATSR	Combined	Interpretation
0	0	0	clear for AATSR and MODIS
1	0	1	probably clear for MODIS, clear for AATSR
2	0	2	probably cloudy for MODIS, clear for AATSR
3	0	3	cloudy for MODIS, clear for AATSR
0	1	4	clear for MODIS, cloudy for AATSR
1	1	5	probably clear for MODIS, cloudy for AATSR
2	1	6	probably cloudy for MODIS, cloudy for AATSR
3	1	7	cloudy for AATSR and MODIS

5. Inversion model

The AATSR aerosol retrieval algorithm has its inheritance in algorithms developed for ATSR-2 for application over ocean (single view) [Veefkind and de Leeuw, 1998] and over land (dual view) [Veefkind et al., 1998]. Both algorithms include multiple scattering and the bi-directional reflectance of the surface. The ATSR-2 algorithms were successfully applied over the east coast of the USA [Veefkind et al., 1998] and the North Atlantic [Veefkind et al., 1999], over Europe [Robles-Gonzalez et al., 2000; Veefkind et al., 2000], over the South Asian continent and the Indian Ocean [Robles-Gonzalez et al., 2006] over Africa [Robles-Gonzalez and de Leeuw, 2008] and over the desert of the UAE [de Leeuw et al., 2005].

The single and dual view algorithms were coupled for the semi-operational processing of large amounts of ATSR-2 data, with good results over Europe for the year 2000. However, the application to AATSR for Europe for the year 2003 appeared less successful, as described in de Leeuw et al. [2007], where also some improvements were proposed. Simultaneously a heavier research version of the model was developed which is described here. It is referred to as the ATSR-DV algorithm and was used in, for example, Kokhanovsky et al. [2008] and Thomas et al. [2007]. Major upgrades include the aerosol models and the number of AOD levels considered, which is reflected in the structure of the LUTs, in addition to some changes in the decision tree. The ATSR-DV algorithm was developed at TNO (Netherlands Organization for Applied Scientific Research, The Hague, The Netherlands)) and in early 2007 it was transferred to the University of Helsinki (UHEL) and the Finnish Meteorological Institute (FMI) in Helsinki, Finland, where it is further developed and applied for scientific studies.

A schematic description of the algorithm is shown in Fig. 5.3. The algorithm comprises two main parts: an inversion part where the actual retrieval is done and a forward part where all necessary *a priori* data are defined and/or computed using a radiative transfer model.

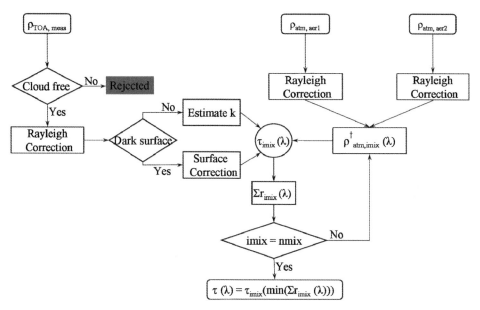

Fig. 5.3. Schematic representation of the ATSR-DV aerosol retrieval algorithm. For a dark surface the single-view algorithm is used which is not considered here since this paper focuses on retrieval over land. The algorithm uses the atmospheric reflectance at TOA measured by AATSR ($\rho_{TOA,\ meas}$) which is compared with the calculated reflection at TOA due to a mixture of two aerosol models ($\rho_{atm,\ aeri}$) for which the reflection is pre-calculated and stored in LUTs. To account for interactions, the TOA reflection is calculated including molecular effects, which in turn is corrected for in the retrieval algorithm before the modeled reflections are mixed ($\rho^{\dagger}_{atm,\ imix}(\lambda)$). The measured reflection for cloud free pixels is also Rayleigh corrected and subsequently a surface correction is applied. The remaining measured and calculated reflections are compared in an iterative loop to determine the optimum mixture *imix*, using all available wavelengths to fit the most appropriate AOD, by minimizing the error function $\Sigma r_{imix}(\lambda)$. Note that the LUT contains a range of AODs for each of the two aerosol models and therefore both are determined in the same step. The aerosol components are selected for the area of interest based on *a priori* knowledge from climatology or other information. The final result is the AOD for the most appropriate aerosol mixture $\tau(\lambda)$, for individual AATSR pixels (nominal is 1×1 km^2).

5.1 Theory

Commonly the reflectance ρ is used in retrieval algorithms, which can be derived from the satellite measured radiance L assuming that the surface acts as a Lambertian reflector and the atmosphere is horizontally uniform using:

$$\rho = \frac{\pi L}{\cos(\theta_0)F_0}, \tag{1}$$

where F_0 is the extraterrestrial solar irradiance and θ_0 is the solar zenith angle.

The reflectance in the visible and near-infrared (NIR) domains measured by AATSR at TOA is used to retrieve aerosol optical properties. To determine the aerosol contribution, $\rho_{aer,meas}(\lambda)$, the measured TOA reflectance at wavelength λ is corrected for the contribution of a Rayleigh atmosphere, $\rho_0(\lambda)$,

$$\rho_{aer,meas}(\lambda) = \rho_{atm}(\lambda) - \rho_0(\lambda). \tag{2}$$

The retrieval procedure is based on two main assumptions:

1. The TOA reflectance due to an external mixture of two aerosol types $\rho_{aer,imix}(\tau_{ref}, \lambda)$ can be approximated as the weighted average of the reflectance TOA of each of the aerosol types [Wang and Gordon, 1994]:

$$\rho_{aer,imix}(\tau_{ref}, \lambda) = \rho_{aer,meas}(\lambda) + \varepsilon_{imix}(\lambda) = n_{aer1}\rho_{aer1}(\tau_{ref}, \lambda) + n_{aer2}\rho_{aer2}(\tau_{ref}, \lambda) \tag{3}$$

where $\rho_{aer1}(\lambda)$ and $\rho_{aer2}(\lambda)$ are the modeled TOA reflectances due to the presence of each aerosol type. n_{aer1} and n_{aer2} are the contributions of each type ($n_{aer1} + n_{aer2} = 1$) and are determined by minimization in the ATSR-DV algorithm. $\varepsilon_{imix}(\lambda)$ accounts for the deviation between the measured, $\rho_{aer,meas}(\lambda)$, and the weighted average, $\rho_{aer,imix}(\lambda)$. Abdou et al. [1997] modified this equation to account for mutual interactions between the aerosol components in the mixture.

2. The reflectance at TOA can be expressed as a linear function of the aerosol optical depth with the reflectance of an aerosol-free atmosphere as an offset [Durkee et al., 1986]:

$$\rho_{atm}(\lambda) = \rho_0(\lambda) + c(\lambda)\tau(\lambda), \tag{4}$$

where $\rho_{atm}(\lambda)$ is the path reflectance due to the presence of aerosols and molecules and $\tau(\lambda)$ is the aerosol optical depth. Veefkind [1999] demonstrated that Eq. (4) is a valid assumption for the region of aerosol optical depths characterizing tropospheric aerosols, i.e. $0.1 < \tau < 0.6$ (see Fig. 5.4). For higher aerosol loads the intercept ρ_0 does not solely represent Rayleigh scattering and is influenced by aerosol particles. However, in practice, TOA reflectances are calculated for discrete AOD values using the LUTs which allows for interpolation assuming linear dependence over smaller segments. It is further noted that the forward model takes multiple scattering into account.

The inversion is based on the comparison between modeled and measured TOA reflectances. To this end, aerosol and molecular contributions are separated by a Rayleigh correction and also ozone absorption is accounted for. Multiple land/atmosphere scattering is estimated assuming that the surface reflectance behaves as a Lambertian reflector. The

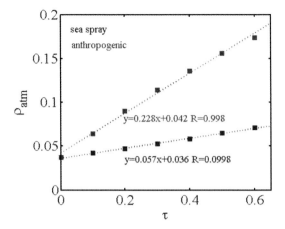

Fig. 5.4. Path reflectance (ρ_{atm}) at 0.55 μm as a function of aerosol optical depth (AOD) for anthropogenic and sea spray aerosol types as defined by WMO [1983] and a sun–satellite geometry defined by: $\Theta_0 = 45°$, $\Theta = 30°$, and $\varphi - \varphi_0 = 90°$. The lines are linear fits through the points, the correlation coefficients are larger than 0.99 for both fits.

inversion procedure can use either a least squares minimization method or a least median square method to fit the modeled reflectance to the satellite measurements. The error function $\Sigma_l \varepsilon_{imix}^2(\lambda_l)$ is minimized to determine the best fit for the mixing ratio:

$$\sum_l \varepsilon_{imix}^2(\lambda_l) = \sum_{l=1}^{3} \left(\frac{\rho_{aer,meas}(\lambda_l) - \rho_{aer,imix}(\tau_{ref}, \lambda_l)}{\rho_{aer,meas}(\lambda_l)} \right)^2, \tag{5}$$

where λ_{13} are the available AATSR wavelengths (0.55, 0.67 and 1.6 μm which are used over land). In cloud-free situations the measured TOA reflectance contains information about both the surface and the atmosphere. For dark surfaces, the surface contribution to the TOA reflectance is computed by reconstructing the bi-directional reflectance of the surface and the aerosol optical depth can be directly determined. Over brighter surfaces, the contributions of the surface and atmospheric reflections to the TOA reflectance need to be separated. In the dual-view algorithm described here, the surface effect is eliminated by taking advantage of the two views of the Along-Track Scanning Radiometer.

5.2 Dual-view algorithm

The dual-view algorithm was developed by Veefkind et al. [1998] for ATSR-2 for aerosol retrieval over land. In the single-view algorithm, used over water, the surface reflectance is assumed isotropic, therefore the difference observed in measured TOA radiance between the two views can easily be explained by a difference in the atmospheric path length. However, over land, the surface reflectance is usually not isotropic and the surface bi-directional reflectance distribution (BRDF) affects the interactions between the surface and the atmosphere. It is, therefore, necessary to use a more complex model to retrieve the atmospheric contribution to the TOA radiance.

Over land the contribution of the surface reflection in the visible to near-infrared domain is mainly driven by the direct contribution, i.e. photons are reflected at the surface and transmitted through the atmosphere. On a pixel-by-pixel approach, we assume that the surface can be approximated as a Lambertian reflector. The TOA reflectance, ρ, for an underlying Lambertian surface, can be written as [Chandrasekhar 1960]:

$$\rho(\lambda) = \rho_{atm}(\lambda) + \frac{\rho_{sfc}(\lambda)}{1 - \rho_{sfc}(\lambda)s(\lambda)} T(\lambda), \tag{6}$$

where $\rho_{atm}(\lambda)$ is the atmospheric path reflectance, $T(\lambda)$ is the atmospheric total transmittance (i.e. the product of the downward and upward transmissions, where upward is due to reflection by the surface), $s(\lambda)$ is the spherical albedo of the atmosphere. The surface reflectance ρ_{sfc} depends both on the wavelength and on the geometry. Flowerdew and Haigh [1995] point out that the angular variation of the surface reflectance is due to the macroscopic structure of the underlying surface, which is of a much larger scale than the wavelength of the incident light. Hence the surface reflectance is a strong function of the wavelength while its shape is in comparison independent of the wavelength. Therefore, in the dual-view algorithm, it is assumed that the ratio between the surface reflectance in the nadir view and the surface reflectance in the forward view, k, is independent of the wavelength:

$$k_\lambda = \frac{\rho_{sfc,f}(\lambda)}{\rho_{sfc,n}(\lambda)} \approx k, \tag{7}$$

where the subscripts f and n represent respectively the forward and the nadir view. For $k = 1$, this approximation corresponds to a Lambertian surface. Using Eqs. (6) and (7) and also an assumption that $\rho_{surf,n}(\lambda)s(\lambda) << 1$, the TOA reflectances in the forward, $\rho_f(\lambda)$, and nadir, $\rho_n(\lambda)$, views, can be written as:

$$\rho_f(\lambda) = \rho_{atm,f}(\lambda) + k\rho_{sfc,n}(\lambda)T_f(\lambda), \tag{8}$$

$$\rho_n(\lambda) = \rho_{atm,n}(\lambda) + \rho_{sfc,n}(\lambda)T_n(\lambda). \tag{9}$$

Using Eqs. (8) and (9) we can eliminate the unknown surface albedo. Then it follows:

$$\frac{\rho_f(\lambda) - \rho_{atm,f}(\lambda)}{kT_f(\lambda)} = \frac{\rho_n(\lambda) - \rho_{atm,n}(\lambda)}{T_n(\lambda)} \tag{10}$$

Assuming that Eq. (4) applies to the path reflectance, $\rho_{atm,f}$, the aerosol optical depth, $\tau(\tau_{ref}, \lambda)$, can be derived using an iterative procedure:

$$\tau^i(\tau_{ref}, \lambda) = \frac{1}{\zeta^i(\tau_{ref}, \lambda)} \left[\frac{\rho_n(\lambda) - \rho_{o,n}(\tau_{ref}, \lambda)}{T_n^{i-1}(\tau_{ref}, \lambda)} - \frac{\rho_f(\lambda) - \rho_{o,f}(\tau_{ref}, \lambda)}{k^i(\tau_{ref})T_f^{i-1}(\tau_{ref}, \lambda)} \right], \tag{11}$$

where

$$\zeta^i(\tau_{ref}, \lambda) = \frac{c_n(\tau_{ref}, \lambda)}{T_n^{i-1}(\tau_{ref}, \lambda)} - \frac{c_f(\tau_{ref}, \lambda)}{k^i(\tau_{ref})T_f^{i-1}(\tau_{ref}, \lambda)}. \tag{12}$$

For most continental aerosol types the aerosol extinction decreases rapidly with wavelength thus the aerosol optical depth at 1.6 μm is small compared to the aerosol optical depth in the visible.

The ratio k^i for the ith iterative step is computed from the 1.6 μm channel:

$$k^i = \frac{\rho_{f,meas}(1.6\ \mu m) - \rho_{atm,f}^{i-1}(1.6\ \mu m)}{\rho_{n,meas}(1.6\ \mu m) - \rho_{atm,n}^{i-1}(1.6\ \mu m)} \tag{13}$$

The iterative procedure is initiated by ignoring the atmospheric contribution at 1.6 μm: $\rho_{f,meas}(\lambda = 1.6\ \mu m) \approx \rho_{sfc,f}(\lambda = 1.6\ \mu m)$. The k-approximation does not apply to the 0.87 μm channel because of the strong reflection by vegetation at this wavelength [Robles-Gonzalez et al., 2000].

The spectral aerosol optical depth, $\tau^i(\tau_{ref}, \lambda)$, is computed using mixtures of two aerosol types. The best aerosol optical depth for the 0.55, 0.67, and 1.6 μm channels and subsequently the best mixing ratio is determined by applying a least squares minimization as explained in Section 5.1.

6.   Forward model

The determination of aerosol optical properties from the measured TOA reflectance is an ill-posed problem because there is insufficient information to constrain for possible solutions. Therefore, radiative transfer calculations using DAK (double adding, KNMI; KNMI is the Royal Netherlands Meteorological Institute where the current version was developed [de Haan et al., 1987]) are performed to derive the Rayleigh and surface contributions, and *a priori* assumptions are made on the aerosol properties. This procedure requires aerosol models describing the aerosol microphysical and optical properties, determined by the chemical composition, to calculate the reflectance at the top of the atmosphere (TOA) using a radiative transfer model.

Reflectances and transmissions for Rayleigh atmospheres and atmospheres containing both gases and aerosols were calculated for a set of wavelengths, aerosol concentrations and geometries (solar zenith angle, viewing zenith angle, relative azimuth angles). Table 5.3 presents the amount of tiepoints used for each dimension. The results are stored in look-up tables (LUT), these tables contain nine calculated variables, and an overview of the variables stored in the LUT as well as their dimensions is given in Table 5.3. By using LUTs, the retrieval algorithm can make full use of the capabilities of DAK to account for bi-directional reflectance and multiple scattering, while the algorithm computational speed is not affected.

DAK makes use of the doubling adding method, a two-step process. Calculations start with a very thin atmospheric layer in which only single-scattering occurs. An identical layer is then introduced, the optical properties of the combined layer are calculated and internal scattering is included. This step is the 'doubling' one. The doubling step will be repeated until the layer has reached the required depth. Multiple scattering is thus taken into account. The 'adding' step is a similar mechanism designed to handle layers with different optical properties. These layers are combined in a single one which will be added to a third layer and so on. The 'adding' is assumed completed when the boundary

Table 5.3. Overview of the layout of the different LUTs. Each variable represents an axis of the multi-dimensional grid. The different parameters contained in each LUT can be found in Table 5.4.

Variable name	Symbol	Units	Number	Entries
Wavelength	λ	nm	4	555, 659, 865, 1600
Surface pressure	P_{sfc}	hPa	1	1013
Cos(solar zenith angle)	μ_0		15	0.15, 0.2, 0.25, ..., 0.65, 0.7, 0.8, 0.9, 1
Cos(viewing zenith angle)	μ		15	0.15, 0.2, 0.25, ..., 0.65, 0.7, 0.8, 0.9, 1
Relative azimuth angle	$\Delta\varphi$	degree	19	0, 10, 20, ..., 180
Aerosol optical depth at 500 nm	τ_{500}	–	10	0.05, 0.1, 0.25, 0.5, 1, 1.5, 2, 2.5, 3, 4

Table 5.4. Variables and dimensions for the data stored in the different LUTs.

Variable name	Symbol	Dimension
Path reflectance	ρ_{atm}	$p_s, \lambda, \mu_o, \mu, \Delta\varphi, \tau_{500}$
Surface downward reflectance	ρ_d	$p_s, \lambda, \mu_o, \mu, \Delta\varphi, \tau_{500}$
Total transmittance	T	$p_s, \lambda, \mu_o, \mu, \tau_{500}$
Total diffuse transmittance	t	$p_s, \lambda, \mu_o, \mu, \tau_{500}$
Diffuse transmittance down	$t_\downarrow$	$p_s, \lambda, \mu_o, \mu, \tau_{500}$
Total transmittance down	T_{tot}	$p_s, \lambda, \mu_o, \mu, \tau_{500}$
Spherical albedo	s	p_s, λ, τ_{500}
Aerosol optical depth	τ	λ, τ_{500}
Single scattering albedo	ω_o	λ

fluxes of the modeled layers converge. In this study, each input parameter was set to achieve a four or five decimal accuracy.

7. Aerosol description

In this section, the microphysics and chemical properties of the aerosols used for the LUTs are described. The aerosol particles are assumed spherical, which allows for the application of a Mie code [Mie, 1908] to compute the optical properties. This assumption is justified because most atmospheric particles are to some degree hygroscopic and thus contain water vapor when the relative humidity is above a certain threshold value (deliquescence point). Due to hysteresis, once the particles have absorbed water vapor, they remain wet unless the relative humidity (RH) drops below a certain value where crystallization occurs. For common atmospheric aerosol particles the crystallization point occurs at very low RH (for further discussion cf. Seinfeld and Pandis [1998] and references cited therein). Hence it is assumed that most atmospheric aerosol particles are in solution. Dust particles are an exception and should be treated as non-spherical particles.

The aerosol number concentrations N_i are distributed over the aerosol particle radii r_i following an n-mode lognormal distribution:

$$\frac{\mathrm{d}N}{\mathrm{d}\ln r} = \sum_{i=1}^{n} \frac{N_i}{(2\pi)^{1/2}\ln\sigma_i} \exp\left(-\frac{(\ln r_i - \ln \bar{r}_{gi})^2}{2\ln^2 \sigma_i}\right), \tag{14}$$

where $\bar{r}_{gi}$ is the geometric mean radius and σ_i is the geometric standard deviation of the ith lognormal mode. In the initial versions of the ATSR-2 retrieval code, the aerosol model used was the Naval Oceanic Vertical Aerosol Model (NOVAM) [de Leeuw et al., 1989] which uses the Navy Aerosol Model (NAM) [Gathman, 1983] as kernel. In this model, the aerosol is described as mixture of sea salt aerosol, water-soluble anthropogenic particles and water-insoluble particles (dust). It obviously works well over the ocean, but also over the western USA and over Europe the anthropogenic aerosol provides an adequate description (e.g., Robles-Gonzalez et al. [2000]). However, for INDOEX (INDian Ocean EXperiment) several other aerosol models were tested, such as the OPAC (Optical Properties of Aerosols and Clouds) database [Hess et al., 1998] and models derived from AERONET [Dubovik et al., 2002]. The latter provided satisfactory results [Robles-Gonzalez et al., 2006]. Over Africa mixtures of more absorbing and less absorbing aerosol were used with good results [Robles-Gonzalez and de Leeuw, 2008]. This shows that aerosol models need to be selected that are most appropriate for the region for which the retrieval is made, possibly also as a function of season. The degree of detail may depend on the application, i.e., for global retrieval obviously less detail can be provided than for scientific studies for which the aerosol models can be tuned to provide optimum results.

With this is mind, look-up tables were computed for four aerosol models, using field campaign measurements and/or literature data:

1. *Continental background*. This model represents the aerosol background in industrial or urban areas, and has a trimodal lognormal distribution. The chemical composition is defined following the work of Heintzenberg [1989] distinguishing between water-soluble inorganic, non-water-soluble inorganic and organic fractions. The complex refractive index is derived from the chemical composition using Maxwell–Garnet mixing rules [Chylek et al., 2000] and the complex refractive index of each main fraction is given by Hess et al. [1998].
2. *Anthropogenic particles*. Monomodal lognormal distribution representing the aerosol particles from anthropogenic origin [Veefkind, 1999].
3. *Sea spray*. Bimodal lognormal distribution [de Leeuw et al., 1989].
4. *Biomass burning*. Aerosol particles originating from biomass burning are represented by this model. The size distribution is represented by a trimodal lognormal distribution [Quinn et al., 2002; Anderson et al., 1996]. The complex refractive index is defined following Haywood et al. [2003].

Detailed information on the aerosol models, such as geometric mean radii, geometric standard deviations, complex refractive indices and the mode fractions are presented in Table 5.5.

Table 5.5. Physical and optical properties of the aerosols models used to build the LUT. The complex refractive index is given for $\lambda = 555$ nm.

Name	Mode	$\overline{r_g}\ \mu m$	$\ln \sigma_g$	fraction	n_r	n_i
Continental	m1	0.010	0.161	0.4	1.44	0.0039
background	m2	0.05	0.217	72	1.44	0.0039
	m3	0.900	0.380	28	1.44	0.0039
Biomass	m1	0.085	0.161	85	1.54	0.0180
burning	m2	0.395	0.217	3	1.54	0.0180
	m3	1.200	0.380	12	1.53	0.0180
Sea spray	m1	0.070	0.460	25	1.38	5.38×10^{-9}
	m2	0.389	0.370	75	1.38	5.38×10^{-9}
Anthropogenic	m1	0.030	0.139	100	1.41	0.00241

8.   Results and evaluation

Fig. 5.5 shows a composite map of the aerosol optical depth at 555 nm over Germany, for October 13, 2005. The spatial resolution is 1×1 km^2 at nadir. The aerosol optical depth derived is low and ranges from 0.07 to 0.23, with a mean aerosol optical depth of 0.15. The observed aerosol optical depth pattern was retrieved using a mixture of continental background aerosol, aerosol originating from human activities and naturally generated aerosol such as wind-blown dust and sea spray. Some higher AOD values are observed in or near cloud areas, likely due to residual cloud contamination, illumination of aerosol from a cloud side (3D effects), or due to enhanced relative humidity in these areas causing swel-

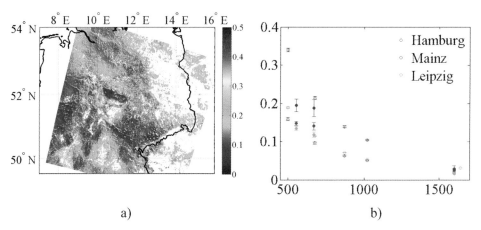

a)   b)

Fig. 5.5. (a) Composite map of the aerosol optical depth derived from AATSR measurements at 0.55 μm over Germany for October 13, 2005, with a 1×1 km^2 resolution. (b) The mean aerosol optical depth retrieved using the dual-view algorithm (filled circles) and from sunphotometer data (open circles) for selected AERONET stations as a function of wavelength (nm). Error bars indicate the spatial standard deviation for the mean aerosol optical depth retrieved within a radius of 2 km around the AERONET site and the temporal standard deviation of the mean aerosol optical depth measured by ground-based sunphotometers [Kokhanovsky et al., 2008].

ling and thus elevated extinction. Statistical methods are tested to remove such pixels or flag them as suspect of cloud influences.

Fig. 5.5(b) presents a comparison between the mean aerosol optical depths retrieved (filled circles) and their measured counterparts (open circles) for Hamburg, Leipzig and Mainz. The results compare favorably with the AERONET measured aerosol optical depth and are within the aerosol optical depth uncertainty of 0.05 [Robles-Gonzales et al., 2006], for ATSR-2.

8.1   Comparison with AARDVARC algorithm: effect of *k*-approximation

The AOD over Germany was retrieved by available instruments and intercomparisons were presented by Kokhanovsky et al. [2007, 2009]. The inter-comparison included the ATSR-DV and Swansea University (UK) AARDVARC (Atmospheric Aerosols Retrieval using Dual_View Angle Reflectance Channels) [Grey et al., 2006] algorithms, which correlate quite well with most data points within the ATSR-DV standard deviation, but also with quite some scattering between individual data points [Kokhanovsky et al., 2009]. The good correlation was to be expected because the same data were used and both algorithms use the dual view. Observed differences in the aerosol optical depth are therefore solely in-duced by differences in the processing schemes. First, the algorithms have independent cloud screening and sampling protocols. Thus, different radiances are ingested by the al-gorithms depending on masking and pixel aggregation. Second, the parameterizations of aerosol chemical and optical properties used to constrain the ill-posed inverse problem contain different assumptions. Third, the atmospheric radiative transfer models used are different, the Swansea University AARDVARC algorithm uses the Second Simulation of the Satellite Signature in the Solar Spectrum (6S) [Vermote et al., 1997] whereas the ATSR-DV algorithm uses DAK [de Haan et al., 1987]. Fourth, the sampling of the look-up tables of atmospheric parameters influences the aerosol retrievals. Finally, the *a priori* constraint used to account for the surface reflectance is not the same. A study by North et al. [1999] indicates that the *k*-approximation constraint (Eq. (7)) does not strictly apply in all cases. In AARDVARC a model is applied which separates the angular effects of the surface into a structural component that depends only on the viewing direction and a spec-tral component that depends only on the wavelength [Grey et al., 2006].

Thus, the variation between the algorithms themselves and the assumptions upon which they are based will result in different estimates of the surface and atmospheric reflectances.

8.2   Comparison with MODIS (Terra)

AATSR on board of ENVISAT and MODIS on board of Terra, overpass the same area within 30 minutes, thus, an exact temporal match is not possible. However, the timespan is small enough that large changes in the aerosol optical depth are unlikely between the two overpasses. Fig. 5.6 presents a comparison with the aerosol optical depth derived using MODIS (Terra, Collection 5). In order to compare the aerosol optical depth, the ATSR-DV retrieved AOD was meshed within the MODIS grid (10×10 km^2). The over-lapping area contains information for 1,283 pixels. The datasets compare favorably, with a homogeneous distribution of the differences. The ATSR-DV algorithm appears to overes-timate the aerosol optical depth in about 75 % of the pixels with respect to MODIS, i.e. the

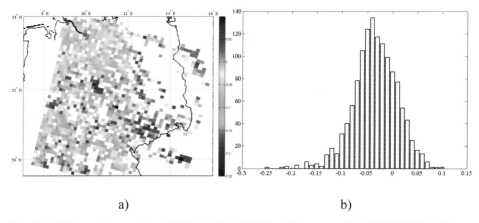

a)   b)

Fig. 5.6. Difference between the AOD derived by MODIS and by means of the ATSR-DV algorithm, at 0.55 μm, over Germany on October 13, 2005. (a) Spatial distribution of the difference (MODIS-AATSR). (b) Histogram of the same distributions.

AATSR values are on average 0.04 high, as derived from the histogram in Fig. 5.6(b). For 69 % of the pixels, the absolute value of difference is within the AATSR standard deviation of 0.05 [Robles-Gonzales et al., 2006].

8.3 Aerosol over Po Valley

Fig. 5.7 shows a composite map of the AOD over the Po valley for September 4, 2004. Strong aerosol optical depth gradients are observed with values varying from about 0.1 near the Alps to about 0.5 in the central Po valley. This variability is caused by the presence of the Alps which form a natural barrier for transport, i.e., aerosols produced in industrialized areas such as Milan and Turin, can only be ventilated through the Po valley to the Adriatic Sea. As observed in the validation study, over Germany for October 13, 2005, an unreasonable increase of the aerosol optical depths occurs at the edges of areas flagged as cloudy. The eastern edge of the composite map shows some unrealistic pattern.

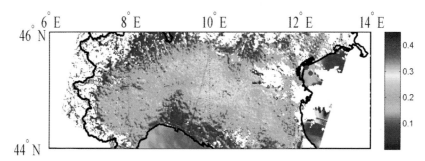

Fig. 5.7. Composite map of the aerosol optical depth derived by means of the ATSR-DV algorithm, over the Po valley, northern Italy, for September 4, 2004. White areas indicate that no data were computed due to cloud occurrences. Red crosses mark the Ispra (45.7°N, 8.6°S), Nicelli (45.4°N, 12.4°S) and Venice (45.3°N, 12.5°S) AERONET sites.

Table 5.6. Direct comparison of aerosol optical depth derived at 0.55 μm for September 4, 2004. The aerosol optical depths retrieved by means of the ATSR-DV algorithm were averaged over 5 km around the ground measurement sites. The AERONET values are averaged over ± 30 minutes of AATSR overpass. The closest MODIS value available within 10 km was used.

Site	AERONET	ATSR-DV	MODIS
Ispra	0.10 ± 0.04	0.26 ± 0.01	0.12
Venice	0.18 ± 0.02	0.15 ± 0.01	–
Nicelli	0.16 ± 0.02	0.17 ± 0.01	0.26

The source of these artifacts, once flagged as sunglint contamination, has not been found. These artifacts seem to be a peculiar feature of this specific scene, as similar artifacts have not been encountered in other processed scenes.

The land–sea transition over the coastline of the Adriatic Sea is smooth and shows little evidence of a sudden change in the aerosol burden. On the other hand, the land–sea transition observed over the coastline of the Ligurian Sea shows a discontinuity with a severe increase of the aerosol optical depth.

Table 5.6 shows a direct comparison with the measured aerosol optical depth for three AERONET ground measurement sites, Ispra (45.7°N, 8.6°S), Nicelli (45.4°N, 12.4°S) and Venice (45.3°N, 12.5°S). The AERONET averaged aerosol optical depths within ±30 minutes of an AATSR overpass are compared to derived aerosol optical depths averaged over 5 km. The average aerosol optical depth derived for Ispra is overestimated by a factor of 2.6. Fig. 5.7 shows that the AOD derived in the area around the site was quite heterogeneous. By reducing the averaging area, and considering only values within 1 km of the ground site measurement, the averaged aerosol optical depth retrieved over the Ispra site is 0.144 ± 0.006, i.e. within experimental error. The averaged aerosol optical depths derived for Nicelli and Venice compare well with the AERONET averaged aerosol optical depths and are within the standard deviation of the AERONET measurements.

8.3 Smoke plumes over Spain

In the previous version of the ATSR-DV algorithm, smoke plumes over the Iberian Peninsula appeared to go largely undetected, resulting in relatively low AOD in the summer of 2003, in spite of the frequent occurrence of wildfires [de Leeuw et al., 2007]. The previous version was designed for semi-operational use and therefore some simplifications had been made to optimize the computing time. Due to the unsuitable and simplified aerosol model, the previous dual-view algorithm was unable to produce solutions for extreme aerosol loads, e.g., smoke plumes, over land. However, over ocean the plumes could be detected well because of the different kind of inversion procedure (single view). This is illustrated with an example for August 11, 2003, where a vast smoke plume over land can be seen in Fig. 5.8(a) over NW Portugal while another plume is extending over the ocean in SW Portugal. As Fig. 5.8(b) shows, the smoke plume over the ocean is clearly visible in the retrieved AOD map whereas over land the algorithm is unable to produce any result for the smoke (white areas in Fig. 5.8(b) designate either cloud or missing (no solution) AOD). Fig. 5.8(c) shows a closer view of the smoke plume advected out

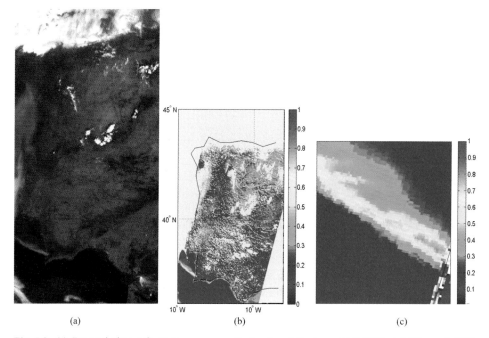

(a)   (b)   (c)

Fig. 5.8. (a) Forward view reflectance map over Portugal and Spain on 11.8.2003 at 0.55 μm. In NW Portugal a large smoke plume appears over land, another plume is extending over the ocean in SW Portugal. (b) Composite map of AOD at 0.55 μm on 11.8.2003 retrieved with the 'old' (TNO semi-operational) algorithm. The wildfire plume over the ocean is clearly visible whereas over land the code was unable to produce any result for the smoke plume. (c) Enlarged view of the retrieved smoke plume over the ocean in SW Portugal.

over the ocean. The structure of the smoke plume is clearly visible with AOD increasing up to about 0.8 towards the source.

The situation described above was not satisfactory and therefore the algorithm was further upgraded after its transfer and installation at the Finnish Meteorological Institute (FMI) and the University of Helsinki (UHEL). The results presented below were obtained with the FMI/UHEL version of the algorithm which was further debugged and tested for a variety of situations with either high (smoke, pollution) or very low (rural remote areas) AOD, and different aerosol components.

The first example is smoke detection over land, following up on the results shown in Figs 5.8(b) and (c). Fig. 5.9 shows the same scene as in Fig. 5.8(b), processed with the upgraded code which clearly shows the smoke plume over land in the western part of Portugal. In addition, the upgraded code shows more structure in the AOD, in particular over areas where higher AOD was expected, such as in the north of Spain. Furthermore, we see a smooth transition between AOD over land and over ocean at the south coast of Spain, mainly due to more reliable retrieval over land surfaces, as indicated by the enhanced AOD (compare Figs 5.8(b) and 5.9) which compares favorably with AERONET data as well (not shown). It is noted that over sea the AOD looks smoother than over land where the AOD looks patchier due to more areas where the retrieval did not provide a solution.

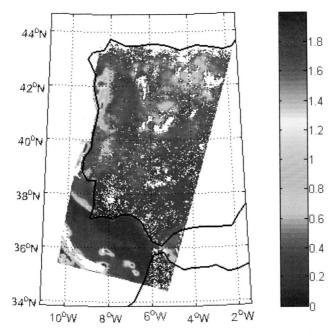

Fig. 5.9. Composite map of AOD at 0.55 μm on 11.8.2003 over Portugal and Spain retrieved with the upgraded ATSR-DV algorithm. The improved algorithm makes it possible to detect smoke over land too.

8.4   AOD over Finland: clean air

Finland is commonly very clean, with predominantly only natural aerosol sources, and hence the AOD is usually very low (less then 0.1). The retrieval of AOD in such cases is a greater challenge than over more polluted areas for which examples were shown earlier in this chapter and in the literature. However, pollution plumes, which are usually transported to Finland from Central and Eastern Europe, when AOD increases to 0.4–0.6, are occasionally observed.

Because an AERONET AOD network was not available in Finland until 2008, when three sites were established, the retrieved AOD is compared with AOD measurements using the PFR (Precision Filter Radiometer) data from the WMO-GAW stations in Sodan-kylä (67.22°N; 26.37°E) and Jokioinen (60.8°N, 23.5°E).

Fig. 5.10 shows an example of the AOD retrieved over Finland on 13 June, 2006. The cloud band over the Barents and Norwegian Seas is taken out by using the cloud mask procedure. Isolated clouds over Kola peninsula and northern Finland are either deleted from the calculations or retrieved as a high AOD, showing the cloud's edge.

Most of the remaining area over Finland was free from clouds during that particular day. The average AOD is about 0.05; higher (up to 0.1) values are observed close to the cloud edges in the north. Lower AOD over the area between isolated clouds can be related to advection of clean Arctic air.

Fig. 5.11 shows a comparison of AOD retrieved from AATSR data with AOD measured using a Precision Filter Radiometer (PFR) in Sodankylä, part of the WMO-GAW network.

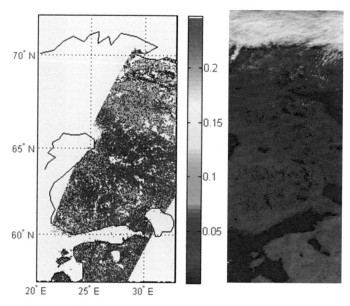

Fig. 5.10. Composite map of AOD (a) derived from AATSR measurement at 0.55 μm, and satellite image (b) over Finland for 13.06.2006.

In view of the very low AOD over Finland, and the associated difficulty in retrieving its value, this comparison is quite satisfactory. There are only seven data points, some of which have a very large standard deviation. The latter were checked against the spatial variability which appears to be high in all these cases, due to the vicinity of clouds causing high AOD variability or strong AOD gradients. This also explains the differences between

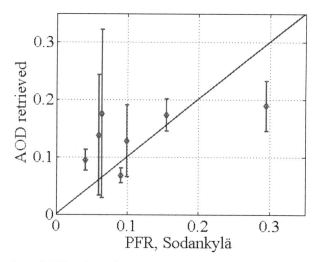

Fig. 5.11. Comparison of AOD retrieved from AATSR data with AOD measured using a Precision Filter Radiometer (PFR) in Sodankylä, part of the WMO-GAW network. The error bars on the AATSR data were obtained from averaging over a 10×10 km^2 area.

both retrievals because the AATSR and PFR measurements were not exactly matched in time. Nevertheless, the trend in the PFR-derived AOD, known as a very accurate instrument, is well-reproduced by the AATSR retrieval. For cases with small spatial variability the AOD is within the previous derived standard deviation of 0.05.

8.5 AOD over China: polluted air

Beijing is the second largest city in China and one of the most polluted cities in the world. The worldwide interest in air quality in Beijing has been growing in recent years because of the Summer Olympic Games in August 2008. Even though many actions have been taken to reduce air pollution in Beijing, the mean aerosol mass concentration is still well above national and international standards and is characterized by a significant fraction of fine particles (e.g., Guinot et al. [2007]). Coal combustion remains a large contributor to Beijing atmospheric pollution. In addition, Beijing is also influenced by emissions from the surrounding densely populated region; in particular, air masses flowing from south and south west import industrial pollution to Beijing. During springtime, the occurrence of so-called dust events can also decrease the air quality significantly.

Fig. 5.12(a) shows an example of AOD retrieval at 0.555 μm wavelength, for April 3rd 2005. The composite map shows AOD-values averaged on a grid of 10×10 km^2, derived from 1×1 km^2 resolution data. White areas with no data are discarded, e.g., because of cloud flags or missing data. The AOD is mainly around 0.5 over the Beijing area. Highest AOD values (0.8–1) are found east and south from Beijing. The lowest AOD-values (0.1) are observed in the west and north, where the AOD pattern shows a clear gradient. This can be partly explained by the topography: Beijing is surrounded by mountains in the north and west. High mountains block transport of pollution originating from emission in the Beijing

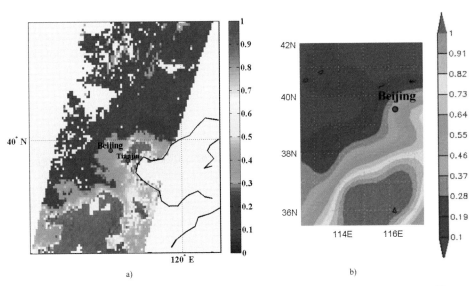

Fig. 5.12. (a) Composite map of AOD derived from AATSR measurements at 0.55 μm over Beijing on April 3, 2005. The map shows AOD-values averaged from the original pixel size of 1×1 km^2 to 10×10 km^2. The patterns are similar to those available from MODIS for the same day (b).

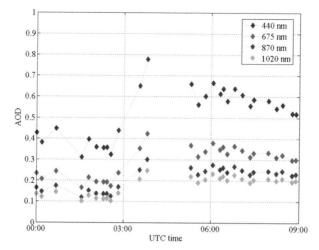

Fig. 5.13. AERONET AOD-measurements for four different wavelengths at Beijing April 3, 2005. The AATSR overpass time was about 2:40 UTC, which corresponds to 10:40 local Beijing time.

area and prevent ventilation. On the other hand, airflow from north or west has to rise over 1,500–2,000 m high mountains before arriving in Beijing. The northern Beijing sector is also considered a relatively clean part, because agriculture is the main activity rather than heavy industry. For this example, back-trajectory analysis shows that the synoptic flow was from the cleaner northern sector.

Fig. 5.12(b) shows the AOD available from MODIS for the same day. The patterns are similar, which provides confidence in the AATSR results.

The AATSR overpass time was about 2:40 UTC which corresponds to 10:40 local Beijing time. The AERONET measurements show a strong increasing trend of AOD during the AATSR overpass time (Fig. 5.13). Therefore the comparison between AATSR and AERONET is not necessary meaningful. However, for other days the comparison has shown good agreement between AERONET and AATSR for Beijing.

9. Conclusion

In this chapter the ATSR-DV algorithm for the retrieval of aerosol properties over land, using data from ATSR-2 on ERS-2 and AATSR on ENVISAT, has been presented. This algorithm resulted from an upgrade and coupling of the single-view and dual-view algorithms developed by Veefkind and de Leeuw [1998] and Veefkind et al. [1998] for ATSR-2. The aerosol properties retrieved with these algorithms are the aerosol optical depth at the available wavelengths (0.55, 0.67, 0.87 and 1.6 μm) and the Ångström coefficient α. The aerosol mixing ratio, for two predefined components, is potentially available as well, as explained in the text and demonstrated in Robles-Gonzalez et al. [2006] and Robles-Gonzalez and de Leeuw [2008]. It is noted that the ATSR-DV algorithm works on individual pixels and averaging over larger areas occurs as a post-processing step. This allows for retrieval of fine detail (spatial resolution 1×1 km^2) at the cost of computer time. Also the accuracy might be improved by first averaging of the original pixels to a large spatial scale and removal of unreliable pixels beforehand.

The cloud-screening protocol developed for the algorithm is found to be satisfactory with the ATSR-DV algorithm and MODIS cloud masks agreeing in 83 % of the conclusive cases. The sensitivity of the cloud-screening protocol for various conditions (i.e., desert, forest fire) remains to be explored.

The aerosol optical depth estimated using the ATSR-DV algorithm has been compared with results from the Swansea University AARDVARC algorithm, MODIS and AERO-NET for an area over Germany. The favorable correlation observed provides confidence in the ability of the ATSR-DV algorithm to retrieve aerosol optical depth. Other applications were presented for the Po valley, northern Italy, wildfire plume identification over the Iberian Peninsula, very low AOD over Finland and polluted air in the Beijing area. In all cases the results compare favorably with ground-based measurements.

It is worth pointing out that the aerosol optical depths are retrieved within the standard deviation of 0.05 over land that was earlier derived by Robles-Gonzales et al. [2006] from comparison of ATSR-2 retrievals with AERONET data.

Land–sea transitions remain an issue in the current implementation of the ATSR-DV algorithm.

Acknowledgments. Research on the retrieval of aerosol properties using satellite-based instruments at TNO has been supported throughout the years by various grants from SRON (The Netherlands Space Research Organization) and NIVR (Netherlands Agency for Aerospace Programmes) in the framework of national user support programs. The ESA (European Space Agency) has contributed ATSR-2 and AATSR data in the framework of AO-257. Further support was received in the framework of various projects sponsored by ESA and the European Commission. We thank the PI investigators and their staff for establishing and maintaining the sites used in this investigation.

References

Abdou, W.A., J.V. Martonchik, R.A. Kahn, R.A. West, and D.J. Diner, 1997: A modified linear-mixing method for calculating atmospheric path radiances of aerosol mixtures. *J. Geophys. Res.*, **102**(D14), 16883–16888.

Anderson, B.E., W.B. Grant, G.L. Gregory, E.V. Browell, J.E. Collins Jr, G.W. Sachse, D.R. Bagwell, C.H. Hudgins, D.R. Blake, and N.J. Blake, 1996: Aerosols from biomass burning over the tropical south Atlantic region: distribution and impacts. *J. Geophys. Res.*, **101**(D19), 24117–24137.

Andreas, E. L., 1992: Sea spray and the turbulent air–sea heat fluxes. *J. Geophys. Res.*, **97**, 11429–11441.

Chandrasekhar, S., 1960: *Radiative transfer.* Oxford University Press, London.

Chylek, P., G. Videen, D. Wally Geldart, J.S. Dobbie, and H.W. Tso, 2000: Light scattering by nonspherical particles. In: *Effective Medium Approximations for Heterogeneous Particles*, 273–308, Academic Press, San Diego, CA.

Chylek, P., V. Srivastava, R.G. Pinnick, and R.T. Wang, 1988: Scattering of electromagnetic waves by composite spherical particles: experiment and effective medium approximations. *Applied Optics*, **27**(12), 2396–2404.

de Haan, J.F., P.B. Bosma, and J.W. Hovenier, 1987: The adding method for multiple scattering calculations of polarized light. *Astronomy Astrophysics*, **183**, 371–391.

de Leeuw, G., K.L. Davidson, S.G. Gathman, and R.V. Noonkester, 1989: Modeling of aerosols in the marine mixed-layer. SPIE Proceedings Volume 1115, 'Propagation Engineering', 287–294.

de Leeuw, G., A.N. de Jong, J. Kusmierczyk-Michulec, R. Schoemaker, M. Moerman, P. Fritz, J. Reid, and B. Holben, 2005: Aerosol retrieval using transmission and multispectral AATSR data. In: Reid, J.S., S.J. Piketh, R. Kahn, R.T. Bruintjes, and B.N. Holben (Eds), *A Summary of First Year Activities of the United Arab Emirates Unified Aerosol Experiment: UAE²*. NRL Report Nr. NRL/MR/7534–05-8899, 105–110.

de Leeuw, G., R. Schoemaker, L. Curier, Y. Bennouna, R. Timmermans, M. Schaap, P. Builtjes, and R. Koelemeijer, 2007: AATSR derived aerosol properties over land. *Proc. 'Envisat Symposium 2007', Montreux, Switzerland, 23–27 April 2007 (ESA SP-636, July 2007), paper 461895.*

Dubovik, O., B.N. Holben, T.E. Eck, A. Smirnov, Y.J. Kaufman, M.D. King, D. Tanré, and I. Slutsker, 2002: Variability of absorption and optical properties of key aerosol types observed in worldwide locations. *J. Atmos. Sciences*, **59**, 590–698.

Durkee, P.A., D.R. Jensen, E.E. Hindman, and T.H. VonderHaar, 1986: The relationship between marine aerosols and satellite detected radiance. *J. Geophys. Res.*, **91**, 4063–4072.

Flowerdew, R.J. and J.D. Haigh, 1995: An approximation to improve accuracy in the derivation of surface reflectance from multi-look satellite radiometers. *Geophys. Res. Lett.*, **22**, 1693–1696.

Forster, P., V. Ramaswamy, P. Artaxo, T. Berntsen, R. Betts, D.W. Fahey, J. Haywood, J. Lean, D.C. Lowe, G. Myhre, J. Nganga, R. Prinn, G. Raga, M. Schulz, and R. Van Dorland, 2007: Changes in atmospheric constituents and in radiative forcing. In: *Climate Change 2007: The Physical Science Basis. Contribution of Working Group I to the Fourth Assessment Report of the Intergovernmental Panel on Climate Change*, Solomon, S., D. Qin, M. Manning, Z. Chen, M. Marquis, K.B. Averyt, M.Tignor, and H.L. Miller (Eds). Cambridge University Press, Cambridge, UK.

Gathman, S.G., 1983: Optical properties of the marine aerosol as predicted by the Navy Aerosol Model. *Opt. Eng.*, **22**, 57–62.

Grey, W.M.F., P.R. North, S. O. Los, and R.M. Mitchell, 2006: Aerosol optical depth and land surface reflectance from multi-angle AATSR measurements: global validation and inter-sensor comparisons. *EEE Transactions Geosciences Remote Sensing*, **44**, 2184–2197.

Guinot, B., H. Cachier, J. Sciare, Y. Tong, W. Xin, and Y. Jianhua, 2007: Beijing aerosol: atmospheric interactions and new trends. *J. Geophys. Res.*, **112**, (D14314), doi:10.1029/2006JD0061848195.

Haywood, J.M., R.S. Osborne, P.N. Francis, A. Keil, P. Formenti, M.O. Andreae, and P.H. Kaye, 2003: The mean physical and optical properties of regional haze dominated by biomass burning aerosol measured from the C-130 aircraft during safari 2000. *J. Geophys. Res., 108 (D13), 8473, doi:10.1029/ 2002JD002226.*

Haywood, J. and O. Boucher, 2000: Estimate of the direct and indirect radiative forcing due to troposphere aerosol: a review. *Rev. Geophys.*, **38**(4), 513–543.

Heintzenberg, J., 1989: Fine particles in the global troposphere a review. *Tellus*, **41, 149–160.**

Hess, M.P., P. Koepke, and I. Schult, 1998: Optical properties of aerosols and clouds: the software package OPAC. *Bull. Amer. Meteor. Soc.*, **79**(5), 831–844.

Koelemeijer, R.B. , P. Stammes, J. W. Hovenier, and J.D. de Haan, 2001: A fast method for retrieval of cloud parameters using oxygen a-band measurements from the global ozone monitoring instrument. *J. Geophys. Res.*, **106**, 3475–3490.

Kokhanovsky, A.A., F.-M. Breon, A. Cacciari, E. Carboni, D. Diner, W. Di Nicolantonio, R.G. Grainger, W.M.F. Grey, R. Höller, and K.-H. Lee, 2007: Aerosol remote sensing over land: a comparison of satellite retrievals using different algorithms and instruments. *Atmospheric Research*, **85**, 372–394.

Kokhanovsky, A.A., R.L. Curier, G. de Leeuw, W.M.F. Grey, K.-H. Lee, Y. Bennouna, R. Schoemaker, and P.R.J. North, 2009: The inter-comparison of AATSR dual view aerosol optical thickness retrievals with results from various algorithms and instruments. *Int. J. Remote Sensing*, in press.

Mie, G., 1908: Beiträge zur Optik trüber Medien, speziell kolloidaler Metallösungen. *Ann. Phys.*, **25**, 377–445.

Myhre, G., F. Stordal, M. Johnsrud, A. Ignatov, M.I. Mishchenko, I.V. Geogdzhayev, D. Tanré, J.L. Deuzé, P. Goloub, T. Nakajima, A. Higurashi, O. Torres, and B.N. Holben, 2004: Intercomparison of satellite retrieved aerosol optical depth over ocean, *J. Atmos. Sci.*, **61**, 499–513.

Myhre, G., F. Stordal, M. Johnsrud, D.J. Diner, I.V. Geogdzhayev, J.M. Haywood, B.N. Holben, T. Holzer-Popp, A. Ignatov, R.A. Kahn, Y.J. Kaufman, N. Loeb, J.V. Martonchik, M.I. Mishchenko, N.R. Nalli, L.A. Remer, M. Schroedter-Homscheidt, D. Tanré, O. Torres, and M. Wang, 2005: Intercomparison of satellite retrieved aerosol optical depth over ocean during the period September 1997 to December 2000. *Atmos. Chem. Phys.*, **5**, 1697–1719.

North, P.R.J., S.A. Briggs, S.E. Plummer, and J. Settle, 1999: Retrieval of land surface bidirectional reflectance and aerosol opacity from ATSR-2 multiangle imagery. *IEEE Trans. Geosci. Rem. Sens.*, **37**, 526–537.

Quinn, P.K., D.J. Coffman, T.S. Bates, T.L. Miller, J.E. Johnson, E.J. Welton, C. Neususs, M. Miller, and P.J. Sheridan, 2002: Aerosol optical properties during INDOEX 1999: means, variability, and controlling factors. *J. Geophys. Res.*, **107**(D19), 8020, doi:10.1029/2000JD000037.

Robles-Gonzalez, C., 2003: Retrieval of aerosol properties using ATSR-2 observations and their interpretation. PhD thesis, University of Utrecht, Utrecht, The Netherlands.

Robles-Gonzalez, C. and G. de Leeuw, 2008:, Aerosol properties over the SAFARI-2000 area retrieved from ATSR-2. *JGR-Atmospheres*, **113**, doi:10.1029/2007JD008636.

Robles-Gonzalez, C., J.P. Veefkind, and G. de Leeuw, 2000: Mean aerosol optical depth over Europe in August 1997 derived from ATSR-2 data. *Geophys. Res. Lett.*, **27**, 955–959.

Robles-Gonzalez, C., G. de Leeuw, R. Decae, J. Kusmierczyk-Michulec, and P. Stammes, 2006: Aerosol properties over the Indian Ocean Experiment (INDOEX) campaign area retrieved from ATSR-2. *J. Geophys. Res.*, **111**, D15205, doi:10.1029/2005JD006184.

Robles-Gonzalez, C. and G. de Leeuw, 2008: Aerosol properties over the SAFARI-2000 area retrieved from ATSR-2. *J. Geophys. Res.*, **113**, doi:10.1029/2007JD008636.

Saunders, R.W. and K.T. Kriebel, 1988: An improved method for detecting clear sky and cloudy radiances from AVHRR data. *Int. J. Remote Sensing*, **9**, 123–150.

Seinfeld, J.H. and S.N. Pandis, 1998: *Atmospheric Chemistry and Physics*. John Wiley, Chichester, UK.

Thomas, G.E., C.A. Poulsen, R.L. Curier, G. de Leeuw, S.H. Marsh, E. Carboni, R.G. Grainger, and R. Siddans, 2007. Comparison of AATSR and SEVIRI aerosol retrievals over the Northern Adriatic. *QJRM*, (ADRIEX special issue), **133**(S1), 85–95 doi: 10.1002/qj.126.

Torres, O., P.K. Bhartia, J.R. Herman, Z. Ahmad, and J. Gleason, 1998: Derivation of aerosol properties from satellite measurements of backscattered ultraviolet radiation: theoretical basis. *J. Geophys. Res.*, **103**(D14), 17099–17110.

Veefkind, J.P., 1999: Aerosol satellite remote sensing. PhD thesis, University of Utrecht, Utrecht, The Netherlands.

Veefkind, J.P. and G. de Leeuw, 1998: A new algorithm to determine the spectral aerosol optical depth from satellite radiometer measurements. *J. Aerosol Sci.*, **29**, 1237–1248.

Veefkind, J.P., G. de Leeuw, and P.A. Durkee, 1998: Retrieval of aerosol optical depth over land using two-angle view satellite radiometry during TARFOX. *Geophys. Res. Letters*, **25**(16), 3135–3138.

Veefkind, J.P., G. de Leeuw, P.A. Durkee, P.B. Russell, P.V. Hobbs, and J.M Livingston, 1999: Aerosol optical depth retrieval using ATSR-2 and AVHRR data during TARFOX. *J. Geophys. Res.*, **104**(D2), 2253–2260.

Veefkind, J.P., G. de Leeuw, P. Stammes, and B.A. Koelemeijer, 2000: Regional distributions of aerosols over lands derived from ATSR-2 and GOME. *Remote Sens. Environ.*, **74**, 377–386.

Vermote, E., D. Tanré, J.L. Deuzé, M. Herman, J.J. Morcrette, 1997: Second simulation of the satellite signal in the solar spectrum, 6S: an overview. *IEEE Transactions Geosciences Remote Sensing*, **35**, 675–686.

Wang, M. and H.R. Gordon, 1994: Radiance reflected from ocean–atmosphere system: synthesis from individual components of the aerosol size distribution. *Applied Optics*, **33**, 7088–7095.

WMO (1983). Cas/radiation commission of iamap meeting of experts on aerosols and their climatic effects. Technical report, WMO, Williamsburg, VA.

6 Aerosol optical depth from dual-view (A)ATSR satellite observations

William M. F. Grey, Peter R. J. North

1. Introduction

1.1 Aerosols and the climate system

Atmospheric aerosol particles play a critical role in the Earth's radiation budget, yet the global radiative forcing by aerosols is widely recognized as a major uncertainty in our understanding of the climate [IPCC, 2007]. The radiative characteristics of aerosol particles are determined by their shape, size, total amount and chemical composition [Kaufman et al., 1997a]. Overall though, aerosols have a cooling effect at the Earth's surface by reducing the amount of solar radiation arriving at the surface below the layer of aerosols in the atmosphere. This cooling effect by aerosols is achieved by increasing the planetary albedo at the top-of-the-atmosphere (TOA) through directly scattering some of incoming sunlight back into space. However, some of the radiation can also be absorbed in the atmosphere by aerosols and reemitted. The volcanic eruption of Mount Pinatubo in 1991 provides an excellent natural experiment to demonstrate the surface cooling effect of aerosols. In the following two years after the eruption the average global surface temperature was reduced by about half a degree Celsius, principally owing to the scattering of sunlight by volcanically enhanced stratospheric sulfate aerosol [Hansen et al., 1992]. In addition, to the direct influence that aerosols have on the climate system, aerosols have an indirect effect on the radiative forcing through their interaction with cloud droplets and influence on cloud albedo.

In addition to ash from volcanic eruptions, aerosols are produced by a myriad of natural and anthropogenic processes from a variety of sources. Naturally occurring aerosols include wind-blown dust from arid and semi-arid regions, and salts from sea-spray and bursting bubbles, for instance. Aerosols caused by human activities include soot and sulfates from fossil fuel burning and heavy industry, and smoke from the burning of vegetation, often as part of land management clearance practices and deforestation (e.g. in the Sahel and Amazon). Anthropogenic aerosols cause a net global cooling that have a combined direct and indirect radiative forcing of -0.4 to -2.7 W m^2, which is of the same order of magnitude as the positive forcing caused by anthropogenic greenhouse emissions estimated as approximately 2.5 W m^2 [IPCC, 2007]. All these different types of aerosols vary in shape, size and chemical composition and hence have different radiative characteristics. For instance, black carbon (soot) has strongly absorbing characteristics, whereas aerosols generally are highly reflective.

Anthropogenic aerosols may have reduced the climate sensitivity by reflecting much of the incoming solar radiation back into space. The climate sensitivity is described as the increase in temperature caused by a doubling of atmospheric carbon dioxide concentrations from the pre-industrial levels of 280 parts per million by volume. However, we cannot rely on aerosols to mitigate the effects of our greenhouse emissions for several reasons. Firstly, it is often the same anthropogenic processes that release both aerosols and carbon dioxide (e.g. coal-powered electricity stations emit carbon dioxide, sulfates and also soot). Secondly, not all aerosols have reflective radiative characteristics. From radiation transfer principles, strongly absorbing anthropogenic aerosols can actually lead to a warming of the troposphere even though the global balance of aerosol forcing is negative. This has been elegantly demonstrated in a recent study by Ramanathan et al. [2007], where it was shown that the brown haze over Asia caused by the release of black carbon in soot from burning fossil fuels had an equal contribution to atmospheric warming as greenhouse gases in this region. Thirdly, the net effect of aerosol on climate forcing depends not only on its own absorption and scattering properties but also on the albedo of the underlying surface. For instance, a layer of aerosol over a dark ocean surface tends to increase the TOA radiance, while over bright surfaces such as deserts or snow the TOA radiance may decrease owing to the greater absorption by the aerosols than the surface. Even non-absorbing aerosols over snow may decrease the TOA reflectivity relative to no atmospheric aerosol contamination. Finally, there has been a reduction in anthropogenic particulate emissions worldwide in part owing to legislation brought into effect since the 1990s in order to reduce harmful sulfate aerosols at the source of industrial output by the European Union and in the United States [Crutzen, 2006]. This was highlighted in a study of satellite data by Mishchenko et al. [2007b] that showed a global reduction in aerosol concentrations since the early 1990s to the present day, leading to a reverse in global dimming. This clearly has serious implications for climate sensitivity, where we may be at an even greater risk of dangerous climate change with reduced atmospheric aerosol concentrations to counter the warming caused by our greenhouse emissions. Geo-engineering solutions for the deliberate manipulation of the Earth's climate such as the controlled emission of aerosols are not currently in the mainstream debate on climate change, but these controversial ideas are being considered as possible solutions even though our understanding of their consequences is poor.

The uncertainty in our understanding of the radiative effects of aerosols stems from the lack of accurate and repetitive measurements at global scales. In addition to improving our knowledge of atmospheric aerosols in general, one of the principal drivers behind the study of atmospheric aerosols is the need to reduce this magnitude of uncertainty in the Earth's atmospheric radiative forcing [Kaufman et al., 2002]. Unlike greenhouse gases which tend to be well mixed in the atmosphere and have long residence times, atmospheric aerosol concentrations are highly dynamic varying both spatially and temporally and as a result need to be continuously monitored at regional and global-scales. Typically, aerosols have short atmospheric residence times of the order of days to weeks before undergoing dry deposition through gravitational settling and turbulent mixing, and wet-deposition through precipitation. Aerosol particles are derived from relatively local sources although they can be distributed between continents and so their distribution and composition are highly variable worldwide.

Aerosols also influence other aspects of the climate system such as the biogeochemical and hydrological cycles. For instance, aerosols affect the total and diffuse radiation fraction at the surface which has consequences on carbon assimilation by vegetation (e.g., Alton et al. [2005]). In addition, large concentrations of aerosols are thought to lead to precipitation suppression, as is the case for dust in the West African Sahel [Rosenfeld et al., 2001]. This is because aerosols act as cloud condensation nuclei around which cloud droplets coalesce. When the concentration of aerosols is high these cloud droplets are smaller and are less likely to result in precipitation. Small droplets are also more effective at reflecting sunlight back into space and trapping longwave terrestrial radiation. Saharan dust aerosol is a major source of iron for the fertilization of phytoplankton in the Atlantic Ocean, possibly leading to the greater uptake of atmospheric carbon dioxide.

1.2 The role of Earth observation

Clearly, understanding the influence of atmospheric aerosols on the Earth's radiation budget is of profound importance. To meet these scientific objectives, satellite observations in combination with comprehensive modeling and field experiments and long-term ground monitoring programs are being carried out. Satellite Earth observations are extremely useful because they provide information that is both frequent and global in coverage. However, some recent studies using satellite data have yielded different conclusions on the global radiative forcing by anthropogenic aerosols (e.g., Bellouin et al. [2005], Chung et al. [2005] and Yu et al. [2005]). This is because the satellite data themselves have different information content with respect to the view angle, spectral channels, spatial and temporal resolution, and polarization [Kokhanovsky et al., 2007]. Moreover, different simplifying assumptions and aerosol retrieval algorithms applied to the same data source can also result in different estimates of aerosol properties. Thus, we should be extremely wary when interpreting satellite datasets and provide information on the limitation to any potential end-users of these products.

The AATSR (Advanced Along-Track Scanning Radiometer) onboard the European Space Agency's (ESA) ENVISAT platform and its predecessor the ATSR-2 (Along-Track Scanning Radiometer-2) are passive optical and thermal instruments that have the potential to make a major contribution to our understanding of atmospheric aerosols. This is because global long-term aerosol retrievals spanning more than a decade can be attained based on very well calibrated radiance measurements.

When interpreting measurements from passive optical instruments the challenge lies in separating out the atmospheric and surface scattering contributions to the satellite signal. The method presented here involves the coupled inversion using a physical model of light scattering that requires no *a priori* knowledge of the land surface. In this chapter, we provide a review of the method developed by North et al. [1999] and subsequently implemented and tested in North [2002] and Grey et al. [2006a, 2006b] for retrieving atmospheric aerosol properties from AATSR and ATSR-2 satellite data. This chapter is based heavily on these papers and we refer the interested reader to these works. The pioneering surface and atmospheric studies from ATSR-2 can be found in Veefkind et al. [1998], Mackay et al. [1998], Flowerdew and Haigh [1996] and Godslave [1996].

1.3 Algorithm development cycle

There are several key stages in the development cycle of an aerosol retrieval algorithm for a satellite instrument and the development of our algorithm has not differed from this generic process. The first stage is the algorithm definition phase during which the algorithm is conceived from a theoretical standpoint, is written up as a computer program and generally tested on a simulated dataset of TOA reflectances. The next stage is the validation phase where satellite data are ingested into the algorithm and the aerosol properties are retrieved over a range of sites worldwide corresponding to ground-based sunphotometers that serve as the benchmark for testing satellite-derived aerosol products. In practice, validation is an on-going process that occurs throughout the development cycle. Once the algorithm has been rigorously tested and proven to provide accurate estimates of aerosol properties, we can apply it within an operational framework for routine retrievals. To process satellite data operationally and routinely the implementation must necessarily be computationally efficient. To provide useful aerosol climatological datasets the entire archive of a satellite instrument needs to be processed. Once the aerosol products have been produced it is then possible to compare with other existing satellite products. During the production and validation of a dataset of aerosol properties a great deal can be learnt about the algorithm and situations where it performs well or poorly. To take account of improvements in scientific algorithms and in the derived geophysical parameters the satellite data archives can be reprocessed several times. This is important because the same version of an algorithm applied to the whole dataset will provide consistent products for climatological studies, otherwise artifacts in the geophysical products will occur that will be the result of the changing of the algorithm. Throughout this chapter we describe each of these stages of algorithm development in more detail.

1.4   Chapter outline

We begin with an outline of remote sensing of aerosols in the context of multi-view-angle passive optical satellite instruments and provide a description of the AATSR instrument. In Section 3, a brief overview of the problem of aerosol retrieval from passive optical remote sensing instruments is presented. This is followed by a description of the multi-view-angle algorithm and processor of Grey et al. [2006a, 2006b] developed for AATSR. In Section 4, we describe the practical operational implementation of the algorithms and the validation that has been carried out against AERONET (Aerosol Robotic Network) sunphotometer ground-based measurements and the inter-comparison with other satellite instruments. We then provide some recommendations for future research. Finally, we summarize the salient points of this chapter.

2.   Remote sensing of aerosols

Satellite remote sensing offers a viable means for routinely measuring aerosols over very large areas. For instance, the AVHRR (Advanced Very High Resolution Radiometer) is used operationally for monitoring aerosols over oceans [Stowe et al., 1997]. Another instrument that has a long-term archive of aerosol measurements is TOMS (Total Ozone Mapping Spectrometer) [Torres et al., 2002; Hsu et al., 1999]. Aerosol estimates from MODIS (Moderate Resolution Imaging Spectroradiometer) have also been retrieved based on the spectral separation of the surface and atmospheric signal [Levy et al., 2007; Remer et al., 2005].

Single-look approaches do not provide all the information on the surface needed for the remote sensing of aerosols [King et al., 1999]. When viewing from a single direction, we must rely on the spectral signature to distinguish atmospheric from ground scattering. Where a target of approximately known reflectance can be identified, such as dense vegetation or a body of water, atmospheric properties at the target location may be estimated on the basis of known correlation of ground reflectance at different wavelengths (e.g., Kaufman et al. [1997b]). For a given set of surface reflectances derived by assuming a certain atmospheric profile when a large number of channels are available it is possible to represent the target reflectance as a linear mixture of an idealized vegetation and soil spectrum, or set of spectra. A number of variations on such methods have been used successfully for aerosol retrieval with Envisat/MERIS (Medium Resolution Imaging Spectrometer) and CHRIS (Compact High Resolution Imaging Spectrometer) On-Board PROBA (Project for on-board Autonomy) [Guanter et al., 2007; Santer et al., 2007; von Hoyningen-Huene et al., 2003]. However, routine application is limited to where such targets are available at the appropriate spatial resolution (i.e. oceans and dark dense vegetation), and accuracy is limited to the level of uncertainty in the *a priori* estimate of target reflectance variation.

A number of instruments, such as POLDER (Polarization and Directionality of the Earth's Reflectances), MISR (Multi-angle Imaging Spectroradiometer), ATSR-2 (Along-Track Scanning Radiometer), and AATSR (Advanced Along-Track Scanning Radiometer) have been developed with the enhanced capability of acquiring near-simultaneous multi-angle observations through different path lengths allowing the atmospheric properties to be inferred. Moreover, future missions such as Nasa's Glory/APS (Aerosol Polarimetry Sensor) and ESA's Sentinel-3/SLSTR (Sea and Land Surface Temperature Radiometer) instruments will also have multi-view-angle capability (see Table 6.1). The APS instrument will have the unique capability of obtaining the phase function. Retrievals of aerosol properties have previously been demonstrated in a number of studies based on multi-look observations of MISR [Martonchik et al., 2004; Diner et al., 1998], POLDER [Roujean et al., 1997] and ATSR-2 [North, 2002; Veefkind et al., 1998; North et al., 1999]. The additional information contained within multi-view-angle observations can potentially lead to improved aerosol estimates compared with conventional single-look instruments [Diner et al., 2005a]. The advantage of the multi-look approach over single-look methods is that assumptions are not required about the land surface spectral properties, thus aerosols can potentially be retrieved over any surface even bright desert scenes [Martonchik et al., 2004]. Satellite remote sensing of aerosol properties have mainly been retrieved using passive optical satellite instruments, but more recently active lidar instruments such as the CALIPSO (Cloud-Aerosol Lidar and Infrared Pathfinder Satellite Observation) instrument that forms part of NASA's A-Train satellite constellation have provided useful complementary information on the profile and height of atmospheric aerosols. Passive optical instruments cannot easily retrieve this information, although a recent study by Kahn et al. [2007] did retrieve the height of smoke plumes from multi-view-angle MISR data.

For passive optical instruments, obtaining accurate information on aerosol properties is also necessary for atmospheric correction of bi-directional reflectances so that quantitative analysis of the land and ocean surface can be performed. The uncertainties in the retrieved biophysical properties such as leaf area index, the fraction of absorbed photosynthetically

Table 6.1. Past, current and future satellite passive optical instruments with multi-view-angle capability

Sensor	Platform	Launch date/ end date	Features
ATSR-2	ERS-2	1995	Dual-view-angle observations; more than a decade of observations; 5–6 day coverage at equator; 1 km spatial resolution. 7 channels with four in optical region at 550, 670, 870 and 1610 nm.
AATSR	Envisat	2002	As ATSR-2
SLSTR	Sentinel-3	2012	As (A)ATSR but wider swath and two additional channels at 1300 and 2250 nm.
CHRIS	PROBA	2001	5 viewing angles, hyper-spectral between 400 and 1000 nm, very high spatial resolution of up to 17 m.
MISR	Terra	1999	9 viewing angles, 6-day coverage at equator; 1 km spatial resolution. Four channels: 446, 558, 672, 866 nm
POLDER	ADEOS-1	11/1996–6/1997	14 viewing angles; polarization; but relatively coarse resolution at 6x7 km resolution. Eight channels: 444, 445, 492, 565, 670, 763, 763, 908 and 861 nm.
	ADEOS-2	4/2003–10/2003	
	Parasol	2004	
APS (Aerosol Polarimetry Sensor)	Glory	2010	~ 250 scattering angles per scene; polarization; 6 km resolution; can obtain phase function. Nine channels: 410, 443, 555, 670, 865, 910, 1,370, 1,610, 2,200 nm

active radiation and albedo will be strongly influenced by the quality of the atmospheric correction of the measured radiances [Guanter et al., 2007].

2.1 Description of AATSR and ATSR-2

The ATSR-2 and AATSR sensors were designed for atmospheric, land and ocean scientific applications. AATSR was launched by the European Space Agency onboard ENVISAT in March 2002 and is one of a series of satellite instruments with the purpose of providing a well calibrated long-term global dataset of satellite data for climate research. Principally designed as a geophysical ocean sensor to measure sea-surface temperature, the instrument has also found applications over land. To all practical intents, the instrument is identical to its predecessor, the ATSR-2 sensor launched in 1995 on ERS-2 (European Remote Sensing), and provides continuity to the ATSR-1 and ATSR-2 datasets. We note here that ATSR-1 does only have thermal channels so this instrument is not used for aerosol retrieval. One of the benefits of these series of missions is that a long time-series of aerosol and other biophysical and geophysical properties spanning more than a decade has been obtained. Moreover, the proposed follow-up SLSTR mission as currently scoped on Sentinel-3 will ensure the continuity of the multi-view-angle data into the future.

One of the characteristics of the AATSR instrument is its ability of acquire two near-simultaneous observations of the same area of the Earth's surface at a view zenith angle of 55° (forward view at the surface) and then approximately 120 seconds later at an angle close to vertical (nadir view) (see Figure 6.1). The observations made in forward view are

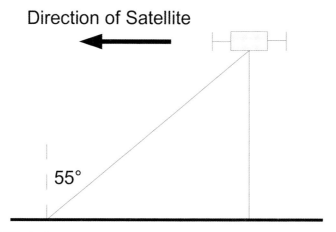

Fig. 6.1. (A)ATSR viewing geometry.

more influenced by atmospheric scattering and absorption than in the nadir view because
the path length is approximately twice that of the nadir view. The swaths are approximately
500 km in width, which means that on average global coverage can be attained every 6
days at the equator. The nominal size of each pixel at nadir is 1 by 1 km. In the forward
view the pixels sizes are larger, but in the AATSR level 1 products the pixels are resampled
to the same size as those in the nadir view. There are seven spectral bands, but only the four
bands in the visible and near-infrared (555, 660, 870 and 1610 nm) are used for aerosol
retrieval in our scheme presented here. These spectral bands are narrow (approximately
20 nm) and avoid atmospheric water vapor absorption regions in the electromagnetic spectrum. Another feature of the (A)ATSR is that it is a well calibrated instrument with an
onboard diffusely reflecting target for calibration of the optical channels, and a blackbody
reference target for the thermal channels.

3.   Model inversion for the retrieval of aerosol optical depth

3.1   Atmospheric radiative transfer

Satellite observations at optical wavelengths consist of solar radiation scattered by both the
atmosphere and the surface back in the direction of the sensor. We need to separate out the
atmospheric and surface scattering components through atmospheric radiative transfer
modeling if we are to obtain accurate estimates of biophysical and geophysical properties.
For remote sensing we can infer the radiative characteristics of aerosol particles from the
measured satellite radiances. The atmospheric and surface contributions to the radiance
measured by a spaceborne sensor at the top of the atmosphere, for a given wavelength and
viewing and illumination geometry, over a Lambertian surface are given by the widely
used equation:

$$L_{toa} = L_0 + \frac{T(\theta_s)T(\theta_v)F_0\cos(\theta_s)\rho_{surf}}{\pi(1 - \rho_{surf}S)}, \tag{1}$$

where L_0 is the atmospheric path radiance due to Rayleigh and aerosol scattering, ρ_{surf} is the surface reflectance, F_0 is the solar radiance at the top-of-the-atmosphere for a directly overhead sun, S is the spherical albedo of the atmosphere, $T(\theta_v)$ is the total transmittance from the surface to the top of the atmosphere in the view direction of the satellite, $T(\theta_s)$ is the total transmittance from the top of the atmosphere to the surface along the path of the incoming solar radiation. In the 6S formulation of the radiative transfer that we employ in our work we use an approximation [Vermote et al., 1997] by assuming that the gaseous absorption layer is above the aerosol scattering and absorption layer such that:

$$L_{toa} = T_g(\theta_s, \theta_v) \left[L_0 + \frac{T(\theta_s)T(\theta_v)F_0 \cos(\theta_s)\rho_{surf}}{\pi(1 - \rho_{surf}S)} \right], \tag{2}$$

where $T_g(\theta_s, \theta_v)$ is the gaseous transmission team. In the non-absorbing regions of the spectrum this assumption has little effect, but at strongly absorbing wavelengths the effect of coupling between molecules and aerosols is much stronger and this approach is not valid [Guanter et al., 2007]. Although the decoupling approach may be valid for some of the gases such as ozone, it is not appropriate for water vapor, therefore in 6S Eq. (2) is modified as:

$$L_{toa} = T_g(\theta_s, \theta_v) \left[L_M + L_A T_g^{H_2O}(\theta_s, \theta_v) + \frac{T_g^{H_2O}(\theta_s, \theta_v)T(\theta_s)T(\theta_v)F_0 \cos(\theta_s)\rho_{surf}}{\pi(1 - \rho_{surf}S)} \right], \tag{3}$$

where L_M is molecular path radiance owing to Rayleigh scattering, L_A is the aerosol path radiance, and $T_g^{H_2O}(\theta_s, \theta_v)$ is the transmittance owing to water vapor only. Moreover, for (A)ATSR the optical bands are away from absorption regions, thus using the decoupling approach as employed in 6S is valid.

All reflectance and radiances terms are a function of the wavelength, solar θ_s and satellite θ_v zenith angles and relative azimuth angle between the sun and viewing position ϕ, as well as the optical properties of the atmosphere and surface. The $1 - \rho_{surf}S$ term refers to multiple scattering between the surface and atmosphere but this term can be neglected for an optically thin atmosphere. We note here that radiances L can also be expressed as reflectances ρ using:

$$\rho = \frac{\pi L}{\cos(\theta_s)F_0}. \tag{4}$$

Over a completely absorbing surface (e.g., over deep dark oceans at infrared wavelengths) the second term in Eq. (1–3) is zero and the measured TOA radiance is due entirely to the atmospheric path radiance. In contrast, over bright land surfaces the surface reflectance term makes a large contribution to the measured TOA radiance, while the atmospheric path radiance contribution is small. This is where the challenge in aerosol remote sensing lies because even a small error in the surface reflectance can lead to a large error in our retrieved aerosol properties.

The atmospheric path radiance L_0 can be separated into molecular L_M and aerosol L_A radiance scattering components such that

$$L_0 = L_M + L_A. \tag{5}$$

However, this standard approach does not include the multiple scattering between the molecules and aerosols. A more realistic representation is given in Rozanov and Kokhanovsky [2005] and the following discussion is based around this study. By taking into account the coupling between molecular and aerosol scattering L_{MA}, we have

$$L_0 = L_M + L_A + L_{MA}. \tag{6}$$

The coupling between aerosols and molecules contributes significantly to the path radiance particularly at blue and ultraviolet wavelengths. At wavelengths of 550 nm or greater coupling is not so large and the error in the retrieved aerosol path radiance will be small, typically below 5 % for solar zenith angles smaller than $60°$ [Rozanov and Kokhanovsky, 2005]. At these solar positions for the AATSR optical wavelengths we do not need to take into account multiple scattering between the molecules and aerosols. However, at solar zenith angles larger than approximately $60°$ this error does become significant and will result in large uncertainties in retrieved estimates of AOD.

It is also pertinent to discuss here the error in our calculations of scattered light intensity by not taking into account the polarization of light. Unpolarized sunlight becomes polarized after interaction with atmospheric molecules and particles. The scalar radiative transfer equation as used in our aerosol retrieval scheme does not take account of polarization, while the vector radiative transfer equation does characterize the polarization of light and its influence on the intensity of reflected radiation detected by a satellite sensor [Rozanov and Kokhanovsky, 2005]. The scalar approach can be used in the infrared, where the molecular contribution is negligible. Errors in the calculated radiance in the UV, where the Rayleigh scattering contribution is strong, can reach 10 % [Rozanov and Kokhanovsky, 2006; Levy et al., 2004; Mishchenko and Travis, 1997], however. Using the scalar radiative transfer equation at AATSR wavelengths (550 nm and above) will result in a small error of the scattered light intensity, but modifications in future work of our aerosol retrieval scheme will make use of the vector radiative transfer equation.

To retrieve the aerosol path radiance we subtract the molecular scattering from the total atmospheric radiance [Rozanov and Kokhanovsky, 2006]. The parameters required to model aerosol radiative effects are aerosol optical depth τ, for a given reference wavelength λ, its spectral dependence, defined by Ångström coefficient a, single scattering albedo ω and phase function $P(\Theta)$, where Θ denotes the scattering angle. The single scattering albedo is equal to the ratio of scattering coefficient σ_s to the extinction coefficient σ_e such that $\omega = \sigma_s/\sigma_e$ where the extinction coefficient is the sum of the scattering and absorption coefficients σ_a, $\sigma_e = \sigma_s + \sigma_a$. These properties are determined by the amount, chemical composition, size and shape of the aerosol particles. For aerosol retrievals we are particularly interested in the aerosol path radiance L_A given as

$$L_A = f(\tau, \omega, P(\Theta), \theta_s, \theta_V, \phi_v, \phi_s), \tag{7}$$

where ϕ_v and ϕ_s are the viewing and solar azimuth angles, respectively. The aerosol path radiance is calculated using the successive orders of scattering approach to take account of multiple scattering [Kotchenova and Vermote, 2007; Kotchenova et al., 2006]. AOD is the principal property derived from satellite observations and is a function of the number and size of particles, and is the extinction of light integrated from the bottom to top of an atmospheric column

$$\tau(\lambda) = \int_0^\infty \sigma_e(z, \lambda) \; dz. \tag{8}$$

The phase function provides a description of the distribution of scattered radiation of the aerosol particles as a function of the scattering angle Θ between the incident and scattering direction. To parameterize the backward and forward scattering direction of incident light we use the asymmetry factor, where a positive asymmetry factor indicates strong forward scatter and a negative asymmetry value indicates scatter in the backward direction. The scattering angle is a function of θ_s, θ_v and $_\phi$ and is calculated by

$$\Theta = \arccos(-\cos(\theta_v)\cos(\theta_s) + \sqrt{(1 - \cos^2(\theta_v))}\sqrt{(1 - \cos^2(\theta_s))}\cos(\phi)). \tag{9}$$

The AOD is the most important parameter for calculating the atmospheric radiative forcing [Chylek et al. 2003]. Climatologically defined aerosol models are used to constrain the other aerosol parameters (e.g., ω and $P(\Theta)$) in Eq. (7) in order to retrieve AOD.

To retrieve estimates of aerosol properties from measured satellite radiances, we need to solve this inverse problem and separate the atmospheric and surface scattering contributions to the observed signal. If the land surface is Lambertian then the differences between the measured radiances from the different viewing positions could be attributed to atmospheric path radiance only. However, all natural surfaces contain some degree of anisotropy at optical wavelengths, therefore it is necessary to consider how the bi-directional reflectance of the land surface changes with the viewing and illumination geometry in order to decouple the atmospheric and surface scattering contributions with any accuracy. For aerosol retrieval the difficulty lies in obtaining a reliable estimate of the land surface reflectance. Over bright land surfaces the problem is particularly challenging because surface scattering dominates the satellite signal. In addition, the land surface is heterogeneous and temporally varying making *a priori* assumptions difficult. Multi-view-angle observations help to constrain the inverse problem since the surface is imaged through different atmospheric path lengths allowing us to infer the atmospheric properties. Also an angular constraint for the surface scattering behavior is provided that can be exploited for the estimation of bi-directional reflectance.

3.2 Model of land surface reflection

In general, it is more challenging to retrieve aerosol properties over land than over ocean. This is because the scattering from the land surface tends to dominate the satellite signal making it difficult to discern the atmospheric scattering contribution to the satellite signal particularly over bright surfaces. In addition, obtaining an accurate model of the land surface is further complicated because bi-directional reflectances are highly variable [North *et al.*, 1999; Veefkind et al., 2000]. In contrast, atmospheric scattering dominates the signal over the ocean. Thus, single-look satellite observations can provide accurate estimates of AOD over ocean [Remer et al., 2005].

We present a multi-view-angle aerosol retrieval and atmospheric correction algorithm over land. Similar approaches have been developed elsewhere, for instance at the Finnish Meteorological Institute and Oxford University and details on both these algorithms can be found in this volume in Curier et al. (Chapter 5) and Thomas et al. (Chapter 7) respectively. The algorithm presented here has been integrated into ESA's GPOD (Grid Processing on Demand, eogrid.esrin.esa.int) high-performance computing facility for operational retrievals of AOD and bi-directional reflectance from AATSR. Moreover, similar approaches have been applied to multi-angle CHRIS and MISR data [Diner et al., 2005b]. The algorithm presented here is based on a simple physical model of light scattering for the dual-angle sampling of the ATSR-2 and AATSR instruments and can be used to separate the surface bi-directional reflectance from the atmospheric aerosol properties without recourse to *a priori* information of the land surface properties, based on an angular constraint [North *et al.*, 1999]. Previous studies have shown that the shape of the surface bi-directional reflectance distribution function (BRDF) is similar at different wavelengths. This is because the scattering elements of the surface are much larger than wavelengths used in the retrieval procedure and so the angular variation of surface reflectance is often dominated by wavelength-independent geometric effects. This has been demonstrated by multi-angle observations from ATSR-2 and AATSR (e.g., Veefkind et al. [1999, 2000]) and MISR (e.g., Diner *et al.* [2005b]). For AATSR, the ratio of surface reflectances at the nadir (where the view zenith angle is close to $0°$) and forward viewing angles (where the view zenith angle is $55°$) is well correlated between bands. Thus

$$\rho(\lambda_i, n) \approx \frac{\rho(\lambda_j, n)}{\rho(\lambda_j, f)} \rho(\lambda_i, f), \qquad (10)$$

where nadir n and forward f refer to forward and nadir view zenith angles of (A)ATSR, respectively, and i and j are any combination of the (A)ATSR optical channels (550, 670, 870, 1630 nm). In the shortwave infrared region, at 1,630 nm there is very little scattering and absorption owing to aerosols, thus correction for gaseous absorption and Rayleigh scattering can be sufficient for atmospheric correction, although this is not the case for large particles like dust or sea-salt with spectrally flat AODs.

North *et al.* [1999] developed this approach further by considering the variation of the diffuse fraction of light with wavelength. Scattering of light by atmospheric aerosols tends to be greater at shorter wavelengths where scattering varies proportionally with $2\pi r_e/\lambda$, where r_e is the effective radius of the scattering elements and λ is the wavelength. This is important to model because the fraction of diffuse to direct radiation influences the anisotropy of the surface. The anisotropy is reduced when the diffuse irradiance is high because the contrast between shadowed and sunlit surfaces decreases. Anisotropy is similarly dependent for bright targets owing to the multiple-scattering of light between the surface elements. The atmospheric scattering elements such as aerosols are comparable in size to the wavelength of light at optical wavelengths. As a result, the effect of atmospheric scattering on the anisotropy will be a function of wavelength and the shape of the BRDF will vary. This is clearly demonstrated in Fig. 6.2 which shows the BRDF from Flight [North, 1996] canopy radiative transfer simulations for Old Jack Pine forest with a solar zenith angle of $45°$ differs for the red (670 nm) and near-infrared (870 nm) channels of the

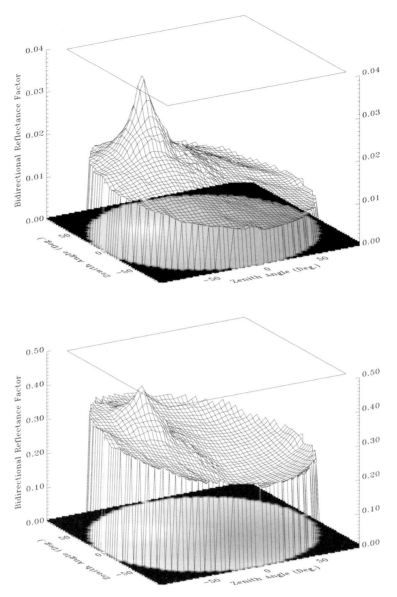

Fig. 6.2. BRDF obtained from Flight simulations for Old Jack Pine forest canopy type with a solar zenith angle of 45° corresponding to the (a) red (670 nm) and (b) near-infrared (870 nm) channels of the (A)ATSR instruments.

(A)ATSR instruments. Taking these effects into account results is a physical model of spectral change with view angle [North et al. 1999]

$$\rho_{\mathrm{mod}}(\lambda, \Omega) = (1 - D(\lambda))v(\Omega)w(\lambda) + \frac{\gamma w(\lambda)}{1 - g}[D(\lambda) + g(1 - D(\lambda))], \qquad (11)$$

where $g = (1 - \gamma)w(\lambda)$, λ is the wavelength, Ω is the viewing geometry (forward or nadir view in the cases of ATSR-2 and AATSR), ρ_{mod} is the modeled bi-directional reflectance, γ is the fraction contributing to higher-order scattering and is set equal to 0.3 [North et al., 1999], D is the fraction of diffuse irradiance, v is a wavelength independent parameter, and w is a parameter independent of the viewing and illumination geometry. The first and second terms refer to direct and diffuse scattering, respectively. The model separates the angular effects of the surface into two components (see also Diner et al., 2005b), a structural parameter v that is dependent only on the view direction, and the spectral parameter w that is dependent only on the wavelength. The free parameters that we want to retrieve through model inversion are $w(\lambda)$ and $v(\Omega)$. If we assume that the influence of diffuse irradiance on surface reflectance is negligible (i.e. $D(\lambda) = 0$) then we get

$$\rho_{mod}(\lambda, \Omega) = v(\Omega)w(\lambda) + g\frac{\gamma w(\lambda)}{1 - g}. \tag{12}$$

Conversely, under completely diffuse illumination conditions, where $D(\lambda) = 1$, we get:

$$\rho_{mod}(\lambda, \Omega) = \frac{\gamma w(\lambda)}{1 - g}. \tag{13}$$

However, these situations are most likely to occur under cloudy conditions where we will be unable to retrieve AOD. By inversion of Eq. (11), this model of surface scattering has been shown theoretically to lead to a tractable inversion method which is potentially more robust than the simple assumption of angular invariance alone [North et al., 1999]. The angular reflectance of a wide variety of natural land surface fits this simple model [North et al., 1999]. In contrast, reflectance that is a mixture of atmospheric and surface scattering does not fit this model well. As a result, the model can be used to estimate the degree of atmospheric contamination for a particular set of reflectance measurements and to find the atmospheric parameters which allow retrieval of a realistic surface reflectance. The minimum inputs into the algorithm are radiances from two bands and two solar and viewing geometries, so it can be applied to all existing multi-view-angle sensors.

3.3 Numerical inversion

In remote sensing we often want to solve the inverse problem, where we need to make inferences about a quantity that cannot be measured directly. The unknown environmental variables $\mathbf{V}$ that we wish to retrieve are a function of a known set of measured satellite radiances $\mathbf{L}_{atm}$ for a range of wavelengths and viewing and solar geometries, such that $\mathbf{V} = f^{-1}(\mathbf{L}_{atm})$ where the forward problem is expressed as $\mathbf{L}_{atm} = f(\mathbf{V})$. Thus, if the geophysical and biophysical properties of the atmosphere and surface are known then we can predict the radiances measured by a remote-sensing instrument. However, it is the unknown geophysical and biophysical properties from the observed radiances that are required.

For retrieving aerosol properties from top-of-atmosphere reflectance data we need to solve the model given in Eq. (1) and expressed again here but in a simpler form as [Chylek et al., 2003]:

$$L_{TOA} = L_M + L_A + T(\theta_v)L_S, \tag{14}$$

where L_{TOA} is the measured radiance, L_M and L_A are the atmospheric molecular and aerosol scattering, respectively and L_S is the surface-leaving radiance. We neglect the multiple scattering between the molecules and aerosols. By neglecting the interaction between the Rayleigh and aerosol scattering, we should be aware that our retrievals of aerosol properties at large solar zenith angles at high latitudes may be unreliable.

In practice we have L_{TOA} but we need to find all the terms in Eq. (14) in order to retrieve the aerosol properties. The land surface model Eq. (11) allows us to obtain L_S from multi-view-angle observations, and L_M is easily defined as a function of surface pressure and climatology. The measured satellite radiances are equal to the atmospheric path radiance once surface-leaving radiance has been taken into account. After correcting for gaseous absorption and Rayleigh scattering, the remaining contribution to the signal is dominated by atmospheric aerosol scattering. Thus, we can find the remaining unknown term L_A, and with a prescribed aerosol model defining the single scattering albedo and phase function we can retrieve AOD.

The inherent difficulty when solving inverse problems is that there are often fewer measurements than model parameters, thus the problem needs to be regularized in order to provide a unique solution. This is where much of the uncertainty in the retrieval of aerosol properties can arise because value-based assumptions are sometimes used to regularize the problem. For instance, we often use an aerosol model to constrain some of the aerosol properties. These can be set as free parameters, which better characterize the atmospheric state than a rigid prescribed aerosol model based on climatology or we can use derived parameters from ground-based observations [Levy et al., 2007].

To regularize the problem so that AOD is the only unknown atmospheric parameter, assumptions are made concerning the other aerosol optical properties including phase function and single-scattering albedo. In practice, a range of models representing generalized aerosol climatologies (smoke, urban, desert dust, continental and maritime) are used to constrain the inverse problem. A summary of these properties for the models used in our scheme are given in Table 6.2. The selection of aerosol model depends on the location and time of year, and the accuracy of the AOD estimates will depend on how well the generalized aerosol model characterizes the actual atmospheric state.

Table 6.2. Optical properties of the five aerosol models used in the 6S radiative transfer code to retrieve AOD and bi-directional reflectance derived using Mie theory.

Aerosol model	Single scattering albedo at 550 nm	Asymmetry parameter at 550 nm	Ångström coefficient 550–630 nm
Biomass (smoke)	0.97	0.68	1.44
Continental	0.89	0.64	1.17
Dust	0.94	0.70	0.30
Maritime	0.99	0.74	0.22
Urban (black carbon)	0.65	0.59	1.34

Our aim is to retrieve AOD from multi-angle top-of-atmosphere (TOA) cloud-free radiances. This is achieved through a coupled inversion of the 6S (Second Simulation of the Satellite Signal in the Solar Spectrum) radiative transfer model of Vermote et al. [1997] and the model of surface scattering given in Eq. (11). The inversion is achieved through iteration of a two-stage numerical process [North et al., 1999], the schematic of which is presented in Fig. 6.3. In this scheme the Powell multi-dimensional minimization routine is nested within the Brent one-dimensional numerical optimization algorithm [Press et al., 1992]. One convenient feature of both these optimization routines is that the partial derivatives are not required. These routines are quick and are generally robust to finding the global minima. However, as with many numerical routines, they may find a local minima instead. The algorithm does not explicitly screen for local minima, but we are confident these cases are rare. In tests using simulated data our forward and inverse modeled estimates of AOD were similar in all cases, indicating that the global minima were always found. Only in situations where the cost function is complex and highly nonlinear does false convergence become a significant problem. In contrast, for a given aerosol model and surface reflectance AOD varies smoothly with measured TOA.

Once a model of the surface scattering is obtained, we can find the atmospheric path radiances by using an atmospheric radiative transfer model. The optimal aerosol properties are found by minimizing the difference between the derived and modeled set of surface reflectances. We use the 6S atmospheric radiative model to iterate through a possible range of AODs and aerosol models. Aerosol models are used to constrain the inverse problem and provide generalized climatologies of other aerosol properties such as single scattering albedo and phase function.

The first stage is to retrieve a set of surface reflectances and estimates of diffuse irradiance given an initial estimate of the atmospheric aerosol model and AOD at 550 nm by inversion of 6S. The best-fit of the two parameters v and w from Eq. (11) for two observations of the same target from different angles at the same wavelength are found using Powell routine. The fraction of diffuse irradiance D is calculated in 6S from the AOD value and aerosol model used for that interaction. In operation this optimization routine is nested within the Brent minimization routine. The interaction between the Brent and Powell routines is shown in Fig 6.3. The second stage uses Brent to iterate through a range of AOD values to converge on the optimum value for AOD at 550 nm. When the best fit of the parameters for a given AOD is found, another iteration is calculated using a different estimate of AOD. This process continues until the interaction between AOD and the two v and ω parameters leads to the overall best-fit solution. Generally, it takes between 10 and 20 Brent iterations to converge on an optimum estimate of AOD. For each iteration, each estimate of AOD results in a different set of surface reflectance values. The dual-view algorithm makes use of the full information content of observations and incorporates a set of eight reflectance values in the four AATSR optical channels and at the two looks. The optimum value of AOD and aerosol model is selected on the basis of best-fit of surface reflectances to the model given by Eq. (11), and is attained by minimizing the error function E_{mod}

$$E_{\mathrm{mod}} = \sum_{\Omega=1}^{2} \sum_{\lambda=1}^{4} \left[\rho_{surf}(\lambda, \Omega) - \rho_{\mathrm{mod}}(\lambda, \Omega) \right]^2, \tag{15}$$

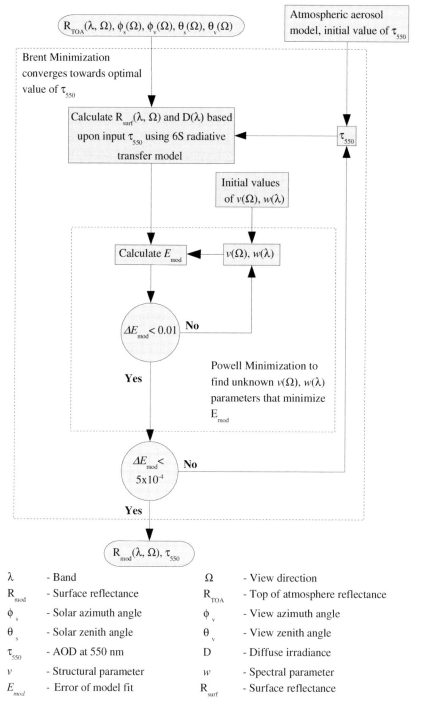

Fig. 6.3. Schematic showing numerical inversion of AOD and surface reflectance retrieval. This is achieved through iteration of a nested two-step numerical optimization process. Reproduced from Grey et al. [2006a], with permission from IEEE.

where ρ_{mod} is the surface reflectance estimated using Eq. (11) based on the best-fit values of the parameters v and w, and ρ_{surf} is the surface reflectance calculated using 6S given the TOA reflectance ρ_{TOA} and the estimated atmospheric profile. Surface reflectance is related to the TOA reflectance by rearranging Eq. (1) [Vermote et al., 1997]

$$\rho_{surf}(\theta_s, \theta_v, \phi_s - \phi_v, \lambda) = \frac{\rho'_{TOA}}{1 + \rho'_{TOA}S}, \tag{16}$$

where θ_s is the solar zenith angle, θ_v is the view zenith angle, ϕ_s is the solar azimuth angle, ϕ_v is the view azimuth angle, S is the atmospheric spherical albedo, and ρ'_{TOA} is

$$\rho'_{TOA} = \frac{\rho_{TOA}(\theta_s, \theta_v, \phi_s - \phi_v, \lambda) - \rho_{atm}(\theta_s, \theta_v, \phi_s - \phi_v, \lambda)}{T(\theta_s)T(\theta_v)}, \tag{17}$$

where ρ'_{atm} is the intrinsic atmospheric reflectance, and $T(\theta_s)$ and $T(\theta_v)$ denote downward and upward transmittance, respectively, and ρ_{TOA} is given by:

$$\rho_{TOA} = \frac{\pi L_{TOA}}{\cos(\theta_s)F_0}. \tag{18}$$

AOD will be retrieved at 550 nm, but values of AOD can be attained at other wavelengths depending on the selected aerosol model.

3.4   Look-up tables of atmospheric parameters

We have developed a robust and computationally efficient method for retrieving estimates of AOD from (A)ATSR that can be applied operationally at regional and global scales. Computational efficiency is achieved by using pre-calculated look-up tables for the numerical inversion of a radiative transfer model of the atmosphere instead of the full inversion approach that is more computationally intensive. Performing inversions using repeated forward model runs of atmospheric radiative transfer models is computationally very expensive. By using look-up tables, the most computationally expensive aspect is completed prior to inversion. Here, we use look-up tables created from forward runs of the 6S atmospheric radiative transfer model. The values within the look-up tables are composed of the fraction of diffuse irradiance, atmospheric transmittance and spherical albedo, allowing us to retrieve the aerosol properties and bi-directional surface reflectance. The parameters in the look-up tables are dependent on the solar and viewing geometry and the atmospheric profile. The look-up tables are constructed at the four visible and near-infrared bands of AATSR in four-dimensions and indexed with AOD at 550 nm (from 0 to 3 at 0.05 intervals), solar zenith angle (from $0°$ to $80°$ at $10°$ intervals), view zenith angle (from $0°$ to $60°$ at $10°$ intervals), and relative azimuth angle (from $0°$ to $180°$ at $20°$ intervals) (see Table 6.3). The look-up tables are generated for the five tropospheric aerosol models (see Table 6.2). Only at the grid points do we have the modeled values in the look-up tables. During operation, the atmospheric parameters are estimated in the look-up tables using piecewise multi-dimensional linear interpolation. In this approach a linear model is applied between neighboring grid points in all dimensions. Mathematically, this can be expressed as

Table 6.3. Summary of look-up tables grid points of atmospheric parameters and indices

Parameter	Minimum	Maximum	Interval
AOD	0	3.0	0.05
Solar zenith angle	$0°$	$80°$	$10°$
View zenith angle	$0°$	$60°$	$10°$
Relative azimuth angle	$0°$	$180°$	$20°$

$$C = C_0 + \sum_{k=1}^{k=4} \partial x_k \frac{\partial C_k}{\partial x_k} \qquad (19)$$

where C is the interpolated parameter, C_0 is the parameter at the reference grid point, and ∂x represents the shift along each of the four parameter vectors k $(\tau, \theta_v, \theta_s, \phi)$ away from the reference grid point towards the other corner grid points. The partial derivatives represent the gradient along each dimension. By using look-up tables there will inevitably be a small decrease in the accuracy of the retrieved measurements compared with performing on-the-fly inversions, but the method gives an approximately 60-fold increase in speed [Grey et al., 2006b].

4.   AATSR AOD retrieval and algorithm validation

4.1   Simulated datasets

To develop and test the AOD retrieval algorithms we often use synthesized Earth observation data derived from coupled surface and atmospheric radiation transfer models. Given the central role that simulated datasets have played in our algorithm development we explain here how these datasets are generated and used but do not present any specific results of our analysis. Our goal here is to explain the purpose and value of these simulated datasets.

There are several reasons why we might want to use simulated datasets. Firstly, if we are developing an operational algorithm for a satellite instrument that has yet to be launched where no data are available, a synthesized dataset configured with the same spectral and viewing characteristics can be used to test and develop the algorithm and potential of the available dataset. For instance, this was the case with MODIS where much of the algorithm development occurred before the satellite instrument was launched. Secondly, simulated data allows us to perform controlled experiments, where we can examine the influence of one particular variable on our retrieval of AOD.

Despite the benefits of using simulated data, we need to be very careful when interpreting the results that are based on simulation studies. To begin with it is very difficult to properly characterize the noise within a particular sensor system. Usually we rely on some statistical random model to generate noise in our synthetic spectra to take account of uncertainties in calibration and so on but it still may not realistically characterize all noise sources. In addition, the atmospheric radiative transfer codes cannot fully characterize the geophysical characteristics of the atmosphere, because they are by nature of models that provide a simplified representation of reality. Thus, there is no substitute for the use of satellite-acquired data for validating algorithms and data characteristics, but synthetic datasets do have their value even if we do need to be cautious in the inferences that we can draw.

A large set of simulated data can be created using coupled surface and atmospheric radiative transfer models for a range of atmospheric and surface conditions. A canopy radiative transfer model such as that of North [1996] can be used to generate a representative set of vegetated surfaces corresponding to forests and grasslands. Similarly ocean and soil surface models can be used. The viewing and illumination conditions correspond to realistic satellite geometries. The generated surface reflectances are then converted to TOA reflectances using an atmospheric radiative transfer model in the forward mode. We select realistic atmospheric parameters for the corresponding land surface using a range of aerosol models (desert, biomass, continental and maritime) and for a range of AODs. Finally statistical noise is added to the TOA reflectances in order to simulate poor calibration, image misregistration and other potential sources of error. The schematic in Fig. 6.4 summarizes how TOA reflectance measurements are generated based on a coupling between the surface and atmospheric radiative transfer models. Once the dataset is generated we can then retrieve aerosol properties from the generated TOA radiances by using the aerosol retrieval algorithms to solve the inverse problem. The algorithms are applied to the simulated TOA reflectances. If the algorithms perform well then the retrieved AOD will be consistent with the AOD used in the forward model of the atmospheric radiative transfer to generate the TOA reflectances. Inversions are performed on both the noise-free and Gaussian noise-added TOA reflectances. Low RMSEs indicate that the algorithm is performing well, when the dynamic range is large.

The initial development of the dual-view AATSR algorithm was performed using simulated (A)ATSR satellite data [North et al., 1999]. This allowed refinement of the performance of the algorithm and testing against other candidate algorithms for controlled

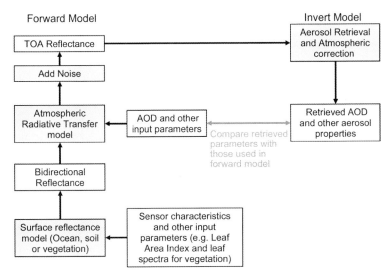

Fig. 6.4. Forward and inverse modeling experiments from simulated TOA reflectance measurements.

experimental scenarios. Recently simulated data have been used to examine the performance of the future Sentinel-3/SLSTR instrument which has a different off-nadir geometry to the AATSR and an additional SWIR channel.

4.2   Comparison against AERONET

AERONET (the Aerosol Robotic Network) is a federated system of Cimel sunphotometers that measure solar and sky radiances from the ground and is dedicated to the long-term monitoring of atmospheric properties worldwide [Holben et al., 1998]. The measurement of the solar irradiance at the ground offers the most direct way of determining aerosol optical depth, so this extensive dataset serves as the principal means of validating the satellite-based retrieval of aerosol properties. The accuracy of AERONET derived AOD is reported to be within 0.02 [Smirnov et al., 2000]. Validation against AERONET establishes limits of applicability and quantitative error in all the main aerosol properties to be derived. The satellite-derived AOD at 550 nm have been compared with ground-based sunphotometer measurements taken within 1 hour of the AATSR image acquisition. AERONET retrieves AOD at multiple wavelengths between 340 and 1,020 m, but does not retrieve AOD at 550 nm; therefore to allow direct comparison with the satellite-derived aerosol estimates, the ground-based sun-photometer measurements were interpolated to 550 nm using

$$\tau(\lambda) = \tau(\lambda_i)\left(\frac{\lambda}{\lambda_i}\right)^{-a}, \tag{20}$$

where τ is the AOD and a is the Ångström exponent, which is inversely related to the size of aerosol particles and describes the variability of AOD with wavelength. The Ångström exponent is calculated from τ at two wavelengths

$$a = \frac{\log_{10}(\tau(\lambda_i)) - \log_{10}(\tau(\lambda_j))}{\log_{10}(\lambda_i) - \log_{10}(\lambda_j)}, \tag{21}$$

where $\tau(\lambda_i)$ and $\tau(\lambda_j)$ are the AODs at wavelengths λ_i and λ_j, respectively. For interpolating to 550 nm we substitute λ_i and λ_j with 440 nm and 670 nm, respectively. Our algorithm can indirectly retrieve a through selection of the best-fit aerosol model, but the reliability of these retrievals have yet to be evaluated.

For comparison with AERONET, AATSR radiances are averaged over a 15 by 15 km area around the location of the corresponding site in order to reduce noise and minimize the effect of co-registration errors between the nadir and forward looks of the AATSR instrument. In addition, identifying and removing cloudy pixels is also a prerequisite for aerosol retrievals. The standard approach to identifying cloud is to apply a set of threshold and statistical tests that were developed by Stowe et al. [1997] and Zavody et al. [2000] to optical and thermal infrared channels. However, these tests are not always effective at removing all cloud over land, partly because they were developed for ocean [Plummer, 2008]. Thus additional cloud tests have been incorporated to ensure that as much cloud as possible is screened out. One such test is the gross cloud threshold that uses the brightness temperatures in the 12 μm channel. The brightness temperature below which we assume cloud varies between 210 and 270 K depending on climatology [Birks, 2007]. Another test based on NDVI is also implemented. Negative NDVI values are considered to be cloud although we cannot distinguish between cloud and snow. Further details on these can be found in Plummer [2008], Birks [2007], Zavody et al. [2000] and Stowe et al. [1997].

The averaged TOA reflectances within the window are calculated from only the cloud-free pixels and are only considered to be cloud-free if more than 40 % of the pixels are not identified as cloud. This represents a cloud-free conductive scheme that results in some clear pixels being incorrectly labeled as cloudy, although there still may sub-pixel cloud contamination in some pixels that pass through the cloud-masking stage.

For validation we acquired more than 200 cloud-free coincident estimates of AOD between the AATSR and AERONET for 19 sites from around the world representing a range of land covers (e.g. deserts, savanna, northern boreal regions, urban areas), aerosol types, and latitudes to examine the effect solar and viewing geometry may have. The overall Pearson's correlation coefficient, for all 19 sites combined is 0.70, although the absolute error of the residuals increases with increasing AOD (see Fig. 6.5). The RMSE (root mean square error) of all the data is 0.16. There is little evidence of systematic error in the estimates of AOD as the mean AOD over all AERONET measurements is 0.27 compared with a mean of the AATSR-derived AOD estimates of 0.26. A site-by-site breakdown summary of statistics is presented in Grey et al. [2006a].

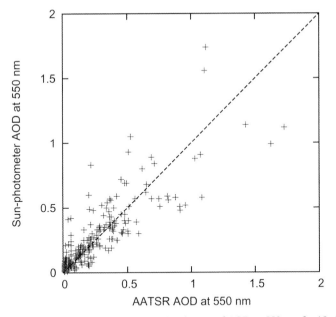

Fig. 6.5. Plot of AERONET versus AATSR derived estimates of AOD at 550 nm for 19 sites worldwide (Alta Floresta (− 9.9° N, 56.0° E), Banizoumbou (13.5° N, 2.7° E), Beijing (40.0° N, 116.4° E), Cart Site (36.6° N, − 97.4° E), Chulalongkorn (13.7° N, 100.5° E), Ilorin (8.3° N, 4.3° E), Jabiru (− 12.7° N, 132.9° E), Kanpur (26.5° N, 80.4° E), Konza (39.1° N, − 96.6° E), Lake Argyle (− 16.1° N, 128.7° E), Lille (50.6° N, 3.1° E), Mexico City (19.3° N, − 99.2° E), Mongu (− 15.3° N, 23.2° E), Oostende (51.2° N, 2.9° E), Ouagadougou (12.2° N, − 1.4° E), Phimai (15.2° N, 102.6° E), Solar Village (24.9° N, 46.4° E), Tinga Tingana (− 29.0° N, 140.0° E), Tomsk (56.5° N, 85.0° E)). The dashed line corresponds to the 1:1 line. Reproduced from Grey et al. [2006a] with permission from IEEE.

Differences between the sunphotometer- and AATSR-derived estimates of AOD may be due to a number of factors including the small time differences between acquisition of sunphotometer measurements and the satellite overpass, undetected sub-pixel cloud contamination, and heterogeneity of the land surface within the 15 by 15 km area of AATSR observations. In addition, scaling between the point measurement of the AERONET and areal measurement of the satellite reflectances integrated over a larger area may also cause differences. The selection of an aerosol model that does not properly characterize the atmospheric scattering may also lead to uncertainties in our retrieved estimates of AOD. For instance, these inversions do not take account of non-spherical dust particles, and improved estimates of AODs can be achieved if non-spherical rather than spherical dust particles are assumed. Future work will look to the improvement of the aerosol models.

4.3 Time series of AOD

The AATSR and ATSR-2 dataset when used together potentially represents a unique 13-year archive of AOD retrievals over land. An example of a much shorter time-series of AATSR derived AOD at 550 nm, from a site in the Amazon at Abracos Hill ($-62.36°$ E, $10.76°$ N) corresponding to AERONET retrievals is given in Fig. 6.6. In general there is good agreement between the AATSR and AERONET estimated AOD at 550 nm, but occasionally there are spikes in the AATSR observations where AOD is overestimated. This is most likely due to the failure of the cloud removal algorithm to identify all cloudy pixels. Thus, during the inversion unscreened cloud is considered as a high concentration of aerosol hence the peaks in the AATSR time-series, when there is no corresponding peak in the AERONET estimates of AOD. In addition, for September 2005 the AATSR estimates of AOD are underestimated compared with the AERONET retrievals. During the months September to November there is large amount of vegetation burning as land is cleared for agricultural uses. Consequently smoke aerosol concentrations are high (typically with AOD greater that 1 at 550 nm) during this period and is clearly identified in both the AATSR and AERONET retrievals. Thus AATSR is useful not only in providing a time-series of observations, but also in providing these globally. Clearly though there is a need to better discriminate clouds over tropical regions.

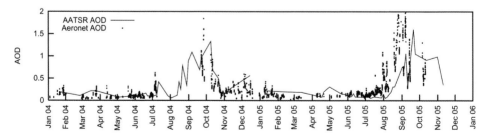

Fig. 6.6. Time-series of AATSR and AERONET AOD at 550 nm from January 2004 to December 2005 at Abracos Hill in the Amazon.

4.4   Satellite inter-comparisons

Aerosol products derived from different satellite sensors should be consistent despite their different characteristics in terms of temporal, spatial, polarization, angular and spectral information content. A range of algorithms have been developed because the instruments have different characteristics, and this needs to be taken into account. For some instruments (e.g., AATSR) several algorithms have been developed. Although these retrieval algorithms are different they should produce consistent values for the same aerosol properties for a given scene. However, inter-comparisons between different aerosol products derived from different passive optical satellite instruments acquired at the same time over the same location have shown not to agree as well as they should. This point has been realized in a recent study by Kokhanovsky et al. [2007] for a scene over Germany that showed that even estimates of the same aerosol optical properties derived from the same sensor and input images were not consistent between the range of algorithms that have been developed. Thus, there is some concern within the satellite aerosol community that this issue needs to be resolved if satellite estimates of aerosol properties are to have any credibility. Scatterplots of satellite estimates of AOD versus AERONET abound in the literature and there is often high correlation (e.g., Grey et al. [2006a], Abdou et al. [2005] and Chu et al. [2002]). Yet when inter-comparisons are made between different aerosol products over large spatial extents agreement tends to be much poorer [Mishchenko et al., 2007a]. This is possibly because the aerosol retrieval algorithms are developed and tested over a relatively small number of AERONET sites and may not be generically applicable to other land covers and aerosol types. Currently, there is poor understanding of why there are such large differences between the different aerosol products and there is a need to reappraise some of the products and to establish exactly why the different estimates of aerosol are not consistent. This will allow us to arrive at product convergence.

The inherent difficulty in retrieving aerosol properties is that there are more free parameters than there are measurements, thus a range of assumptions need to be made in order to regularize the problem. How these assumptions are derived is often value-based. Therefore, differences shown are due only to different processing schemes and it is important to understand the reasons behind these discrepancies [Kokhanovsky et al., 2007]. There are a number of steps within the processing scheme where differences can arise. Firstly, the different algorithms have independent cloud masking and resampling strategies. Thus, different radiances are ingested at the fundamental algorithm level depending on masking and pixel aggregation. Secondly, the variation between the algorithms themselves and the assumptions upon which they are based will result is different estimates of the surface reflectance. Thirdly, the selection of aerosol models used to regularize the inverse problem is value-based and contains different parameterizations of aerosol optical properties such as single scattering albedo. Fourthly, a range of atmospheric radiative transfer models have been developed (e.g., 6S and MODTRAN (MODerate resolution atmospheric TRANsmission)) and different aerosol retrieval schemes use different radiative transfer models some using vector codes others scalar. Finally, the sampling of the look-up tables of atmospheric parameters will influence the aerosol retrievals. Moreover, inter-sensor comparisons have the additional difficulty that there can be small time differences between image acquisition owing to different overpass times. However this cannot explain the poor agreement between the MERIS- and AATSR-derived aerosol estimates of AOD because both instruments are onboard the same platform and therefore imaging the surface simultaneously [Kokhanovsky et al., 2007].

Thus, it is no surprise that the different aerosol products vary as much as they do given that so many assumptions are required at different stages in the processing scheme. An in-depth understanding of each of the steps in the processing schemes will lead to better performance and also convergence of the various algorithms. Good agreement between aerosol retrievals from different satellite datasets and algorithms will enhance confidence in remotely sensed estimates of aerosol properties within climate studies.

4.5 Operational implementation

In order to test the aerosol retrieval look-up table scheme at global-scales, approximately 400 AATSR ATS-TOA-1P striplines for September 2004 were processed for AOD and bi-directional surface reflectance. The results are a spatially and temporally composited mosaic with a pixel size of 5 arcminutes (0.1°) as shown in Figs 6.7 and 6.8. In order to generate a global-scale product of AOD from the input AATSR ATS-TOA-1P products the following pre-processing steps are performed.

Firstly, we perform cloud screening using the scheme of Plummer [2008] and Zavody et al. [2000]. The AOD retrieval algorithm is applied to integrated pixels within a 9 by 9 km window in this case. This provides us with a resolution of approximately 0.1 degrees at the equator and is approximately the same as the MODIS (10 km) and MISR (17 km) AOD products. When compositing the images we need to strike a balance between providing as much coverage as possible, but not integrating over too long a time period because aerosols are temporally and spatially dynamic. Given the swath width of (A)ATSR this works out at about a month, but even over this time period there are regions where no cloud-free data are available, for instance over tropical regions, and so neither AOD nor surface reflectance can be retrieved.

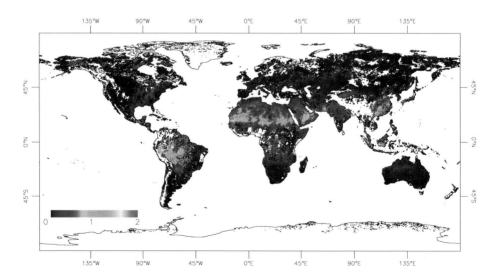

Fig. 6.7. A global composite of AATSR-derived AOD at 550 nm for September 2004 over land.

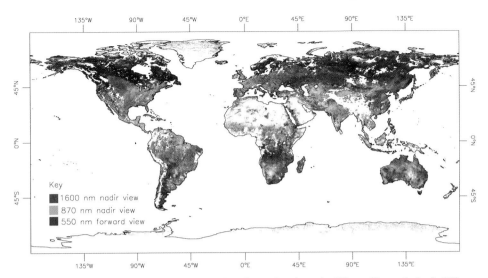

Fig. 6.8. A global surface reflectance composite of AATSR channels 550 nm (forward view), 870 nm (nadir view), and 1,630 nm (nadir view) displayed in green, red and blue, respectively.

Fig. 6.7 shows that during September 2004 high levels of AOD were present over the Sahel in West Africa. This is because continental-wide dust plumes are common at this time of year owing to the strong harmattan winds. The largest sources of dust are the Bodélé depression in northern Chad and in northern Mauritania and Mali [Koren et al., 2006]. In general the dust is blown from the east towards the Atlantic Ocean. The high reflectivity of the Sahara in the optical region and the lack of contrast between the surface and overlying aerosol makes this one of the most challenging environments over which to retrieve aerosol properties. Approximately 10^9 kg of dust is emitted from the Sahara each year [IPCC, 2007], thus understanding and monitoring aerosols in this region is particularly important. Another large dust source is in southern Saudi Arabia. Throughout this discussion, we do recognize the limitation of using a spherical model assumption for dust particles, but comparisons of our retrieval using this approach do compare well against AERONET retrievals even if there is a small overestimate [Grey et al., 2006b], and our aerosol model is not ideal. Another potential source of error is that the retrieved AOD is influenced by the background surface reflectance. This is particularly apparent over the Sahara desert, where the shapes of surface geomorphological features can be traced in spatial patterns of the AOD estimates.

Other regions that show high concentrations of AOD are the Amazon and South Western China. In the Amazonian Basin during September there is considerable vegetation-burning for agricultural land clearance. Over many parts of China there is a thick layer of haze resulting from anthropogenic emissions [Li et al., 2007], and this can in part explain the high level of AOD retrieved from AATSR over this region.

The corresponding global surface reflectance false color composite for AATSR is shown in Fig 6.8. These are the retrieved reflectances based on the estimated AOD. It is possible to reconstruct the original TOA radiances from the surface reflectances and AOD. One of the key problems is cloud, where over some parts of the globe no surface data are available for the whole month, and in some locations there is cloud contamination.

5. Avenues for future research

There are a range of standard improvements that can be made to the retrieval scheme including the retrieval of other aerosol properties and AOD at different wavelengths, the use of more aerosol models that better characterize the atmospheric profile, including the explicit representation of non-spherical particles, and the use of look-up tables with more parameters (e.g., water vapor, ozone and surface pressure). Here, we identify two avenues of future work, the generation of a long-term record of AOD from (A)ATSR, and the convergence of different AOD products from different satellite instruments and algorithms.

5.1 A global product of AOD from AATSR and ATSR-2

Our operational multi-look AOD retrieval approach is currently being implemented within ESA's GPOD (Grid Processing on Demand) high-performance computing facility for global retrievals of AOD and bi-directional reflectance from ATSR-2 and AATSR. The aerosol retrieval code is applied to the (A)ATSR ATS-TOA-1P striplines. To improve efficiency and because of the misregistration between corresponding pixels in the forward and nadir views of AATSR, AOD is retrieved on a coarser grid of 10 by 10 pixels. Surface reflectance is retrieved for every 1 km pixel corresponding to the top-of-atmosphere observations.

We do not make assumptions concerning the aerosol climatology, thus we calculate AOD separately five times using different aerosol models with their parameters summarized in Table 6.2. Therefore, five sets of AOD, bi-directional reflectances and model best-fit errors are produced. The optimal AOD and aerosol model are selected on the basis of the minimization of the error of best-fit; however, all five sets of aerosol estimates and errors are retained for future experiments.

The software has been written in the C programming language and is tested on a standard desktop with a 2.8 GHz processor and 1 GB RAM under Fedora Linux, and is computationally efficient both in terms of processing requirements and memory usage. In tests performed on this system, it generally takes less than 5 minutes to retrieve aerosol optical depth and surface reflectance from a single orbital AATSR ATS-TOA-1P stripline for one aerosol model. For each stripline we apply five aerosol models to each stripline, thus the total processing time for each stripline is less than 30 minutes even after dealing with the input/output and format conversion issues. To derive AOD and surface reflectance globally over several years, we intend to process all of the ATS-TOA-1P products in the archive. Although the software is computationally efficient, the large volume of AATSR data (approximately 20,000 striplines at 14 Tb) in the archive makes this a substantial processing requirement.

The principal output products from the processor will be the surface bi-directional reflectance at all the (A)ATSR optical channels and both looks, aerosol optical depth at 550 nm at 10 by 10 km, and the error of the best fit of the model inversion. Also monthly and daily global composites of AOD at 0.10 degrees greed will be produced. The result will be a 12-year time-series of global estimates of surface and atmospheric aerosol properties. We do not explicitly retrieve the Ångström exponent which is an extremely useful variable from which particle size can be inferred. Large particles tend to have more spectrally flat AOD variation with wavelength, whereas smaller particles have greater variability. However, the Ångström exponent can be retrieved indirectly from the selection of the best-fit aerosol model. In addition to providing operational retrievals of aerosol and surface properties users will be able eventually to process the (A)ATSR data for their particular regions and time periods of interest through a web interface at eogrid.esrin.esa.int.

5.2 Convergence of aerosol products

The aerosol remote sensing community needs to make a concerted collaborative effort to arrive at convergence of the best estimates of aerosol properties (Kokhanovsky and de Leeuw, 2009). In order to use these products operationally for assimilation into climate models for instance, there needs to be much better consistency between the different aerosol products and understanding of uncertainty. Thus, an inter-comparison experiment is required. Greater consistency between different aerosol products will give us greater confidence in our satellite retrieved aerosol properties. Up to now groups have developed independently their own systems for the operational retrievals of AOD from space. There exist at least three independent AATSR retrievals and three MERIS algorithms. Both instruments are placed on the same satellite platform and observe the same area at the same time with MERIS having larger swath with and spatial resolution (0.3 km as compared to 1 km for AATSR). Agreement between the retrieved AODs is not as good as might be expected. We need to understand why these differences arise at each stage in the various processors.

6. Summary

Owing to the complexity of the atmospheric and surface coupled radiative transfer problem we are required to make a number of assumptions in our parameterizations and model development. To begin with the invariance of the BRDF with wavelength under direct illumination conditions is a critical assumption in the surface model development. This effect has been widely documented and quantified and shown to be reliable over a number of land surface types (e.g., North et al. [1999], Veefkind et al. [1998] and Flowerdew and Haigh [1996]). The atmospheric radiative transfer problem is extremely complex and we do make some simplifying assumptions. The key assumptions we make that probably have greatest influence on the accuracy of our aerosol retrievals are the not account for the polarization of light, limited climatologies of aerosol types, no multiple scattering of radiation between aerosol and gases, assumed spherical dust particles, no water absorption at AATSR wavelengths, and constant surface pressure.

Of course, we can always improve our algorithms by increasing their complexity but we also have to consider the additional computational burden required. But it is only through development of the first-generation aerosol retrieval algorithms that we learn of the sensitivity and limitations of our assumptions and improvements that we can make to subsequent versions of our algorithms.

The dual-view capability of the (A)ATSR instruments have the potential to make a major contribution to atmospheric aerosol science, particularly over land, where retrieval of aerosol properties is notoriously difficult. Combined with this is the development of an algorithm that exploits this multi-view-angle capability allowing robust estimates of AOD without recourse to prescribed information on the land surface. The algorithm solves the inverse problem of separating out the atmospheric and surface scattering contributions to the observed signal at the satellite level. The scheme makes use of numerical minimization routines to converge on optimal estimates of atmospheric properties and surface reflectances. To allow for rapid inversion, the retrieval code uses pre-calculated look-up tables of atmospheric coefficients derived from forward runs of the 6S atmospheric radiative transfer model. The good agreement between the AATSR- and AERONET-derived estimates of AOD in addition to the ability of this approach to derived estimates even over bright desert surfaces is a promising development. There exists now a 13-year archive of observations from these missions, allowing us to potentially explore this climate dataset for long-term trends. Moreover, the proposed Sentinel-3/SLSTR mission will ensure the continuity of this powerful climate dataset. Despite this, there are still some issues that need to be resolved particularly with the poor agreement between the different AATSR algorithms' estimates of aerosol properties. The lack of consistency between the various aerosol products is of major concern. In addition, the approach presented here is limited to the retrieval of AOD at 550 nm. Further developments with respect to the look-up tables of atmospheric properties are required. Also we do not explicitly retrieve the Ångström exponent. However, the Ångström exponent can be retrieved indirectly from the selection of the best-fit aerosol model, but in future algorithm developments we will retrieve this directly. Considerations of aerosol particle shape will lead to improved estimates of aerosol properties over dust sources, for instance. Finally, the synergistic use of both MERIS multispectral and AATSR multi-view-angle datasets will lead to major improvements in retrieved aerosol properties from space. The premise of synergy is that the additional information from multiple instruments will allow us to better constrain the radiative transfer inverse problem. This may result in more aerosol parameters being retrieved with less uncertainty compared with existing single-instrument algorithms.

Acknowledgments. William Grey was supported by the UK Natural Environment Research Council. We would like to thank the AERONET principal investigators for establishing and maintaining the sites. We also gratefully acknowledge the European Space Agency and the NERC Earth Observation Data Centre for providing the ATSR-2 and AATSR data. Thanks go to Suzanne Bevan for providing several of the figures.

References

Abdou, W.A., D.J. Diner, J.V. Martonchik, C.J. Bruegge, R.A. Kahn, B.J. Gaitley, K.A. Crean, L.A. Remer, and B. Holben, 2005: Comparison of coincident Multiangle Imaging Spectroradiometer and Moderate Resolution Imaging Spectroradiometer aerosol optical depths over land and ocean scenes containing Aerosol Robotic Network sites, *Journal of Geophysical Research, 110*, D10S07, doi:10.1029/2004JD004693.

Alton, P.B., P.R.J. North, J. Kaduk, and S.O. Los, 2005: Radiative transfer modeling of direct and diffuse sunlight in a Siberian pine forest. *Journal of Geophysical Research*, **110**, D23209, doi:10.1029/2005JD006060.

Bellouin, N., O. Boucher, J. Haywood, and M.S. Reddy, 2005: Global estimation of aerosol direct radiative forcing from satellite measurements. *Science*, **438**, 1138–1141.

Birks, A.R., 2007: Improvements to the AATSR IPF relating to Land Surface Temperature Retrieval and Cloud Clearing over Land, Technical Report. Rutherford Appleton Laboratory.

Chu, D.A., Y.J. Kaufman, C. Ichoku, L.A. Remer, D. Tanré, and B.N. Holben, 2002: Validation of MODIS aerosol optical depth retrieval over land. *Geophysical Research Letters*, **29**, 12, doi:10.1029/2001GL013205.

Chung, C.E., V. Ramanathan, D. Kim, I. Podgorny, 2005: Global anthropogenic aerosol direct forcing derived from satellite and ground-based observations. *Journal of Geophysical Research*, **110**, D24207, doi:10.1029/2005JD006356.

Chylek, P., B. Henderson, and M. Mishchenko, 2003: Aerosol radiative forcing and the accuracy of satellite aerosol optical depth retrieval. *Journal of Geophysical Research*, **108**, D24, 4764, doi:10.1029/2003JD004044.

Crutzen, P.J. 2006: Albedo enhancement by stratospheric sulfur injections: a contribution to resolve a policy dilemma? *Climatic Change*, **77** (3–4), 211–219.

Diner, D., J. Beckert, T. Reilly, C. Bruegge, J. Conel, R. Kahn, J. Martonchik, T. Ackerman, R. Davies, S. Gerstl, H. Gordon, J.-P. Muller, R. Myneni, P. Sellers, B. Pinty, and M. Verstraete, 1998: Multi-angle Imaging SpectroRadiometer (MISR) instrument description and experiment overview. *IEEE Transactions on Geoscience and Remote Sensing*, **36**(4), 1072–1087.

Diner, DJ. B.H. Braswell, R. Davies, N. Gobron, J.N. Hu, Y.F. Jin, R.A. Kahn, Y. KKnyazikhin, N. Loeb, J.-P. Muller, A.W. Nolin, B. Pinty, C.B. Schaaf, G. Seiz, and J. Stroeve, 2005a: The value of multiangle measurements for retrieving structurally and radiatively consistent properties of clouds, aerosols, and surfaces. *Remote Sensing of Environment*, **97**(4), 495–518.

Diner, D., J. Martonchik, R. Kahn, B. Pinty, N. Gobron, D. Nelson, and B. Holben, 2005b: Using angular and spectral shape similarity constraints to improve MISR aerosol and surface retrievals over land. *Remote Sensing of Environment*, **94**, 155–171.

Dubovik, O., A. Smirnov, B. Holben, M. King, Y. Kaufman, T. Eck, and I. Slutsker, 2000: Accuracy assessments of aerosol optical properties retrieved from AERONET sun and sky radiance measurements. *Journal of Geophysical Research*, **105**, 9791–9806.

Flowerdew, R.J., and J.D. Haigh, 1996: Retrieval of aerosol optical thickness over land using the ATSR-2 dual-look satellite radiometer. *Geophysical Research Letters*, **24**(4), 351–354.

Godslave, C. 1996: Simulation of ATSR-2 optical data and estimates of land surface reflectance using simple atmospheric corrections. *IEEE Transactions on Geoscience and Remote Sensing*, **34**(5), 1204–1212.

Grey, W, P. North, and S. Los, 2006a: Computationally efficient method for retrieving aerosol optical depth from ATSR-2 and AATSR data. *Applied Optics*, **45**, 2786–2795.

Grey, W.M.F., P.R.J. North, S.O. Los, and R.M. Mitchell, 2006b: Aerosol optical depth and land surface reflectance from multi-angle AATSR measurements: global validation and inter-sensor comparisons. *IEEE Transactions on Geoscience and Remote Sensing*, **44**(8), 2184–2197.

Guanter, L., M. Del Carmen Gonz\'{a}lez-Sanpedro, J. Moreno, 2007: A method for the atmospheric correction of ENVISAT/MERIS data over land targets. *International Journal of Remote Sensing*, **28**(3–4), 709–728.

Hansen, J., A. Lacis, R. Ruedy, M. Sato, 1992: Potential climate impact of Mount-Pinatubo eruption. *Geophysical Research Letters*, **19**(2), 215–218.

Holben, B.N., Y.J. Kaufman, T.F. Eck, I. Slutsker, and D. Tanré 1998: AERONET – A federated instrument network and data archive for aerosol characterization. *Remote Sensing of Environment*, **66**, 1–16.

Holzer-Popp, T., M. Schroedter, and G. Gesell, 2002: Retrieving aerosol optical depth and type in the boundary layer over land and ocean from simultaneous GOME spectrometer and ATSR-2 radiometer measurements, 1, Method description. *Journal of Geophysical Research*, **107**, doi:10.1029/2001JD002013.

Hsu, N., J. Herman, O. Torres, B.N. Holben, D. Tanré, T.F. Eck, A. Smirnov, B. Chatenet, and F. Lavenu, 1999: Comparisons of the TOMS aerosol optical thickness: results and applications. *Journal of Geophysical Research*, **104**(D6), 6269–6279.

Intergovernmental Panel on Climate Change, 2007: *Climate Change 2007: The Physical Science Basis*, Contribution of Working Group I to the Fourth Assessment Report, Cambridge, UK Cambridge University Press.

Kahn, R.A., W.H. Li, C. Moroney, D.J. Diner, J.V. Martonchik, E. Fishbein, 2007: Aerosol source plume physical characteristics from space-based multiangle imaging. *Journal of Geophysical Research-Atmospheres*, **112**(D11), D11205. doi:10.1029/2006JD007647.

Kaufman, Y.J., D. Tanré, H.R. Gordon, T. Nakajima, J. Lenble, R. Frouin, H. Grassl, B.M. Herman, M.D. King, P.M. Teillet, 1997a: Passive remote sensing of tropospheric aerosol and atmospheric correction for the aerosol effect. *Journal of Geophysical Research*, **102**,(D14), 16815–16830.

Kaufman, Y.J., A. Wald, L. Remer, B.-C. Gao, R.-R. Li, and L. Flynn, 1997b: The MODIS 2.1-μm channel – correlation with visible reflectance for use in remote sensing of aerosol. *IEEE Transactions on Geoscience and Remote Sensing*, **35**(5), 1286–1298.

Kaufman, Y.J., D. Tanré, and O. Boucher, 2002: A satellite view of aerosols in the climate system. *Nature*, **419**, 215–223.

King, M., Y.J. Kaufman, D. Tanré, and T. Nakajima, 1999, Remote sensing of tropospheric aerosols from space: past, present, and future. *Bulletin of the American Meteorological Society*, **80**, 2229–2259.

Kokhanovsky, A.A., F.-M. Breon, A. Cacciari, E. Carboni, D. Diner, W. Di Nicolantonio, R.G. Grainger, W.M.F. Grey, R. Holler, K.-H. Lee, P.R.J. North, A. Sayer, G. Thomas, W. von Hoyningen-Huene, 2007: Aerosol remote sensing over land: satellite retrievals using different algorithms and instruments. *Atmospheric Research*, **85**, 372–394, doi:10.1016/j.atmosres.2007.02.008.

Kokhanovsky, A.A., G. de Leeuw, 2009: Determination of atmospheric aerosol properties over land using satellite measurements, Bull. American Met. Society, in press.

Kokhanovsky, A.A., L. Curie, G. de Leeuw, W. M. F. Grey, K.-H. Lee, Y. Bennouna, R. Schoemaker, P. R. J. North, 2009: The inter-comparison of AATSR dual view aerosol optical thickness retrievals with results from various algorithms and instruments. *International Journal of Remote Sensing, in press*.

Koren, I., Y.J. Kaufman, R. Washington, M.C. Todd, Y. Rudich, J.V. Martins, and D. Resefdeld, 2006: The Bod\'{e}l\'{e} depression: a single spot in the Sahara that provides most of the dust to the Amazon forest, *Environmental Research Letters*, **1**, 014005, doi:10.1088/1748-9326/1/1/014005.

Kotchenova, S.Y., E.F. Vermote, 2007: Validation of a vector version of the 6S radiative transfer code for atmospheric correction of satellite data. Part II. Homogeneous Lambertian and anisotropic surfaces. *Applied Optics*, **46** (20): 4455–4464.

Kotchenova, S.Y., E.F. Vermote, R. Matarrese, F.J. Klemm, 2006: Validation of a vector version of the 6S radiative transfer code for atmospheric correction of satellite data. Part I: Path radiance. *Applied Optics*, **45**(26),6762–6774.

Levy, R.C., L.A. Remer, Y.J. Kaufman, 2004: Effects of neglecting polarization on the MODIS aerosol retrieval over land. *IEEE Transactions on Geoscience and Remote Sensing*, **42**(11): 2576–2583.

Levy, R.C., L.A. Remer, O. Dubovik, 2007: Global aerosol optical properties and application to Moderate Resolution Imaging Spectroradiometer aerosol retrieval over land, *Journal of Geophysical Research-Atmospheres*, **112**(D13), D13210, doi:10.1029/2006JD007815.

Li, Z., X. Xia, M. Cribb, W. Mi, B. Holben, P. Wang, H. Chen, S-C. Tsay, T.F. Eck, F. Zhao, E.G. Dutton, and R.E. Dickerson, 2007: Aerosol optical properties and their radiative effects in northern China. *Journal of Geophysical Research*, **112**, D22S01, doi:10.1029/2006JD007382.

Mackay, G, M. Steven, and J. Clark, 1998: An atmospheric correction procedure for the ATSR-2 visible and near-infrared land surface data. *International Journal of Remote Sensing*, **19**(5), 2949–2968.

Martonchik, J., D. Diner, R. Kahn, B. Gaitley, and B. Holben, 2004: Comparison of MISR and AERONET aerosol optical depths over desert sites. *Geophysical Research Letters.* **31**, L16102, doi:10.1029/2004GL019807.

Mishchenko, M,I, and L.D. Travis, 1997: Satellite retrieval of aerosol properties over the ocean using polarization as well as intensity of reflected sunlight. *Journal of Geophysical Research – Atmospheres.* **102**(D14), 16989–17013.

Mishchenko, M.I., I.V. Geogdzhayev, B. Cairns, B.M. Carlson, J. Chowdhary, A.A. Lacis, L. Liu, W.B. Rossow, L.D. Travis, 2007a: Past, present, and future of global aerosol climatologies derived from satellite observations: a perspective, *Journal of Quantitative Spectroscopy and Radiative Transfer*, **106**(1–3), 325–347.

Mishchenko, M.I., I.V. Geogdzhayev, W.B. Rossow, B. Cairns, B.E. Carlson, A.A. Lacis, L. Liu, L.D. Travis, 2007b: Long-term satellite record reveals likely recent aerosol trend. *Science*, **315**(5818), 1543, doi:10.1126/science.1136709.

North, P.R.J., 1996: Three-dimensional forest light interaction model using a Monte Carlo method. *IEEE Transactions on Geoscience and Remote Sensing*, **34**(5), 946–956.

North, P.R.J., 2002: Estimation of aerosol opacity and land surface bidirectional reflectance from ATSR-2 dual-angle imagery: operational method and validation. *Journal of Geophysical Research*, **107**(D12), doi:10.1029/2000JD000207.

North, P.R.J., Briggs, S.A., Plummer, S.E. and Settle, J.J., 1999: Retrieval of land surface bi-directional reflectance and aerosol opacity from ATSR-2 multi-angle imagery. *IEEE Transactions on Geoscience and Remote Sensing*, **37**(1), 526–537.

Plummer, S.E., 2008: The GLOBCARBON cloud detection system for the Along Track Scanning Radiometer (ATSR) sensor series. *IEEE Transactions on Geoscience and Remote Sensing* (in press).

Press, W.H., S.A. Teukolsky, W.T. Vetterling, and B.P. Flannery, 1992: *Numerical Recipes in C, The Art of Scientific Computing*, 2nd edn. Cambridge, UK: Cambridge University Press.

Ramanathan, V., M.V. Ramana, G. Roberts, D. Kim, C. Corrigan, C. Chun, D. Winker, 2007: Warming trends in Asia amplified by brown cloud solar absorption. *Nature*, **448**, 575–578, doi:10.1038/nature06019.

Remer, L.A., Y. J. Kaufman, D. Tanré, S. Mattoo, D. A. Chu, J. V. Martins, R.-R. Li, C. Ichoku, R.C. Levy, R. Kleidman, T.F. Eck, E. Vermote, and B.N. Holben, 2005: The MODIS aerosol algorithm, products and validation. *Journal of the Atmospheric Sciences*, **62**, 947–973.

Rosenfeld, D., Y. Rudich, R. Lahav, 2001: Desert dust suppressing precipitation: a possible desertification feedback loop. *Proceedings of the National Academy of Sciences*, **98**(11), 5975–5980, doi:10.1073/pnas.101122798.

Roujean, J.-L., D. Tanré, F.-M. Brecon, and J.-L. Deuze, 1997: Retrieval of land surface parameters from airborne POLDER bidirectional reflectance distribution function during HAPEX-Sahel. *Journal of Geophysical Research*, **102**(D10), 201–218.

Rozanov, V.V., and A.A. Kokhanovsky, 2005: On the molecular-aerosol scattering coupling in remote sensing of aerosol from space. *IEEE Transactions on Geoscience and Remote Sensing*, **43**(7), 1536–1541.

Rozanov, V.V., and A.A. Kokhanovsky, 2006: The solution of the vector radiative transfer equation using the discrete ordinates technique: selected applications. *Atmospheric Research*, **79**, 241–265.

Santer, R., D. Ramon, J. Vidot, E. Dilligeard, 2007: A surface reflectance model for aerosol remote sensing over land. *International Journal of Remote Sensing*, **28**(3–4), 737–760.

Smirnov, A., B.N. Holben, T.F. Eck, O. Dubovik, and I. Slutsker, 2000: Cloud-screening and quality control algorithms for the AERONET database. *Remote Sensing of Environment*, **73**, 337–349.

Stowe, L., A. Ignatov, and R. Singh, 1997: Development, validation, and potential enhancements to the second-generation operational aerosol product at the national environmental satellite data, and information service of the national oceanic and atmospheric administration. *Journal of Geophysical Research-Atmospheres*, **102**(D14–16), 923–934.

Thomas, G.E., S.H. Marsh, S.M. Dean, E. Carboni, R.G. Gainger, C.A. Poulsen, R. Siddans, B.J. Kerridge, 2007: An optimal estimation retrieval scheme for (A)ATSR, AOPP Memorandum 2007.1.

Torres, O., P.K. Bhartia, J.R, Herman, A. Sinyuk, P. Ginoux, and B. Holben, 2002: A long-term record of aerosol optical depth from TOMS observations and comparison to AERONET measurements. *Journal of the Atmospheric Sciences*, **59**(3), 398-413.

Veefkind, J.P., G. de Leeuw, and P.A. Durkee (1998). Retrieval of aerosol optical depth over land using two-angle view satellite radiometry during TARFOX. *Geophysical Research Letters*, **25** (16), 3135–3138.

Veefkind, J.P., G. de Leeuw, P.A. Durkee, P.B. Russell, P.V. Hobbs, and J.M. Livingston, 1999: Aerosol optical depth retrieval using ATSR-2 and AVHRR data during TARFOX, *Journal of Geophysical Research-Atmospheres*, **104**(D2), 2253–2260.

Veefkind, J.P., G. de Leeuw, P. Stammes, R.B.A. Koelemeijer, 2000: Regional distribution of aerosol over land derived from ATSR-2 and GOME. *Remote Sensing of Environment*, **74**(3), 377–386.

Vermote E.F., D. Tanré, J.L. Deuze, M. Herman, J.J. Morcrette, 1997: Second simulation of the satellite signal in the solar spectrum, 6S: an overview. *IEEE Transactions on Geoscience and Remote Sensing*, **35**(3), 675–686.

von Hoyningen-Huene W., M. Freitag, J.B. Burrows, 2003: Retrieval of aerosol optical thickness over land surfaces from top-of-atmosphere radiance. *Journal of Geophysical Research*, **108**(D9), 4260, doi:10.1029/2001JD002018.

Yu, H., Y.J. Kaufman, M. Chin, G. Feingold, L.A. Remer, T.L. Anderson, Y. Balkanski, N. Bellouin, O. Boucher, S. Christopher, P. DeCola, R. Kahn, D. Koch, N. Loeb, M.S. Reddy, M. Schulz, T. Takemura, and M. Zhou, 2005: A review of measurement-based assessment of aerosol direct radiative effect and forcing. *Atmospheric Chemistry and Physics*, **5**, 7647–7768, doi:10.1029/2005JD006356.

Zavody, A.M., C.T. Mutlow, and D.T. Llewellyn-Jones, 2000: Cloud clearing over the ocean in the Processing of data from the Along-Track Scanning Radiometer (ATSR). *Journal of Atmospheric and Oceanic Technology*, **17**, 595–615.

7 Oxford-RAL Aerosol and Cloud (ORAC): aerosol retrievals from satellite radiometers

Gareth E. Thomas, Elisa Carboni, Andrew M. Sayer, Caroline A. Poulsen, Richard Siddans, Roy G. Grainger

1. Introduction

This chapter describes an optimal estimation retrieval scheme for the derivation of the properties of atmospheric aerosol from top-of-atmosphere (TOA) radiances measured by satellite-borne visible-IR radiometers. The algorithm makes up part of the Oxford-RAL Aerosol and Cloud (ORAC) retrieval scheme (the other part of the algorithm performs cloud retrievals and is described in detail elsewhere [by Watts *et al.*] [37]).

The following sections will describe three separate versions of the ORAC algorithm. The first is the original ORAC aerosol retrieval algorithm, which has already been applied in producing global aerosol datasets from ATSR-2, AATSR and SEVIRI measurements (brief descriptions of these instruments are given in Section 2), through the GRAPE and GlobAEROSOL projects. This algorithm makes use of visible and near-infrared channels and assumes the Earth's surface acts as a Lambertian reflector.

The second version improves on the original Lambertian ORAC, by implementing a new forward model which uses a bi-directional reflectance distribution function (BRDF) to describe the surface reflectance. This forward model is more accurate and also allows the use of multiple views of the same scene (as are produced by the (A)ATSR instruments) to be incorporated into the retrieval.

Thirdly, a version of ORAC is described which makes use of thermal infrared channels, which greatly improve the detection of lofted dust above desert surfaces.

Finally, example results produced with all three versions of the ORAC algorithm will be presented and compared, both with each other, ground-based measurements and other satellite aerosol products.

2. Instrument descriptions

2.1 The ATSR-2 and AATSR instruments

The second and third generation Along-Track Scanning Radiometers (ATSR-2 and Advanced ATSR) were launched on the ESA polar orbit satellites ERS-2 and ENVISAT in 1995 and 2002, respectively. As the instruments are essentially the same in their operation, with the only major difference being the bandwidth available for data transfer, they can be described together.

The primary design goal of the ATSR instruments is the measurement of sea-surface temperature, with a secondary objective of ATSR-2 and AATSR being the determination of land surface and vegetation properties. ORAC makes use of the atmospheric component of the ATSR signal, which is considered contamination in its primary and secondary roles.

Both ATSR-2 and AATSR have seven spectral channels centered at 0.55, 0.67, 0.87, 1.6, 3.7, 10.7 and 12 μm. The instruments use a dual-view system, with a continuously rotating scan mirror directing radiation from two apertures and two onboard blackbody calibration targets onto the radiometer. One viewing aperture produces a scan centered on the nadir direction, while the other views the surface approximately 900 km ahead of the satellite (at a viewing angle of 55° from the nadir). This continuous scanning pattern produces a nadir resolution of approximately 1 × 1 km with a swath width of 512 pixels. The dual-view system is one of the great strengths of the ATSR instruments, as it allows the atmospheric and surface contributions to the TOA radiance to be more effectively decoupled than is possible with a single view. This offers much improved accuracy in both derived surface and atmospheric parameters. In addition, the instruments are designed to be self-calibrating, with two integrated, thermally controlled blackbody targets for calibration of the thermal channels, as well as an opal visible calibration target (illuminated by sunlight) for the visible/near-IR channels.

Due to bandwidth limitations on the ERS-2 satellite, ATSR-2 is usually run in a 'narrow swath' mode over the oceans, which produces a swath of only 256 pixels in some of the visible channels (with the 0.55 μm channel being the most commonly effected, followed by 0.67 μm). In addition, although ATSR-2 is still operational, the ERS-2 satellite developed a pointing problem in October 2001, which means that post-2001 data from the instrument has to have a geo-location correction applied before it can be used. Additionally, in June 2003, the data tape recorder on ERS-2 failed, with the result that ATSR-2 data from this date is only available while the satellite is within range of a data downlink ground station. ATSR-2 ceased operating in February 2008, when the scan mirror mechanism failed.

ERS-2 and ENVISAT are in similar polar orbits with periods of approximately 100 minutes. Both ATSR-2 and AATSR nominally provide global coverage every 6 days.

2.2   The SEVIRI instrument

The Spinning Enhanced Visible and InfraRed Imager (SEVIRI) is a line-scanning radiometer and has been the primary instrument onboard the European geostationary[1] meteorological satellites since Meteosat-8 began operation. The satellite was launched in 2003 and the first data were available in early 2004. In April 2007 Meteosat-9 took over as the primary operational satellite.

SEVIRI provides data in four visible and near-infrared channels and eight infrared channels with a resolution of 3 km at the sub-satellite point. The channels used in the analysis presented here are the 0.67, 0.87 and 1.6 μm in the visible and near-infrared, with the 10.8 and 12.0 μm being used from the thermal infrared. A key feature of SEVIRI is its ability to continuously image the Earth every 15 minutes. This allows the tracking of fast-moving aerosol events, such as dust storms, which offers a great advantage over polar orbiting instruments. Also, although SEVIRI lacks the dual-view capability of the ATSR instruments, in conditions of relatively stable aerosol loading, the change in solar elevation

---

[1]   The Meteosat sub-satellite point is at 0° longitude, just to the east of the African coastline.

throughout the day can be utilized to provide repeated views under different angles of illumination. The difference in the surface BRDF and aerosol scattering phase function can then be used to decouple the aerosol and surface signals in a way analogous to the dual-view system. The main disadvantage of this instrument is the susceptibility of the larger field-of-view to contamination by cloud.

3.   The ORAC forward model

The core of the ORAC retrieval algorithm is the forward model, which uses radiative transfer code to predict the radiance observed at the satellite as a function of aerosol properties, using assumptions about the atmospheric state and the reflectance of the Earth's surface. For the sake of numerical efficiency, ORAC makes use of two forward models: firstly a full radiative transfer model (referred to here simply as the forward model, FM), which attempts to accurately account for all relevant physical processes effecting the measurement, is run 'off-line' to produce look-up tables of total atmospheric reflectance and transmission for the plausible range of viewing geometries and aerosol states. These look-up tables are then used to produce TOA radiances during a retrieval run using a simple arithmetic expression, known as the fast forward model (Fast-FM). This section details the aerosol FM used with ORAC, which is the same for both the Lambertian and BRDF surface reflectance versions of the retrieval scheme, as well as describing the extensions needed to incorporate thermal infrared channels into the retrieval scheme.

The FM can itself be thought of as consisting of three separate elements:
1. A model of aerosol scattering and absorption.
2. A model of atmospheric gas absorption.
3. Radiative transfer code to produce TOA radiance based on the output of the first two models, Rayleigh scattering and viewing geometry.

3.1   Aerosol scattering and absorption

In a given location, atmospheric aerosols are characterized by their morphology, concentration, size distribution, chemical composition (which determines their complex refractive index), and their vertical profile. With knowledge of these properties, the required radiative characteristics may be approximated by assuming the particles are spherical and applying Mie theory [20].

The aerosol optical depth, τ, is the primary quantity obtained from ORAC. It is defined as:

$$\tau(\lambda) = \int_0^\infty \beta_e(z, \lambda) \mathrm{d}z = \int_0^\infty (\beta_s(z, \lambda) + \beta_{-a}(z, \lambda)) \mathrm{d}z \tag{1}$$

The total extinction coefficient, β_e, is defined as the sum of the extinction due to absorption, β_a, and scattering, β_s. The vertical profile of β_a and β_s along with the scattering phase function, $P(\theta)$, which determines the angular distribution of the scattered radiation, and the degree of polarization as a function of scattering angle, fully describe the aerosol radiative characteristics. Other convenient ways of defining aerosol optical properties are the single scattering albedo, ω_o, which is the ratio of β_s to β_e, and the asymmetry parameter, g, which is the integral of $P(\theta)$ over all possible scattering angles ($0 \leq \theta \leq 180°$), weighted by $\cos \theta$ (i.e. it is the first moment of the phase function). For a given aerosol model (shape, size, and refractive index), β_e is proportional to the aerosol concentration while $P(\theta)$ is not.

Mie theory shows that the extinction coefficient is given by:

$$\beta(z, \lambda) = \int_0^\infty Q_e(z, m, x)\pi r^2 \, n(z, r)\mathrm{d}r,$$

(2)

where Q_e is the Mie extinction efficiency factor, and is dependent on the Mie size parameter $x = 2\pi r/\lambda$, and the refractive index of the particles ($m = m_r + im_i$), $n(r)$ is the number size distribution.

The lognormal distribution is the most suitable representation for characterizing the size distribution of the atmospheric aerosols [4]. The distribution, in terms of number density as a function of radius $n(r)$, is described by its median radius (r_m), standard deviation (σ) of $\ln r$, and total number density (N_0):

$$n(r) = \frac{N_0}{\sqrt{2\pi}} \frac{1}{\sigma r} \exp\left[-\frac{(\ln r - \ln r_m)^2}{2\sigma^2}\right]$$

(3)

The primary source of aerosol properties used in the retrieval is the OPAC (Optical Properties of Aerosols and Clouds) database [11]. The database provides optical (most importantly, the complex refractive index as a function of wavelength) and physical properties (such as the size distribution and vertical distribution) for a set of aerosol components from which representative aerosol types can be built.

The quantity used to define the size of the aerosol particles in ORAC is the effective radius, defined as the the ratio of the third and second moments of the size distribution:

$$r_e = \frac{\int_0^\infty r^3 n(r) \, \mathrm{d}r}{\int_0^\infty r^2 n(r) \, \mathrm{d}r}$$

(4)

In order to produce radiance look-up tables from this database the scattering properties of each aerosol type are calculated. Scattering properties are calculated for the central wavelength of each channel across a range of effective radii from 0.02 to 20 μm. Two assumptions are made during this step:

- That the radiative properties of the aerosol are constant across the width of each instrument channel. As the features of aerosol extinction spectra are very broad in comparison with gas features this is a reasonable approximation.
- Assumptions must be made in determining both the form of the aerosol size distribution and how its shape varies with changing aerosol effective radius. To model aerosol distributions with different effective radii to those prescribed by the OPAC database, the relative concentration of the different-sized aerosol components which make up each aerosol class are changed. For example, if the effective radius needs to be decreased, the relative concentration of the smallest component of the aerosol (the accumulation mode) will be increased, while the larger components will be decreased.

If the required effective radius is equal to that given by the smallest or largest component of a given aerosol type, then the type effectively becomes a single component aerosol. If the size is outside of this range, then the mode radius of the smallest/largest components is shifted (while keeping the width of the component's distribution constant). Clearly, in such situations, the accuracy of the model can be called into question, so we are relying on the prescribed effective radius being relatively close to that found in the real world. It should also be pointed out that in the case of very small aerosol particles, the composition of the particles become less important in determining their scattering effects, since they will act more like Rayleigh scatterers.

These scattering properties are then used to generate a vertical profile of aerosol extinction and phase function, based on vertical profiles of number density, N:

$$N(z) = N(0)\exp(-z/Z), \tag{5}$$

where z is the height and Z is a scale height, defined by the aerosol type. For each layer at which the aerosol distribution is defined, the extinction coefficient, single scattering albedo and the coefficients of a Legendre expansion of the scattering phase function are calculated for each instrument channel and over 20 logarithmically spaced effective radii between 0.01 and 10 μm.

3.2 Modelling atmospheric gas absorption

Once aerosol scattering properties have been calculated, gas absorption over the instrument band passes is calculated in terms of an optical depth, and convolved with the instrument filter transmission functions, using MODTRAN [1]. MODTRAN provides tropical, mid-latitude summer and winter, subarctic summer and winter, and US Standard Atmosphere climatological atmospheres for the following gases: H_2O, CO_2, O_3, N_2O, CO, CH_4, plus single profiles for: HNO_3, NO, NO_2, SO_2, O_2, N_2, NH_3 and the heavy molecules (CFCs). ORAC look-up tables are generated using the mid-latitude summer atmosphere only. This simplification can be made as gas absorption is weak compared to aerosol extinction in the visible and the (A)ATSR channels are free from strong absorption features of gases which show large spatial and temporal variability (most notably, H_2O). Although the 1.6 μm channel of SEVIRI is slightly affected by H_2O, the effect on aerosol retrievals has been found to be negligible.

3.3 Modeling atmospheric transmission and reflectance

The final step in the FM is the prediction of atmospheric transmission and bi-directional reflectance, based on the aerosol phase functions and gas optical depth calculated in the previous two steps. The ORAC FM uses the DIscrete Ordinates Radiative Transfer (DISORT) software package [34] to perform this step.

The DISORT algorithm solves the equation for the transfer of monochromatic light at wavelength λ as described by the equation

$$\mu \frac{dL_\lambda(\tau_\lambda, \mu, \phi)}{d\tau} = L_\lambda(\tau_\lambda, \mu, \phi) - L_\lambda^S(\tau_\lambda, \mu, \phi), \tag{6}$$

where $L_\lambda(\tau_\lambda, \mu, \phi)$ is the intensity along direction μ, ϕ (where μ is the cosine of the zenith angle and ϕ is the azimuth angle) at optical depth τ_λ measured perpendicular to the surface of the medium. $L_\lambda^S(\tau_\lambda, \mu, \phi)$ is the source function.

It should be noted that DISORT still makes some important approximations, which can limit its accuracy in certain circumstances. The most important of these are:

- It assumes a plane parallel atmosphere, which makes it inapplicable at viewing or zenith angles above approximately $80°$, where the curvature of the Earth has a significant influence on radiative transfer.
- It is a one-dimensional model, so cannot reproduce the effects of horizontal gradients in the scattering medium. This is important where strong gradients exist, such as near cloud edges.
- It does not model polarization effects and hence cannot be used to model measurements made by instruments which are sensitive to polarization and does not take the polarization introduced into the diffuse component of radiance by Rayleigh scattering.[2]

DISORT is provided with the aerosol-scattering properties defined by the Mie scattering calculations and the gas absorptions defined by MODTRAN and a series of 20 logarithmically spaced aerosol optical depths (defined at a wavelength of $0.55\,\mu m$) between 0.008 and 5.6. Although DISORT has the ability to include a surface of arbitrary reflectance 'below' the modeled atmosphere, no surface reflectance is included at this step. Rather, the transmission and reflectance of the atmosphere alone is computed for both direct beam and diffuse radiation sources separately. These calculations produce five look-up tables for each aerosol type/channel combination:

- Bidirectional reflectance of the atmosphere, from the top of the atmosphere, $R_{BD}(\theta_0, \theta_v, \phi)$.
- Diffuse reflectance of the atmosphere to diffuse radiance, from the bottom of the atmosphere, R_{FD}.
- Diffuse transmission of an incident beam, $T_{BD}^\downarrow(\theta_0)$.
- Direct transmission of the beam, $T_{DB}^\downarrow(\theta_0)$, or $T_{DB}^\uparrow(\theta_v)$.
- Total transmission in the viewing direction, $T^\uparrow(\theta_v)$.

Here, a $\downarrow$ denotes transmission from the top to the bottom of the atmosphere, while $\uparrow$ indicates the reverse. θ_0, θ_v and ϕ indicate a dependence on the solar zenith, viewing zenith and relative azimuth angles, respectively. Each of these files contains tabulated transmission or reflectance (depending on the file) values for each of the twenty effective radii, nine $0.55\,\mu m$ optical depths and sun/satellite geometry (specified by 20 equally spaced zenith angles and 11 equally spaced azimuth angles).

Effects of molecular absorption and Rayleigh scattering are included by adjustment of the layer's optical depth and the particle's single scattering albedo and phase function with the following:

$$\tau = \tau_a + \tau_R + \tau_g, \tag{7}$$

$$\omega = \frac{\tau_R + \omega_a \tau_a}{\tau_g + \tau_R + \tau_a}, \tag{8}$$

---

[2] It should be noted that DISORT has now be superseded by the vectorized VDISORT code [31], which does include polarization effects. However, this code has yet to be implemented in the ORAC scheme.

$$P(\theta) = \frac{\tau_a \omega_a P_a(\theta) + \tau_R P_R(\theta)}{\tau_a \omega_a + \tau_R}, \tag{9}$$

where τ_a, τ_R and τ_g are the contributions to the total optical depth τ due to aerosol scattering, Rayleigh scattering and gaseous absorption within each layer respectively. The aerosol single scattering albedo is denoted ω_a.

For each layer bounded by lower and upper pressure levels p_l and p_u, respectively and ground-level pressure p_0, τ_R is calculated from

$$\tau_R = \frac{\tau_{RT}[p_l - p_u]}{p_0}, \tag{10}$$

where τ_{RT}, the wavelength-dependent Rayleigh scattering optical depth for a column of atmosphere extending from the ground surface to the top of the atmosphere, is obtained from [15]:

$$\tau_{RT}(\lambda) = \frac{p_0}{p_s} \times \frac{1}{117.03\lambda^4 - 1.316\lambda^2}, \tag{11}$$

where p_s is the standard pressure ($p_s = 1013.25$ hPa), p_0 is the ground pressure in hPa and λ is in μm.

4. Surface reflectance

Of crucial importance in the retrieval of aerosol properties from 'near-nadir' visible/near-infrared satellite measurements (i.e. measurements in which the Earth's surface contributes to the measured radiances) is an accurate description of the surface reflectance. Both the Lambertian and BRDF surface reflectance versions of ORAC retrieve the surface reflectance in addition to the aerosol optical depth and effective radius, however, it is still necessary to have accurate *a priori* knowledge of it.

The methodology used to produce an *a priori* surface reflectance differs between measurements made over sea or land. Over the sea a surface reflectance model based on the method presented by Koepke [17] is used. This model includes upwelling radiance from volume scattering within the water itself [22], specular reflections from the wind-roughened surface (as modeled by the Cox and Munk method [4, 5]) and reflection from white caps [21, 23]. The model uses ECMWF reanalysis wind fields to determine wave statistics and white cap coverage and can also make use of chlorophyll concentrations and gelbstoff loading from MERIS products. A detailed description of the model is given by Sayer [30].

Over land the MODIS[3] land surface bidirectional reflectance product [14] is used to define the *a priori* surface reflectance. The product consists of a set of three parameters for the MODIS AMBRALS (Algorithm for Modelling Bidirectional Reflectance Anisotropies of the Land Surface) surface reflectance model [36], which itself consists of three simple reflectance kernels for different surface types:

---

[3] MODerate resolution Imaging Spectrometer.

- Isotropic kernel. Lambertian reflectance, for which the kernel is $\equiv 1$.
- Ross-thick kernel, $K_{Rt}(\theta_0, \theta_v, \phi)$. Parameterizes densely packed, randomly oriented reflectors, such as leaves.
- Li-sparse kernel, $K_{Li}(\theta_0, \theta_v, \phi)$. Parameterizes the shadowing effects of isolated large objects, such as scattered trees.

The three coefficients, p_{iso}, p_{vol} and p_{geo} for the isotropic, Ross-thick and Li-sparse kernels respectively, provided by the BRDF product weight these models to reproduce the atmospherically corrected bi-directional surface reflectance observed by MODIS over a 16-day period.

The MODIS BRDF product has as specified uncertainty of ± 0.02 in the white sky albedo derived from the AMBRALS model coefficients. Validation work [13, 28] has shown this to be a reasonable estimate of the true accuracy of the product in general, although accuracy can decrease for scenes with a highly heterogeneous surface.

Since the Ross-thick and Li-sparse kernels are both dependent only on the solar and viewing directions, the AMBRALS model can be written in the form:

$$R_{SBD} = p_{iso} + K_{Rt}(\theta_0, \theta_v, \phi)p_{vol} + K_{Li}(\theta_0, \theta_v, \phi)p_{geo}. \tag{12}$$

These coefficients can also be combined to form either a black-sky albedo:

$$R_{SLB} = p_{iso} + \left(b_{bs1} + b_{bs2}\theta_0^2 + b_{bs3}\theta_0^3\right)p_{vol} + \left(c_{bs1} + c_{bs2}\theta_0^2 + c_{bs3}\theta_0^3\right)p_{geo}, \tag{13}$$

or a white-sky albedo:

$$R_{SLW} = p_{iso} + b_{ws}p_{vol} + c_{ws}p_{geo}, \tag{14}$$

where the quantities b_{ws}, c_{ws}, b_{bs1}, etc. are constant coefficients published by the MODIS BRDF team [32].

Sunglint

A major problem encountered in making nadir satellite measurements is the specular reflection of sunlight off the ocean surface, usually referred to as sunglint. Sunglint has two detrimental impacts:

1. The TOA signal becomes dominated by the directly reflected radiance from the surface.
2. Consequently, the Lambertian surface reflectance approximation (usually valid for the ocean surface) becomes wholly inadequate. For these reasons, a retrieval using the Lambertian Fast-FM (see Section 5) will not produce accurate results in regions of sunglint and they must be masked out, resulting in significant loss of data. The BRDF forward model is able to successfully model the radiative transfer in sun-glint regions. However, it becomes highly dependeant on accurate modeling of the surface reflectance (which in turn depends on the accuracy of the assumed surface wind conditions), due to the relatively small contribution of the atmosphere to TOA radiance in these regions.

5.   The Lambertian fast forward model

The Fast-FM uses the transmission and reflectance look-up tables produced by the FM to predict a top-of-atmosphere radiance using the scheme shown diagrammatically in Fig. 7.1. The solar beam is incident on the atmosphere and the first contribution to the TOA observed radiance is the direct bi-directional reflectance, R_{BD}, of the atmosphere. Transmission through the atmosphere is partly by direct transmission of the beam, $T_{DB}^{\downarrow}$, and partly by diffuse transmission of scattered radiance, $T_{BD}^{\downarrow}$. As an underlying Lambertian surface is assumed, any preferred directionality of the radiance is lost on reflection and these transmitted terms can be combined to give the total transmission downward through the atmosphere, $T^{\downarrow} = T_{BD}^{\downarrow} + T_{DB}^{\downarrow}$. Radiation reflected by the surface (with reflectance R_s) is partially transmitted by the atmosphere into the viewing direction. This transmission, again with the direct and diffuse components combined, is denoted by $T^{\uparrow}$. The atmosphere also reflects downwards (with reflectance R_{FD}) so there is a set of multiple reflections and transmissions giving rise to a series of rapidly decreasing contributions to the TOA reflectance. This process is represented in the following equation:

$$R(\theta_0, \theta_v, \phi) = R_{BD}(\theta_0, \theta_v, \phi) + T^{\downarrow}(\theta_0)R_s T^{\uparrow}(\theta_v) \tag{15}$$
$$+ T^{\downarrow}(\theta_0)R_s^2 T^{\uparrow}(\theta_v)R_{FD}$$
$$+ T^{\downarrow}(\theta_0)R_s^3 T^{\uparrow}(\theta_v)R_{FD}^2$$
$$+ \dots .$$

This expression can be simplified to give

$$R(\theta_0, \theta_v, \phi) = R_{BD}(\theta_0, \theta_v, \phi) + T^{\downarrow}(\theta_0)R_s T^{\uparrow}(\theta_v)\left(1 + R_s R_{FD} + R_S^2 R_{FD}^2 + \dots\right) \tag{16}$$

which, in turn, can be simplified further in terms of a geometric series limit,

$$R(\theta_0, \theta_v, \phi) = R_{BD}(\theta_0, \theta_v, \phi) + \frac{T^{\downarrow}(\theta_0)T^{\uparrow}(\theta_v)R_s}{1 - R_s R_{FD}}. \tag{17}$$

It is this equation that is used to calculate the top-of-atmosphere radiances seen by the satellite.

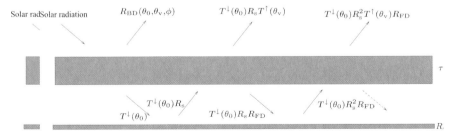

Fig. 7.1. Atmosphere surface interactions.

5.1 Forward model gradient

The gradient of the forward model $(\partial y/\partial x)$ where y is a radiance measurement in a single channel and x is one of the state variables is required for the following two purposes:

1. The gradient with respect to parameters which are to be derived from the measurements (state parameters) is a vital quantity for the inversion of the non-linear reflectance model by the Levenberg–Marquardt algorithm.
2. The gradient with respect to parameters which might be considered known and not part of the inversion procedure (model parameters), e.g., surface reflectance spectral shape, is used to judge the sensitivity to these parameters and thus to estimate their contribution to the retrieval error.

If Eq. (17) is rewritten as

$$R = R_{\mathrm{BD}} + S, \tag{18}$$

then without reproducing the algebra it can be simply stated that the gradient of the model with respect to optical depth or effective radius is given by:

$$\frac{\partial R}{\partial x} = R'_{\mathrm{BD}} + S\left(\frac{T^{\downarrow}T^{\uparrow\prime} + T^{\downarrow\prime}T^{\uparrow}}{T^{\downarrow}T^{\uparrow}} + \frac{R_{\mathrm{S}}R'_{\mathrm{FD}}}{1 - R_{\mathrm{S}}R_{\mathrm{FD}}}\right) \tag{19}$$

where all $'$ indicate $\partial/\partial x$ and x is either τ or r_{e}.

The gradient with respect to surface reflectance is given by:

$$\frac{\partial R}{\partial R_{\mathrm{s}}} = \frac{T^{\downarrow}T^{\uparrow}}{\left(1 - R_{\mathrm{s}}R_{\mathrm{FD}}\right)^2}. \tag{20}$$

6.   The BRDF fast forward model

Although the approximation of a Lambertian surface reflectance is reasonable for many surface types, it can be grossly inaccurate in some situations. For this reason a new Fast-FM has been developed for the ORAC retrieval system that, while using the same reference forward model and look-up tables as the Lambertian forward model, no longer makes this assumption. Dropping the assumption of a Lambertian surface reflectance has two consequences on the derivation of the Fast-FM:

1. The surface reflectance can no longer be described by one value. Rather, three values are required:
    *   A bi-direction reflectance, R_{SBD}, is needed to characterize the reflection of the direct solar beam into the viewing angle. This is a function of both solar and viewing angles.
    *   An equivalent black-sky albedo, or hemispherical reflectance, R_{SLB}, is needed to characterize the diffuse reflection of the direct beam over the whole hemisphere. This is only a function of the solar angle.
    *   An equivalent white-sky albedo, or bi-hemispherical reflectance, R_{SLW}, is needed to characterize the reflection of diffuse downwelling radiation. This is independent of solar and viewing angles, since it assumes reflection is isotropic.
2. The combination of the direct transmission of the solar beam, $T^{\downarrow}_{\mathrm{DB}}$, and the diffuse transmission of the scattered radiance from the solar beam, $T^{\downarrow}_{\mathrm{BD}}$, into a single term

$T^{\downarrow}$ is no longer possible, because the direct beam and diffuse radiance are subject to different reflectances at the ground.

This means Eq. (15) becomes:

$$R(\theta_0, \theta_v, \phi) = R_{BD}(\theta_0, \theta_v, \phi) + T_{DB}^{\downarrow}(\theta_0)R_{SBD}(\theta_0, \theta_v, \phi)T_{DB}^{\uparrow}(\theta_v) \tag{21}$$

$$+ T_{DB}^{\downarrow}(\theta_0)R_{SLB}(\theta_0)\left(T^{\uparrow}(\theta_v) - T_{DB}^{\uparrow}(\theta_v)\right)$$

$$+ T_{BD}^{\downarrow}(\theta_0)R_{SLW}T^{\uparrow}(\theta_v)$$

$$+ T_{DB}^{\downarrow}(\theta_0)R_{SLB}(\theta_0)R_{FD}R_{SLW}T^{\uparrow}(\theta_v)$$

$$+ T_{BD}^{\downarrow}(\theta_0)R_{SLW}R_{FD}R_{SLW}T^{\uparrow}(\theta_v)$$

$$+ T_{DB}^{\downarrow}(\theta_0)R_{SLB}(\theta_0)R_{FD}R_{SLW}R_{FD}R_{SLW}T^{\uparrow}(\theta_v)$$

$$+ T_{BD}^{\downarrow}(\theta_0)R_{SLW}R_{FD}R_{SLW}R_{FD}R_{SLW}T^{\uparrow}(\theta_v)$$

$$+ \ldots.$$

Here the term $T_{DB}^{\downarrow}(\theta_0)R_{SBD}(\theta_0, \theta_v, \phi)T_{DB}^{\uparrow}(\theta_v)$ is the direct reflection of the solar beam into the viewing angle at the surface and thus uses the transmission of the direct beam for both the downward and upward paths through the atmosphere. $T_{DB}^{\downarrow}(\theta_0)R_{SLB}(\theta_0)\left(T^{\uparrow}(\theta_v) - T_{DB}^{\uparrow}(\theta_v)\right)$ is the diffuse reflection of the direct beam, i.e. it is the radiance seen at the satellite due to the direct beam being diffusely scattered by the ground. Note that for this term, the upwelling transmission is the total transmission minus the direct beam transmission, since we are only interested in the diffuse component. Together, these two terms are equivalent to the $T_{DB}^{\downarrow}(\theta_0)R_S T^{\uparrow}(\theta_v)$ term present in the Lambertian forward model. Notice also that we require the diffuse-only transmission for the upwelling radiation from the black-sky ($R_{SLB}(\theta_0)$) term.

The term $T_{BD}^{\downarrow}(\theta_0)R_{SLW}T^{\uparrow}(\theta_v)$ approximates the reflection of the diffusely transmitted solar radiation into the viewing direction. Ideally, this would be split into two terms: $T_{BD}^{\downarrow}(\theta_0)R_{SLW}T_{BD}^{\uparrow}(\theta_v) + T_{BD}^{\downarrow}(\theta_0)R_{SLB}(\theta_v)T_{DB}^{\uparrow}(\theta_v)$, where the first gives the upward diffuse transmission to the satellite and the second gives the direct transmission. By setting $R_{SLB}(\theta_v) = R_{SLW}$ these two terms combine to give the above expression and $R_{SLB}(\theta_v)$ can be eliminated from the equation. This greatly simplifies the formulation of the forward model and reduces the number of values which must be propogated through the retrieval by one. The rest of the terms in Eqn.(21) are multiple surface–atmosphere reflections, analogous to those which appear in the Lambertian forward model.

This model makes two simplifying assumptions:

1. The combining of the upwelling direct and diffuse components of the reflected diffusely transmitted solar radiation described above amounts to the approximation that the surface acts as a Lambertian reflector when diffusely illuminated. That is to say that the surface will appear the same from all viewing directions if lit by a purely diffuse source.
2. We also assume that when taken as a pair, the surface and atmosphere act as Lambertian reflectors, so that any directionality left the reflected beam ($T_{DB}^{\downarrow}(\theta_0)R_{SBD}(\theta_0, \theta_v, \phi)$) is lost in that proportion which is reflected back towards the ground.

Following on from this approximation, after the first surface-atmosphere pair of reflections, the radiation has lost all directionality, and thus the white sky albedo is used for

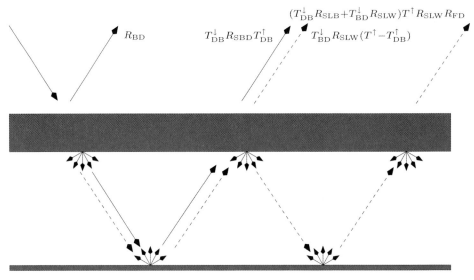

Fig. 7.2. Atmospher–surface interactions using the simplified BRDF model.

subsequent surface reflections, and the total atmospheric transmission $(T^\uparrow(\theta_v))$ for the final upward transmission (Fig. 7.2).

Again, Eq. (21) can be simplified:

$$R(\theta_0, \theta_v, \phi) = R_{BD}(\theta_0, \theta_v, \phi) + T^\downarrow_{DB}(\theta_0)R_{SBD}(\theta_0, \theta_v, \phi)T^\uparrow_{DB}(\theta_v) \qquad (22)$$
$$- T^\downarrow_{DB}(\theta_0)R_{SLB}(\theta_0)T^\uparrow_{DB}(\theta_v)$$
$$+ \left(T^\downarrow_{DB}(\theta_0)R_{SLB}(\theta_0) + T^\downarrow_{BD}(\theta_0)R_{SLW}\right)T^\uparrow(\theta_v)$$
$$\left(1 + R_{SLW}R_{FD} + R^2_{SLW}R^2_{FD} + \dots\right).$$

Applying the same series limit as before, our new reflectance value is given by:

$$R(\theta_0, \theta_v, \phi) = R_{BD}(\theta_0, \theta_v, \phi) + T^\downarrow_{DB}(\theta_0)(R_{SBD}(\theta_0, \theta_v, \phi) - R_{SLB}(\theta_0))T^\uparrow_{DB}(\theta_v) \qquad (23)$$
$$+ \frac{\left(T^\downarrow_{DB}(\theta_0)R_{SLB}(\theta_0) + T^\downarrow_{BD}(\theta_0)R_{SLW}\right)T^\uparrow(\theta_v)}{1 - R_{SLW}R_{FD}}.$$

6.1 Retrieving the surface reflectance with the BRDF forward model

Using the Lambertian surface forward model, the magnitude of the surface reflectance (in particular, the albedo at $0.55\,\mu m$) is retrieved. The spectral shape of the surface reflectance (i.e. the ratios between the reflectance in different channels) is fixed by the *a priori* surface albedo (from the MODIS white-sky albedo over land and from the surface reflectance model over the ocean), but the magnitude of this reflectance is allowed to vary. When using the BRDF forward model we are faced with having three separate reflectances for each channel. How then do we include the surface reflectance as a retrieved parameter?

As has been described already, the black-sky albedo describes the amount of radiation scattered into the entire hemisphere for a single incoming beam at a given zenith angle. Hence it can be derived from the BRDF:

$$R_{\mathrm{SLB}}(\theta_0) = \frac{\int_0^{2\pi} \int_0^{\pi/2} R_{\mathrm{SBD}}(\theta_0, \theta_{\mathrm{v}}, \phi) \cos\theta_{\mathrm{v}} \sin\theta_{\mathrm{v}} \, \mathrm{d}\theta_{\mathrm{v}} \, \mathrm{d}\phi}{\int_0^{2\pi} \int_0^{\pi/2} \cos\theta_{\mathrm{v}} \sin\theta_{\mathrm{v}} \, \mathrm{d}\theta_{\mathrm{v}} \, \mathrm{d}\phi} \tag{24}$$

$$= \frac{1}{\pi} \int_0^{2\pi} \int_0^{\pi/2} R_{\mathrm{SBD}}(\theta_0, \theta_{\mathrm{v}}, \phi) \cos\theta_{\mathrm{v}} \sin\theta_{\mathrm{v}} \, \mathrm{d}\theta_{\mathrm{v}} \, \mathrm{d}\phi.$$

Similarly the white-sky albedo is the amount of light scattered over the entire hemisphere from isotropic diffuse downwelling radiance. It can be calculated by integrating R_{SLB} across all solar zenith angles:

$$R_{\mathrm{SLW}} = \frac{\int_0^{\pi/2} R_{\mathrm{SLB}}(\theta_0) \cos\theta_0 \sin\theta_0 \mathrm{d}\theta_0}{\int_0^{\pi/2} \cos\theta_{\mathrm{v}} \sin\theta_{\mathrm{v}} \, \mathrm{d}\theta_0} = 2 \int_0^{\pi/2} R_{\mathrm{SLB}}(\theta_0) \cos\theta_0 \sin\theta_0 \, \mathrm{d}\theta_0. \tag{25}$$

It is clear from these two equations that a small change in any one of the three surface reflectance values will result in a proportional change in the other two, since a constant can simply be moved outside the integral.

Examining Eq. (12), (13) and (14) for the calculation of bi-directional reflectance, black-sky albedo and white-sky albedo from the MODIS BRDF product, it can be seen that, for a given pixel, we have three linear equations of the form

$$R = p_{\mathrm{iso}} + c_1 p_{\mathrm{vol}} + c_2 p_{\mathrm{geo}}. \tag{26}$$

Hence the reflectances calculated using these expressions also scale linearly.

The ORAC BRDF retrieval is set up to treat the white-sky albedo as the retrieved parameter, with the bi-directional and black-sky albedo values being derived from it, as the white-sky albedo is independent of the viewing geometry.

6.2 Derivatives of the forward model expression

The derivative of Eq. (23) with respect to optical depth or effective radius can be shown to be

$$\frac{\partial R}{\partial x} = R'_{\mathrm{BD}} + (R_{\mathrm{SBD}} - R_{\mathrm{SLB}}) \left(T_{\mathrm{DB}}^{\downarrow} T_{\mathrm{DB}}^{\prime\uparrow} + T_{\mathrm{DB}}^{\prime\downarrow} T_{\mathrm{DB}}^{\uparrow} \right) \tag{27}$$

$$+ \frac{\left(T_{\mathrm{DB}}^{\downarrow} R_{\mathrm{SLB}} + T_{\mathrm{BD}}^{\downarrow} R_{\mathrm{SLW}} \right) R_{\mathrm{SLW}} T^{\uparrow} R'_{\mathrm{FD}}}{(1 - R_{\mathrm{SLW}} R_{\mathrm{FD}})^2}$$

$$+ \frac{\left(T_{\mathrm{DB}}^{\downarrow} R_{\mathrm{SLB}} + T_{\mathrm{BD}}^{\downarrow} R_{\mathrm{SLW}} \right) T'^{\uparrow} + T^{\uparrow} \left(R_{\mathrm{SLW}} T_{\mathrm{BD}}^{\prime\downarrow} + R_{\mathrm{SLB}} T_{\mathrm{DB}}^{\prime\downarrow} \right)}{1 - R_{\mathrm{SLW}} R_{\mathrm{FD}}},$$

where all $'$ indicate $\partial/\partial x$ and x is either τ or r_{e}.

The derivative with respect to surface reflectance requires that we express the derivatives of R_{SBD} and R_{SLB} in terms of a derivative of R_{SLW}. Since R_{SBD} and R_{SLB} both depend linearly on R_{SLW} for a given viewing geometry, we can write:

$$\frac{\partial R}{\partial R_{SBD}} = \frac{\partial R}{\partial R_{SLW}} \frac{\partial R_{SLW}}{\partial R_{SBD}} = \frac{1}{\alpha} \frac{\partial R}{\partial R_{SLW}} \tag{28}$$

$$\frac{\partial R}{\partial R_{SLB}} = \frac{\partial R}{\partial R_{SLW}} \frac{\partial R_{SLW}}{\partial R_{SLB}} = \frac{1}{\beta} \frac{\partial R}{\partial R_{SLW}}, \tag{29}$$

and the derivative can then be expressed as:

$$\frac{\partial R}{\partial R_{SLW}} = T_{DB}^{\downarrow}(\alpha\beta)T_{DB}^{\uparrow} + \frac{T_{DB}^{\downarrow}\beta T^{\uparrow} + T_{BD}^{\downarrow}T^{\uparrow}}{1 - R_{SLW}R_{FD}} + \frac{\left(T_{DB}^{\downarrow}R_{SLB} + T_{BD}^{\downarrow}R_{SLW}\right)T^{\uparrow}}{(1 - R_{SLW}R_{FD})^2}. \tag{30}$$

7.   The thermal infrared forward model

In general aerosol has a relatively small impact on the TOA radiance in the thermal infrared, as the particles are generally small enough to act as Rayleigh scatters at these wavelengths (i.e. the aerosol signal becomes lost in the Planck curve of the atmosphere). However, large particles, such as wind-blown dust, can have a significant thermal infrared signature. This fact has been widely used to develop dust indices, such as the Saharan Dust Index (SDI) [19] used with SEVIRI. The potential of thermal information on wind-blown dust is especially great in situations where the lofted dust is above surfaces with a similar composition. In such cases the contrast between the aerosol and surface in the visible/near-infrared is particularly poor. However, thermal channels show a strong contrast, as the lofted dust is almost universally at a significantly lower temperature than the surface.

In order to take advantage of this potential, a version of the ORAC aerosol retrieval which can utilize thermal window channels,[4] has been developed. Using the thermal channels complicates the retrieval scheme in four main ways:

1. A separate Fast-FM is required for thermal channels, as their signal is dominated by thermal emission from the Earth's surface and atmosphere, rather than scattering and reflection of solar radiation.

2. In window channels, the thermal signal is dominated by the surface temperature and emissivity. Thus, these parameters must also be accounted for in the retrieval.

3. In order to accurately model the TOA brightness temperature, the thermal emission of the atmosphere without aerosol loading (i.e. clear-air) must also be modeled. This involves further radiative transfer calculations and requires accurate knowledge of the temperature structure and trace-gas concentrations of the atmosphere.

4. Since the thermal emission of the aerosol depends on its altitude, this must be accounted for in the retrieval.

---

[4] Channels for which the signal from trace gas absorption and emission is low and the signal is dominated by the thermal emission from the ground and atmosphere.

In order to model the clear-sky brightness temperature, RTTOV [25, 29] is used in conjunction with ECMWF reanalysis data for the scene of interest. The ECMWF fields used are the surface temperature and pressure, and profiles of temperature, humidity and ozone. ECMWF data are available at 6-hour intervals, and are linearly interpolated to the measurement time of the satellite. Clearly, the lack of any trace gas information in the ECMWF data, aside from water vapor and ozone, means this approach will only be valid for instrument channels where there is little signal from gas absorption or emission lines (i.e. the analysis is limited to making use of window channels). RTTOV provides the temperature at 43 layers, the transmission from the surface to each layer and from each layer to the TOA, as well as the upwelling and downwelling clear-sky radiance at each level, for each pixel in the satellite image to analysed.

In the calculation of aerosol look-up tables, it is necessary to extend the calculation of transmission and reflectance of the aerosol layer to diffuse radiation to the thermal wavelengths, as well as calculating its emissivity (values for the direct beam transmission and bi-directional reflectance are not required, as there is no direct beam component in the thermal radiative transfer). Thus, aerosol optical properties are required which extend into the thermal infrared.

Additionally the dependence of the thermal signal on the height distribution of the aerosol must be accounted for. In the generation of the look-up tables, the aerosol is modelled as a single infinitesimally thin layer, for which the transmission, reflectance and emissivity are computed. In the retrieval, the height of this layer is variable (i.e. the layer height is a retrieved parameter), and is assumed to be in thermal equilibrium with the surrounding atmosphere. This approximation will only be valid in certain circumstances (i.e. where a single elevated aerosol layer exists). However, in the case of wind-blown dust, this is a common scenario (see [18] for example). Simulations have shown that grossly different aerosol height distributions (such as boundary layer aerosol below an elevated layer or aerosol across a broad height range) lead to errors in retrieved parameters which are similar to, or significantly smaller than the expected error due to uncertainty in the measured radiances.

Derivation of the thermal Fast-FM follows a similar pattern to that of the shortwave Fast-FM. If we define:

- Upwelling TOA radiance, $I^\uparrow$.
- Transmission of the atmosphere above the aerosol layer, T_{al}.
- Upwelling TOA radiance from the atmosphere above the aerosol layer, $I_{al}^\uparrow$.
- Downwelling radiance at the top the aerosole layer from the overlying atmosphere, $I_{al}^\downarrow$.
- Transmission of the atmosphere below the aerosol layer, T_{bl}.
- Upwelling radiance at the bottom of the aerosol layer neglecting multiple reflection, between the layer and underlying surface, $I_{bl}^\uparrow$.
- Black body radiance at of the aerosol layer, B_l.
- Emissivity of the aerosol layer, ε_l.
- Reflectance of the aerosol layer, R_l.
- Surface reflectance, R_s.

The TOA thermal intensity can then be expressed as

$$I^\uparrow = B_l \varepsilon_l T_{al} + I_{al}^\downarrow R_l T_{al} + I_{al}^\uparrow + I_{bl}^\uparrow T_l T_{al} + I_{bl}^\uparrow R_l T_{bl}^2 R_s T_l T_{al} + I_{bl}^\uparrow R_l^2 T_{bl}^4 R_s^2 T_c T_a l + \dots . \quad (31)$$

Here the first term is the thermal emission from the aerosol layer itself, the second term gives the contribution from downwelling radiance reflected by the aerosol layer and the third and fourth terms are the contribution from thermal emission of the atmosphere above and below the aerosol layer, respectively. The remaining terms account for multiple reflections between the surface and the aerosol layer. As with the shortwave forward model, this expression can be simplified by factorizing and applying a geometric series limit:

$$I^\uparrow = B_1\varepsilon_1 T_{al} + I^\downarrow_{al}R_1 T_{al} + I^\uparrow_{al} + I^\uparrow_{bl}T_1 T_{al}\left(1 + R_1 T^2_{bl}R_s + R^2_1 T^4_{bl}R^2_s + \ldots\right), \qquad (32)$$

leading to the expression

$$I^\uparrow = B_1\varepsilon_1 T_{al} + I^\downarrow_{al}R_1 T_{al} + I^\uparrow_{al} + \frac{I^\uparrow_{bl}T_1 T_{al}}{1 - R_1 T^2_{bl}R_s}. \qquad (33)$$

Note that the value $I^\uparrow_{bl}$ includes both the atmospheric transmission below the aerosol layer and emission from the surface:

$$I^\uparrow_{bl} = I^{\uparrow(atm)}_{bl} + B_s\varepsilon_s T_{bl}, \qquad (34)$$

where B_s is the blackbody radiance of the surface and ε_s is its emissivity.

If we neglect multiple scattering between the surface and aerosol layer (which is a reasonable approximation at thermal wavelengths), the later terms in Eq. (31) and (32) become zero and the denominator of Eq. (32) becomes unity.[5] Applying this approximation and substituting (34), we get:

$$I^\uparrow = B_1\varepsilon_1 T_{al} + I^\downarrow_{al}R_1 T_{al} + I^\uparrow_{al} + B_s\varepsilon_s T_{bl}T_1 T_{al} + I^{\uparrow(atm)}_{bl}T_1 T_{al}. \qquad (35)$$

In order to account for the height dependence of the aerosol thermal emission and the surface-temperature dependence, both of these quantities are retrieved by the thermal algorithm. The state vector for the thermal retrieval thus consists of:

- Aerosol optical depth at 0.55 μm.
- Aerosol effective radius.
- Surface albedo at 0.55 μm.
- Aerosol layer height.
- Surface temperature.

Due to the greatly extended spectral range encompassed when thermal channels are included in the retrieval, accurate knowledge of the aerosol spectral refractive index is even more essential than in a visible near-infrared algorithm. A case study carried out by Carboni et al. [3], showed that dust-like aerosol models currently available from the literature [6, 11] could not be used to reproduce radiances observed by SEVIRI over the Sahara Desert and Atlantic Ocean during a dust storm event. The errors in aerosol optical depth and effective radius from retrievals using these refractive indices were completely dominated by this discrepancy. More success was achieved using refractive indices from in-house spectral measurements of Saharan dust samples [24], however, there remain discre-

---

[5] Note that this is in effect the same as assuming a zero surface reflectance.

pancies that show the need for future improvement of the characterization of aerosol and the Earth's surface in the IR.

8.   The retrieval algorithm

All three of the Fast-FMs described in the previous section fit into the same basic ORAC retrieval algorithm, with minor changes to deal with differing input variables (such as the necessity of dealing with three surface reflectance values for each pixel for a retrieval using the BRDF forward model). The algorithm is built around the optimal estimation framework described by Rogers [26, 27]. If we define the vector made up of the retrieved parameters to be the state vector, $\mathbf{x}$, and the a vector of the measurements, $\mathbf{y}$, then the probability density function of the state subject to the measurements is defined, by application of Bayes' theorem and Gaussian statistics, to be

$$-2 \ln P(\mathbf{x}|\mathbf{y}) = (\mathbf{y} - \mathbf{F}(\mathbf{x}))\mathbf{S}_\varepsilon^{-1}(\mathbf{y} - \mathbf{F}(\mathbf{x})) + (\mathbf{x} - \mathbf{x}_a)\mathbf{S}_a^{-1}(\mathbf{x}\mathbf{x}_a). \tag{36}$$

Here $\mathbf{F}(\mathbf{x})$ is the forward function (i.e. the function which maps the state parameters to measurements, which we approximate with a forward model $\mathbf{f}(\mathbf{x})$), $\mathbf{S}_\varepsilon$ is the measurement error covariance matrix, $\mathbf{x}_a$ is the *a priori* state vector and $\mathbf{S}_a$ is the *a priori* error covariance matrix. Together $\mathbf{x}_a$ and $\mathbf{S}_a$ denote our best guess at the state before the measurement is made and the precision of this guess. The retrieval problem is, therefore, that of finding the minimum value of Eq. (36), which is known as the cost function (i.e. maximizing the probability of $\mathbf{x}$ subject to $\mathbf{y}$), which is known as the cost function.

ORAC uses the Levenberg–Marquardt numerical optimization to perform this minimization. This is an iterative procedure, whereby, if the number of measurements in $\mathbf{y}$ is m, and there are n state parameters, $\mathbf{x}$ is incremented by

$$\mathbf{x}_{i+1} = \mathbf{x}_i + \left(\mathbf{S}_a^{-1} + \mathbf{K}_i^T\mathbf{S}_\varepsilon^{-1}\mathbf{K}_i + \gamma\mathbf{D}_n\right)^{-1}\left[\mathbf{K}_i^T\mathbf{S}_\varepsilon^{-1}(\mathbf{y} - \mathbf{F}(\mathbf{x})) - \mathbf{S}_a^{-1}(\mathbf{x} - \mathbf{x}_a)\right], \tag{37}$$

where $\mathbf{K}$ is the weighting function matrix, γ is variable parameter, $\mathbf{D}_n$ is a $n \times n$ diagonal scaling matrix and the i subscript denotes values for the current iteration. $\mathbf{K}$ is a $m \times n$ matrix, with each column containing the derivative of the forward model with respect to each state parameter, i.e.

$$k_{i,j} = \frac{\partial f_i(\mathbf{x})}{\partial x_j}. \tag{38}$$

Thus, for a linear system, we could write $\mathbf{y} = \mathbf{K}\mathbf{x}$.

The parameter γ is the key to the efficiency and robustness of the Levenberg–Marquardt algorithm. If $\gamma \to \infty$, Eq. (37) tends to the step given by the steepest descent algorithm, which will always lie in the direction of the local 'downhill' gradient and is therefore very robust. If $\gamma \to 0$, however, the algorithm behaves like Gauss–Newton iteration, which, although less numerically robust than steepest descent, will provide an exact solution to a linear problem in one iteration. The procedure for determining the value of γ is to start with a fairly small value (so the initial iteration will resemble Gauss–Newton), then at each iteration:

- If, as a result of the step suggested by Eq. (37), the cost function increases, do not update the state vector and increase γ.
- If the cost function is decreased by a step, update the state vector and decrease γ for the next step.

ORAC uses a factor of 10 for increasing and reducing γ. The scaling matrix, $\mathbf{D}_n$, is used to ensure that the state parameters are of similar magnitude, in the interests of numerical stability.

This iterative procedure is continued until either a convergence criterion is satisfied, or a maximum number of iterations is exceeded (in the former case the retrieval is said to have converged, while the later case can generally be rejected as a failed retrieval). ORAC uses the change in the cost function between iterations to determine whether the algorithm has converged – a negligible change in cost between iterations indicates that the retrieval is no longer improving the fit between measurements and forward model.

The optimal estimation framework offers two main advantages over more ad hoc retrieval algorithms:

1. *A priori* information is explicitly included in the retrieval in a way which is consistent with the way measurement information is included.
2. Rigorous error propagation, including the incorporation of forward model and forward model parameter error, is built into the system, providing extra quality control and error estimates on the retrieved state.

Error estimates for the retrieved state can be calculated by applying

$$\hat{\mathbf{S}} = \mathbf{S}_a^{-1} + \mathbf{K}_i^T \mathbf{S}_\varepsilon^{-1} \mathbf{K}_i \tag{39}$$

after the final iteration, where $\hat{\mathbf{S}}$ is the covariance of the retrieved state. If there is a known limitation in the forward model, due to approximations or incomplete modeling of the relevant physics, this can be accounted for in the retrieval as forward model error described by a covariance matrix $\mathbf{S}_{fm}$. Uncertainty in parameters on which the forward model depends, but which are not retrieved (for instance, the height distribution of aerosol), can also be included in the retrieval as forward model parameter error. These extra error terms are combined with the measurement error:

$$\mathbf{S}_y = \mathbf{S}_\varepsilon + \mathbf{S}_{fm} + \mathbf{K}_p \mathbf{S}_p \mathbf{K}_p^T, \tag{40}$$

where $\mathbf{S}_p$ is the covariance matrix describing the uncertainty in the forward model parameters and $\mathbf{K}_p$ is the weighting function which maps this error into measurement space (i.e. it is analogous to the $\mathbf{K}_i$ matrix used above). The new measurement covariance $\mathbf{S}_y$ then replaces $\mathbf{S}_\varepsilon$ in Eq. (36) to (39).

9.   Aerosol speciation

Although the ORAC algorithm does not directly retrieve any information on the composition of the aerosol, except the change in mixing state implied by the retrieval of effective radius (see Section 3.1), it is still possible for the system to provide some indication of the aerosol type present in a given scene. This capability is achieved by running the retrieval repeatedly using a different predefined aerosol class each time. The resulting set of aerosol retrievals can be merged into a single 'speciated' product by comparing the retrieval cost function for each of the aerosol classes used, weighted by *a priori* knowledge of the likely aerosol type at that particular location.

For example, over the mid-Atlantic Ocean, the dominant aerosol would be expected to be maritime. However, during periods of agricultural and forest burning in central and southern Africa, a biomass aerosol class will often provide a significantly better fit to the measurements (as indicated by a lower value of the cost function), indicating the presence of outflow from the African fires.

Although this method has been found to be somewhat successful, it is subjective in nature (relying on an ad hoc *a priori* weighting of the cost function comparison) and can only be considered as an indicative measure of the actual aerosol composition. Also, the method is only sensitive to aerosol types which appear significantly different in the measurement channels. For instance, it has been found that when using (A)ATSR or SEVIRI visible channels, maritime aerosol and desert dust are difficult to distinguish except at very high optical depths.

10.   Example results

The following sections give examples of aerosol properties derived using the different versions of ORAC described in the previous sections.

10.1 Single-view retrieval from AATSR

Fig. 7.3 shows monthly means of optical depth, effective radius and speciation for September 2004, from AATSR, as retrieved using the Lambertian Fast-FM described in Section 5. Nadir view AATSR radiances have been cloud cleared and then averaged onto a 10 km sinusoidal grid [33] prior to retrieval. Cloud clearing over the ocean used the ESA operational cloud flag [10], while over the land a custom method, which uses a threshold on normalized difference vegetation index values derived from the 0.55, 0.67 and 0.87 μm channels, as well as thresholds on the radiances in these channels themselves [2] was used. The retrieval has been run using five separate aerosol classes:

- continental clean [†]
- desert dust [†]
- maritime clean [†]
- urban [†]
- biomass burning [‡]

where a [†] indicates the aerosol class originates from the OPAC database [11] and [‡] originates from the work of Dubovik et al. [9]. The monthly mean was performed on a $1° \times 1°$ latitude/longitude grid and the data has had the following quality-control criteria applied:

- The retrieval must have converged.
- The final value of the retrieval cost function must be below a threshold value.
- The retrieval must have converged to a state within the bounds of the LUTs.
- The retrieved surface reflectance at 550 nm must be less than 0.2.

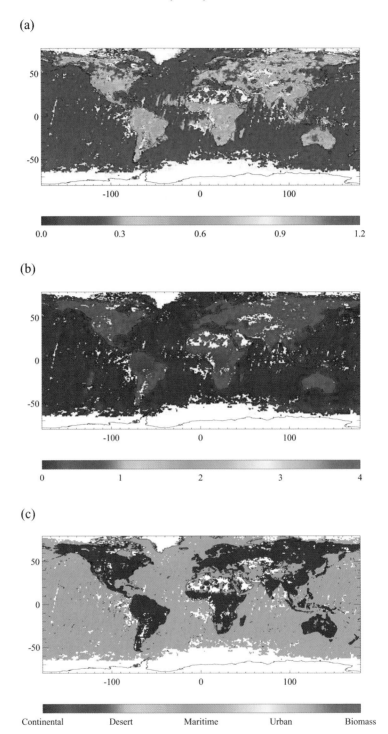

Fig. 7.3. Global mean (a) 0.55 μm aerosol optical depth, (b) effective radius in μm and (c) speciation from AATSR for the month of September 2004.

- Over land pixels the fraction of measurement pixels within a given sinusoidal grid cell flagged as cloud must be less than 50%.[6]

The first thing to note from Fig. 7.3 is the limitations introduced by the narrow swath of AATSR. Since ORAC uses the visible wavelength channels of AATSR, only data from the descending (daylight) half of the orbit can be used. Even with a month of data, there remain areas where cloud cover and sunglint mean there is no coverage, and the retrieved fields still show evidence of the orbital pattern. Taking this into account, the optical depth field shows the expected patterns, with reasonable values. However, the limitations of the single view using a Lambertian surface reflectance are clear, especially over land. Generally, retrievals over land surfaces show slightly higher than expected optical depths and there is a clear step change in both optical depth and effective radius along coastal boundaries. It is also notable that some regions which would be expected to show high aerosol loading, such as China, are not particularly conspicuous in the optical depth field. The optical depth over the oceans appears to be much more as expected, although a very low value of effective radius has been retrieved. Speciation (Fig. 7.3(c)) shows that the expected aerosol class is derived for most pixels, although there is clear evidence of the difficulty in distinguishing desert dust and maritime aerosol using the AATSR channels. Also, there is little indication of the biomass burning plume visible in the optical depth field off the coast of Africa.

These conclusions are supported by Fig. 7.4, which displays the results a comparison between AERONET optical depths and AATSR retrievals for the same month as in Fig. 7.3. Fig. 7.5 shows the locations of the AERONET stations used in the comparisons presented in this chapter. The stations used have been limited to those known to provide measurements which are representative of their surrounding area, based on recommendations by S. Kinne [16]. These comparisons relate the mean AERONET optical depth within 30 minutes of the satellite overpass with the mean satellite optical depth within 20 km of the AERONET site. This averaging is done to try and ensure that similar airmasses are being measured by the satellite and ground-based instruments and is based on the procedure used for MODIS aerosol validation [12]. Each AERONET station used in the comparison has been assigned a unique coloured symbol, with stations within similar geographical regions having the same color. Stations which are in open sea or coastal sites are denoted with open symbols, while solid symbols are used for inland sites. The ORAC data used for comparison with AERONET have had the same filters applied as those applied for the monthly mean products, with the addition of a threshold of the variability of the retrieved optical depth within the 20 km spatial window, which was included to remove cases where high spatial variablity in aerosol loading could introduce significant sampling biases to the comparison.

The comparison with AERONET data clearly shows that the retrieval is performing relatively poorly over land surfaces. The high optical depth bias of the single-view Lambertian ORAC results suggested by the monthly mean plot is confirmed, and the results show a great deal of scatter, with a correlation of 0.4 between the two datasets. The results

---

[6] The scheme averages all pixels cloud-free within each sinusoidal grid cell, to give a cloud free radiance. However, the cloud flag applied to AATSR over the land was found to leave some residual cloud contamination. This was not found to be a problem over ocean pixels.

(a)

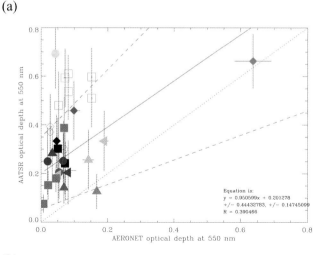

(b)

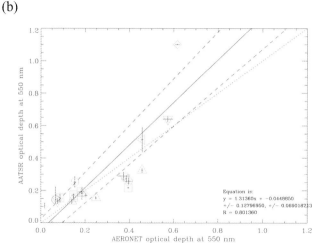

Fig. 7.4. Scatterplots of AERONET optical depth at 550 nm versus coincident AATSR retrievals for September 2004 from the single-view Lambertian surface retrieval. The results are a composite of five separate aerosol classes, with the class of each point being determined by comparing the retrieval costs. Error bars indicate the standard deviation of the values that went into the spatial or temporal averaging. Each plot includes a least-absolute-deviation fit line (solid line) and its equation, plus uncertainty estimates on this fit (dashed lines), with the uncertainties in the fitted parameters included below the equation. The one-to-one line (dotted) is also included for comparison, and the correlation coefficient for the plot, R is given. Plot (a) shows results for inland AERONET stations. Plot (b) shows results for coastal stations, where only satellite results over the ocean are compared.

for ocean pixels are far more encouraging. Here there is no clear bias in the results and there is much less scatter, with the correlation being 0.8.

These results show that the single-view Lambertian retrieval is very dependent on an accurate *a priori* description of the surface reflectance. The higher reflectance and much greater anisotropicity (and hence greater uncertainty in its reflectance) of the land surface, when compared to the ocean, results in relatively poor retrieval performance over the land.

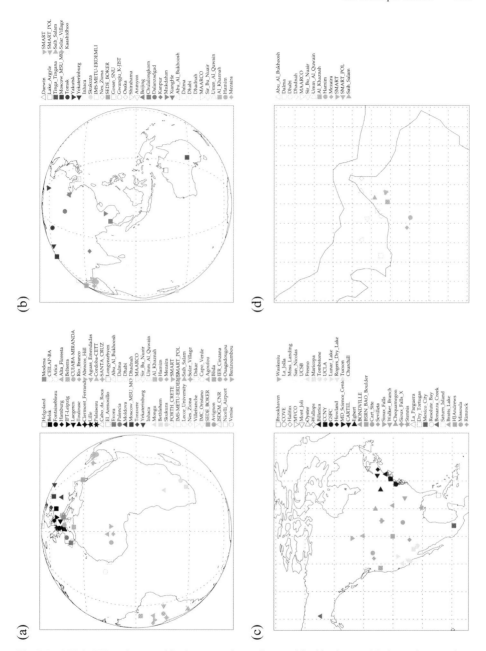

Fig. 7.5. AERONET stations used in the comparisons discussed in this chapter. (a) shows those stations within the SEVIRI observation disk, (b) shows stations within the Eastern Hemisphere, North American stations appear in (c) and those in Arabia are shown in (d).

10.2 Dual-view retrieval from AATSR

This section presents results equivalent to those presented in Section 10.1, but using the BRDF surface reflectance version of ORAC incorporating both views of the AATSR instrument. Fig. 7.6 can be exactly compared to Fig. 7.3 and shows a marked improvement in the retrieved aerosol properties. The optical depth field shows far less evidence of elevated optical depths over land pixels and is much more continuous across coastal boundaries. Areas of high optical depth are evident, but are generally in regions where high aerosol loading is to be expected – such as Southern and Eastern Asia). Optical depths in desert regions are also more believable than in the single-view Lambertian results. The effective radius field shows features that are also expected such as larger particles in regions of desert outflow, and smaller particles associated with polluted regions. Speciation also shows evidence of improvement over the single-view Lambertian results, with significant amounts of biomass-burning aerosol being detected in regions where it might be expected (the Amazon Basin and central/southern Africa) as well as clear biomass plume extending across the Alantic (which even shows evidence of aging – becoming more like background continental aerosol as it extends towards South America). However, there is still evidence of the difficulty in distinguishing desert dust and maritime aerosol over the sea and at high latitudes the speciation over the sea breaks down somewhat (with a scattering of different aerosol types being retrieved). The reason for this is not clear; however it could due to the phase functions of the various aerosol types producing similar spectral responces across the AATSR channels at high solar zenith angles.

Comparison with AERONET data (Fig. 7.7) offers further evidence of the improved quality of these results with those discussed in Section 10.1. In this instance all of the comparisons (land and sea) have been plotted together and show a very high degree of correlation (over 0.95) and very little bias. This level of correlation is extremely impressive, given the widely different methods and sampling used to make the measurements.

Comparing the results from this section and Section 10.1 shows the great improvement given by the BRDF surface reflectance Fast-FM. This improvement is due to three factors:

1. The surface reflectance is better modeled. This is particularly important for anisotropic land surfaces and in regions of the ocean effected by sunglint, where the Lambertian approximation breaks down.
2. The retrieval is able to make use of the AATSR dual-view measurement system. Not only does this double the number of measurements available to the retrieval but, since each surface pixel is viewed twice through differing atmospheric paths, the constraint on both the surface reflectance and aerosol properties being consistent between views allows the retrieval to decouple their effects much more effectively than with a single view.
3. The inclusion of four more measurements provides enough information for the retrieval of the surface reflectance in each channel independently, greatly reducing the reliance on an accurate *a priori* knowledge of it.

Although it is difficult to disentangle these three effects, the results do strikingly demonstrate the strength of the ATSR dual-view system.

(a)

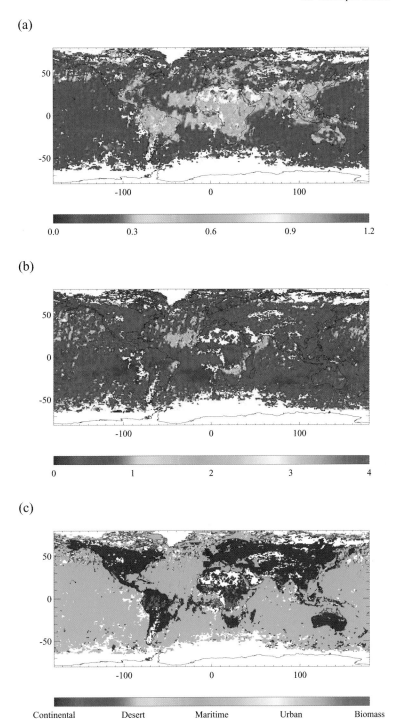

(b)

(c)

Fig. 7.6. Global mean (a) 0.55 μm aerosol optical depth, (b) effective radius and (c) speciation from AATSR for the month of September 2004. The data were produced by the BRDF version of ORAC, using both nadir and forward views on a ∼ 10 km sinusoidal grid. The monthly mean was performed on a 1° × 1° latitude/longitude grid.

(a)

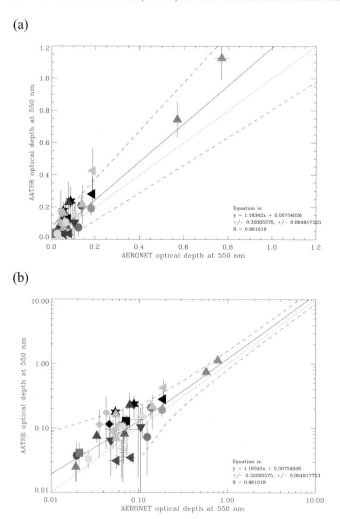

(b)

Fig. 7.7. Scatterplots of AERONET optical depth at 550 nm versus coincident AATSR retrievals for September 2004 from the dual-view BRDF retrieval. The results are a composite of five separate aerosol classes, with the class of each point being determined by comparing the retrieval costs. Both plots show the results for all AERONET stations, with (b) using a logarithmic scale for clarity. See Fig. 7.5 for a definition of the plotting symbols.

10.3 Retrieval from SEVIRI using the BRDF Fast-FM

Fig. 7.8 shows the mean 0.55 μm optical depth retrieved from SEVIRI data over September 2004, as given by ORAC using the BRDF forward model. The production of this composite followed a very similar method as used for the AATSR monthly means discussed in the previous two sections. The same retrieval grid and aerosol classes were used, and very similar quality controls were applied. The composite has been formed using two SEVIRI measurements per day, one at 10:15 and one at 16:15 UT. Although SEVIRI does not

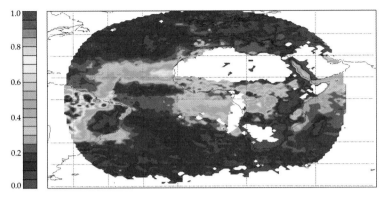

Fig. 7.8. Global mean 0.55 μm aerosol optical depth from SEVIRI for the month of September 2004. The data were produced by the BRDF version of ORAC, on a $\sim$ 10 km sinusoidal grid. The monthly mean was performed on a $1° \times 1°$ latitude/longitude grid. Speciation has been carried out in the usual way, from the same set of five aerosol classes used for the AATSR retrievals.

provide a dual-view measurement, the BRDF forward model still provides a better description of the surface reflectance than the Lambertian approximation does. One of the largest difficulties to be overcome when applying the ORAC scheme to SEVIRI data is that of cloud-contamination. This is largely due to the larger pixel size of the SEVIRI instrument when compared to the (A)ATSR sensors, which results in relatively large number of cloud contaminated pixels being flagged as clear by the operational EUMETSAT SEVIRI cloud flag. Although post-retrieval quality control has removed the vast majority of residual cloud contamination, there is still some evidence of contaminaton in the very high average optical depths seen in the Amazon basin.

Fig. 7.9 shows that, as far as agreement with AERONET is concerned, the SEVIRI retrieval lies somewhere between the AATSR single-view Lambertian and dual-view BRDF results. Agreement over the ocean is excellent, with a correlation of 0.79 and very little bias. For land-based stations, the best-fit line shows a strong negative bias. However, it can be seen that this is mostly caused by a bias against a single AERONET station – Mongu (denoted by yellow triangles). This station lies in an arid region of southern Africa (see Fig. 7.5(a)) that is characterized by a high surface reflectance. This suggests that the SEVIRI retrieval is prone to underestimating aerosol optical depth in regions where the surface component dominates the TOA signal. The negative bias in the SEVIRI results apparent when compared against another desert AERONET site, Blida (denoted by brown squares), supports this conclusion. Furthermore, if Fig. 7.8 is compared to Fig. 7.6, it can be seen that the SEVIRI-based optical depth does tend to be lower than the AATSR value in arid regions. Although the SEVIRI results have had a stringent surface reflectance threshold applied to them (resulting in the regions of missing data over the Sahara, Arabia and southern Africa), these results suggest that a stronger quality-control criterion may be needed.

If one neglects the Mongu points in Fig. 7.9, the agreement between SEVIRI and AERONET is again excellent. Indeed, even with the Mongu points included in the calculation, the correlation for inland AERONET stations is actually higher than that for the ocean/coastal sites, at 0.82. This high correlation is largely due to excellent agreement

(a)

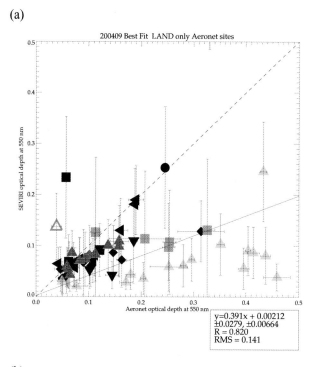

(b)

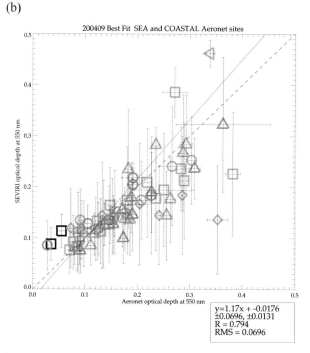

Fig. 7.9. Scatter-plots of AERONET optical depth at 550 nm versus coincident SEVIRI retrievals for September 2004. Plot (a) shows results for inland AERONET stations. Plot (b) shows results for coastal stations, where only satellite results over the ocean are compared. See Fig. 7.5 for a definition of the plotting symbols.

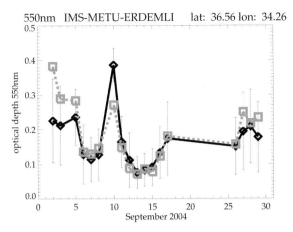

Fig. 7.10. A time-series of 0.55 μm optical depth from SEVIRI retrievals (black) and AERONET sun-photometer (orange). Only retrievals over land pixels were considered in computing the SEVIRI optical depth.

between the ORAC retrieval and AERONET over Europe. An example of this is shown in Fig. 7.10, where time series of collocated AEROSOL optical depth from SEVIRI and the IMS-METU-ERBEMLI AERONET station are given. In almost all cases the two measurements agree within the error bars.

10.4 Retrieval using the thermal infrared forward model

At the time of writing, work on including a thermal infrared forward model in the ORAC aerosol retrieval was still at a fairly early stage. However, the retrieval has been applied to a limited quantity of SEVIRI data from March 2006. This month was marked by a very large dust storm in the Sahara, which resulted in large plumes of dust being blown south, towards the Ivory Coast. There the lofted dust encountered a strong westerly airflow, which carried it out into the Atlantic at extremely high concentrations. Fig. 7.11 shows the SEVIRI data collected on March 9, 2006, at the height of the event. The false color image clearly shows the dust plume extending in a northwesterly direction across the Atlantic, but also demonstrates the problem encountered when trying to retrieve lofted dust over a desert surface using a visible/near-infrared retrieval scheme – it is very difficult to distinguish the lofted dust from the background surface over the desert in Fig. 7.11(a).

Fig. 7.11(b) shows the 550 nm optical depth retrieved using ORAC with the thermal forward model enabled. In this instance the aerosol class has been assumed to be desert dust (based on the optical properties of Peters et al. [24]), and the maximum optical depth covered by the LUTs has been extended from 2 to 5. The retrieval has been performed on a 10-km sinusoidal grid. The dust plume extending into the Atlantic is very clear, and shows extremely high optical depths, which in some places exceed 5. The optical depth field also shows the plume extending along the southern edge of the Sahara, and shows elevated optical depths (exceeding 1) extending north into the Sahara itself.

It must be noted that Fig. 7.11 also shows evidence of the need for further development of the thermal-infrared retrieval. The differentiation of heavy dust loading and cloud is a

(a)

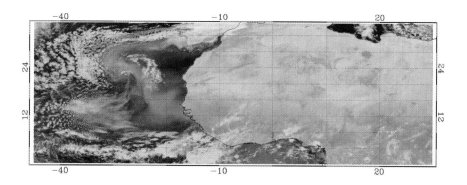

(b)

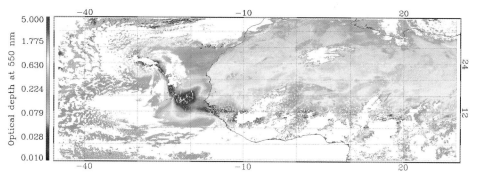

Fig. 7.11. An example of retrieved optical depth making use of thermal channels. Plot (a) shows a false-color image from SEVIRI taken on March 9, 2006, at approximately 12:12UT, during a large dust storm event in the southern Sahara. Plot (b) shows the retrieved optical depth field for this scene. White areas in plot (b) indicate the presence of cloud or failed retrievals (due, for example, to the optical depth being outside the 0.01 to 5.0 optical depth range).

difficulty common among aerosol retrieval schemes, and it is a problem with the thermal-infrared version of ORAC as well. The optical depth field also shows features which are in reality clearly associated with the surface. This is due to both limitations in the description of the surface and the applicability of the assumed aerosol properties. Steps are being taken to address these issues, namely:

- The description of the surface reflectance in the thermal-infrared retrieval will be improved by incorporating the visible/near-infrared BRDF Fast-FM.
- The use of non-spherical scattering code in the calculation of LUTs for non-spherical aerosol classes (such as desert dust) is being investigated.
- Further measurements of Saharan aerosol samples will be undertaken.
- It is hoped that a surface BRDF product derived from SEVIRI measurements will become available in the future [35], which will provide an improved description of the surface for the retrieval.

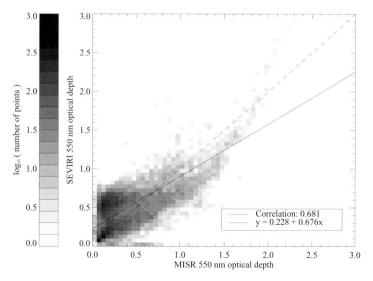

Fig. 7.12. Scatter between collocated SEVIRI and MISR optical depths over the Saharan region for the period March 5–30, 2006. The SEVIRI results, all of which are for 12:12UT, have been retrieved with the thermal-infrared version of ORAC on a 10 km sinusoidal grid, which has then been interpolated onto the MISR level 2 optical depth grid. The 1-to-1 line is indicated by the dashed line, while the solid line gives the best-fit line (least absolute deviation fit, weighted by the error in both datasets), the equation of which is also given.

Despite these limitations, however, these results do provide a good overall indication of aerosol optical depth field. Fig. 7.12 shows a comparison between collocated thermal-infrared ORAC and MISR (Multi-angle Imaging SpectroRadiometer) retrievals. Due to its multiple-view observation system, MISR is regarded as one of the most reliable satellite-based aerosol products, particularly over bright surfaces like deserts [8]. The agreement between the two datasets is very close, especially considering the difference in measurement time between the two (MISR observes at approximately 11:30 local solar time, which means only MISR swaths which lie near the east coast of the Saharan region will be temporally close to the SEVIRI measurement).

11.   Conclusion

This chapter has detailed the ORAC aerosol retrieval algorithm in its various forms. ORAC is a retrieval scheme based around the optimal estimation framework [26, 27] that uses a plane parallel radiative transfer, using assumed aerosol properties, to forward model the TOA radiance measured by satellites. The basic ORAC algorithm provides a framework for the retrieval of aerosol properties from visible/infrared radiometers. As demonstrated by the variety of different ORAC variants described, the generic nature of the algorithm makes it suitable for a wide range of instruments, provided they meet the following criteria:

- The measured signal is not polarization-dependent.
- The measured signal is not strongly effected by molecular Rayleigh scattering, as the variation of this signal with topography height and atmospheric state is not modeled.

- The channels used in the retrieval are fairly free of molecular absorption by atmospheric trace gases.
- Instrument and solar zenith angles must be less than $\sim 80°$ for the plane parallel radiative transfer used in the forward model to be reliable.

The algorithm can also be easily modified to make use of different forward models, a feature which has allowed the development of the BRDF, ATSR dual-view and thermal-infrared versions of the algorithms presented.

The example results presented demonstrate that the ORAC algorithm shows good agreement with ground-based measurements of aerosol optical depth, as well as with current operational satellite aerosol products. The quality of the products combined with the adaptability of the algorithm, makes the ORAC algorithm unique in its ability to quickly adapt to a range of different instruments. This, combined with the strengths inherent in an optimal estimation retrieval scheme, namely:

- statistically rigorous error propagation and error estimates on all retrieved quantities,
- the inclusion of *a priori* information in a statistically consistent way,

make ORAC a strong addition to the stable of satellite aerosol retrieval algorithms.

References

1. A. Berk, L.S. Bernstein, G.P. Anderson, P.K. Acharya, D.C. Robertson, J.H. Chetwynd, and S.M. Adler-Golden. MODTRAN cloud and multiple scattering upgrades with application to AVIRIS. *Remote Sens. Environ.*, **65**:367–375, 1998.
2. A. Birks. Improvements to the AATSR IPF relating to land surface temperature. technical note, European Space Agency, 2004.
3. E. Carboni, G.E. Thomas, R.G. Grainger, C.A. Poulsen, R. Siddans, D. Peters, E. Campmany, A. M. Sayer, H.E. Brindley. Retrieval of aerosol properties from SEVIRI using visible and infrared channels. *Proceedings of EUMETSAT/AMS Conference*, Amsterdam, 24–28, 2007.
4. C.S. Cox, W.H. Munk. Measurement of the roughness of the sea surface from photographs of the Sun's glitter. *J. Opt. Soc. Am.*, **44**:838–850, 1954.
5. C.S. Cox, W.H. Munk. Statistics of the sea surface derived from Sun glitter. *J. Mar. Res.*, **13**:198–227, 1954.
6. G.A. dÁlmeida, P. Koepke, E.P. Shettle. *Atmospheric Aerosols: Global Climatology and Radiative Characteristics*. A Deepak Publishing, Hampton, Virginia, 1991.
7. C.N. Davies. Size distribution of atmospheric particles. *J. Aerosol Sci.*, **5**:293–300, 1974.
8. D.J. Diner, L. Di Girolamo, A. Nolin. Preface to the MISR Special Issue. *Remote Sens. Environ.*, **107**:1–2, 2007.
9. O. Dubovik, B. Holben, T.F. Eck, A. Smirnov, Y.J. Kaufman, M.D. King, D. Tanre, I. Slutsker. Variability of absorption and optical properties of key aerosol types observed in worldwide locations. *J. Atmos. Sci.*, **59**:590–608, 2002.
10. European Space Agency. *ESA AATSR Product Handbook*, Issue 2.2, 2007.
11. M.P. Hess, P. Koepke, I. Schult. Optical properties of aerosols and clouds: the software package OPAC. *Bull. Am. Met. Soc.*, **79**:831–844, 1998.
12. C Ichoku, D.A Chu, S. Mattoo, Y.J. Kaufman, L.A. Remer, D. Tanre, I. Slutsker, B.N. Holben. A spatio-temporal approach for global validation and analysis of MODIS aerosol products. *Geophys. Res. Lett.*, **29**(12.1616), 2002.
13. Y. Jin, C.B. Schaaf, C.E. Woodcock, F. Gao, X. Li, A.H. Strahler. Consistencey of MODIS surface BRDF/albedo retrievals: 2. Validation. *J. Geophys. Res.*, **108**(D5), 2003. doi:10.1029/2002JD002804.
14. Y. Jin, C.B. Schaaf, C.E. Woodcock, F. Gao, X. Li, A.H. Strahler, W. Lucht, S. Liang. Consistency of MODIS surface BRDF/albedo retrievals: 1. Algorithm performance. *J. Geophys. Res.*, **108**(D5), 2003. doi:10.1029/2002JD002803.

15. C.G. Justus, M.V. Paris. Modelling solar spectral irradiance and radiance at the bottom and top of a cloudless atmosphere. *J. Appl. Meteorol.*, **24**:193–205, 1985.

16. S. Kinne. Private communication, 2007.

17. P. Koepke. Effective reflectance of oceanic whitecaps. *Appl. Opt.*, **23**:1816–1824, 1984.

18. J.-F. Léon, D. Tanré, J. Pelon, Y.J. Kaufman, J.M. Haywood, B. Chatenet. Profiling of a Saharan dust outbreak based on a synergy between active and passive remote sensing. *J. Geophys. Res.*, **108**(D18), 2003. doi:10.1029/2002JD002774.

19. C.J. Merchant, O. Embury, P. Le Borgne, B. Bellec. Saharan dust in nighttime thermal imagery: detection and reduction of related biases in retrieved sea surface temperature. *Remote Sensing Env.*, **104**:15–30, 2006.

20. G. Mie. Beiträge zur Optik trüber Medien, speziell kolloidaler Metallösungen. *Ann. Phys.*, **25**:377–445, 1908.

21. E.C. Monahan, I. Ó Muircheartaigh. Optimal power-law description of oceanic whitecap coverage dependence on wind speed. *J. Phys. Oceanogr.*, **10**:2094–2099, 1980.

22. André Morel, Louis Prieur. Analysis of variations in ocean color. *Limnol. Oceanogr.*, **22**:709–722, 1977.

23. J.M. Nicolas, P.Y. Deschamps, R. Frouin. Spectral reflectance of oceanic whitecaps in the visible and infra-red: aircraft measurements over open ocean. *Geophys. Res. Lett.*, **28**:4445–4448, 2001.

24. D.M. Peters, R.G. Grainger, G.E. Thomas, R.A. McPheat. Laboratory measurments of the complex refractive index of Saharan dust aerosol. *Geophys. Res. Abstracts*, **9**, 2007.

25. P.J. Rayer. Fast transmittance model for satellite sounding. *Appl. Opt.*, **34**:7387–7394, 1995.

26. C.D. Rodgers. Retrieval of atmospheric temperature and composition from remote measurements of thermal radiation. *Rev. Geophys. and Space Phys.*, **14**:609–624, 1976.

27. C.D. Rodgers. *Inverse Methods for Atmospheric Sounding: Theory and Practice*. World Scientific, 2000.

28. J.G. Salomon, C.B. Schaaf, A.H. Strahler, F. Gao, Y. Jin. Validation of the MODIS bidirectional reflectance distribution function and albedo retrievals using combined observations from the AQUA and TERRA platforms. *IEEE Trans. Geosci. Remote Sens. Environ.*, **83**(6):1555–1565, 2006.

29. R.W. Saunders. RTTOV-7 science and validation report. RTTOV documentation, NWP SAF, 2002. http://www.metoffice.com/research/interproj/nwpsaf/rtm.

30. A. M. Sayer. A sea surface albedo model suitable for use with AATSR aerosol retrieval, version 2. Technical memorandum, Atmospheric, Oceanic and Planetary Physics, University of Oxford, 2007. http://www.atm.ox.ac.uk/main/research/technical/2007.2.pdf.

31. F.M. Schulz, K. Stamnes, F. Weng. Vdisort: An improved and generalized discrete ordinate method for polarized (vector) radiative transfer. *J. Quant. Spectrosc. Radiat. Transfer*, **61**(1):105–122, 1999.

32. C.L. Shaaf. *MODIS BRDF/Albedo Product (MOD43B) User's Guide*. Boston University, 2004. http://www-modis.bu.edu/brdf/userguide/index.html.

33. J.P. Snyder. Map projections – a working manual. Professional Paper 1395, U.S. Geological Survey, 1987.

34. K. Stamnes, S.C. Tsay, W. Wiscombe, K. Jayaweera. A numerically stable algorithm for discrete-ordinate-method radiative transfer in multiple scattering and emitting layered media. *Appl. Opt.*, **27**:2502–2509, 1988.

35. S. Wagner, Y. Govaerts, A. Lattanzio, P. Watts. Simultaneous retrieval of aerosol load and surface reflectance using MSG/SEVIRI observations. *Geophys. Res. Abstracts*, **9**, 2007.

36. W. Wanner, A.H. Strahler, P. Lewis B. Hu, J.-P. Muller, X. Li, C.L. Barker Schaaf, M.J. Barnsley. Global retrieval of bidirectional reflectance and albedo over land from EOS MODIS and MISR data: theory and algorithm. *J. Geophys. Res.*, **102**(D14):17,143–17,161, 1997.

37. P.D. Watts, C.T. Mutlow, A.J. Baran, A.M. Zavody. Study on cloud properties derived from Meteosat Second Generation observations. ITT 97/181, EUMETSAT, 1998.

8 Benefits and limitations of the synergistic aerosol retrieval SYNAER

Thomas Holzer-Popp, Marion Schroedter-Homscheidt, Hanne Breitkreuz, Dmytro Martynenko, Lars Klüser

1. Introduction

Air pollution by solid and liquid aerosol particles suspended in the air is one of the major concerns in developed countries because of potential health impact of increasing numbers of nano-particles in particular from diesel engines (see, e.g., Pope et al. [2002] and Stedman [2004]), as well as in developing countries with their high particle concentrations in the air. Furthermore, windblown dust can also act as carrier for long-range transport of diseases, e.g., from the Sahara to the Caribbean or Western Europe [Pohl, 2003], or even around the globe [Prospero et al., 2002]. Also well known in principle are direct (by reflecting light back to space) and several indirect (e.g., by acting as cloud condensation nuclei) climate effects of aerosols, although large uncertainties exist in the exact values of the forcing [IPCC, 2007]. Finally, the highly variable atmospheric aerosol load has a major impact on satellite observations of the Earth's surface that need to be atmospherically corrected for quantitative analysis and on the solar irradiance which is exploited in solar energy applications (aerosols are the determining factor in clear-sky conditions). In all these cases an estimation of the type of aerosols is required for an accurate quantitative assessment. For example, Kaufman, et al. [2002] point out, that the absorption behavior of particles (mainly soot and minerals) needs to be known in order to assess their total direct and indirect climate effects. This is because strongly absorbing particles can regionally reverse the sign of the aerosol direct forcing from cooling to heating or suppress cloud formation. Therefore, attempts have been made to extend satellite aerosol retrieval beyond observation of the spatial–temporal distribution patterns to estimate the type of aerosols.

In recent years the satellite monitoring capabilities in particular to derive maps of aerosol optical depth (AOD) have increased tremendously. A good overview of different satellite retrieval principles to derive AOD is presented in Kaufman et al. [1997a] and a review of achieved AOD retrieval capabilities is given in Kaufman et al. [2002]. Examples of satellite retrieval of additional aerosol optical properties include the Ångström coefficient, (e.g., AATSR dual view [Veefkind et al., 1999]) and the separation into fine and coarse mode aerosols (e.g., MODIS multispectral collection 5 [Levy et al., 2007]; fine mode AOD only by POLDER polarized multispectral measurements [Deuzé et al., 2001]). Further examples of aerosol characterization use a choice from predefined aerosol types (e.g., MISR multi-angle [Kahn et al., 2005]), the single scattering albedo (MODIS deep blue [Hsu et al., 2004]), or derived quantities such as particle number concentrations (e.g., parameterization based on MERIS multispectral measurements von Hoyningen-Huene et al., 2003; Kokhanovsky et al., 2006]).

Another approach that has been used to extract additional aerosol properties (beyond AOD) is the SYNAER (SYNergistic AErosol Retrieval) method which was developed to exploit data from a combination of a radiometer (ATSR-2, Along-Track Scanning Radiometer 2) and a spectrometer (GOME, Global Ozone Monitoring Experiment) onboard a single platform (ERS-2, European Remote Sensing Satellite) to provide a multispectral retrieval ranging from deep blue to red bands with two different spatial scales [Holzer-Popp et al., 2002a]. A first-case study validation and retrieval of 15 coincidences of these ERS-2 data with AERONET observations showed a multispectral AOD accuracy of 0.1 at three visible wavelengths [Holzer-Popp et al., 2002b].

The SYNAER method is further detailed in this chapter, which is divided into six sections. Section 2 gives an overview of the SYNAER sensors and an analysis of its information content with regard to the aerosol type including realistic noise in the retrieval. The chapter then gives a comprehensive description of the SYNAER methodology and details the major advances of the new SYNAER methodology developed during the transfer to ENVISAT in Section 3: extension of the aerosol component database with moderately absorbing soot and mineral dust, two additions to the cloud screening, and an improved dark field method. Section 4 contains a validation of the SYNAER algorithm with the modifications comparing the results of AOD at 440, 550 and 670 nm for 42 orbits of summer 2005 against AERONET measurements. In addition a first 4-month dataset derived with the SYNAER method from ENVISAT sensors AATSR and SCIAMACHY is discussed. The chapter concludes with a discussion and outlook to further validation and a long time series by combining ENVISAT and METOP retrievals with SYNAER in Section 5.

2.   SYNAER: exploited satellite instruments and information content analysis

2.1 Exploited sensors

ATSR-2 and GOME have been acquiring data from onboard the European platform ERS-2 since 1995. The two instruments simultaneously observe the same area on the globe. ATSR-2 measures Earth reflected radiances in five spectral bands centered at 0.55, 0.67, 0.87, 1.6, and 3.7 μm (at 3.7 μm also including emitted terrestrial radiation) with bandwidths of 25 to 66 nm and brightness temperatures in two thermal channels at 11 and 12 μm. All observations are taken under two viewing angles (nadir and approximately 55° forward) with a ground resolution of approximately 1.1 km^2 at nadir. The main application of ATSR-2 is the retrieval of sea-surface temperature with high accuracy [Zavody et al., 1995].

GOME observes near-nadir reflection from the Earth in the range from 240 to 790 nm with a spectral resolution of 0.2 nm to 0.4 nm and a pixel size of either 320 × 40 km^2 or 80 × 40 km^2. The latter pixel size was selected during the entire 9-months commissioning phase and since July 1997, it has operated in this coarser resolution mode 3 days per month. Its principal goal is the monitoring of stratospheric ozone but other stratospheric trace gases can be also measured [Burrows et al., 1999]. Both instruments measure the solar illumination regularly. Consequently, reflectances $R = \pi \cdot L/\mu_0 \cdot E_0$ can be calculated, where L is the reflected light radiance and E_0 is the solar irradiance. This significantly reduces calibration errors as compared to the use of Just calibrated radiances. The cross-correlation of spectrally and spatially integrated reflectances measured by both instruments was found to satisfy high accuracy requirements with deviations less than 4 % of absolute reflectance [Koelemeijer et al., 1997].

The Advanced Along-Track Scanning Radiometer (AATSR) and the Scanning Imaging Absorption Spectrometer for Atmospheric Cartography (SCIAMACHY) are flown on the European platform ENVISAT which was launched on March 1 in 2002. AATSR has spatial and spectral characteristics very similar to ATSR-2 but the data are provided in a simpler format (one data file contains one entire orbit). SCIAMACHY has a nadir and an additional limb viewing mode. Its nadir pixel size is 60×30 km^2 and it has an extended spectral range covering 240 to 1750 nm completely and 1940 to 2380 nm partly with a spectral resolution between 0.2 and 1.5 nm. In the latest calibration version the cross-correlation of spectrally and spatially integrated reflectances measured by both instruments and against another radiometer MERIS (Medium Resolution Imaging Spectrometer) onboard ENVISAT was found to satisfy high accuracy requirements with deviations of the order of 1 % [Kokhanovsky et al. 2007].

2.2 Analysis of the information content

It is important to understand the information content of remotely sensed observations and, therefore, how different observations contribute to a retrieval algorithm. By observing reflectances at different wavelengths, the size and quality of measurements can be improved. However, there is a point where adding further observations has a negligible effect. Further, with the vast increase in satellite data, it is not possible to include all available observations in retrievals. Therefore, it is necessary to select an optimal subset of the observations such that the important information is retained. In the case of satellite retrievals, there is a complicated relationship between observed and retrieved variables. The information content determines how many linearly independent pieces of information are contained in a set of observations. This not only depends on the observations, but on the algorithm in which they are used, for example on the radiative transfer model and on the errors in the observations.

The SYNAER method consists of two parts (see Section 3.1 for more detail). In the first part the radiometer data are used to retrieve aerosol optical depth and surface reflectance for selected aerosol types. In the second part the spectrometer data are then used to select the most plausible aerosol type. As the estimation of the aerosol type is the most innovative part of SYNAER, this section provides an analysis of the information content of the second SYNAER part with regard to aerosol composition [Holzer-Popp et al. 2008]. Consequently, the focus of this analysis is on exploiting the spectrometer measurements explicitly using the results of the first retrieval step, namely aerosol optical depth at 0.55 μm and surface reflectance at 0.55, 0.67 and 0.87 μm for each aerosol mixture. In the analysis of the information content seven basic components (water-soluble, water-insoluble, sea salt accumulation and coarse mode, anthropogenic soot, biogenic soot and mineral transported) were used to define a set of 40 mixtures (see Table 8.2, but for insoluble components INSL, INSO, MITR, MILO only those with high absorption, i.e. INSO, MITR, are used), which was then applied to radiative transfer calculations of simulated SCIAMA-CHY spectra using the same radiative transfer code.

Information theory or communication theory is concerned with the information content, i.e. the independent degrees of freedom, contained in a measurement. Information theory was first used by electrical engineers to design better telecommunication systems, but now has a wide variety of applications. In particular, concepts from information theory have been applied to satellite retrieval studies (e.g., Rodgers [2000]). In satellite retrieval studies, it is useful to obtain a single number as a quantitative measure of the information content.

One of the methods used to examine the information content is the singular value decomposition (SVD). SVD is a useful tool to identify the dominant or important part of the observations. This allows us to identify the number of parameters which can be retrieved from the observations and the variables which can be determined. Generally, the number of observations does not equal the degrees of freedom because the observations are not independent and there is usually high correlation between instrument bands. For any remote measurement, the measured quantity, y, is some vector-valued function F of the unknown state vector x, and of some other set of parameters b excluded from the state vector, considering also the experimental error term ε:

$$y = F(x, b) + \varepsilon, \tag{1}$$

where $y \in R^m$ is the measurements vector of dimension m, $x \in R$ is the state vector of dimension n, b is the vector containing all the other parameters necessary to define the radiative transfer from the atmosphere to the spacecraft, $F : R^n \rightarrow R^m$ is the forward model that describes the physics of the measurements that map from the state space to the measurements space and $\varepsilon \in R^m$ is the measurement error vector.

The measurement vector for SYNAER retrieval consists of simulated spectra for 40 different aerosol mixtures for a given surface type. The state vector consists of 40 elements corresponding to the different aerosol mixtures and 12 elements corresponding to predefined SYNAER surface types: savanna, pine forest, bog, pasture, La Mancha, plowed field, and Hildesheimer Börde after Kriebel [1977] and Köpke and Kriebel [1987]; vegetation, water, soil, sand, and snow after Guzzi et al. [1998]. For the purpose of information content, error analysis and the inversion procedure it is necessary to linearize the forward model around a reference state x_0:

$$y - F(x_0) = \frac{\partial F}{\partial x}(x - x_0) + \varepsilon = K(x - x_0) + \varepsilon, \tag{2}$$

where K is the weighting function matrix of dimension $m \times n$. Each element of K is the partial derivative of a forward model element with respect to a state vector element:

$$K_{ij} = \frac{\partial F(x)_i}{\partial x_j}; \forall i = 1 \ldots, m, \ \forall j = 1 \ldots, n. \tag{3}$$

The act of measurement maps the state space into the measurement space according to the forward model. Conversely, a given measurement could be the result of a mapping from anywhere in the state space. For this reason it is necessary to have some prior information about the state, which can be used to constrain the solution.

The information content is condensed into the degrees of freedom for signal (DFS). DFS can be interpreted as the number of independent linear combinations of the state vector that can be independently retrieved from the measurements. It is given by:

$$DFS = \sum \frac{\lambda_i^2}{1 + \lambda_i^2},\tag{4}$$

where λ_i are the singular values of $\boldsymbol{K}' := (\boldsymbol{S}_\varepsilon)^{-1/2}\boldsymbol{K}(\boldsymbol{S}_a)^{1/2}$. $\boldsymbol{S}_\varepsilon$ and $\boldsymbol{S}_a$ are the measurement covariance and a priori covariance matrices. The measurement covariance matrix $\boldsymbol{S}_\varepsilon$ has a diagonal form, with diagonal element $\sigma_\varepsilon^2 = 0.000001$ (i.e. the relative error of measured reflectance spectrum values, which are typically around 0.1, is about 1 %). The *a priori* covariance matrix $\boldsymbol{S}_a$ has a block diagonal form, because there is no correlation in retrieval of surface type and retrieval of aerosol type. The first non-diagonal part of $\boldsymbol{Sa}$ is the explicitly calculated covariance between percent contributions for 40 aerosol mixtures from Table 8.2. The second part of the block diagonal matrix $\boldsymbol{S}_a$ (albedo part) is diagonal, since there is no correlation between different channels for the pre defined 12 SYNAER albedo spectra. The diagonal values for all albedo surfaces, except for the 'snow' case, are equal to 0.0001 ($\sigma_A = 0.01$). For the much brighter 'snow' type $\sigma_A^2 = 0.0016$ is chosen ($\sigma_A = 0.04$). These values are taken in correspondence with the albedo values for the different surface types.

Fig. 8.1(a) shows example spectra of top-of-atmosphere reflectance which were used for this analysis over six surface types and the 40 mixtures for a typical retrieval condition: 'vegetation', 'fresh snow', 'pine forest', 'open water', 'bare soil' and, as a special case, for 'black surface' type, which corresponds to a numerical model of absolutely dark surface. Fig. 8.1(b) describes the dependence of DFS on aerosol optical depth, whereas Fig. 8.1(c) shows the DFS dependence on both sun elevation angle and AOD. The non-monotonous growth of DFS with the sun elevation angle is supposed to be due to the combination of the various phase functions of the basic aerosol components. Those values for typical retrieval conditions as depicted in Fig. 8.1(b) are shown in Fig. 8.1(c) by blue squares. With increasing AOD, the increase of the DFS values is relatively fast (e.g., at AOD = 0.07, DSF is at 3.3). This means that the SYNAER aerosol type retrieval shows meaningful results also for small values of AOD and already at AOD = 0.15 the curve in Fig. 8.1(b) reaches its saturation at about DSF = 4. The offset value of the DFS of 2 at AOD = 0 (i.e. no aerosol content and thus no aerosol signal) is supposed to correspond to the surface brightness and AOD (which are provided from the first SYNAER step). The results of the second step in the SYNAER retrieval are obviously also dependent on the surface type over which the retrieval was made (Fig. 8.2). An analysis was made for the same set of six different surface types as in Fig. 8.1(a). Obviously, the retrieval does not work properly over bright surfaces, such as snow or desert, due to larger absolute noise. Here DFS has approximately a value of 2; i.e. no additional information content for the aerosol type. The maximum values of DFS are over vegetation and water pixels. This agrees well with the choice of surface type for the dark field method. But also for sparsely vegetated surfaces (see the soil example, where DFS is 3 in Fig. 8.2 versus DSF = 4 for vegetation) the information content for the aerosol type is smaller but not negligible (i.e. 2).

In summary, this analysis shows that DFS for the aerosol type (after eliminating the DFS offset of 2) exhibits a variation from 0 to 3.5. These values of DFS correspond only to the

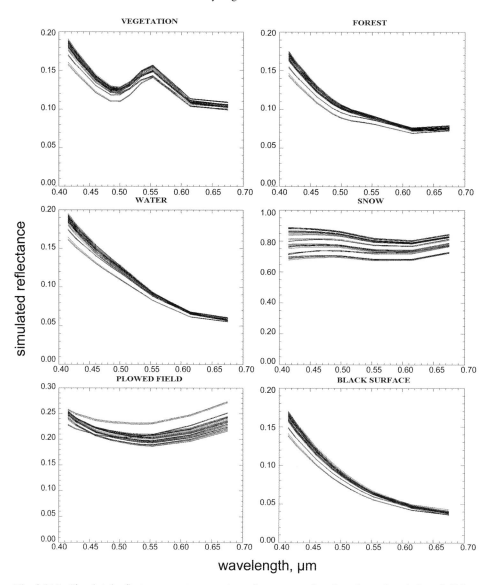

Fig. 8.1(a). Simulated reflectance spectra over six surface types as function of wavelength from 0.415 to 0.675 μm and for a typical retrieval condition with solar zenith angle 42.5° and aerosol optical depth (AOD) at 0.55 μm of 0.35.

determination of the aerosol type, whereas the offset of 2 degrees of freedom is related to the first part of SYNAER, where AOD and surface brightness are retrieved from the radiometer measurements. It should be noted that, based on the retrieved surface reflectances at 0.55, 0.67, and 0.87 μm in this first part of SYNAER, two vegetation indices are calculated, which are then used to select the surface type, i.e. one of 12 predefined spectral form curves (while the absolute surface reflectance values are normalized to the retrieved

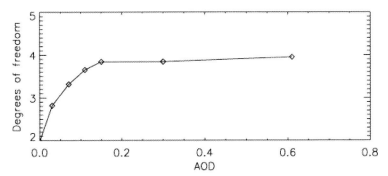

Fig. 8.1(b). Degrees of freedom of SYNAER exploiting the 10 wavelengths from 0.415 to 0.675 μm as function of aerosol optical depth (AOD) at 0.55 μm for a typical retrieval condition with solar zenith angle 42.5° and surface type 'vegetation' (from Holzer-Popp et al. [2008]).

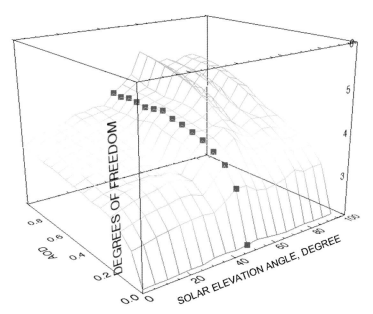

Fig. 8.1(c). Degrees of freedom of SYNAER exploiting the 10 wavelengths from 0.415 to 0.675 μm as function of aerosol optical depth (AOD) at 0.55 μm and solar zenith angle for surface type 'vegetation'. The blue squares indicate the line of a typical retrieval condition shown in Fig. 8.1(b) at solar zenith angle of 42.5° (from Holzer-Popp et al. [2008]).

values at the three wavelengths; see Section 3.7 for more detail). Thus, in the second SYNAER part, the spectra simulations are only done for one selected surface spectrum and accordingly only 1 degree of freedom is needed. It is thus theoretically proven that SYNAER can determine more than two independent aerosol properties in addition to AOD and surface brightness. Also this analysis provides a deeper insight into favorable conditions and limitations of the aerosol type retrieval with SYNAER (surface, sun elevation, AOD).

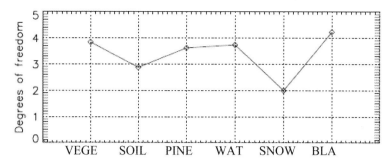

Fig. 8.2. Degree of freedom of SYNAER exploiting the 10 wavelengths from 0.415 to 0.675 μm for aerosol optical depth AOD at 0.55 μm $= 0.35$ and solar zenith angle 42.5° for different surface types: VEGE $=$ vegetation, SOIL $=$ bare soil, PINE $=$ pine forest, WAT $=$ open water, SNOW $=$ fresh snow, BLACK $=$ black surface (from Holzer-Popp et al. [2008]).

3.   The retrieval method

3.1   Method overview

The synergistic aerosol retrieval method delivers aerosol optical depth (AOD) and an estimation of the type of aerosols in the lower troposphere over both land and ocean by exploiting a combination of a radiometer and a spectrometer. In retrieving AOD the free tropospheric and stratospheric aerosol concentration are kept constant at background conditions, whereas the boundary layer aerosol concentration and type and a possible dust layer are varied. The type of aerosol is estimated as percentage contribution of representative components from an extension of the OPAC (Optical Parameters of Aerosols and Clouds; [Hess, et al. [1998]) dataset to AOD in the boundary layer. The high spatial resolution including thermal spectral bands of the radiometer permits accurate cloud detection. The SYNAER aerosol retrieval algorithm comprises then two major parts: step 1, detailed in sections 3.5 and 3.6 (a dark field method exploits single wavelength radiometer reflectances to determine aerosol optical depth and surface reflectance over automatically selected and characterized dark pixels for a set of 40 different predefined boundary layer aerosol mixtures) and step 2, described in Section 3.7 (after spatial integration to the larger pixels of the spectrometer these parameters retrieved in the first step are used to simulate spectra for the same set of 40 different aerosol mixtures with the same radiative transfer code). A least-squares fit of these calculated spectra at 10 wavelengths to the measured spectrum delivers the correct AOD value and – if a uniqueness test is passed – the most plausible spectrum and its underlying aerosol mixture. The entire method uses the same aerosol model of basic aerosol components, each of them representing optically similar aerosol species. These basic components are externally mixed into 40 different aerosol types meant to cover a realistic range of atmospheric aerosol masses. Also the underlying radiative transfer code (SOS; [Nagel et al., 1978], extended after Popp [1995]) is consistently used throughout all retrieval steps. Fig. 8.3 gives an overview of the SYNAER method with more detailed descriptions (based on the original publications) given in the rest of Section 3. For clearer reference we name the original SYNAER version 1.0 [Holzer-Popp et al., 2002a] and the improved version 2.0 [Holzer-Popp et al., 2008].

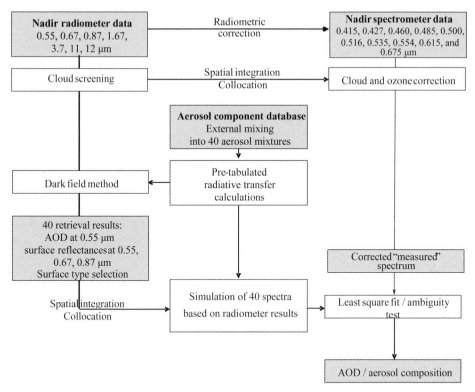

Fig. 8.3. Flowchart of major processing steps in the SYNAER retrieval method. SYNAER is based on three processing columns: radiometer reflectances enable cloud detection, aerosol optical depth retrieval over dark fields and surface albedo retrieval as input to spectrometer simulations. Measured spectra from a contemporary spectrometer are corrected for cloud and ozone effects and then used to select the most plausible aerosol type. The calculations are conducted for a set of typical atmospheric aerosol mixtures which are composed from a basic set of aerosol components. The result of this synergistic procedure is the aerosol optical depth and the aerosol type.

3.2   Radiative transfer model and data preparation

3.2.1 Radiative transfer model

All radiative transfer calculations are conducted with a plane-parallel iterative code (successive orders of scattering, SOS; Nagel et al. [1978]) which includes full multiple scattering. The atmosphere is distributed in at least 15 vertical layers up to 30 km which are subdivided if the total optical thickness of one layer becomes larger than 0.1. This guarantees an accurate modeling of multiple scattering. To correct gas absorption by ozone, oxygen and water vapor in atmospheric window channels, an exponential sum fitting method has been integrated into SOS [Popp, 1995] based on absorption cross-sections from LOWTRAN7 [Kneizys et al., 1988]. Ozone columns are taken from the spectrometer measurements, whereas the total water vapor content is fixed at a climatologic average of 2.3 g cm^{-2} (only the radiometer near-infrared measurements are somewhat affected by it).

In order to account for the strong forward peak of aerosol scattering for large particles a delta-peak approximation is combined with an expansion of the phase function into Legendre polynomials up to the order of 450 (order of expansion is selected automatically to meet an accuracy criterion of 1 %). For a fast application in the aerosol retrieval pre-calculated radiative transfer tables are used. Spectral reflectances are then calculated from bi-quadratic polynomials as a function of surface albedo and aerosol optical depth. Vertical profiles of pressure, temperature and aerosols are based on the mid-latitude summer atmo-sphere [McClatchey et al., 1972]. The aerosol extinction profile within the complete boundary layer is multiplied by a factor to vary the integrated boundary layer aerosol op-tical depth. The bi-directionality of surface reflectance is taken into account in step 1 of the retrieval (aerosol optical depth retrieval with ATSR-2 over dense dark vegetation) by using a normalized bi-directional reflectance distribution function (BRDF) for pine forest [Krie-bel, 1977]. As bi-directional behavior is typically defined by the density of vegetation cover this normalized BRDF is suitable for dense dark vegetation. Furthermore it was shown by [Popp, 1995] that using any average realistic normalized BRDF results in fewer errors than assuming Lambertian reflectance behavior. The spectrometer measurements over different surface types are simulated as Lambertian reflectors because the large pixel size makes a weak anisotropic behavior of near-nadir observations plausible.

3.2.2 Preparation of measured data

First, normalized radiance values of the radiometer channels at 550, 670, 870 and 1,650 nm are converted into reflectances and the cloud-detection algorithm APOLLO is applied (see Section 3.3). By use of a global land–water mask data set in APOLLO (which was derived from the World Database 2 of 1960) and a simple median filter pixels in the neighborhood of clouds (which might be affected by cloud shadow or brightening [Nikolaeva et al., 2005]) and in coastal areas (where underwater reflectance or suspended sediments and chlorophyll prevail and might mislead the retrieval over water) are flagged.

For the GOME reflectance spectra a 'jump correction' [Slijkhuis, 1999] is applied using information from the polarization measurement devices (PMD) of GOME, which are read out with an integration time much smaller than the detector arrays and synchronized with the first detector pixel. For SCIAMACHY so far the operational geo-located reflectance spectra (level 1c product) are obtained. The examples presented in this chapter were de-rived using version 5.4 spectra. As the absolute radiometric calibration of both spectro-meters shows significant errors, for each pixel the relative difference of co-located spectro-meter pixels (spectrally integrated over the response function of the radiometer pixel) with spatially integrated radiometer pixels are calculated to determine a pixel-wise correction factor which is then applied to the entire visible spectrum of GOME and SCIAMACHY. Also for both spectrometers (GOME and SCIAMACHY), the total ozone amount (opera-tional GOME and SCIAMACHY level 2 products) for each pixel is used to correct the ozone absorption inside the Chappuis spectral region. To reduce perturbations with high spectral variability (e.g., the Ring effect, a second-order Raman scattering) spectral reflec-tances at all exploited wavelengths are integrated over windows of 2 nm.

3.2.3 Cloud correction of the spectrometer measurements

The simulated spectrometer measurements do not contain any cloud contribution. Therefore, the fit between simulated and measured spectra needs a cloud corrected 'measured' spectrum. This is very significant as only a few percent of spectrometer pixels are totally cloud-free. The basic idea for cloud correction is the synergistic use of the radiometer and spectrometer information (see Holzer-Popp et al. [2002a] for more detail).

The radiometer cloud detection (see *Section* 3.3) discriminates between fully cloudy, partly cloudy and cloud-free ATSR-2 pixels and retrieves the cloud fraction of the partly cloudy (1×1 km^2) pixels. Through summation of fully cloudy and partly cloudy radiometer pixel reflectances inside one spectrometer pixel, the reflectance of the cloudy spectrometer pixel part is separated from the reflectances of the cloud-free spectrometer pixel part. For partly cloudy radiometer pixels their retrieved cloud cover fraction retrieved is taken into account. The cloudy part of the spectrum seen by the spectrometer can be taken as equal to the radiometer-integrated cloudy reflectance based on the assumption that the underlying surface within the spectrometer pixel is, on average, homogeneous. The difference between the integrated reflectance of all radiometer pixels and of only the cloudy radiometer pixels gives the necessary correction term for the measured spectrum. This difference is calculated for all three radiometer wavelengths at 0.56, 0.67 and 0.87 μm. Additionally, an extrapolation of the correction term at the three AATSR wavelengths to a subset of 10 wavelengths in the spectrometer spectral range, which are exploited in the SYNAER method (see Section 3.7), is performed with a second-order fit for non-vegetated surfaces. For vegetated surfaces the influence of the vegetation on a cloud spectrum is modeled with a normalized sun-angle-dependent structure curve that describes how the difference between cloudy and cloud-free spectra is typically modified through vegetation.

After application of this approach, the cloudy part of the spectrometer spectrum is known without any assumption of cloud type or cloud microphysics. Knowing the cloud fraction inside the spectrometer pixel from the summation of the radiometer cloud information, a linear mixing of the cloudy and cloud-free parts of the spectrum into a mixed spectrum can be assumed, neglecting nonlinear 3D effects of clouds which are still an unknown issue of ongoing research. This then delivers the cloud-corrected spectrum.

It was shown in Holzer-Popp et al. [2002a] that this approach represents the cloud influence very well for non-vegetated surfaces in the wavelength range between 400 and 800 nm. It should be noted that the gas absorption bands cannot be fitted, but this does not matter as SYNAER exploits only wavelengths selected especially for low gas absorption (see Section 3.7). The underlying simulations were performed for several cloud types (cumulus, stratus, altostratus, nimbostratus, stratocumulus and cirrus), for cloud top heights varying between 0.5 and 10 km, for solar zenith angles between 0 and 80°, for several aerosol types (urban, rural, maritime, tropospheric, desert as described in MODTRAN; Kneizys et al. [1996]) and several surface types used in SYNAER (see Section 3.7). The ozone amount was always set to zero due to the fact that in SYNAER a separate ozone correction using GOME ozone columns is performed before the cloud correction. The absolute difference between the second-order fit and simulated cloud spectrum was typically less than 0.02 reflectance units (at 400 nm, where largest errors occurred). The procedure was validated for GOME pixels with up to 35 % cloud coverage;

for high cloud fractions it is expected that the inaccuracy of this approach dominates the total inaccuracy and spoils the entire retrieval procedure.

3.3   Accurate cloud detection and its improvements

Accurate cloud detection is an important prerequisite for aerosol retrieval. The well established APOLLO (AVHRR Processing Scheme Over CLoud Land and Ocean; Saunders and Kriebel [1988], Kriebel et al. [1989] and Kriebel et al. [2003]) software was adapted for ATSR-2 and AATSR data. Five radiometer channels which correspond to AVHHR spectral bands (0.67 and 0.87, 3.7, 11 and 12 μm) are used for cloud detection and the 1.6 μm channel is used for separating clouds from snow. APOLLO yields cloud fraction, four cloud layers and cloud optical parameters. Few errors occur along the coast due to the limited accuracy in geo-location through mapping to a land–sea mask, but these coast pixels are not used in SYNAER anyway. Because snow pixels are too bright, they cannot be used for aerosol retrieval (see Section 3.5) and they are also flagged as forbidden values. APOLLO also determines pixels affected by sunglint through a combination of a visible reflectance threshold and a geometric calculation of the potential sunglint area. The capability of retrieving cloud cover in boxes of approximately 1 km^2 provides a significant strength to SYNAER because it reduces the erroneous aerosol detection due to the presence of sub-pixel clouds significantly. It even allows the exploitation of partly cloudy spectrometer pixels as described in Section 3.2.3.

Unfortunately, adapting the cloud-screening scheme to the ATSR radiometers has two shortcomings, which have to be accounted for in order to derive an accurate cloud mask for aerosol retrievals. First, heavy aerosol load over oceans (mainly mineral dust, less frequent smoke plumes from wildfires) is classified as 'cloudy' by APOLLO and these AATSR pixels are then not used for the retrieval of AOD in SYNAER, leading to slightly reduced AOD values in the dust belts. The second shortcoming is an improper detection of shallow cumulus cloud cover over land due to a simple temperature threshold test for the rejection of cloudy pixels in order to not classify desert surfaces as low clouds. Thus, in some obviously cloudy AATSR scenes no clouds are detected, and those much too bright pixels are used for the AOD derivation, leading to high AOD over land. Both these shortcomings of the APOLLO cloud-detection scheme applied to the ATSR radiometers essentially require corrections to the cloud-screening procedure, which were implemented in SYNAER version 2.0 [Holzer-Popp et al., 2008].

3.3.1 Corrected mis-classification of mineral dust over ocean

Heavy dust plumes in the Atlantic region are usually embedded in an air layer often called Saharan Air Layer (SAL), which is described in detail, e.g., by Wong and Dessler [2005]. The main characteristic properties of this SAL are being warm, dry and well mixed. Thus 11 μm brightness temperatures of dust-laden pixels are well above 273 K. So this brightness temperature value is chosen as a first threshold, which prevents cool mixed-phase or ice clouds from being taken into account for the further analysis, together with scenes in polar regions. This condition can also be met by thin cirrus or semi-transparent clouds, so another criterion for the following dust discrimination scheme is the cloud type 'low cloud' determined by the original APOLLO tests.

Dunion and Velden [2004] use images of the Geostationary Operational Environmental Satellites (GOES) to track the SAL across the Atlantic Ocean. They discriminate the 'SAL-strength' by means of the brightness temperature difference (BTD) between 11 μm and 12 μm, arguing that non-SAL BTD values are well above $+5$ K. The SAL-strength analysed by Dunion and Velden [2004] does not directly represent the aerosol load or optical depth and also is sensitive to cloud screening.

The BTD alone does not seem to be an appropriate measure to discriminate mis-classified clouds (not shown), but it is a good method to select AATSR pixels which have to undergo further inspection. In this APOLLO improvement scheme a slightly more conservative BTD threshold of $+2$ K is used, because heavy aerosol plumes can be shown to inhibit values well below this threshold (and low aerosol loads seem not to be mis-classified by APOLLO). Evan et al. [2006] use different BTD thresholds in different stadiums of their dust-detection algorithm with the minimal threshold being -0.5 K and the maximum BTD value, for which dust classification remains possible, being $+3.5$ K. For the purpose of saving computing time a single BTD threshold of $+2$ K has been chosen here for the initial test.

Furthermore two reflectance thresholds are applied to pre-select possible dusty mis-classified pixels: they have to inhibit 1.6 μm reflectances below 0.2 and 0.67 μm reflectances below 0.3. Brighter pixels classified as cloudy by APOLLO remain unchanged.

Thus the pre-selection scheme of possible mis-classified pixels consists of the following tests:

$$T_{11\mu m} > 273 \text{ K and } R_{1.6\mu m} < 0.2 \text{ and } R_{0.67\mu m} < 0.3 \text{ and } T_{11\mu m} - T_{12\mu m} \leq 2 \text{ K} \quad (5)$$

Pixels classified as low-cloud covered by APOLLO, for which these tests apply, can still be cloud-contaminated or aerosol-laden. The discrimination between clouds and mineral dust can be achieved by means of the ratio of reflectance at 1.6 μm and 0.67 μm due to the higher reflectance of water clouds at 1.6 μm compared to mineral dust.

Fig. 8.4 shows AATSR reflectance values at 1.6 μm and 0.67 μm for a desert dust outbreak scene from March 9, 2006, off the western coast of North Africa. The analysis includes a box area of 1,000 $\times$ 512 AATSR pixels. For 0.67 μm reflectance greater than about 0.1 one can clearly distinguish two different regimes in the scatterplot. The lower branch, colored blue in the figure, represents pixels which can be identified as dust-laden by visual inspection (not shown). The upper branch, colored red, can be identified as definitely cloud covered in RGB images of the scene.

The discrimination between both branches does not exactly follow the reflectance ratio of 1. Dust discrimination requires the condition

$$\frac{R_{1.6\mu m} + 0.035}{R_{0.67\mu m}} < 1 \quad (6)$$

to be met. The additional constant, being 0.035 for AATSR, originates from a best-fit test and can be shown to differ slightly for other sensors than AATSR (e.g., it is 0.03 for SEVIRI onboard the MSG satellite).

Furthermore, Fig. 8.4 shows that discrimination between dust and cloud by means of the reflectance ratio is not possible for 0.67 μm reflectance below about 0.1, values corresponding to moderate to low dust load or thin low-level clouds.

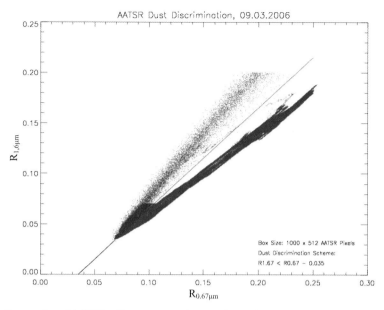

Fig. 8.4. Scatter plot of AATSR reflectances at 1.6 μm against reflectances at 0.67 μm for a scene from 09.03.2006 showing the differentiation of heavy dust load in blue and shallow stratocumulus clouds in red (from Holzer-Popp et al. [2008]).

For pixels, which meet the conditions in (5) and inhibit the 0.67 μm reflectance below 0.1, an even more conservative BTD threshold of 0 K is applied to discriminate mis-classified pixels, as described in Huang et al. [2006] and also in Evan et al. [2006], while the reflectance ratio test is not applied for those pixels.

3.3.2 Shallow convection over land

The original adapting of APOLLO to AATSR data includes a temperature threshold test scheme for the exclusion of very bright desert surfaces from cloud detection. In this scheme, cloudy pixels having 11 μm brightness temperatures above a scene-dependent threshold between 285 K and 305 K are rejected (from cloud detection) if they inhibit 0.67 μm reflectance below 0.6. This test is included into the APOLLO scheme to determine whether a pixel is cloudy or shows bright desert surface. Without this test, APOLLO classifies many desert areas as 'cloudy' due to their high shortwave reflectance. On the other hand, inclusion of this test rejects pixels, which obviously show low-level convective cloud fields with warm cloud-top temperatures somewhere in the temperature range of the thresholds. This makes the application of another test over land necessary, which accounts for these cloud fields and flags them 'cloudy'.

Only pixels classified as 'cloud-free' by APOLLO after the temperature rejection tests within the APOLLO scheme are regarded for the additional tests described below. Furthermore the pixels have to inhibit 11 μm brightness temperatures in the range 285–305 K, which covers the range of possible thresholds for the temperature rejection test. Pixels having 11 μm brightness temperatures above the highest possible threshold value of

305 K remain flagged cloud-free, following the original APOLLO strategy for desert surfaces.

As a second test, the ratio of reflectance at 1.6 μm and 0.87 μm has to be above 0.65 and below 1.0. This somewhat arbitrary threshold accounts for the near-equality of those reflectances for dense water clouds and desert surfaces which often show reflectance ratios well above 1.0 (not shown).

As the described conditions alone are not enough to distinguish between desert surface and low cloud fields, pixels meeting the above conditions are re-classified as 'cloudy', if either of the conditions

$$R_{0.87\mu m} > 0.25 \text{ and } R_{0.67\mu m} > 0.25 \text{ and } T_{11\mu m} - T_{12\mu m} \geq 1.25 \text{ K} \qquad (7)$$

are met or if

$$R_{0.87\mu m} > 0.4 \text{ and } R_{0.67\mu m} > 0.4 \text{ and} -0.5 < T_{11\mu m} - T_{12\mu m} < 1.25 \text{ K} \qquad (8)$$

is true.

The threshold of 0.4 for 0.67 μm and 0.87 μm reflectance is adapted from Rosenfeld and Lensky [1998] for pixels with BTD lower than 1.25. This BTD test split value of 1.25 K results from the APOLLO algorithm. So, for BTD values greater than 1.25 a slightly lower reflectance threshold of 0.25 can be chosen, which follows the cloud-detection method of Kaufman and Fraser [1997]. Actually, those authors use a threshold 0.2 for 0.67 μm reflectance, combined with a difference in brightness temperatures between 3.7 μm and 11 μm of greater than + 8 K. A slightly higher threshold has been chosen here because in difference to Kaufman and Fraser [1997] pixels having 11 μm brightness temperatures warmer than 290 K are also included and no brightness temperature difference between 3.7 μm and 11 μm is used. Only the combination of both additional tests enables proper discrimination between rejected desert surfaces and low cloud fields within the APOLLO cloud detection scheme.

Fig. 8.5 shows an AATSR scene with a large number of obviously mis-classified cloudy pixels. The left-hand side of the image shows an RGB composite image, in which the low-level cloud field easily can be observed. On the right-hand side the APOLLO cloud mask is shown. Green pixels show cloud-free land and white pixels show clouds detected by the original APOLLO scheme. A clear disagreement between the cloud detection and the clouds seen in the composite occurs. The red pixels show pixels classified as cloud-covered by the improved APOLLO scheme, while not by the original one. A great improvement of cloud detection is obvious and clearly shows its necessity.

The improved APOLLO cloud classification, extended by the tests described above, has been tested by visual inspection with 39 different AATSR orbits of the years 2006 and 2007 during all seasons with many scenes including low-level convection and heavy dust plumes. These tests show clear improvements of the APOLLO cloud screening for SYNAER, which should clearly lead to a reduced bias in AOD both over land and over ocean.

It has to be noted that there can still be some very warm or moderately bright low-level clouds which go undetected by the APOLLO screening procedure. On the other hand also slightly more desert pixels, showing no clear evidence of being cloudy, are flagged cloud covered with the improved scheme, leading to slightly too high mean cloud cover in desert

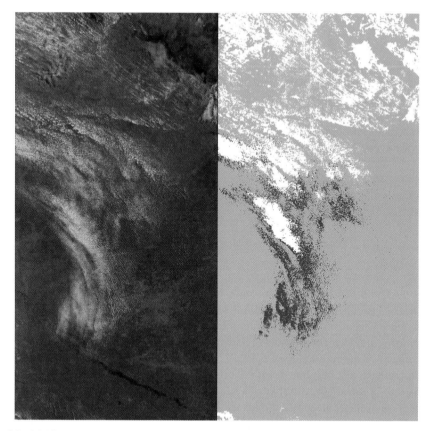

Fig. 8.5. AATSR scene over land on March 8, 2006, showing APOLLO misclassifications: left, RGB composite of bands at 12 μm, 0.87 μm and 0.67 μm; right, corrected cloud mask. Green pixels are cloud-free land, white pixels are clouds detected by the original APOLLO scheme and red pixels show clouds only detected by the additional cloud detection scheme (from Holzer-Popp et al. [2008]).

regions. But compared to the benefits of additional detection of low-level clouds these misclassifications of potentially cloud-free desert pixels can surely be accepted in the case of SYNAER.

3.4 Aerosol model and its upgrade

3.4.1 The SYNAER aerosol model

The aerosol model consists of three vertical layers with different optical properties. Aerosol optical depth and surface albedo values for the radiometer pixels are derived for 40 boundary layer aerosol mixtures because the retrieved values depend strongly on the aerosol type and relative humidity. The 40 boundary layer aerosol mixtures are constructed using the external mixing approach [World Climate Program, 1986], which allows the mixing of arbitrary aerosol types from a set of basic components. Each mixture is defined by percentage contributions of the components to the boundary layer aerosol optical depth.

For large particles with a large specific extinction coefficient, only small percentages of AOD at 550 nm are used in conformity with observations [Köpke et al., 1997]. However, in the case of a desert dust outbreak (mixtures number 15–17, and 35–37 in Table 8.2) a dominating contribution of large particles is allowed. The basic components are defined by their complex index of refraction, a lognormal size distribution, and humidity dependence for hygroscopic particles. Through Mie calculations the optical properties of each component are derived and each component summarizes an ensemble of optically similar particles. For SYNAER version 1.0 [Holzer-Popp et al. 2002a] six components (water soluble, water insoluble, sea salt accumulation and coarse modes, soot, and mineral transported) from the OPAC database (Optical Properties of Aerosols and Clouds [Hess et al., 1998]) were used. OPAC was compiled on the basis of a vast number of measurements and the experiences and thorough evaluation of earlier aerosol climatologies in order to describe the typical composition of the global aerosol with a representative but small set of basic components (Global Aerosol Data Set, GADS; Köpke et al., [1997]).

All particles are assumed to be spherical so that their optical properties can be calculated with Mie theory. Non-sphericity of the particle shape is not taken into account as the sensitivity of near-nadir measurements of both the spectrometer and the radiometer is expected to be too low for retrieving information on particle shape. However, for large particles (minerals, sea salt) this might be a source of error, which has to be investigated further [Dubovik et al., 2006]. A future option to overcome this limitation is the use of a specific database for spheroids (e.g.,[Dubovik et al. [2006]). It is noted that these components are meant to describe a representative set of typical parts of the global aerosol load, since the optical properties (spectral extinction coefficient, spectral absorption fraction and spectral phase function) are assumed to exist only in a limited number of combinations due to their link to the microphysical properties.

Above the boundary layer, free tropospheric and stratospheric aerosol optical depth at 550 nm are set to 0.025 and 0.005, respectively [World Climate Program, 1986], with a free tropospheric and stratospheric background aerosol type as used in LOWTRAN7 [Kneizys et al., 1988]. The values of 0.025 and 0.005 were recommended by an expert panel on the basis of a large number of observations for the purpose of defining a mean standard atmosphere above the boundary layer. This means that the conversion of boundary layer aerosol optical depth values into total AOD values at 550 nm is simply achieved by adding 0.03. This approach is justified for the investigated period since 1995 up to now (no major volcanic eruption after 1992). The height of the boundary layer is varied from 2 to 6 km to account for the typical vertical extent of different aerosol types. It should be noted that for version 1.0 in the case of a desert dust outbreak the 'boundary layer' in SYNAER merges two sub-layers (the desert dust outbreak which is above a background aerosol layer) into one boundary layer reaching up to 4 km with water-soluble and transported mineral particles. This approach was adopted as the lower layer is partly masked by the mineral dust layer above it. Extreme cases of desert dust outbreaks reaching as high as 8 km were therefore not covered by this model. Up to now, the surface level has been fixed at sea level (a future option is the use of an elevation database such as GTOPO30). Due to humidity-dependent components (water-soluble, sea salt) two models with 50 % and 80 % relative humidity have been included for each aerosol mixture. The same set of 40 mixtures is applied to the radiometer aerosol optical depth and albedo retrieval, and the spectrometer simulations.

3.4.2 Updates to the original SYNAER aerosol model

Table 8.1 summarizes the relevant microphysical properties and optical characteristics derived from Mie calculations. In SYNAER version 2.0 [Holzer-Popp et al., 2008], on the basis of more recent campaigns and AERONET data exploitation, some specific updates have been implemented.

The original soot component was split into two components for strongly absorbing diesel soot (DISO) more representative for industrial areas and weakly absorbing biomass burning soot (BISO). The optical properties of strongly absorbing diesel soot were taken from Schnaiter, et al. [2003], while optical properties for soot from biomass burning cases, as for example in Amazonian, South American cerrado, African savanna and boreal regions were adopted from Dubovik et al. [2002]. As size distributions measured in Schnaiter et al. [2003] were similar to the OPAC database, the size distribution described in the

Table 8.1. Optical and microphysical characteristics of basic components used for external mixing of in SYNAER (new components are highlighted in italics), where the size distribution is defined by

$$\frac{dN(r)}{d\log r} = \frac{N}{\sqrt{2\pi} \cdot \log\sigma} \exp\left[-\frac{1}{2}\left(\frac{\log r - \log r_0}{\log\sigma}\right)^2\right]$$

Component	Species	Complex refract. index at 550 nm	Geometric mean radius of lognormal size distribution	Geometric standard deviation of lognormal size distribution	Particle density	Extinction coefficient for 1 particle per cm^3 at 550 nm [km^{-1}]	Single scattering albedo at 550 nm	Literature source
			r_0 [µm]	σ	ρ[g/cm^{-1}]			
WASO, rH = 70%	Sulfate/ nitrate	1.53 – 0.0055 i	0.028	2.24	1.33	7.9 e-6	0.981	Hess et al., 1998
INSO	Mineral dust, high hematite content	1.53 – 0.008 i	0.471	2.51	2.0	8.5 e-3	0.73	Hess et al., 1998
INSL	Mineral dust, low hematite content	1.53 – 0.0019 i	0.471	2.51	2.0	8.5 e-3	0.891	Dubovik et al., 2002
SSAM, rH = 70%	Sea salt, accumulation mode	1.49 – 0 i	0.378	2.03	1.2	3.14 e-3	1.0	Hess et al., 1998
SSCM, rH = 70%	Sea salt, coarse mode	1.49 – 0 i	3.17	2.03	1.2	1.8 e-1	1.0	Hess, et al. 1998
BISO	Biomass burning soot	1.63 – 0.036 i	0.0118	2.0	1.0	1.5 e-7	0.698	Dubovik et al., 2002
DISO	Diesel soot	1.49 – 0.67 i	0.0118	2.0	1.0	7.8 e-7	0.125	Schnaiter et al., 2003
MITR	Transported minerals, high hematite content	1.53 – 0.0055 i	0.5	2.2	2.6	5.86 e-3	0.837	Hess et al., 1998
MILO	Transported minerals, low hematite content	1.53 – 0.0019 i	0.5	2.2	2.6	5.86 e-3	0.93	Dubovik, et al. 2002

OPAC database are used also for the DISO and BISO components replacing now the original OPAC SOOT component.

For mineral dust, an additional component (MILO, mineral dust with low absorption) was introduced in order to take dust sources with lower hematite content into account. Moulin et al. [2001] discuss earlier measurements [Patterson et al., 1997], used for example in OPAC, being conducted in regions with large hematite content. As hematite is a strongly absorbing material, small amounts can change the optical properties of atmospheric dust significantly. Recent measurements (e.g., Schnaiter et al., [2003], Moulin et al. [2001], and Sinyuk et al. [2003]) in Bahrain, Cape Verde, Sahara and Saudi Arabia (as regions with low hematite content) show a reduced imaginary part of the refractive index between 0.001 and 0.002 compared to the OPAC value of 0.0055. Larger hematite concentrations can be found only in restricted areas such as the Sahel area, northern India and

Table 8.2. Predefined external aerosol mixtures of the basic components (Table 8.1) which are used in SYNAER

No.		Name	Rel. hum. [%]	Vert. prof. [km]	Component contributions to aerosol optical depth (AOD) at 0.55 μm [%]								
					WASO	INSO	INSL	SSAM	SSCM	BISO	DISO	MITR	MILO
1	21	Pure water-soluble	50/80	2	100				50/80				
2	22	Continental	50	2	95	5	5						
3	23				90	10	10						
4	24				85	15	15						
5	25	Maritime	50/80	2	30			70					
6	26				30			65	5				
7	27				15			85					
8	28				15			75	10				
9	29	Polluted water-soluble	50/80	2	90						10		
10	30				80						20		
11	31	Polluted continental	50	2	80	10	10				10		
12	32				70	10	10				20		
13	33	Polluted maritime	50/80	2	40			45	5		10		
14	34				30			40	10		20		
15	35	Desert outbreak	50	2–4	25							75	75
16	36			3–5	25							75	75
17	37			4–6	25							75	75
18	38	Biomass burning	50/80	3	85					15			
19	39				70					30			
20	40				55					45			

WASO = water-soluble, INSO = insoluble, INSL = insoluble/low hematite, SSAM = sea salt accumulation mode, SSCM = sea salt coarse mode, BISO = biomass burning soot, DISO = diesel soot, MITR = mineral transported, MILO = mineral transported/low hematite.
For mixture number N and mixture number $N + 20$ alternative humidity (50 % or 80 %) or mineral composition (INSO or INSL, MITR or MILO), respectively is chosen.

eastern Australia [Claquin et al., 1999]. Also, measurements from the GOES-8 satellite optimally reproduce ground measurements of mineral dust concentrations if an imaginery part of 0.0015 is assumed [Wang et al., 2003].

As the insoluble component in OPAC (INSO) is modeled with the identical refractive index as the mineral transported component MITR, also an insoluble component with low absorption (INSL) was introduced. Both, MILO and INSL have the same size distributions as the OPAC components MITR and INSO, respectively. In the case of a desert dust outbreak, the lowest aerosol layer of up to 4–6 km is now modeled as two distinct sub-layers representing a dust layer above background aerosols, as they occur in nature.

Table 8.2 shows the updated definition of the 40 mixtures used in the SYNAER retrieval method version 2.0. The set of 40 mixtures is meant to model all principally existing aerosol types and allow for some variability in the composition of each type. This set of mixtures has proven to provide a fit in the GOME spectra retrieval which is in many cases at a 1 % noise level. Values in the table show the vertical profile, relative humidity in the boundary layer and the percentage contribution to the optical depth at 0.55 μm of the respective components. Two groups of 20 mixtures, each are applied where either relative humidity or the absorption of the mineral component are altered. Alternative values are marked with grey boxes: for example, mixture number 1 has 50 % relative humidity and mixture number 21 has 80 % relative humidity; mixture number 2 has a 5 % insoluble (large absorption) component, whereas mixture number 22 has a 5 % insoluble (low absorption) contribution to the optical thickness at 550 nm.

3.5   Dark field selection and characterization

3.5.1 Dark field selection and characterization over ocean and its improvement

Dark water pixels are selected only over deep cloud-free ocean using the water mask and cloud filter described in Section 3.3 as shadow (near clouds) and coastal water pixels may potentially be very dark or affected by ocean color effects. In version 1.0 an average fixed albedo value of 0.015 at 870 nm was assumed for the aerosol retrieval over these pixels. The rather high near-infrared albedo value of 0.015 was chosen to account for some minor diffuse reflection from the water surface. In version 2.0 the near infrared ocean surface albedo is allowed to vary between 0.005 and 0.035, depending on a 'chlorophyll index'

$$\mathrm{CVI} = (R_{0.55\mu m}) - R_{0.67\mu m}/(R_{0.55\mu m} + R_{0.67\mu m}), \tag{9a}$$

which is determined by the ratio of the reflectances at 0.67 and 0.55 μm. The near-infrared ocean surface albedo at 0.87 μm is then calculated with an experimentally derived function as

$$R_{0.86\mu m} = 0.005 + 0.15(\mathrm{CVI} - 0.05); \ R_{0.86\mu m} \in [0.005, 0.035] \tag{9b}$$

Sunglint is also excluded based on APOLLO results. It is evident that a more sophisticated treatment would require a wind-speed driven adjustment of the ocean albedo and accounting for varying chlorophyll absorption and whitecaps. However, this was not included to avoid another complexity in the algorithm. The case studies which have been conducted so far show some agreement between simulated and measured spectra over water so that for the large spectrometer pixels the accuracy of this rather crude open ocean treatment can be accepted for global studies. An evaluation of this approach needs to be conducted, in particular far away from any land surfaces, by future validation with data from ship measurements (as AERONET ocean stations are always located on islands or coasts). This approach is suitable over open ocean, but over case 2 coastal waters difficulties are expected. For this reason, SYNAER excludes coastal water pixels by applying a near-coast filter before the retrieval.

3.5.2 The SYNAER dark field selection and characterization over land

Dark radiometer pixels over land are selected on the basis of the 1.6 μm channel and in version 1.0 also using the reflected contribution to the 3.7 μm channel. It is assumed that the aerosol effect at these wavelengths can be neglected for most aerosol types. For large concentrations of large particles this assumption is not valid, but it seems that the interplay of scattering and absorbtion effects (and maybe non-sphericity of the particles) of these aerosols in the mid-infrared spectral region (1.0–6.0 μm) allows the application of the method even in these cases despite violating the basic principle. This has also been in-dicated by Kaufman et al. [2000]. Thus, in general all cloud-/snow-free pixels with a 1.6 μm (or 3.7 μm) reflectance below given thresholds are selected. To reject sub-pixel snow and ice, open water on bare soil pixels or small inland waters, which are also dark in the 1.6 μm channel but brighter in the visible [King et al., 1999], a second criterion is applied in combination with the 1.6 μm channel test using the Normalized Difference Vegetation Index

$$\text{NDVI} = (R_{0.87\mu m} - R_{0.67\mu m})/(R_{0.87\mu m} + R_{0.67\mu m}), \tag{10}$$

which must be larger than a minimum value. This minimum value was dynamically chosen in a box of 64 × 64 pixels in SYNAER version 1.0. In SYNAER version 1.0 for the pixels selected through this scheme the dark field albedo values at 0.67 μm could then be esti-mated from the 1.6 μm and 3.7 μm reflectances by dividing them by a conversion factor of 3.7 and 0.8, respectively. The value of 3.7 at 1.6 μm had been selected on the basis of Landsat-5 TM case studies [Holzer-Popp et al., 2002a], whereas the value of 0.8 at 3.8 μm had been derived experimentally with GOME/ATSR-2 test pixels.

The correlation of reflectances of vegetated pixels in the visible and mid-infrared chan-nels can be understood by similar characteristics of vegetation and soil in both wavelength regions: vegetation reflectivity decreases due to chlorophyll absorption in the visible and due to absorption by liquid water in the mid-infrared. Wet soil has a lower reflectance in the visible due to trapping of light and again in the mid-infrared due to liquid water absorption. Other influences such as surface roughness, shadows, and inclination effects decrease the reflectance values across the entire solar spectrum [Kaufman and Remer, 1994]. Kaufman et al. [1997b] suggest a similar selection scheme based on the 2.1 μm and 3.8 μm channels of the MODIS sensor onboard the TERRA-1 platform.

The conversion factor from the IR-wavelengths to the VIS-wavelengths is just a sta-tistical average value and therefore not explicitly valid for all pixels and situations. This results in a significant scatter of the retrieved AOD values even for neighboring pix-els. Therefore, a minimum number of adjacent pixels had to be exploited during the dark field selection to allow for an appropriate averaging. Especially in areas with sparse dark fields singular outlier values must be rejected by comparison to the nearest available dark fields.

To deal with these requirements the selection scheme is a stepwise process. Dark fields are grouped into three different classes with increasing surface albedo, i.e. decreasing retrieval accuracy. Threshold values for the NDVI and $R_{3.7}$ are selected in boxes of 64×64 ATSR-2 pixels. The threshold values were chosen as an optimal compromise between finding a suitable number of dark fields in different climate zones and restricting the detection to the darkest available pixels in the visible channel. For the selection of dark pixels an iterative scheme is used within boxes of 25×25 pixels. Only dark pixels of the lowest level available, i.e. most accurate, class are exploited.

3.5.3 Improved dark field selection and characterization over land

More detailed analysis with larger data amounts of the ENVISAT sensors showed significant scatter of the dark field selection and characterization of the earlier version which leads to significant numbers of outliers with large errors of the retrieved AOD values. In SYNAER version 2.0 therefore the dark field scheme was altered [Holzer-Popp et al., 2008]. Now, no use of the 3.7 μm reflectances is made (as the retrieval accuracy over these dark fields versus the 1.6 μm reflectances is not well understood since the 3.7 μm radiance consists of a thermally emitted and a solar reflected signal, which cannot be separated accurately in all cases) and a more sophisticated estimation of the surface reflectances in the 1.6 μm channel is now applied. In order to retrieve AOD with an accuracy of 0.1 the surface albedo of the treated dark field should be known with an accuracy of 0.01 (see, e.g., Holzer-Popp et al. [2002a]). To achieve this accuracy in an automatic retrieval procedure over land for AATSR, dark fields are selected from a combination of absolute thresholds for the NDVI (top of the atmosphere) and the reflectance $R_{1.6\mu m}$ in the mid-infrared at 1.6 μm. Best values set in the retrieval are now NDVI > 0.5 and $R_{1.6\mu m} < 0.23$. Dark fields are now grouped into four different classes with increasing surface albedo and decreasing vegetation amount, i.e. decreasing retrieval accuracy. These classes are defined by following threshold values:

$$
\begin{aligned}
&\text{Class 1}: \quad 0.01 < R_{1.6\mu m} < 0.17 \text{ and NDVI} > 0.65, \\
&\text{Class 2}: \quad 0.01 < R_{1.6\mu m} < 0.17 \text{ and NDVI} > 0.50, \\
&\text{Class 3}: \quad 0.17 < R_{1.6\mu m} < 0.23 \text{ and NDVI} > 0.65, \\
&\text{Class 4}: \quad 0.17 < R_{1.6\mu m} < 0.23 \text{ and NDVI} > 0.50.
\end{aligned}
\tag{11}
$$

For these dark field pixels the surface reflectance over land at 0.67 μm is then estimated by a linear regression with the reflectance at 1.6 μm. Similar to the latest update of the MODIS retrieval algorithm Collection 5 (regression between 2.2 μm and 0.67/0.49 μm becomes vegetation-dependent in Collection 5; Levy et al. 2007]) this regression shows a dependence on the vegetation amount. Fig. 8.6 shows the regression of surface reflectance at 0.67 μm versus top-of-atmosphere reflectance at 1.6 μm for intervals of 0.05 in NDVI (top of the atmosphere). For this plot 2474 AATSR dark fields were analyzed, where an AERONET sunphotometer measurement was available within 60 minutes and 50 km from the respective AATSR pixel, where the AOD at 0.55 μm was below 0.1 and where the scattering angle was between $140°$ and $160°$. The surface reflectance was derived by atmo-

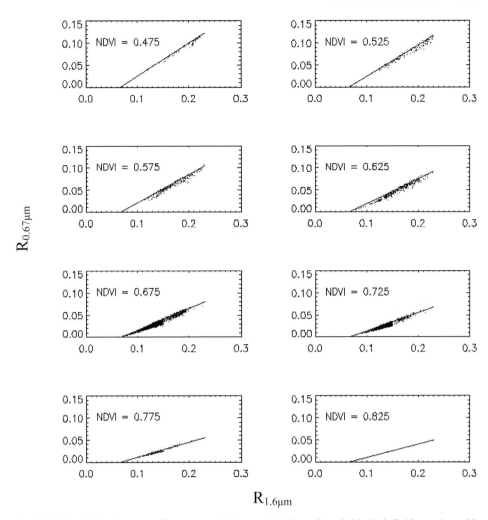

Fig. 8.6. Correlation between reflectances at 1.6 μm and 0.67 μm for suitable dark fields varying with normalized vegetation index NDVI. Vegetation surfaces with NDVI > 0.45 and R1.6 < 0.23 were chosen and are plotted for NDVI intervals of 0.05. The figure shows results based on 2,474 automatically selected dark fields from 42 ENVISAT orbits in 2005 with aerosol optical depth AOD < 0.1 and scattering angles between 140° and 160°. Regression lines in each plot show the dependence function which is used in the SYNAER retrieval (from Holzer-Popp et al. [2008]).

spheric correction with the SYNAER radiative transfer algorithm and the respective AERONET AOD value. This analysis leads then to following regression function between top-of-atmosphere reflectances $R_{1.6\mu m}$ and surface reflectances $R_{0.67\mu m}$:

$$R_{0.67\mu m} = aR_{1.6\mu m} + b + c \tag{12}$$

with

$$a = -1.5\mathrm{NDVI} + 1.5,$$
$$b = 0.1\mathrm{NDVI} - 0.1,$$
$$c = 0.1(\cos\psi - \cos 150°), \quad \text{for } \theta < 150°,$$

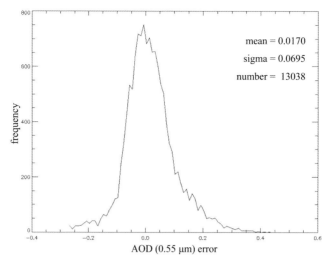

Fig. 8.7. Absolute error histogram of the retrieved aerosol optical depth (AOD) at 0.55 μm for real dark field pixels of ENVISAT-AATSR in summer 2005 against coincident AERONET measurements in the vicinity of up to 50 km (from Holzer-Popp et al. [2008]).

where θ is the scattering angle and NDVI at top of the atmosphere is used. The third term is not shown in Fig. 8.6 as the few extreme values of the scattering angle were not included in this analysis, but it was found necessary to improve the AOD retrieval for a small number of cases with scattering angles close to $120°$. Finally, dark fields where the estimated surface reflectance at 0.67 μm exceeds 0.085 are not used in the retrieval.

In support of the regression function extracted by optimizing AOD agreement with nearby AERONET observations, Fig. 8.7 shows the results of applying the regression function to dark fields with all AOD values. This histogram of retrieved AOD errors against AERONET ground-based measurements (up to 50 km away) for the real dark field pixels confirms the NDVI- and scattering-angle-dependent regression function. The achieved standard deviation (0.089) and bias (0.017) are quite satisfactory given the mis-distance/mis-time to the AERONET station of up to 50 km and 60 min.

This new regression function is used to determine the surface reflectance in the visible as main precondition for retrieving aerosol optical depth. One drawback of using a regression based on NDVI is the fact, that the NDVI values themselves are independent of AOD in the first order but do exhibit a residual dependence on the aerosol optical depth (typically decreasing with increasing AOD). To overcome this dependence, a one-step iteration is conducted, where the preliminary retrieved AOD value is used to adjust the NDVI value and subsequently the same regression function is applied again to calculate a corrected visible surface reflectance, which is then exploited for a corrected AOD retrieval. The function derived empirically to provide optimal agreement of retrieved AOD after iteration against AERONET observations is:

$$\text{NDVI}_{corrected} = \text{NDVI} + 0.25\text{AOD}_{preliminary}/\mu_0, \tag{13}$$

where μ_0 is the cosine of the solar zenith angle.

3.6 Retrieving AOD and surface albedo for 40 aerosol types

Using the estimated surface albedo and pre-tabulated radiative transfer calculations the top-of-atmosphere (TOA) reflectance over dark field pixels is inverted into values of aerosol optical depth at 0.67 μm (over land) and 0.87 μm (over ocean). In the inversion TOA reflectances are interpolated between the pre-tabulated values by a second-order polynomial as a function of surface albedo and aerosol optical thickness. These calculations are conducted for all 40 different boundary layer aerosol mixtures described above. The extended OPAC database includes a well-defined spectral dependence of the aerosol extinction coefficient. This allows an easy, and for a certain aerosol mixture, unique conversion to other wavelengths by multiplying a wavelength-dependent factor. Thus, the retrieval can be conducted at wavelengths which are best suited due to darkest surfaces, most accurate albedo characterization, and highest aerosol sensitivity. With the well-defined spectral dependence of the extinction for each aerosol mixture these values are then converted to the reference wavelength of 0.55 μm.

Even after the rather complex scheme of selecting an optimum set of dark fields described in Section 3.5, a significant scatter of the AOD values at adjacent dark fields remains. Therefore, an interpolation and smoothing procedure is applied. The basic assumption behind this procedure is that aerosol variability has, in most cases, a scale of many kilometers. In SYNAER version 1.0 in boxes of 4 × 4, 16 × 16, and 64 × 64 pixels all aerosol optical depth values which are outside a 1-σ interval are excluded if enough pixels are available for the statistics (10, 40, 100). In SYNAER version 2.0 this outlier screening is not applied due to the larger number of spectrometer pixels. For the remaining pixels a 25 × 25 radiometer box average was calculated as $AOD_{box} = 0.4\ AOD_{box-mean} + 0.6\ AOD_{box-minimum}$ in version 1.0, since it was found that a simple average calculation results in AOD values which were too large. This indicated that some of the dark fields which yielded higher AOD values were still not fulfilling the underlying assumptions. Following King et al. [1999] who report similar problems (and therefore use only those results between 10 % and 40 % of a histogram of retrieved dark field AOD values in a 10 × 10 box of MODIS), SYNAER version 2.0 also estimates the 25 × 25 pixel box AOD as average of only the retrieval results which fall into the 10 % to 50 % range of the box AOD histogram. This averaging is applied only to the lowest dark field class with highest accuracy – only if fewer than five dark fields are available in the averaging box, additional dark fields of the next class are added subsequently.

If no dark fields are available inside a box the average is found by inverse distance-weighted interpolation between the nearest adjacent boxes which contain dark fields. In this interpolation the radius is increased until at least three quadrants around the missing value contain AOD results. Then, aerosol optical depth values for every individual pixel are calculated by means of an inverse distance-weighted interpolation function between these (now regularly distributed) box average values.

Using this pixel mask of aerosol optical depth a pixel-wise atmospheric correction (EXACT, EXact Atmospheric Correction by accurate radiative Transfer; Popp [1995]) can be conducted for all cloud-free pixels which results in surface albedo pixel-masks at the radiometer wavelengths 0.55, 0.67 and 0.87 μm. EXACT conducts atmospheric correction by fully taking into account multiple scattering effects of aerosols and molecules, as well as ozone, water vapor and aerosol absorption. It is based on the same radiative transfer code

(SOS) as the other parts of the SYNAER method and the same aerosol model. The aerosol optical depth for each pixel is taken from the dark pixel derived AOD mask for each of the 40 mixtures. By knowing the aerosol type, its properties can be transferred from 0.55 μm to the other ATSR-2 channels at 0.67 and 0.87 μm. Because EXACT is applied for 40 aerosol mixtures it yields 40 sets of tri-spectral atmospherically corrected surface albedos.

In summary, the first retrieval step (radiometer retrieval) yields the aerosol optical depth (stored at 0.55 μm), and the surface albedo values at 0.55, 0.67, and 0.87 μm for all cloud-free pixels. It should be kept in mind that the degree of darkness and density of the dark fields impose some limitation on the accuracy of this approach. But, it was found that a sufficient number of well-characterized dark fields can be found in many locations around the globe. The set of these parameters is derived 40 times for the different 40 aerosol mixtures each with known spectral dependence of aerosol extinction, absorption and phase function.

3.7   Selecting the most plausible aerosol type

At this point surface and aerosol results for all 40 aerosol mixtures are available from step 1 of the retrieval but the decision of the best suitable aerosol mixture is still open. A bi-spectral (0.55 and 0.67 μm) 'ATSR-only' method failed as the 0.55 μm channel has a much weaker correlation to the mid-infrared spectral range due to strong vegetation influences. Attempts to use the two viewing angles of ATSR-2 were not successful as the co-registration of both views (which is done by the data distributor without using a digital elevation model) showed significant errors in structured terrain and thus the two measurements did not contain the same target. Therefore, the simultaneously measured spectrometer spectra are used in step 2 of the retrieval in order to exploit their additional spectral information. It is not really expected that 40 different mixtures can be separated but this large number is tested to guarantee an accurate simulation of the truth in the observed atmospheric spectra.

First, the radiometer results are co-located and spatially integrated to match the larger spectrometer pixels. From the accuracies of geo-locations specified for both instruments an error of the respective surface area of 1/2 pixel along the edge of the GOME pixel can be assumed yielding a relative error of $\frac{1}{2} \cong 4\%$. This agrees well with an assessment of the radiometric relative accuracy between ATSR-2 and GOME instruments, which was investigated by Koelemeijer et al. [1997] and found to be 2% and 4% at 0.55 and 0.67 μm, respectively. In the latest calibration version of SCIAMACHY (version 6.1) the cross-correlation of spectrally and spatially integrated reflectances measured by SCIAMACHY and AATSR instruments (and against another radiometer MERIS onboard ENVISAT) was even found to show deviations on the order of 1% [Kokhanovsky et al., 2007].

A surface albedo spectrum is needed for the extrapolation of the surface spectrum to the blue and UV spectral range where no radiometer channels are available. Its form shape is chosen out of 12 available data sets [Guzzi et al., 1998; Köpke and Kriebel, 1987]; see the plots of plate 3 in Holzer-Popp et al. [2002a]) and adapted to absolute values of the derived radiometer surface albedos at the three radiometer wavelengths. The selection of the surface type is based on brightness (snow in the green channel brighter than 0.25, water in the near-infrared channel darker than 0.03), NDVI and a green-to-red ratio GRR = $R_{0.55\mu m}/R_{0.67\mu m}$ which are calculated from the radiometer-derived three spectral surface albedo values. In the selection scheme NDVI and GRR are compared to the values from the 10 signature curves (except water and snow) to choose the type with the closest coincidences of both values. The surface albedo value at 0.33 μm is kept fixed for each type. This means that the choice of one of the 12 surface spectral curves defines only details of the spectral form of the surface albedo signature in the spectrum. However, the absolute values (except for 0.33 μm) are exactly defined by the corrected radiometer values at 0.55, 0.67, and 0.87 μm and interpolated linearly in between. This adjustment scheme is applied to both land and water dominated spectrometer pixels.

In SYNAER version 2.0 a different extrapolation scheme for vegetation types (PINE FOREST and PASTURE from Köpke and Kriebel [1987], VEGETATION from Guzzi et al. [1998]) is applied. Following the MODIS retrieval scheme, where the blue surface reflectance is estimated as 50 % of the red surface reflectance, the reflectance at 0.50 μm is calculated as $R_{0.500\mu m} = 0.5\ R_{0.67\mu m}$. Then wavelengths below 0.50 μm are linearly extrapolated from the 0.67 and 0.50 μm surface reflectances, whereas larger wavelength surface reflectances are interpolated between the 0.50, 0.55 and 0.67 μm bands.

Using the aerosol spectral information GOME or SCIAMACHY spectra can then be simulated for the 40 mixtures with the pre-tabulated radiative transfer calculations and the extrapolated surface spectra based on the radiometer surface reflectance values. By a simple unweighted least-squares fit of the 40 simulations to the cloud-corrected 'measured' spectrum the simulated spectrum which fits best is selected. This spectrum is then assumed to represent the real (surface and atmospheric) conditions in the pixel, and its aerosol optical depth and type of aerosol are taken as the retrieval result. This procedure is conducted at 10 wavelengths which have been selected based on their high aerosol sensitivity and low uncertainty due to instrument errors, surface brightness and gas absorption: 0.415, 0.427, 0.460, 0.485, 0.500, 0.516, 0.535, 0.554, 0.615, and 0.675 μm. Wavelengths above 0.700 μm are not utilized because the simulation often does not fit the bright vegetation peak. Wavelengths shorter than 0.415 μm are not used as they suffer from a strong sensitivity to the boundary layer height for absorbing aerosols.

Application of SYNAER to a large number of pixels yields a number of cases, where all simulated spectra are completely different from the measured spectrum. As a quality check spectrometer reflectances corresponding to radiometer channels are therefore calculated and compared in order to detect large deviations which indicate perturbed pixels. Furthermore, pixels where the fit error is larger than an error due to an assumed 3 % accuracy of the spectral reflectances are also excluded for both aerosol optical depth and type retrieval. These problems may be due to instrument errors, to 3D scattering from broken clouds or inhomogeneous cloud tops inside or near the spectrometer pixel, or to remaining inaccuracies in the radiometric calibrations of both instruments. Finally, an ambiguity test is applied which compares the fit error with the mean deviation between the spectra of the 40 simulated mixtures. Only those spectrometer pixels where the fit is more accurate than this variability are exploited in terms of the type of aerosol. Thus, especially for pixels with low aerosol contents, and therefore low differences between the 40 simulated spectra, the aerosol type cannot be derived as proven theoretically in Section 2.2.

4.   Validation and applications

4.1   Validation status and its limitations

Ground-based photometer measurements are widely used to determine the accuracy of AOD satellite retrievals. However, the natural spatio-temporal variability of aerosol distributions is often not considered, leading to misinterpretations of the significance of such comparisons. This is why the determination of the representativeness of single ground measurement stations – as an indicator for the local variability in AOD and their aptitude for being used as ground truthing station – is of high interest for both satellite- and ground-based retrieval of aerosols.

A possible method for quantifying the natural spatial variability of any given parameter are variogram analyses: they express the variability of a quantity, e.g., AOD, as measured at different locations but approximately at the same time, seen as a function of the distance between two locations considered. In Fig. 8.8 the result of such a variogram analysis [Holzer-Popp et al., 2008] is shown, where the RMSE of AOD at 0.55 μm is plotted as a function of the distance between the ground stations involved. Here all available AERONET ground stations in Europe (squares), the USA (triangles) and the Middle East region including Saudi Arabia (crosses) for 2003 to 2005 are included, allowing only high quality level-2 ground data for the analysis. All measurements within ± 30 minutes are considered for each RMSE value and each pair of ground stations, which are then grouped into bins of 50 km for legibility purposes.

It has to be pointed out that due to the regional distribution of the AERONET stations the database for the first values, at a distance of 0 to 50 km, is very small. For all regions considered the 'natural variability offset' of the curve, i.e. the variability of AOD within a very small region, is around 0.05. This value of atmospheric noise should always be kept in mind, as the optimal accuracy which can be reached when comparing ground to satellite measurements.

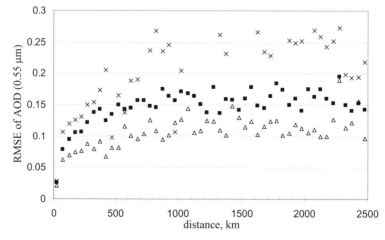

Fig. 8.8. Absolute RMSE of the measured aerosol optical depth (AOD) at 0.55 μm for Europe (squares), the USA (triangles) and the Middle East region (crosses); based on AERONET ground measurements for the years 2003–2005 (from Holzer-Popp et al. [2008]).

For larger distances the average variability generally increases with increasing distance between the locations of ground measurements. However, there are differences for the various regions analyzed: the European stations reach an accuracy of 0.1 at approximately 200 km distance between two ground measurement stations, whereas the US locations can be spaced apart as far as 500 km to reach the same natural variability. In the Middle East this distance amounts to approximately 100 km only, signalizing a rather small representativeness of AOD ground measurements. This means that depending on the geographic location different natural variabilities of aerosol measurements have to be taken into account when determining the accuracy limits of satellite-based AOD retrieval.

Table 8.3. AERONET ground stations used for the inter-comparisons

	Latitude	Longitude	Country	Comment
Avignon	43.93° N	4.88° E	France	rural
Belsk	51.84° N	20.79° E	Poland	rural
Blida	36.51° N	2.88° E	Algeria	dust
Cabauw	51.97° N	4.93° E	Netherlands	rural
Cairo	30.08° N	31.29° E	Egypt	urban
Carpentras	44.08° N	5.06° E	France	rural
Dunkerque	51.04° N	2.37° E	France	Urban, coast
Erdemli	36.57° N	34.26° E	Turkey	Maritime
Evora	38.57° N	7.91° W	Portugal	rural
Forth Crete	35.33° N	25.28° E	Greece	coast
Fontainebleau	48.41° N	2.68° E	France	rural
Ispra	45.80° N	8.63° E	Italy	industrialized
Karlsruhe	49.09° N	8.43° E	Germany	industrialized
Kishinev	47.00° N	28.81° E	Moldova	urban
Laegeren	47.48° N	8.35° E	Switzerland	urban
Lannion	48.73° N	3.46° E	France	coast
Lille	50.61° N	3.14° E	France	urban
Minsk	53.00° N	27.50° E	Belarus	urban
Modena	44.63° N	10.95° E	Italy	industrialized
Mongu	15.25° S	23.15° E	Zambia	fires
Oostende	51.23° N	2.93° E	Belgium	coast
Palaiseau	48.70° N	2.21° E	France	urban
Palencia	41.99° N	4.51° W	Spain	rural
Haute Provence	43.94° N	5.71° E	France	rural
Rome	41.84° N	12.64° E	Italy	industrialized
Saada	31.63° N	8.16° E	Morocco	dust
Santa Cruz de	28.47° N	16.25° W	Spain	Maritime, island
Skukuza	24.99° S	31.59° E	South Africa	rural
The Hague	52.11° N	4.33° E	Netherlands	industrialized/maritime
Toravere	58.26° N	26.46° E	Estonia	rural
Toulon	43.14° N	6.01° E	France	coast

First inter-comparisons of the new SYNAER-ENVISAT version 2.0 results to ground-based sunphotometer measurements of the spectral aerosol optical depth from NASA's Aerosol Robotic Network (AERONET) at 39 coincidences were conducted [Holzer-Popp et al., 2008]. A list of the AERONET stations which were included in this inter-comparison is shown in Table 8.3. These validation cases have a moderately dark surface albedo (below 0.20 at 0.67 μm) and a fit error better than 0.01 (which is equivalent to a few percent noise in the spectra) and show a good agreement with correlations above 0.80, bias values less than 0.02 and standard deviations of 0.10 (0.13, 0.09) at 0.55 (0.44, 0.67) μm as shown in Fig. 8.9(a), 8.9(b) and 8.9(c). In this inter-comparison AOD values at 550 nm for AERONET observations (which are not directly measured) were obtained applying a power law function of AOD values at 500 and 670 nm. This indicates a correct assessment of the amount and type (namely the spectral dependence of extinction) of aerosol. Through error propagation of the natural variability of at least 0.05 (coincident with the SYNAER pixel size of 60×30 km^2) from the variogram analysis of Fig. 8.8 a standard deviation for the SYNAER retrieval only of 0.08 at 0.55 μm can be deduced. This ground-based validation comprised data from Europe and Africa in several climate zones spread over 3 months in the summer season of 2005. A similar case study validation of SYNAER version 1.0 with a smaller number of 15 data pairs of AERONET and the predecessor satellite instruments ATSR-2/GOME onboard ERS-2 showed a similar agreement [Holzer-Popp et al. 2002b]. Furthermore, a comparison of monthly mean results from SYNAER 1.0 and other

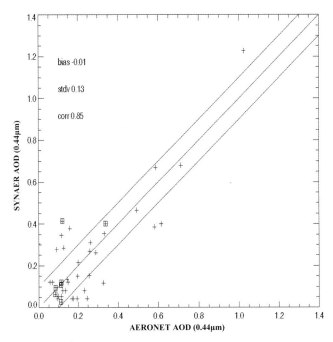

Fig. 8.9(a). Scatter plot of SYNAER version 2.0 versus AERONET aerosol optical depth at 0.44 μm. It should be noted that the synergistic exploitation of AATSR + SCIAMACHY is applied to the large $(60 \times 30$ km$^2)$ spectrometer pixels. Land pixels are denoted as +, whereas coastal and ocean pixels are denoted as ⊞ (from Holzer-Popp et al. [2008]).

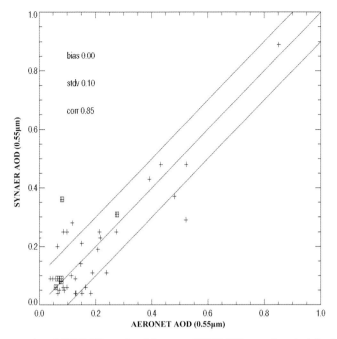

Fig. 8.9(b). Scatter plot of SYNAER version 2.0 versus AERONET aerosol optical depth at 0.55 μm. It should be noted that the synergistic exploitation of AATSR + SCIAMACHY is applied to the large (60×30 km^2) spectrometer pixels. Land pixels are denoted as +, whereas coastal and ocean pixels are denoted as ⊞ (from Holzer-Popp et al. [2008]).

satellite aerosol retrievals as well as AERONET stations over ocean [Myhre et al. 2005] showed a qualitative agreement with the other datasets for a number of cases.

4.2   First datasets

Since the SYNAER processor started routine operations in June 2005, daily aerosol products have been produced since then (but with gaps due to input failure or processing errors) – the continuing processing leads to a further evolving dataset. Additionally, reprocessing with version 2.0 has been conducted with data of the months July to November 2003. Reprocessing of the entire dataset with the new SYNAER version 2.0 has been started recently. This large dataset will then be used for extended statistically significant validation and assessment of seasonal and geographical distribution of bias and noise in the SYNAER results. From the validation efforts conducted so far and the theoretical information content analysis preliminary application limitations were extracted, which are given in Table 8.4. As can be seen in the examples of daily SYNAER 1.0 observations with ENVISAT in Fig. 8.10 of June 25, 2005, these limitation factors lead to a significant reduction of the data coverage. The instrument scan patterns (alternating nadir – limb pattern of SCIAMACHY; swath width limit of 512 km of AATSR) themselves lead to a revisit period over one location of about 12 days. Cloudiness, bright surface (deserts, snow cover) and low sun cause a further reduction of available observations especially in semi-arid regions and in winter times. The images in Fig. 8.10 show the typical summer coverage

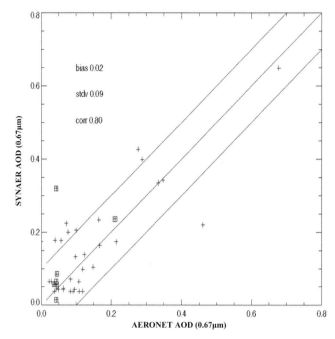

Fig. 8.9(c). Scatter plot of SYNAER version 2.0 versus AERONET aerosol optical depth at 0.67 μm. It should be noted that the synergistic exploitation of AATSR + SCIAMACHY is applied to the large (60 × 30 km²) spectrometer pixels. Land pixels are denoted as +, whereas coastal and ocean pixels are denoted as ⊞ (from Holzer-Popp et al. [2008]).

Table 8.4. Limitation factors for SYNAER application

Limitation factor	Limiting effect	Preliminary threshold
Surface albedo (0.67 μm)	Total aerosol and type sensitivity decreases with increasing brightness of a ground surface	Below 0.20 for SCIAMACHY pixel Below 0.085 for AATSR dark fields
Cloud fraction	Total aerosol and type sensitivity decreases with increasing cloudiness	Below 0.35
Aerosol optical depth (AOD) at 0.55 μm	Type retrieval becomes ambiguous for low aerosol loading	Above 0.15
Solar zenith	Neglected effects of non-spherical atmosphere leads to increasing AOD error	Below 75°
Coarse mode particles	Neglected effects of non-spherical particles lead to increasing AOD error	Mineral dust components, sea salt

and resolution with two to three orbits per day over Europe. As one feature in this example, elevated aerosol optical depth over the Iberian peninsula can be seen, which is mainly due to increased water-soluble and soot aerosols, as is frequently the case for wildfires over southern Europe in the summer months. Examples of the complete coverage of the currently available SYNAER-ENVISAT dataset (version 1.0) are discussed in Holzer-Popp et

ENVISAT
Aerosol Optical Depth 550 nm

Jun 25, 2005
Europe

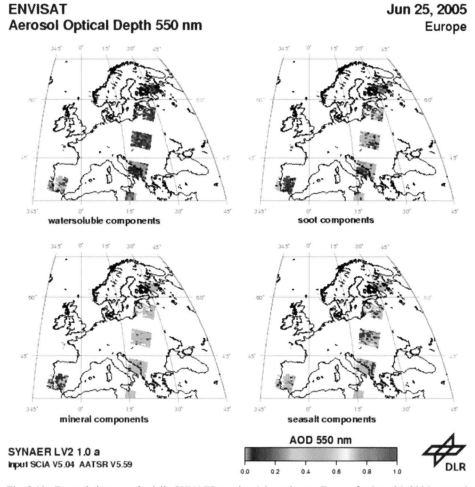

Fig. 8.10. Example images of a daily SYNAER version 1.0 result over Europe for June 25, 2005: aerosol optical depth composition into the 4 major components – clearly biomass burning can be seen over the Iberian peninsula in the WASO and SOOT components.

al. [2008] where several well-known features can already be clearly seen. The datasets shown in Holzer-Popp et al. [2008] give the 4-month average values for the period July–October 2003 on a $5° \times 5°$ grid. In this time period a reasonable coverage in this grid is achieved (as opposed to the earlier ERS-2 coverage, which needed one year of data for a similar pixel number on a 5×5 degree grid). Although validation of this aerosol composition dataset is solely based on the indirect validation through spectral AOD, in the total aerosol optical depth and the aerosol component maps distinct features can be seen which are very plausible. The major ones are the tropical biomass burning regions in Africa and South America, the sub-tropical desert regions (Sahara, Namib/Kalahari, Arabia, South America) and the biomass burning plume over the Atlantic from southern Africa and peaks of the soot component in the biomass burning regions and the mineral components around the desert areas. No retrieval is possible with this method inside the Sahara due to the bright surface.

5. Discussion and conclusions

The SYNAER method provides a retrieval, which exploits the optical measurements made from two sensors in one retrieval algorithm. One significant drawback from this synergistic application is the limited spatial resolution of 60×30 km^2 and the weak temporal repetition frequency of 12 days at the equator in cloud-free conditions when applied to ENVISAT sensors. On the other hand the gain lies in the joint exploitation of 10 spectrometer and four radiometer spectral channels. Consecutively, the information content in this retrieval enables, under favorable conditions (dark and even moderately bright surfaces if there are few dark spots suitable for the 1 km AOD retrieval within the larger spectrometer pixel, high sun), the independent retrieval of not only the aerosol loading (aerosol optical depth), but also its composition.

In this chapter a comprehensive description of SYNAER version 2.0 is provided and in particular the improvements of the SYNAER method versus the original version 1.0 are described and their application with the new sensor pair SCIAMACHY and AATSR is demonstrated and validated. Given the pixel size, the retrieval accuracy of around 0.1 at $0.55\,\mu$m (or 0.08 after atmospheric noise deduction) is satisfactory, but there is room for improvement. Further analysis of the reasons for the remaining error (in addition to the scaling versus the AERONET point measurements) will be conducted. Potential underlying reasons are neglecting non-sphericity of particles, not account for the polarization characteristics of scattered light, residual cloud contamination, and the scatter in the surface reflectance correlation between the $1.6\,\mu$m and $0.67\,\mu$m channels of AATSR. Furthermore, the representativity of an AERONET station for a SCIAMACHY pixel area with large terrain or pollution variance can be absolutely absent.

The theoretical analysis of the information content in the second step of the methodology (aerosol type retrieval) yielded up to five degrees of freedom (two for surface albedo and AOD introduced from the first retrieval step and three for the aerosol type) and thus supports the conclusion that an estimation of the aerosol composition is becoming feasible under favorable conditions with this method. Under typical conditions over vegetated surfaces outside winter periods two degrees of freedom are available for information on the aerosol type. As two independent parameters (e.g., spectral extinction gradient and absorption) are needed to differentiate monomodal aerosol distributions, this would be sufficient to characterize the major aerosol types. At least one more degree of freedom is needed for bimodal aerosol distributions, which is then only feasible for the best possible conditions. Further analysis is required to interpret the information content with regard to these different parameters. For example, the apparent absorption sensitivity in the method (being able to differentiate water-soluble and soot components) may also be provided by the integrating of different scales of the two instruments (1 km and 60/30 km) and thus by averaging dark and bright pixels, which is sensitive to the nonlinearity in radiative transport.

SYNAER includes an accurate cloud detection for the radiometer. A method to cope with partly cloud covered spectrometer pixels by use of the spatially better resolved radiometer measurements was developed by model calculations and proven indirectly through successful SYNAER retrieval. It assumes a linear mixing of cloudy and cloud-free signals, neglecting nonlinear 3D-effects in inhomogeneous clouds.

SYNAER is based on physical understanding and uses accurate (but scalar) modeling of radiative transfer through the atmosphere and scattering by aerosol particles. However, the final selection of thresholds, box sizes and conversion factors for finding and characterizing dark fields relies to some extent on an empirical optimization that was conducted with about 2,600 GOME pixels and corresponding ATSR-2 measurements of about 220 frames. The major goal in this optimization procedure was to find the best balance between computing costs (e.g., number of aerosol mixtures, number of wavelengths, box size for interpolation) and sufficient information content to account for a realistic variability of aerosols and the surface (e.g., density and degree of darkness of dark fields in most climate zones).

SYNAER relies on two major basic assumptions which impose limitations. First, the dark field approach based on the 1.6 μm channel, together with the NDVI (and the 3.7 μm channel in version 1.0), limits the applicability in semi-arid areas. Secondly, the aerosol model includes several simplifications, which might lead to failure of SYNAER under certain conditions.

The dark field approach is based on the physically plausible correlation between mid-infrared and visible reflectances of vegetated and some non-vegetated surfaces and the assumption that aerosol scattering decreases significantly in the mid-infrared region. The assumption of the correlation was supported by both the investigation of some Landsat-5 scenes (SYNAER version 1.0) and the validation results of SYNAER (version 1.0 and 2.0). It should be noted, that aerosol monitoring over land with MODIS relies on comparable basic assumptions. The exploitation of a new wavelength sub-region around 1.6 μm was found to have more restrictions than the 2.2 μm channel and a suitable dark field scheme was developed and improved. On the one hand, the remaining aerosol extinction in this channel is stronger by a factor of about 2 for small particles which reduces retrieval accuracy for high aerosol loading. On the other hand, snow and inland water are also dark at this channel. Therefore, the vegetation index is used as a second criterion to stabilize the dark field selection and a vegetation-dependent surface brightness estimation was developed. Thus now four major dark field approaches were demonstrated: three exploiting mid-infrared to visible correlations based on: 2.2 μm [Kaufman et al., 1997b], 3.7/3.8 μm [Kaufman et al., 1997b; Holzer-Popp et al., 2002a], and 1.6 μm channels together with the NDVI (with fixed correlation factor in Holzer-Popp et al. [2002a] and with vegetation dependant correlation in Holzer-Popp et al. [2008]), and the fourth one (the oldest) assuming fixed dark field albedos for targets selected from the 0.87 μm channel and the vegetation index [Kaufman and Sendra, 1989]. Future applications and comparisons between these approaches and with other case-study aerosol retrievals will be valuable in identifying weaknesses and strengths of the various methods.

The SYNAER aerosol model assumes horizontal aerosol variability at scales above 10 km and a simplified layer structure neglecting disconnected additional aerosol layers in the troposphere (except dust). It includes only spherical particles (which can be treated with Mie calculations). An external mixing of representative components is assumed. These components describe ensembles of optically equivalent (but chemically different) particles with their fixed combinations of extinction, absorption, phase function, vertical profiles and size distribution in several size modes based on measured aerosol particle mixtures. A set of 40 mixtures was chosen as representative for real variability in order to keep calculation times within a reasonable limit. Particle swelling due to increased humidity was taken into account with two relative humidity values of 50 % and 80 % for all mixtures. Improved understanding of aerosols was used to extend the original soot and dust components to two different components representing different size distributions and mixing modes of combustion and biomass burning soot particles and for low or high hematite content. Evidently, the component-based mixing approach allows the separation of these basic components but not the detection of individual molecule species which would be of interest for chemical investigations. This is self-evident due to the fact that the optical features, and thus the impact on the measured reflectances are almost equal for the different particle species in one aerosol component.

Further assumptions are used in the method: SYNAER has a rather high technical complexity, since it exploits synergistic measurements of two sensors. It relies on the high radiometric accuracy and correlation of these two instruments to extract the rather weak aerosol signals from the observations. SYNAER relies on reference signatures of several surface types which are adjusted to absolute values on the basis of the radiometer albedo measurements. In the UV and blue spectral region the simulation of spectrometer spectra over water relies solely on radiometer-derived surface albedo values at 0.55, 0.67, and 0.87 μm and one spectral signature of open ocean water, whereas variations in chlorophyll content are estimated with a simple parameterization and wind-driven diffuse glint is not accounted for. A plane-parallel scalar radiative transfer model is used, limiting the applicability to data with solar elevations above 15 degrees. In the range of 0.415–0.500 μm the lack of knowledge of the vertical aerosol profile and the neglecting of polarization effects in the radiative transfer calculations is another possible source of error for absorbing aerosols [Rozanov and Kokhanovsky, 2006]. Therefore, future work will be done on the retrieval of vertical aerosol profiles from UV spectra. Radiative transfer calculations indicate that it could be feasible to retrieve aerosol optical depths for the boundary layer, free troposphere, and stratosphere or the height of the boundary layer. The possibility to include new effective optical properties (e.g., non-sphericity or internal mixing) in the scalar radiative transfer calculations if this becomes available shall be investigated in the future.

As with all satellite retrieval algorithms for aerosols SYNAER is limited by a mathematically ill-posed system, which makes several assumptions and simplifications necessary. These are in addition to the limited sampling, the limitation to spherical particles (Mie scattering), the dependence on the predefined aerosol mixtures, the decreasing information content with brighter surfaces, and the general optical remote sensing limitations for low sun and high cloud fraction. One future element in SYNAER is still the application of the theoretical analysis to each pixel geometric and surface condition. Here a good compromise between larger coverage (also over moderately bright surfaces) and decreasing information content needs to be established in the future with at least one complete global annual dataset.

As validation of the retrieved aerosol composition is extremely difficult due to a lack of equivalent ground-based data, only an indirect validation approach through multispectral AOD measurements has been used so far. Validation of satellite-derived information with this large pixel size and low spatial–temporal coverage is further impeded due to atmospheric noise and the limited number of ground-based stations. It is also important to understand, that this type of satellite retrieval depends critically on the aerosol model chosen and its limitations or its complexity. In the end, the retrieval of aerosol composition must therefore be considered as a way of interpreting of the optical measurements. But, it is the conviction of the authors that the plausible results shown in this chapter encourage further work in this direction.

The validation against AERONET stations includes several cases where the representativeness of the ground-based station for the SYNAER pixel is weak. Examples are Fontainbleau (at the edge of the mega-city Paris), Ispra (at the edge of an Alpine valley), and Erdemli (at the coast with high mountains behind). In all these cases the local AOD regime at the station differs significantly from the regional AOD. One extreme case at Tenerife island was excluded by the ambiguity test of SYNAER. Here two stations (Izana at 2,367 m above sea level with an AOD at 0.55 μm of 0.39 and Santa Cruz at sea level with an AOD at 0.55 μm of 0.71) fall into 1 SYNAER pixel (AOD at 0.55 μm = 0.94) thus highlighting the possible variability inside a SYNAER pixel for an extreme case of a desert dust outbreak and the subsequent limitation for the AERONET inter-comparison.

Validation of the derived aerosol composition requires further work. One planned approach will use EMEP mass speciation fractions to determine the presence of soot (elemental and organic carbon), mineral dust and sea salt and inter-compare with the SYNAER composition. Other possibilities lie in the inter-comparison to model and other satellite datasets (e.g., MODIS fine/coarse mode, future MISR aerosol composition product).

The application potential of SYNAER ranges from improving satellite trace gas retrievals over data assimilation into atmospheric chemistry transport models for climate research and air quality monitoring and forecasting to service applications such as accurate calculation and forecast (through assimilation into a forecast model) of solar irradiance for solar energy applications (see, for example, Breitkreuz et al. [2007]). Finally, the estimation of the aerosol composition provides one critical piece of information (among vertical profile/boundary layer height, humidity/aerosol type) for a systematic conversion of AOD into near-surface mass concentrations (PM values; e.g., Holzer-Popp and Schroedter-Homscheidt [2004]), which is the key quantity for regulatory purposes.

Due to the need for overlap of two sensors with very different scan patterns, the coverage of SYNAER/ENVISAT is still quite weak, providing approximately one synergistic observation every 12 days if no clouds occur. This leads to the need for large integration grid boxes or time periods. The potential for independent estimation of the aerosol type can be shown, but daily monitoring applications are only feasible by assimilation into atmospheric chemistry models. A further improved coverage (every 1–2 days globally) will be achieved with the transfer to equivalent sensors AVHRR and GOME-2 onboard the operational meteorological METOP platform. A prototype is already available, but needs further adjustments for the instrument characteristics and calibration. Finally, by integrating SYNAER results from ERS-2, ENVISAT and METOP there is a perspective to achieve a long-term record of AOD and composition ranging from 1995–2020. It was always in the light of this final goal, that specific instrument characteristics such as the ATSR dual-view or the SCIAMACHY mid-infrared bands or limb observations were not exploited to ensure application of SYNAER to all three satellite datasets. However, the different pixel sizes and sampling will need thorough assessment when integrating these three datasets.

SYNAER/ENVISAT has been implemented for operational processing at the German Remote Sensing Data Center within the ESA GSE PROMOTE (Protocol Monitoring for the GMES Service Element; see also http://www.gse-promote.org) and delivers daily near-real-time observations (within the same day) and an evolving archive of historic data. SY-NAER data are stored at the World Data Center for Remote Sensing of the Atmosphere (http://wdc.dlr.de).

Acknowledgments. We are thankful to ESA for providing the satellite observations from ERS-2 and ENVISAT platforms, which are exploited with the SYNAER method through announcement of opportunity projects PAGODA-2 (AO3.218) and SENECA (AO ID-106) as well as for funding the validation and application of SYNAER to ENVISAT as part of the ESA GMES Service Element PROMOTE (Stage 2). The development of the SYNAER method was partly supported by the EC FP5 project Heliosat-3 (NNK5-CT-2000-00322). Furthermore we appreciate the independent validation dataset from the AERONET ground-based sunphotometers provided by Brent Holben and all involved AERONET PIs.

References

Breitkreuz, H., Schroedter-Homscheidt, M., Holzer-Popp, T., 2007: A case study to prepare for the utilization of aerosol forecasts in solar energy industries, *Solar Energy*, **81**, 1377–1385.

Burrows J.P., et al. 1999: The Global Ozone Monitoring Experiment (GOME): mission concept and first scientific results, *J. Atmos. Sci.*, **56**, 151–175.

Claquin, T., Schulz, M., Balkanski, Y.J., 1999: Modelling the mineralogy of atmospheric dust sources, *J. Geophys. Res.*, **104**, D18, 22243–22356.

Deuzé L., Bréon, F. M., Devaux, C., 2001: Remote sensing of aerosols over land surfaces from POLDER-ADEOS – 1 polarized measurements, *J. Geophys. Res.*, **106**, D5, 4913–4926.

Dubovik, O., Holben, B., Eck, T.F., Smirnov, A., Kaufman, Y.J., King, M.D., Tanre, D., Slutsker, I., 2002: Variability of absorption and optical properties of key aerosol types observed in worldwide locations, *J. Atmos. Sci.*, **59**, 590–608.

Dubovik O., Sinyuk A., Lapyonok T., Holben B.N., Mishchenko M., Yang P., Eck T.F., Volten H., Muñoz O., Veihelmann B., van der Zande W.J., Leon J.-F., Sorokin M., Slutsker I., 2006: Application of spheroid models to account for aerosol particle nonsphericity in remote sensing of desert dust, *J. Geophys. Res.*, **111**, D11208, doi:10.1029/2005JD006619.

Dunion, J.P., Velden, C.S., 2004.: The impact of the Saharan Air Layer on Atlantic tropical cyclone activity, *Bull. Am. Meteorol. Soc.*, **90**, 353–365.

Evan, A.T., Heidinger, A.K., Pavalonis, M.J., 2006: Development of a new over-water Advanced Very High Resolution Radiometer dust detection algorithm, *Int. Journ. Rem. Sens.*, **27**, 3903–3924.

Guzzi, R., Burrows, J., Cattani, E., Kurosu, T., Cervino, M., Levoni, C., Torricella, F., 1998: GOME cloud and aerosol data products algorithm development – Final Report, ESA Contract 11572/95/NL/CN.

Hess, M., Köpke, P., Schult, I., 1998: Optical properties of aerosols and clouds: the software package OPAC, *Bull. Am. Met. Soc.*, **79**, 831–844

Holzer-Popp, T., Schroedter-Homscheidt, M., 2004.: Satellite-based background concentration maps of different particle classes in the atmosphere, C. A. Brebbia (Ed.), *Air Pollution XIII*, WIT Press, Southampton.

Holzer-Popp, T., Schroedter, M., Gesell, G., 2002a: Retrieving aerosol optical depth and type in the boundary layer over land and ocean from simultaneous GOME spectrometer and ATSR-2 radiometer measurements, 1, Method description, *J. Geophys. Res.*, **107**, D21, AAC16-1-AAC16-17.

Holzer-Popp, T., Schroedter, M., Gesell, G., 2002b: Retrieving aerosol optical depth and type in the boundary layer over land and ocean from simultaneous GOME spectrometer and ATSR-2 radiometer measurements, 2, Case study application and validation, *J. Geophys. Res.*, **107**, D24, AAC10-1AAC10-8.

Holzer-Popp, T., Schroedter-Homscheidt, M., Breitkreuz, H., Kl\'{user, L., Martynenko, D., 2008: Synergistic aerosol retrieval from SCIAMACHY and AATSR onboard ENVISAT, *Atmospheric Chemistry and Physics Discussions*, **8**, 1–49.

Hsu, N.C., Tsax, S.-C., King, M., 2004: Aerosol properties over bright-reflecting source regions, *IEEE Trans. Geosci. Rem. Sens.*, **42**, 3.

Huang, J., Minnis, P., Lin, B., Wang, T., Yi, Y., Hu, Y., Sun-Mack, S., Ayers, K., 2006: Possible influences of Asian dust aerosols on cloud properties and radiative forcing observed from MODIS and CERES, *Geophys. Res. Let.*, **33**, L06824, doi:10.1029/2005GL024724.

IPCC, 2007: *Climate Change 2007: Synthesis Report.* Contribution of Working Groups I, II and III to the Fourth Assessment Report of the Intergovernmental Panel on Climate Change, Pachauri, R.K, and Reisinger, A. (Eds), IPCC, Geneva.

Kahn, R.A., Gaitley, B.J., Martonchik, J.V., 2005: Multiangle Imaging Spectroradiometer (MISR) global aerosol optical depth validation based on 2 years of coincident Aerosol Robotic Network (AERONET) observations, *J. Geophys. Res.*, **110**, D10S04.

Kaufman, Y.J., Fraser, R.S., 1997: The effect of smoke particles on clouds and climate forcing. *Science*, **277**, 1636–1639.

Kaufman, Y. J., Remer, L., 1994: Remote sensing of vegetation in the mid-IR: The 3.75 μm channels, *IEEE Trans. Geosci. Rem. Sens.*, **32**, 672–683.

Kaufman Y.J., Sendra, C., 1989: Algorithm for automatic corrections to visible and near-IR satellite imagery, *Int. J. Rem. Sens.*, **9**, 1357–1381.

Kaufman, Y.J., Tanre, D., Gordon, H.R., Nakajima, T., Lenoble, J., Frouin, R., Grassl, H., Herman, B.M., King, M.D., Teillet, P.M., 1997a: Passive remote sensing of tropospheric aerosol and atmospheric correction for the aerosol effect, *J. Geophys. Res.*, **102**, 16815–16830.

Kaufman, Y.J., Tanre, D., Remer, L.A., Vermote, E.F., Chu, A., Holben, B.N., 1997b: Operational remote sensing of tropospheric aerosol over land from EOS moderate resolution imaging spectrometer, *J. Geophys. Res.*, **102**, 17051–17067.

Kaufman Y.J., Karnieli A., Tanre D., 2000: Detection of dust over deserts using satellite data in the solar wavelengths, *IEEE Trans. Geosci. Rem. Sens.*, **38**, 525–531.

Kaufman, Y.J., Tanre, D., Boucher, O., 2002: A satellite view of aerosols in the climate system, *Nature*, **419**, 215–223.

King M.D., Kaufman, Y.J., Tanre, D., Nakajima, T., 1999: Remote sensing of tropospheric aerosols from space: past, present, and future, *Bull. Am. Met. Soc.*, **80**, 2229–2259.

Kneizys, F.X., Shettle, E.P., Abreu, L.W., Chetwynd, J.H., Anderson, G.P., Gallery, W.O., Selby, J.E.A., Clough, S.A., 1988: *User Guide to Lowtran 7*, AFGL-TR-88-0177.

Kneizys, F.X, Abreu, L.W., Anderson, G.P., Chetwynd, J.H., Shettle, E.P., Berk, A., Bernstein, L.S., Robertson, D.C., Acharya, P., Rothmann, L.S., Selby, J.E.A., Gallery, W.O., Clough, S.A., 1996: The MODTRAN $2/3$ Report and LOWTRAN 7 MODEL, Contract No. F19628-91-C-0132, Philipps Laboratory, Hanscom AFB, USA.

Koelemeijer, R.B.A., Stammes, P., Watts, P.D., 1997: Intercomparison of GOME and ATSR-2 reflectivity measurements, *Proc. 3rd ERS Symposium on Space at the Service of our environment*, Florence, ESA SP-414.

Kokhanovsky, A.A., von Hoyningen-Huene, W., Burrows, J.P., 2006: Atmospheric aerosol load as derived from space, *Atmos. Res.*, **81**, 176–185.

Kokhanovsky, A.A., Bramstedt, K., von Hoyningen-Huene, W., Burrows, J.P., 2007: The intercomparison of top-of-atmosphere reflectivity measured by MERIS and SCIAMACHY in the spectral range of 443–865 nm, *IEEE Trans. Geosci. Rem. Sens. Letters*, **4**, 293–296.

Köpke, P., Kriebel, K.T., 1987: Improvements in the shortwave cloud free radiation budget accuracy, *J. Clim. Appl. Meteorol.*, **26**, 374–409.

Köpke P., Hess, M., Schult, I., Shettle, E.P., 1997: Global aerosol data set, Max-Planck-Institut f\'{ur Meteorologie, Report No. 243.

Kriebel, K.T., 1977: Reflection properties of vegetated surfaces: Tables of measured spectral biconical reflectance factors, Wiss. Mitt., 29, Universit\'{at M\'{unchen.

Kriebel, K.T., Saunders, R.W., Gesell, G., 1989: Optical properties of clouds derived from fully cloudy AVHRR pixels, *Beiträge zur Physik der Atmosphäre, 8, 723–729.*

Kriebel, K.T., Gesell, G., Kästner, M., Mannstein H., 2003: The cloud analysis tool APOLLO: Improvements and validation, *Int. J. Rem. Sens.*, **24**, 2389–2408.

Levy, R.C., Remer, L.A., Mattoo, S., Vermote, E.F., Kaufman, Y.J., 2007: Second-generation operational algorithm: retrieval of aerosol properties over land from inversion of Moderate Resolution Imaging Spectroradiometer spectral reflectance, *J. Geophys. Res.*, **112**, doi:10.1029/2006JD007811.

McClatchey, R.A., Fenn, R.W., Selby, J.E.A., Volz, F.E., Garing, J.S., 1972: *Optical Properties of the Atmosphere*, AFCRL-72-0497, AFGL, Cambridge, MA, USA.

Moulin, C., Gordon, H.R., Banzon, V.F., Evans, R.H., 2001: Assessment of Saharan dust absorption in the visible from SeaWiFS imagery, *Journ. Geophys. Res.*, **106**, D16, 18239–18249.

Myhre, G., Stordal, F., Johnsrud, M., Diner, D.J., Geogdzhayev, I.V., Haywood, J.M., Holben, B., Holzer-Popp, T., Ignatov, A., Kahn, R., Kaufman, Y.J., Loeb, N., Martonchik, J., Mishchenko, M.I., Nalli, N.R., Remer, L.A., Schroedter-Homscheidt, M., Tanre, D., Torres, O., Wang, M., 2005: Intercomparison of satellite retrieved aerosol optical depth over ocean during the period September 1997 to December 2000, *Atm. Chem. Phys*, **5**, 1697–1719.

Nagel, M.R., Quenzel, H., Kweta, W., Wendling, P.,1978, *Daylight Illumination-Color-Contrast-Tables*, Academic Press, New York.

Nikolaeva, O.V., Bass, L.P., Germogenova, T.A., Kokhanovsky, A.A., Kuznetsov, V.S., Mayer, B., 2005: The influence of neighboring clouds on the clear sky reflectance studied with the 3-D transport code RADUGA, *J. Quant. Spectr. Rad. Transfer*, **94**, 405–424

Patterson, E.M., Gillete, D.A., Stockton, B.H., 1997: Complex index of refraction between 300 and 700 nm for Saharan aerosol, *J. Geophys. Res.*, **82**, 3153–3160.

Pohl, O., 2003: News scan: disease dustup, *Sci. Am.*, **7**, 10–11.

Pope III, C.A., Burnett, R.T., Thun, M.J., Calle, E.E., Krewski, D., Ito, K., Thurston, G.D., Lung Cancer, 2002: Cardiopulmonary mortality, and long-term exposure to fine particulate air pollution, *J. Am. Med. Ass.*, **287**, 1132–1141.

Popp, Th., 1995: Correcting atmospheric masking to retrieve the spectral albedo of land surfaces from satellite, *Int. J. Rem Sens.*, **16**, 3483–3508.

Prospero, J.M., Ginoux, P., Torres, O., Nicholson, S., Gill, T., 2002: Environmental characterization of global sources of atmospheric soil dust identified with the NIMBUS 7 Total Ozone Mapping Spectro-meter (TOMS) absorbing aerosol product, *Rev. of Geophys.*, 10.1029/2000RG000095.

Rodgers, C.D., 2000: *Inverse Methods for Atmospheric Sounding, theory and practice*, University of Oxford, UK.

Roger J.C., Vermote E.F., 1997: Computations and use of the reflectivity at 3.75 μm from AVHRR thermal channels, *Rem. Sens. Environ.*, **64**, 103–114.

Rosenfeld, D., Lensky, I.M., 1998: Satellite-based insights into precipitation formation processes in continental and maritime convective clouds, *Bull. Am. Meteorol. Soc.*, **79**, 2457–2476.

Rozanov, V.V., Kokhanovsky, A.A., 2006: The solution of the vector radiative transfer equation using the discrete ordinates technique: selected applications, *Atmos. Res.*, **79**, 241–265.

Saunders, R.W., Kriebel, K.T., 1988: An improved method for detecting clear sky and cloudy radiances from AVHRR data, *Int. Journ. Rem. Sens.*, **9**, 123–150.

Schnaiter, M., Horwath, H., Möhler, O., Naumann, K.-H., Saathoff, H., Schöck, O.W., 2003: UV-VIS-NIR spectral optical properties of soot and soot-containing aerosols, *J. Aerosol Sci.*, **34**, 1421–1444.

Sinyuk, A., Torres, O., Dubovik, O., 2003: Combined use of satellite and surface observations to infer the imaginary part of refractive index of Saharan dust, *Geophys. Res. Lett.*, **30**, 2, 1081, doi 10.1029/2002GL016189.

Slijkhuis S., 1999: *GOME Data Processor Extraction Software User's Manual*, Doc. No. ER-SUM-DLR-GO-0045, Issue 1, 4. August 1999, available from ESA or via http://auc.dfd.dlr.de/GOME/index.html.

Stedman, J.R., 2004: The predicted number of air pollution related deaths in the UK during the August 2003 heatwave, *Atmos. Env.*, **38**, 1087–1090.

Veefkind, J.P., de Leeuw, G., Durkee, P.A., Russell, P.B., Hobbs, P.V., Livingston, J.M., 1999: Aerosol optical depth retrieval using ATSR-2 data and AVHRR data during TARFOX, *J. Geophys. Res.*, **104** (D2), 2253–2260.

von Hoyningen-Huene, W., Freitag, M., Burrows, J.B., 2003: Retrieval of aerosol optical thickness over land surfaces from top-of-atmosphere radiance, *J. Geophys. Res.*, **108**, D94260.

Wang, J., Christopher, S.A., Reid, J.S., Maring, H., Savoie, D., Holben, B.H., Livingston, J.M., Russel, P., Yang, S.K., 2003: GOES 8 retrieval of dust aerosol optical thickness over the Atlantic Ocean during PRIDE, *J. Geophys. Res.*, **108**, 8595–8609, doi: 10.1029/2002JD002494.

Wong, S., Dessler, A.E., 2005: Suppression of deep convection over the tropical North Atlantic by the Saharan Air Layer, *Geophys. Res. Lett.*, **32**, L09808, doi:10.1029/2004GL022295.

World Climate Program (WCP), 1986: *A Preliminary Cloudless Standard Atmosphere for Radiation Computation*, WCP-112, WMO/TD No. 24, Boulder, CO.

Zavody, A.M., Mutlow, C.T., Llewellyn-Jones, D.T., 1995: A radiative transfer model for sea surface temperature retrieval for the Along Track Scanning Radiometer (ATSR), *J. Geophys Res.*, **100**, 937–952.

9 Retrieval of aerosol properties over land using MISR observations

John V. Martonchik, Ralph A. Kahn, David J. Diner

1. Introduction

Global and regional mapping of aerosol properties, including column amount, particle type and effective size, is of great interest for environmental and climate studies. Increased aerosol production results in decreased insolation (a direct aerosol effect), mitigating the rise in global surface air temperature caused by enhanced concentrations of greenhouse gases, though on different spatial and temporal scales [Charlson et al., 1992; Kiehl and Briegleb, 1993; Andreae, 1995]. Indirect aerosol effects include an alteration of cloud particle properties (size, single scattering albedo) that can modify cloud scattering properties, lifetimes, and precipitation amount. Although oceans cover the majority of the Earth's surface, land areas are the source of most aerosols and essentially all anthropogenic production. As a consequence, there is a tendency to find the largest aerosol optical depth values over land and coastal ocean. Monitoring of these vast areas on a frequent, global basis can be effectively accomplished only by means of space-based instruments. However, land and coastal waters generally have complex, heterogeneous, and *a priori* unknown surface reflectance characteristics, making the retrieval of aerosol properties in these locations particularly troublesome. The problem stems from the fact that the radiance measured at the top of the atmosphere (TOA) is a mixture of two components – radiance scattered solely by the atmosphere and radiance produced by multiple surface–atmosphere scattering interactions which is eventually transmitted up through the atmosphere to space. The successful retrieval process must be able to separate and explicitly describe these two radiance components, resulting in a determination of both the aerosol properties and the surface reflectance characteristics.

The retrieval algorithm strategy used to decouple these two radiance components from the TOA measurements will depend on the operational capabilities of the particular instrument. For those instruments that observe a given scene at multiple visible/near-infrared wavelengths but only at a single sun angle and view angle (typically, low Earth orbit scanners, e.g., the MODerate resolution Imaging Spectroradiometer (MODIS) and the MEdium Resolution Imaging Spectrometer (MERIS)), the aerosol retrieval process at these wavelengths always requires the absolute surface reflectance or its spectral variation at the different wavelengths to be prescribed (e.g., Chu et al. [2002]; Kaufman et al. [2002]; Santer et al. [1999]; Ramon and Santer [2001]; Hsu et al. [2006]). However, imaging instruments capable of viewing a scene at multiple wavelengths and at multiple view or sun angles over a sufficiently short time period so that atmospheric conditions can be assumed constant are much more flexible. This chapter describes how such measurements from the Multi-angle Imaging SpectroRadiometer (MISR) on NASA's Terra space platform are used to determine aerosol properties over land, unbiased by any assumptions about surface absolute reflectance or spectral reflectance characteristics.

2.   MISR specifications and operation

MISR [Diner et al., 1998] is one of five instruments (among them, MODIS) observing the Earth on the polar-orbiting Terra platform at an altitude of 705 km (Fig. 9.1). Its nominal data collection mode uses nine cameras, each fixed at a particular view zenith angle in the along-track direction and having four spectral bands (446, 558, 672 and 866 nm) with a cross-track ground spatial resolution of 275 m or 1.1 km. The fore–aft cameras are paired in a symmetrical arrangement and acquire multispectral images with nominal view angles relative to the Earth's surface of $0°$, $26.1°$, $45.6°$, $60.0°$ and $70.5°$, labeled An, Af/Aa, Bf/Ba, Cf/Ca and Df/Da, respectively. The letters A–D denote the lens design, each lens system progressing from shortest focal length (A cameras) to longest (D cameras) to preserve cross-track spatial resolution, and the letters 'n', 'f' and 'a' indicate nadir, forward and aftward views, respectively. Thus, for example, designation Ca is the label for the camera pointing aftward at nominal zenith angle $60°$. This camera configuration allows a scene to be imaged by all cameras in the four bands within a span of 7 minutes, i.e., nearly simultaneously. There are two modes in which the instrument can take data. Global Mode is the nominal operational mode, whereby data are taken at 275 m resolution in 12 channels (all nine cameras in the red band and the nadir camera in the other three bands) and at 1.1 km resolution in the remaining 24 channels. Local Mode is a specialized mode, designed to cover prescribed target areas at 275 m resolution in all 36 channels (nine cameras in four spectral bands) and traversing a distance along track about equal in length to the swath width. The swath width of the eight off-nadir cameras is 413km and that of the nadir camera is 378 km. The polar orbit allows complete coverage between latitudes $82°$ N and S every nine days, characterized by minimal observations of a scene at the equator and in-

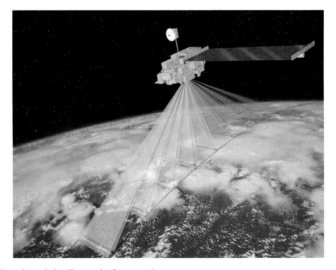

Fig. 9.1. MISR onboard the Terra platform.

creasing multiple observations of a scene with increasing latitude. After the 36 channels of imaging data are radiometrically calibrated, georectified (i.e., geo-located and co-registered) and averaged to a uniform resolution of 1.1 km, they are analyzed to determine aerosol properties over both land and ocean at a resolution of 17.6×17.6 km (16×16 1.1 km size array). This lower-resolution region allows for considerable flexibility in the way retrievals are performed, while still providing useful information on local, regional and synoptic scales.

3.   Aerosol retrieval data set requirements

The remote sensing of aerosol properties using space-based observations is a severely underdetermined problem. The at-launch MISR aerosol retrieval algorithm built upon earlier work by making use of the multi-angle data to remove much of the ambiguity [Martonchik et al. 1998a, 2002]. Since then considerable experience has been gained on how best to retrieve aerosol properties using the MISR multispectral, multi-angle data [Diner et al., 2005]. The strategy is based on a few basic assumptions, some physical constraints and other considerations:

(a) Aerosols are assumed to be laterally homogeneous with the 17.6×17.6 km region at the surface, expanding to approximately 17.6×46 km (the area contained within the view of the two $70.5°$ view angle cameras (Df and Da) at an altitude of 5 km. With this assumption the different camera-dependent effective path lengths observed through the atmosphere vary in a predictable way.

(b) Retrievals are performed by comparing observed radiances with pre-computed model radiances calculated for a suite of aerosol compositions and particle size distributions that cover a range of expected natural conditions. This allows the aerosol retrieval to be computationally efficient. However, no geographical constraints on aerosol types are applied.

(c) A fit function formalism is used to assess the magnitude of the residuals in the comparison of the measured and the model radiances. All models that meet the fit function acceptance criteria are reported as acceptable solutions to the retrieval process.

(d) For land surfaces which are reflectively heterogeneous within a given 17.6 km region no assumptions are made concerning the magnitude or explicit angular shape of the region's spectral bi-directional reflectance factors (BRFs). Instead, in the initial stages of the retrieval it is assumed that the BRF angular shape does not change significantly from one MISR band to another [Diner et al., 2005]. This assumption winnows down the number of possible pre-computed aerosol models as solutions to the retrieval process. For the remaining models additional retrieval processing requires that their BRFs produce a TOA radiance describable in terms of empirical orthogonal functions (EOFs), derived directly from the multi-angle imagery.

(e) Because the retrieval algorithm requires an assumption of atmospheric horizontal homogeneity, no retrieval are performed over land when the surface topography is determined to be complex. This restriction tends to rule out any retrievals over mountainous terrain. Additionally, all 1.1 km sub-regions within the 17.6 km region that are considered to be contaminated by clouds are not used in the aerosol retrieval process.

3.1 MISR aerosol models

To realistically constrain the MISR retrievals, it is advantageous to make reasonable use of what is currently known about the types and characteristics of aerosols found in the troposphere. In general tropospheric aerosols fall into a small number of compositional categories that include sea spray (salt), sulfate/nitrate particles, mineral dust, biomass burning particles, and urban soot. The aerosol information used by the MISR retrieval algorithms is based on these categories, convolved with the expected sensitivity of MISR data to particle properties under good but not necessarily ideal viewing conditions [Chen et al., 2008; Kahn et al., 2001; Kalashnikova and Kahn, 2006]. The aerosol optical models used are described in the Aerosol Climatology Product (ACP), which is composed of two parts, an Aerosol Physical and Optical Properties (APOP) file and a tropospheric Aerosol Mixture file. The APOP file has detailed information on the microphysical and scattering characteristics of 21 different aerosol single composition particle types, called components, of which only eight are used in the MISR operational Version 22 retrieval process (see Table 9.1). All components are modeled using lognormal particle size distributions, characterized by the median radius r_m, the standard deviation σ of the natural log of the radius r, and the total number of particles N:

$$n(r) = \frac{N}{(2\pi)^{1/2} r\sigma} \exp[-\frac{(\ln r - \ln r_m)^2}{2\sigma^2}]. \tag{1}$$

Table 9.1. MISR aerosol components

Aerosol Type	r_m (μm)	σ	Real refractive index	Imaginary refractive index	Effective radius (μm)	Single scattering albedo
1: spherical	0.03	0.501	1.45	0.00	0.06	1.000
2: spherical	0.06	0.531	1.45	0.00	0.12	1.000
3: spherical	0.12	0.560	1.45	0.00	0.26	1.000
6: spherical	1.0	0.642	1.45	0.00	2.80	1.000
8: spherical	0.06	0.531	1.45	0.0147	0.12	B: 0.911 G: 0.900 R: 0.885 N: 0.853
14: spherical	0.06	0.531	1.45	0.0325	0.12	B: 0.821 G: 0.800 R: 0.773 N: 0.720
19: dust	0.5	0.405	B: 1.50 G: 1.51 R: 1.51 N: 1.51	B: 0.0041 G: 0.0021 R: 0.00065 N: 0.00047	0.21	B: 0.919 G: 0.977 R: 0.994 N: 0.997
21: dust	1.0	0.693	1.51	B: 0.0041 G: 0.0021 R: 0.00065 N: 0.00047	3.32	B: 0.810 G: 0.902 R: 0.971 N: 0.983

B = blue, G = green, R = red, N = Near IR MISR bands

Table 9.2. MISR retrieval aerosol mixtures

Model number	Components	Mixture fractional optical depth (green band)
1–10: bimodal	1 and 6	Range: 1.0 (1) and 0.0 (6) to 0.2 (1) and 0.8 (6)
11–20: bimodal	2 and 6	Range: 1.0 (2) and 0.0 (6) to 0.2 (2) and 0.8 (6)
21–30: bimodal	3 and 6	Range: 1.0 (3) and 0.0 (6) to 0.2 (3) and 0.8 (6)
31–40: bimodal	8 and 6	Range: 1.0 (8) and 0.0 (6) to 0.2 (8) and 0.8 (6)
41–50: bimodal	14 and 6	Range: 1.0 (14) and 0.0 (6) to 0.2 (14) and 0.8 (6)
51–53: trimodal	2, 6 and 19	Range: 0.72 (2) and 0.08 (6) and 0.20 (19) to 0.16 (2) and 0.64 (6) and 0.20 (19)
54–56: trimodal	2, 6 and 19	Range: 0.54 (2) and 0.06 (6) and 0.40 (19) to 0.12 (2) and 0.48 (6) and 0.40 (19)
57–59: trimodal	2, 6 and 19	Range: 0.36 (2) and 0.04 (6) and 0.60 (19) to 0.08 (2) and 0.32 (6) and 0.60 (19)
60–62: trimodal	2, 6 and 19	Range: 0.18 (2) and 0.02 (6) and 0.80 (19) to 0.04 (2) and 0.16 (6) and 0.80 (19)
63–66: trimodal	2, 19 and 21	Range: 0.40 (2) and 0.48 (19) and 0.12 (21) to 0.40 (2) and 0.12 (19) and 0.48 (21)
67–70: trimodal	2, 19 and 21	Range: 0.20 (2) and 0.64 (19) and 0.16 (21) to 0.20 (2) and 0.16 (19) and 0.64 (21)
71–74: bimodal	19 and 21	Range: 0.8 (19) and 0.2 (21) to 0.2 (19) and 0.8 (21)

Note: Numbers in parentheses in fractional OD column indicate component label.

The six spherical particle types in Table 9.1 range in size from small to large with the medium size also taking on a range of absorbing strengths. Note that components labeled 2, 8, and 14 have identical size distributions but differ in their imaginary refractive indices, allowing them to simulate either clean or polluted aerosols or biomass burning particles.

The scattering properties for the spherical particles were computed using Mie theory whereas those for the two dust components were computed using the discrete dipole approximation and the T-matrix technique [Kalashnikova et al., 2005]. During the retrieval process 74 distinct aerosol models are created as mixtures of these components, using information detailed in the Aerosol Mixture file of the ACP (see Table 9.2). They simulate the more complex aerosol compositions in the troposphere, i.e., the bimodal and trimodal combinations that are typically found in nature. The first 50 models have bimodal distributions and are comprised of only spherical particles, simulating aerosol types found in rural to urban regions, including biomass burning and sea spray. The remaining 24 models contain at least one dust component. Models 51 through 62 have trimodal distributions where two modes are fine and coarse, non-absorbing spherical particles and the third mode is the medium dust component. Models 63 through 70 also have trimodal distributions but with a single mode of fine, non-absorbing spherical particles mixed with the medium and large dust components. Finally, models 71 through 74 are all dust bimodal mixtures composed of the medium and large dust components.

3.2 Radiative transfer look-up tables

In addition to predetermined aerosol models, another major feature of the MISR aerosol retrieval strategy is the use of the look-up table (LUT) to obtain most of the radiative transfer (RT) parameters needed by the algorithms. Whether retrieving aerosols over ocean or land, the fundamental process involves comparing measured TOA radiances with those derived from a coupled atmosphere/surface RT model. To accommodate the timing requirements of analyzing the large amount of data accumulated on a daily basis and the modeling of complex RT processes needed by the algorithms, the necessary RT parameters have been pre-computed for those aerosol components contained in the ACP and the results stored in the Simulated MISR Ancillary Radiative Transfer (SMART) dataset.

The atmospheric structure of the RT model is an aerosol layer, defined by its base, top and scale heights and optical depth, imbedded in a Rayleigh scattering layer defined by its scale height and optical depth (see Table 9.3). The RT calculations were performed using a scalar (non-polarized) code based on a matrix operator technique [Grant and Hunt, 1968], which accounts for all orders of multiple scattering. The final RT calculations include a correction for Rayleigh polarization effects, two Rayleigh scattering amounts (used to interpolate to a elevation-dependent pressure), and a fixed, standard atmosphere water vapor amount that affects only slightly the radiance in MISR's near-IR band (effective water optical depth ~ 0.005). No ozone is included since the MISR observations are corrected for effects of this gas prior to the use of the SMART dataset.

The SMART dataset contains a variety of computed RT parameters needed for retrieving aerosol properties over both ocean and land and also retrieving land-surface reflection and biophysical properties [Martonchik et al., 1998a, b]. Each RT parameter is a function of the aerosol component type, optical depth and, depending on the parameter, one or more of the view, sun geometry angles, namely view zenith angle, solar zenith angle and the azimuth angle difference between the view and sun directional vectors. To prevent the size of the dataset from being prohibitively large, reasonable restrictions were implemented concerning key variables. For example, the optical depth for each aerosol component at each wavelength was limited to a maximum value of 3.0. Also, the range of the view zenith angles in the SMART dataset is appropriate to cover only those angles observed by each of the nine MISR cameras along an orbital path. Likewise, the solar zenith angle in the dataset is limited to a value of $78.5°$, beyond which the plane-parallel atmosphere approximation starts to break down. However, the needed angular geometry in the SMART dataset is on a sufficiently fine grid to allow for linear interpolation of the RT parameters with no significant loss of computational accuracy. Finally, requiring that the RT calculations be carried out for the individual aerosol components only and not the individual aerosol models, results in the minimal size LUT needed to perform aerosol

Table 9.3. SMART atmospheric layer structure

Layer type	Layer base height (km)	Layer top height (km)	Layer scale height (km)
Rayleigh	0.0	705.	8.0
Components 1–14	0.0	10.0	2.0
Components 19 and 21	3.0	6.0	10.0

retrievals. To transition from RT parameters defined for aerosol components to parameters needed for mixture models, a modified form of linear mixing [Abdou et al., 1997] is applied on-the-fly at the desired retrieval locations. This technique is a modification of the standard linear mixing approach (see, e.g., Wang and Gordon [1994]) and provides a much more accurate result than standard linear mixing in situations where particles with substantially different absorption characteristics are present.

4.   Methodology for aerosol retrieval over land

The quality of the MISR aerosol retrieval product over land has improved considerably since it became operational. Much of the initial quality increase was due to improved filtering of the input top-of-atmosphere TOA data (e.g., using better cloud screening techniques) to provide clear sky conditions and leaving the actual mechanics of the retrieval algorithm virtually unchanged. This retrieval algorithm was based on a principal component analysis of the multispectral, multi-angle TOA MISR data from which empirical orthogonal functions (EOFs) in view angle were obtained to describe the directional reflectance properties of the surface [Martonchik et al., 1998a]. After three years of operation and considerable scrutiny of the retrieval results a major design change was implemented in January 2003 [Diner et al., 2005] This upgrade left the original algorithm intact, serving instead as an additional step in the processing chain. It functions as an independent retrieval algorithm, selecting its own set of acceptable aerosol models based on similarity of the angular shape of the spectral surface directional reflectance properties associated with each model. This subset of acceptable models is then used as a filtered input to the original EOF-based retrieval algorithm, a technique that considerably improves the quality of the final retrieval result. The theory behind these two retrieval steps, both requiring multi-angle data as a prerequisite for their implementation, is summarized in the following sections.

4.1   Basics of the surface reflectance angular shape similarity retrieval algorithm

This algorithm has its heritage in pioneering work performed in implementing the Along-Track Scanning Radiometer (ATSR)-2 experiment. ATSR-2 is a conical scanning radiometer that measures TOA radiance at two view zenith angles, one near nadir and the other at approximately $55°$. Flowerdew and Haigh [1995] suggested that the ratio of the surface reflectance at the two angles should be nearly spectrally invariant because the surface scattering elements are much larger than the wavelengths of the scattered light. Aerosol retrieval algorithms based on this idea, created by a number of researchers using ATSR-2 data [Flowerdew and Haigh, 1996; Veefkind et al., 1998; North et al., 1999; North, 2002], yielded good comparisons with aerosol results from ground-based sunphotometers. The success of this technique prompted its adaptation to the multi-angle observations of MISR [Diner et al., 2005].

In the MISR adaptation, the surface directional reflectance is described in terms of the hemispherical-directional reflectance factor (HDRF), which is the directional surface reflectance, illuminated by both direct and diffuse ambient irradiance, ratioed to the reflectance from an ideal Lambertian target, illuminated under the same conditions [Shaepman-

Strub et al, 2006; Martonchik et al., 2000]. However, a more accurate description of the surface reflectance, and one which is in better agreement with the physics of angular shape invariance, is the bi-directional reflectance factor, defined analogously to that of the HDRF but the surface illumination is restricted to direct (i.e., parallel beam) irradiance only [Shaepman-Strub et al., 2006; Martonchik et al., 2000]. Assuming surface spectral BRFs have strict angular shape invariance, the corresponding spectral HDRFs generally would not have this condition (see, e.g., [Shaepman-Strub et al., 2006]). This is due to the downward diffuse radiance having different directional properties across the visible spectrum, interacting with the surface whose reflectance can be strongly dependent on illumination angle. Both strongly wavelength-dependent Rayleigh scattering and large aerosol optical depths contribute to the angular variability and strength of the diffuse radiance. The rationale for the MISR approach of using the HDRF instead of the BRF is threefold. First, it is more complex, involving additional time and computer resources, to remove the effects of the diffuse field to arrive at the surface spectral BRFs [Martonchik et al.,1998b]. This would have to be done for every aerosol model being tested in this stage of the retrieval process, and covering a wide range of optical depths. Second, it is not clear how precise the spectral angular shape invariance principle is for natural surfaces. For example, it seems intuitive that the surface albedo would have some influence on the invariance principle. Multiple scattering taking place within the surface structure should have a greater effect on reflectance in spectral bands with high surface albedo compared to reflectance in spectral bands with low albedo, so the angular distribution of the reflected radiance would be expected to differ from that of single scattering from the surface. Lastly, the goal of this MISR algorithm is to find and eliminate in an efficient and effective manner those aerosol models and optical depths that are grossly at odds with the angular shape invariance principle. This has been achieved with remarkable success in the MISR operational aerosol retrieval process, in spite of the identified simplifications [Diner et al., 2005].

The spectral reflectance shape invariance is expressed in the algorithm as

$$r_\lambda(-\mu,\mu_0\phi - \phi_0) = a_\lambda f(-\mu,\mu_0\phi - \phi_0), \tag{2}$$

where r_λ is the spectral HDRF at wavelength λ, f is the invariant (wavelength-independent) angular shape function and a_λ is a wavelength-dependent scaling parameter. The independent parameters include cosine of the view zenith angle $-\mu$, cosine of the solar zenith angle μ_0 and relative azimuth angle $\phi - \phi_0$. An explicit but approximate expression for r_λ can be written as

$$r_\lambda(-\mu,\mu_0,\phi - \phi_0) = \frac{L_\lambda^{\mathrm{MISR}}(-\mu,\mu_0,\phi - \phi_0) - L_\lambda^{\mathrm{atm}}(-\mu,\mu_0,\phi - \phi_0)}{[\exp(-\tau_\lambda/\mu) + t_\lambda^{\mathrm{diff}}(-\mu)]L_\lambda^{\mathrm{lam}}(\mu_0)}, \tag{3}$$

where $L_\lambda^{\mathrm{MISR}}$ is the TOA radiance, L_λ^{atm} is the atmospheric path radiance, τ_λ is the total atmospheric optical depth, $t_\lambda^{\mathrm{diff}}$ is the upward diffuse, azimuthally integrated, atmospheric transmittance and L_λ^{lam} is the upward radiance at the surface reflected from a Lambertian target. Dividing r_λ by the camera average value $\langle r_\lambda \rangle_{\mathrm{cam}}$, this normalized HDRF, $r_{\mathrm{camnorm},\lambda}$ is given by

$$r_{\mathrm{camnorm},\lambda}(i) = \frac{r_\lambda(i)}{\langle r_\lambda(i) \rangle_{\mathrm{cam}}} = \frac{[L_\lambda^{\mathrm{MISR}}(i) - L_\lambda^{\mathrm{atm}}(i)]/[\exp(-\tau_\lambda/\mu_i) + t_\lambda^{\mathrm{diff}}(i)]}{\langle [L_\lambda^{\mathrm{MISR}} - L_\lambda^{\mathrm{atm}}]/[\exp(-\tau_\lambda/\mu) + t_\lambda^{\mathrm{diff}}] \rangle_{\mathrm{cam}}}, \tag{4}$$

where the explicit camera geometry, $-\mu, \mu_0, \phi\phi_0$, is replaced by the camera index i. All the atmospheric parameters in this equation are MISR aerosol model RT parameters contained in the SMART dataset. It is presumed that, if the correct aerosol model has been selected, $r_{\text{camnorm},\lambda}$ will be wavelength independent, as stipulated by Eq. (2). For a given aerosol model the wavelength independence condition may be measured by means of a fit function, X^2_{angular}, defined as

$$X^2_{\text{angular}}(\text{model}, \tau_{\text{green}}) = \frac{\sum_\lambda w_\lambda \sum_i q_i [r_{\text{camnorm},\lambda}(i) - \langle r_{\text{camnorm}}(i)\rangle_\lambda]^2}{(0.05)^2 \sum_\lambda w_\lambda \sum_i q_i}, \qquad (5)$$

where $\langle r_{\text{camnorm}}\rangle_\lambda$ is the spectral band average of $r_{\text{camnorm},\lambda}$, w_λ are band weights and q_i are camera weights. As implemented, bands 1–4 (blue, green, red, and near-IR) are assigned w_λ values of 4, 3, 2, and 1, respectively, indicating higher weights for the shorter wavelengths. Similarly, the camera weights, q_i, are set to the view angle reciprocal cosine, $1/\mu$, taking advantage of the greater atmospheric sensitivity in the oblique views.

An alternative view of the surface constraint implied by Eq. (2) is to work with the normalized HDRF, $r_{\text{bandnorm},\lambda}$, defined as the HDRF, r_λ, divided by its band averaged value $\langle r\rangle_\lambda$,

$$r_{\text{bandnorm},\lambda}(i) = \frac{r_\lambda(i)}{\langle r(i)\rangle_\lambda}. \qquad (6)$$

In this formulation the normalized HDRF should be independent of angle. For a given aerosol model, this independence can be measured as a fit function, X^2_{spectral}, defined as

$$X^2_{\text{spectral}}(\text{model}, \tau_{\text{green}}) \frac{\sum_\lambda w_\lambda \sum_i q_i [r_{\text{bandnorm},\lambda}(i) - \langle r_{\text{bandnorm},\lambda}\rangle_{\text{cam}}]^2}{(0.05)^2 \sum_\lambda w_\lambda \sum_i q_i}, \qquad (7)$$

where $\langle r_{\text{bandnorm},\lambda}\rangle_{\text{cam}}$ is the camera average value of $r_{\text{bandnorm},\lambda}$ and weights w_λ and q_i, as defined by Eq. (5).

The aerosol model and optical depth that minimizes either Eq. (5) or Eq. (7), in principle, would be the 'best estimation' of the actual aerosol conditions. The algorithm allows a weighted linear combination, X^2_{shape}, of the two fit functions, i.e.,

$$X^2_{\text{shape}}(\text{model}, \tau_{\text{green}}) = \beta X^2_{\text{angular}} + (1 - \beta) X^2_{\text{spectral}} \qquad (8)$$

in which the parameter β is set to 0.5.

4.2   Basics of the principal components retrieval algorithm

This algorithm has undergone little change since its introduction as the algorithm of choice for operationally retrieving aerosol properties over land using MISR data [Martonchik et al., 1998a, 2002]. Like the surface reflectance shape similarity retrieval algorithm described in the previous section, this algorithm is intrinsically multi-angle in nature and can only be used with this type of observational data. It assumes that the atmosphere is laterally homogeneous over the region of retrieval interest (currently 17.6×17.6 km for the standard MISR aerosol product) and requires that there be sufficient surface contrast contained within the 256 1.1×1.1 km sub-regions defining this region. Under these conditions a principal component analysis (PCA) (see, e.g., Preisendorfer [1988]) in the surface directional reflectance properties can be performed on the MISR TOA radiance data, which results in constraints on the atmospheric path radiance (and the associated aerosol models).

For a given sub-region located at x, y the MISR radiance field $L_{x,y,\lambda}^{\text{MISR}}$ at wavelength λ can be written as

$$L_{x,y,\lambda}^{\text{MISR}}(-\mu,\mu_0,\phi-\phi_0) = L_\lambda^{\text{atm}}(-\mu,\mu_0,\phi-\phi_0) + L_{x,y,\lambda}^{\text{surf}}(-\mu,\mu_0,\phi-\phi_0), \qquad (9)$$

where L_λ^{atm} is the atmospheric path radiance (independent of x, y) and $L_{x,y,\lambda}^{\text{surf}}$ is that component of the TOA radiance which contains all the surface–atmosphere interactions. When the TOA radiance of a particular sub-region within the region is selected as a bias radiance, $L_{\text{bias},\lambda}^{\text{MISR}}$, and this radiance is subtracted from every other sub-region TOA radiance in the region, a scatter matrix C_λ can be constructed which is not explicitly dependent on the unknown atmospheric path radiance L_λ^{atm}:

$$\begin{aligned} C_\lambda(i,j) &= \sum_{x,y} [L_{x,y,\lambda}^{\text{MISR}}(i) - L_{\text{bias},\lambda}^{\text{MISR}}(i)][L_{x,y,\lambda}^{\text{MISR}}(j) - L_{\text{bias},\lambda}^{\text{MISR}}(j)] \\ &= \sum_{x,y} [L_{x,y,\lambda}^{\text{surf}}(i) - L_{\text{bias},\lambda}^{\text{surf}}(i)][L_{x,y,\lambda}^{\text{surf}}(j) - L_{\text{bias},\lambda}^{\text{surf}}(j)] \end{aligned} \qquad (10)$$

Here, Eq. (9) was used, again replacing the camera angular parameters $-\mu$, μ_0, $\phi - \phi_0$ with a single camera index i or j. The camera index nominally runs from 1 through 9, each identified with an individual camera, but the indexing range can be less if certain cameras are not available (e.g., due to cloud contamination). The summation is over the usable sub-regions within the region, where a usable sub-region is defined as having clear sky radiances for each available camera view This requirement implies a usable sub-region location mask in x, y which is identical for each camera. As a choice for a bias sub-region, the MISR operational algorithm Version 22 and earlier uses the darkest sub-region as observed by the nadir (An) camera in the green band. Following standard principal component procedure, the eigenvectors (aka EOFs) and eigenvalues of matrix C_λ are then used to explicitly describe $L_{x,y,\lambda}^{\text{surf}}$, i.e.

$$L_{x,y,\lambda}^{\text{surf}}(i) = L_{\text{bias},\lambda}^{\text{surf}}(i) + \sum_n A_{x,y,\lambda}^n f_{n,\lambda}(i), \qquad (11)$$

where $f_{n,\lambda}$ are the eigenvectors and $A_{x,y,\lambda}^n$ are expansion coefficients for the sub-region location at x, y. The contribution of an individual eigenvector in describing the angular shape of $L_{x,y,\lambda}^{\text{surf}}$, is determined by the relative size of its eigenvalue. The eigenvectors are ordered such that their corresponding eigenvalues decrease monotonically. Therefore, only those eigenvectors with eigenvalues greater than a prescribed magnitude are used in the summation in Eq. (11) to effectively describe $L_{x,y,\lambda}^{\text{surf}}$. From Eq. (9), the MISR radiance $L_{x,y,\lambda}^{\text{MISR}}$ can now be written as

$$L_{x,y,\lambda}^{\mathrm{MISR}}(i) = L_{\lambda}^{\mathrm{atm}}(i) + L_{\mathrm{bias},\lambda}^{\mathrm{surf}}(i) + \sum_{n} A_{x,y,\lambda}^{n} f_{n,\lambda}(i). \tag{12}$$

Averaging $L_{x,y,\lambda}^{\mathrm{MISR}}$ over all the usable sub-regions,

$$\langle L_{\lambda}^{\mathrm{MISR}}(i) \rangle_{x,y} = L_{\lambda}^{\mathrm{atm}}(i) + L_{\mathrm{bias},\lambda}^{\mathrm{surf}}(i) + \sum_{n} \langle A_{\lambda}^{n} \rangle_{x,y} f_{n,\lambda}(i) = L_{\lambda}^{\mathrm{atm}}(i) + \sum_{n} B_{\lambda}^{n} f_{n,\lambda}(i). \tag{13}$$

Note that $L_{\mathrm{bias},\lambda}^{\mathrm{surf}}$ is incorporated into the eigenvector summation term, under the assumption that it can be adequately expressed in that representation. Experience using a PC analysis on MISR data generally indicates that only the first two eigenvectors in this summation are needed to express 95 % of the variation in $L_{x,y,\lambda}^{\mathrm{surf}}$ in Eq. (11), as dictated by the eigenvalues.

Equation (13) embodies the PCA aspect of the aerosol retrieval algorithm. On the left side of the equation is the average MISR radiance over a 17.6 km region. It is fitted via the expression on the right side, where $L_{\lambda}^{\mathrm{atm}}$ is the atmospheric path radiance for an aerosol model previously selected as a successful retrieval candidate using the surface reflectance shape similarity algorithm, and the coefficients B_{λ}^{n} are determined by a least-squares fit:

$$B_{\lambda}^{n} = \sum_{i} [\langle L_{\lambda}^{\mathrm{MISR}}(i) \rangle_{x,y} - L_{\lambda}^{\mathrm{atm}}(i)] f_{n,\lambda}. \tag{14}$$

This expression for B_{λ}^{n} is particularly simple and easily computed due to orthonormality of the eigenvectors $f_{n,\lambda}$. Defining $L_{\lambda}^{\mathrm{model}}$ as the collection of terms on the right side of Eq. (13), a fit function, X_{het}^{2}, is calculated as

$$X_{\mathrm{het}}^{2}(\mathrm{model}, \tau_{\mathrm{green}}) = \frac{\displaystyle\sum_{\lambda} \sum_{i} v_{\lambda}(i) \frac{[\langle L_{\lambda}^{\mathrm{MISR}}(i) \rangle_{x,y} - L_{\lambda}^{\mathrm{model}}(i)]^{2}}{\sigma_{\mathrm{het},\lambda}^{2}(i)}}{\displaystyle\sum_{\lambda} \sum_{i} v_{\lambda}(i)}. \tag{15}$$

where the weight $v_{\lambda} = 1$ if a valid value of $\langle L_{\lambda}^{\mathrm{MISR}} \rangle_{x,y}$ exists; otherwise $v_{\lambda} = 0$. The parameter $\sigma_{\mathrm{het},\lambda}$ is defined as

$$\sigma_{\mathrm{het},\lambda}(i) = 0.05 \ \max[\langle L_{\lambda}^{\mathrm{MISR}}(i) \rangle_{x,y}, 0.04(E_{0,\lambda}/\pi)], \tag{16}$$

where $E_{0\lambda}$ is the solar irradiance at wavelength λ. Multiplying radiance by π and dividing by the TOA solar irradiance defines an equivalent reflectance.

After testing all 74 aerosol models and traversing the total range of model optical depth (0.0 to 3.0 in the green band) for each, the retrieval process in a region is considered to be successful if there is at least one model at an optical depth (green band) greater than zero, which has a minimum X_{het}^{2} less than a prescribed upper limit. If more than one model passes this minimum X_{het}^{2} criterion, the retrieved effective aerosol particle properties and associated spectral optical depths are estimated through an appropriate model averaging procedure [Martonchik et al., 1998a].

5. MISR cloud screening over land

Cloud screening is a discrimination process whereby pixels in an image are classified as being either contaminated by cloud or clear of cloud. This is a critical step in the aerosol retrieval process both over land and ocean, striving to provide assurance that only clear pixels are used in the analysis. Methods for identifying clouds are generally based on radiance threshold, atmospheric model, or statistical techniques making use of spectral and textural features in the imagery (see, e.g., Rossow et al. [1985], Goodman and Henderson-Sellers [1988] and Rossow [1989]). Radiance threshold techniques work on an individual pixel basis, using predefined single or multiple thresholds applied to radiometrically based observables, to differentiate between clear and cloudy pixels. Atmospheric model techniques use one or more spectral radiance measurements as input to an atmospheric radiative transfer or structure model and retrieve a physical quantity such as cloud optical thickness or altitude. The pixels are then classified as clear or cloudy based on thresholds in the retrieved quantity. Statistical techniques use groups of adjacent pixels and methods based on spatial coherence between adjacent pixels, neural networks, maximum likelihood decision rules, and clustering routines.

MISR operational processing generates three separate and unique cloud masks, each employing a markedly different discrimination technique, and all three masks are used to select the clear pixels to be used by the aerosol retrieval algorithm.

5.1 Radiometric camera-by-camera cloud mask (RCCM)

This cloud mask is generated for each of the nine MISR cameras at 1.1 km resolution and is based on the radiance threshold technique. This is a challenging problem for MISR given the small number of spectral channels available, none of which have wavelengths longer than 1 μm. As a result, only a few cloud detection observables can be constructed just using simple arithmetic operations on the camera radiances as practically mandated by time constraints on the operational process. The few observables that are used by the RCCM to determine clear versus cloudy conditions depends on whether the observations are made over water or land. In the case of cloud detection over water the two observables are the TOA BRF in band 4 (near IR at 866 nm) at 1.1 km resolution, R_4, and the standard deviation of the 4×4 array of 275 m band 3 (red at 672 nm) BRFs within a 1.1 km area, σ_3. Here, the band 4 TOA BRF can be written as

$$R_4(i) = \pi L_4^{\mathrm{MISR}}(i)/\mu_0 E_{0,4} \tag{17}$$

where $E_{0,4}$ is the band 4 solar irradiance. The dependence of R_4 on camera is explicitly expressed by the camera index i. Two observables are used over land, namely the D parameter and the σ_3 parameter. The D parameter [Di Girolamo and Davies, 1995] is defined as

$$D(i) = \frac{|\mathrm{NDVI}(i)|^b}{\langle R_3(i) \rangle^2} \tag{18}$$

where $\langle R_3 \rangle$ is the average of the 275 m TOA BRFs in band 3 within a 1.1-km pixel area and NDVI is the Normalized Difference Vegetation Index defined as

$$\text{NDVI}(i) = \frac{R_4(i) - R_3(i)}{R_4(i) + R_3(i)}. \tag{19}$$

The parameter b in Eq. (18) is chosen so as to maximize the distinction between clear and cloudy pixels [DiGirolamo and Davies, 1995]. The choice of b also tends to maximize the spatial variability of clear-sky D-values, thus allowing statistical cloud-detection techniques to be effective. It, however, depends on the underlying surface type. The surface classification used by the RCCM is that of the Cloud Screening Surface Classifications (CSSC) dataset, which is based on the WE1.4D version of Olsen's global ecosystem database [NOAA-EPA Global Ecosystems Database Project, 1992], and has the surface divided into $158°$ contiguous surface types. The other observable, σ_3, is a type of standard deviation defined as

$$\sigma_3(i) = \sqrt{\frac{1}{n}\sum_{p=1}^{n}[R_{3,p}^{275}(i) - \langle R_3^{275}(i)\rangle]^2} \tag{20}$$

where $\langle R_3^{275}\rangle$ is the mean value of the usable number n of 275 m band 3 BRFs, $R_{3,p}^{275}$, of the 4×4 matrix comprising a 1.1 km pixel.

The thresholds associated with the two observables are a function of surface type, solar zenith angle, relative azimuth angle and view zenith angle. They are derived using an automated threshold selection algorithm on MISR TOA data that is suitably fast for use in operational processing [Yang et al., 2007]. There are three thresholds for each of the two observables with those of the D parameter updated seasonally for land classes since surface conditions can vary. The thresholds for spring 2008, for example, are different from those derived for spring 2007, to account for inter-annual variation. Threshold T_1 divides a high-confidence cloudy condition (CloudHC) from a low-confidence cloudy condition (CloudLC), T_2 divides CloudLC from a low-confidence clear-sky condition (ClearLC), and T_3 divides ClearLC from a high-confidence clear-sky condition (ClearHC). Failure for any reason to retrieve any of these four conditions results in a label of no retrieval (NR). Thus, the RCCM contains the classification of each 1.1 km pixel for each camera labeled with one of these five designations.

5.2   Stereoscopically derived cloud mask (SDCM)

This is a single 1.1-km resolution mask (unlike the camera-dependent RCCM), labeling a pixel as cloudy based on a retrieval of the altitude of the radiance-reflecting layer (surface or cloud) within that pixel. Due to the multiple camera configuration of MISR, cloud-motion vectors and cloud-top heights can be determined using a purely geometrical technique that requires locating the same cloud features at different viewing angles. To this end, fast stereo-matching algorithms have been developed to perform this camera-to-camera image matching automatically on an operational basis [Moroney et al., 2002]. Cloud-top height is generally obtained with an accuracy of ± 562 m, even over snow and ice, for clouds that are optically thick enough to have identifiable features in multiple, angular images. This accuracy limitation stems from the lack of sub-pixel acuity in the stereo-matching algorithms. In the operational default case of matching with the nadir and Af/Aa cameras, a one pixel difference in the calculated disparity (observed difference in feature location between cameras) leads directly to a 562 m height difference. Other camera combinations can theoretically produce better height accuracies but they tend to be more computationally expensive on an operational basis. The influence of winds is an important consideration in the stereo-matching process since the perceived parallax (relative shifting of feature locations in two camera views with respect to the ground) can be a combination of true parallax due strictly to geometry and an actual physical shift in a cloud location due to winds. Therefore, the first step in computing reflecting layer heights is to retrieve the wind vector at that height, but it is a difficult process that only succeeds about half the time. For those areas where the wind retrieval succeeded and passed its quality assessment, a wind-corrected height is calculated, but this field usually contains large gaps in coverage. So, a less accurate second version is also calculated, where wind effects are ignored but having almost complete coverage. It is this 'without winds' height field product that is used to generate the WithoutWinds SDCM used in the aerosol-retrieval process.

The 'WithoutWinds' reflecting layer height field provides the information to label pixels, using a five-label height classification which parallels the five-label cloud/clear classification of the RCCM. This reflecting layer height mask labels the reflecting layer as being above the surface with high confidence (AboveSurfHC), above the surface with low confidence (Above SurfLC), near the surface with low confidence (NearSurfLC), near the surface with high confidence (NearSurfHC), or no retrieval (NR), based on various image-matching quality factors associated with the stereo-matching process. The final step in generating the SDCM is a comparison of the RCCM with the reflecting layer height mask to validate, but never override, the height mask classification, resulting in the SDCM labels of CloudHC, CloudLC, NearSurfaceLC, NearSurfaceHC, and NR. Since the SDCM is derived solely from the retrieved height of image features (without incorporating, for example, radiometric information), it cannot be determined with sufficient confidence that a pixel is clear, only that the retrieved reflecting layer height is near the surface.

5.3   Angular signature cloud mask (ASCM)

This is a single 1.1-km resolution mask (like the SDCM), labeling a pixel as cloudy or clear depending on an analysis of its band-differenced angular signature (BDAS) [Di Girolamo and Davies, 1994; Di Girolamo and Wilson, 2003]. A BDAS is the TOA radiance difference in two sufficiently separated, non-absorbing, visible solar spectral bands as a function of view angle. For MISR cloud detection purposes the BDAS can be defined as

$$BDAS(i) = R_1(i) - R_{3,4}(i), \tag{21}$$

where R_1 is the TOA BRF in band 1 (blue at 446 nm). R_4 is used in Eq. (21) when observing over snow- and ice-free and R_3 is used over snow and ice-free land. The basis of the technique is that the BDAS is sensitive to the relative concentration of Rayleigh scattering to the total reflectance. Bands 1 and 4 differ by almost a factor of 15 in their sensitivity to Rayleigh scattering, while bands 1 and 3 differ by more than a factor of 5. Since the contribution of Rayleigh scattering decreases in the presence of high clouds, the resulting changes in the BDAS enable detection of such clouds. Over much of the snow- and ice-free land surface, if bands 1 and 4 were used by the BDAS, its sensitivity to clouds would generally be hampered by the large difference in land reflectance in these bands. This is why bands 1 and 3 are used by the BDAS for these types of surface conditions. This band choice has the advantage of reducing the difference in surface spectral reflectance but also the disadvantage of reducing the difference in the Rayleigh scattering contribution to the two spectral TOA BRFs. For snow and ice surface conditions, however, the use of bands 1 and 4 by the BDAS works extremely well in detecting clouds since the surface reflectances in these bands are comparable. This capability provides needed relief from the difficulties experienced by the RCCM when performing under snow and ice surface conditions.

The BDAS technique, like other cloud-detection techniques, requires a threshold that distinguishes between clear and cloudy pixels. The observable in this case is Ψ, defined as

$$\Psi(i,j) = \frac{[\text{BDAS}(i) - \text{BDAS}(j)]}{\theta(i) - \theta(j)}, \tag{22}$$

where θ is the scattering angle. Ψ is simply the slope of the BDAS with scattering angle, determined by the selection of two cameras i and j. The two cameras nominally used in the MISR processing are the fore camera pair, Cf and Df, or the aft pair, Ca and Da, the choice depending on whether the observations are made in the Northern or Southern Hemisphere, respectively [Di Girolamo and Wilson, 2003] This distinction is made because maximum slope sensitivity is obtained with the BDAS between cloud and surface if the radiation is scattered predominately in the forward direction, i.e., if the C and D cameras have a scattering angle $\theta \leq 90°$. There are three thresholds associated with the observable Ψ, similar in concept to those used by the RCCM. Threshold Ψ_1 divides a high-confidence cloudy condition (CloudHC) from a low confidence cloudy condition (CloudLC), Ψ_2 divides CloudLC from a low confidence clear-sky condition (ClearLC), and Ψ_3 divides ClearLC from a high-confidence clear-sky condition (ClearHC). Implementation of Eq. (22) and associated thresholds on a 1.1-km pixel generates the ASCM product. Failure for any reason to retrieve any of these four cloud/clear conditions results in a label of no retrieval (NR) in the ASCM. Like the RCCM thresholds, the ASCM thresholds also depend on view zenith angle, relative azimuth angle, solar zenith angle and surface type but, unlike the RCCM thresholds, they are currently static.

5.4 Aerosol retrieval cloud mask

The aerosol retrieval process uses these three cloud masks to decide which 1.1-km resolution pixels are suitable for use by the retrieval algorithm. When over ice- and snow-free land, only the RCCM and SDCM masks are considered. For this case Table 9.4 shows the matrix used to decide the classification of a pixel for a given camera. A label of Cloud eliminates that pixel being used in the retrieval. When over snow- or ice-covered terrain all three cloud masks are used to decide the classification of a pixel. Table 9.5 shows the pixel classification matrix for this case.

The processing of MISR imagery using cloud masks is actually the first step in a three-step process to determine which pixels are considered suitable as input to the aerosol retrieval algorithm. These masks are generated external to the aerosol retrieval process, as

Table 9.4. Ice/snow-free land cloud classification matrix

		RCCM				
		No Retrieval	CloudHC	CloudLC	ClearLC	ClearHC
	No Retrieval	Clear	Clear	Clear	Clear	Clear
SDCM	CloudHC	Clear	Cloud	Cloud	Clear	Clear
	CloudLC	Clear	Cloud	Cloud	Clear	Clear
	NearSurfaceLC	Clear	Clear	Clear	Clear	Clear
	NearSurfaceHC	Clear	Clear	Clear	Clear	Clear

Table 9.5. Ice/snow-covered land cloud classification matrix

ASCM = No Retrieval		RCCM				
		No Retrieval	CloudHC	CloudLC	ClearLC	ClearHC
SDCM	No Retrieval	Cloud	Cloud	Cloud	Clear	Clear
	CloudHC	Cloud	Cloud	Cloud	Cloud	Cloud
	CloudLC	Cloud	Cloud	Cloud	Cloud	Cloud
	NearSurfaceLC	Clear	Cloud	Cloud	Clear	Clear
	NearSurfaceHC	Clear	Cloud	Cloud	Clear	Clear

ASCM = CloudHC or CloudLC		RCCM				
		No Retrieval	CloudHC	CloudLC	ClearLC	ClearHC
SDCM	No Retrieval	Cloud	Cloud	Cloud	Cloud	Cloud
	CloudHC	Cloud	Cloud	Cloud	Cloud	Cloud
	CloudLC	Cloud	Cloud	Cloud	Cloud	Cloud
	NearSurfaceLC	Cloud	Cloud	Cloud	Cloud	Cloud
	NearSurfaceHC	Cloud	Cloud	Cloud	Cloud	Cloud

ASCM = ClearLC or ClearHC		RCCM				
		No Retrieval	CloudHC	CloudLC	ClearLC	ClearHC
SDCM	No Retrieval	Clear	Cloud	Cloud	Clear	Clear
	CloudHC	Cloud	Cloud	Cloud	Cloud	Cloud
	CloudLC	Cloud	Cloud	Cloud	Cloud	Cloud
	NearSurfaceLC	Clear	Cloud	Cloud	Clear	Clear
	NearSurfaceHC	Clear	Cloud	Cloud	Clear	Clear

part of the MISR Level 2 TOA/Cloud Product. But two additional data-filtering steps are initiated within the aerosol retrieval process itself to further eliminate contaminated pixels. These are the angle-to-angle smoothness and correlation evaluations.

5.5 Angle-to-angle smoothness evaluation

This is a test on each 1.1-km sub-region within the 17.6-km region to insure that the radiance field is 'smooth' as a function of view angle [Martonchik et al., 2002]. Besides additional clouds somehow missed by the MISR cloud masks, other undesirable features, e.g., glitter, may be detected by this test. It is applied separately to the four forward-viewing directions and nadir, and to the four aftward-viewing directions and nadir, and for each spectral band. The smoothness test consists of fitting the radiance values to a polynomial with one less degree of freedom than the number of cameras and within a specified tolerance. Failure of any of the eight smoothness tests causes the sub-region to be eliminated (all bands and all cameras) from the aerosol retrieval process.

5.6 Angle-to-angle correlation evaluation

The correlation test is designed to detect features, e.g., small clouds, which result in poor correlation of the radiance spatial distribution within a 1.1-km sub-region from one view angle to another [Martonchik et al., 2002]. It makes use of the 275 m data in band 3 (red), routinely available for all nine cameras in MISR Global Mode operation, producing a 4×4 array of radiances for each sub-region. The template used to correlate with each camera view 4×4 array is a simple average of the nine MISR individual camera arrays. If the correlation fails to fall within a prescribed tolerance for any camera, the sub-region (all bands and all cameras) is eliminated from the aerosol retrieval process.

It may seem that the use of three cloud masks and two angle-to-angle radiance evaluation tests to detect clouds and other undesirable features is more than sufficient to guarantee that only clear-sky pixels are used in the aerosol retrieval process. But operational results for aerosol optical depth, for example, show that 'blunders' do occasionally occur, indicating that current cloud filtering procedures are still not completely foolproof, both in the sense that some cloud, especially thin, uniform cirrus, may leak through, and that in some cases, thick smoke or dust aerosol is identified as cloud.

6. Aerosol retrieval results using MISR

The Atmospheric Data Center at NASA Langley has been generating MISR aerosol products over both ocean and land since the start of the EOS Terra mission. Retrieval results are available as a Level 2 product that retains the native swath geopositions at the original 17.6 km resolution and also in a spatially and temporally gridded version by latitude and longitude at a coarser resolution of $0.5°$ (Level 3). Fig. 9.2 shows a year and season matrix of Level 3 aerosol optical depth data covering the region of the African continent. Aerosols in northern Africa are primarily dust, and in central and southern Africa are primarily from biomass burning, though these components are frequently mixed in the atmospheric column. Note the good spatial coverage of aerosols over the bright Sahara desert as well as elsewhere in Africa. An overall consistency in the aerosol patterns from year to year for a given season is apparent.

A large amount of work has gone into assessing the quality of the MISR aerosol product, over both land and ocean. This validation effort has relied heavily on the use of sunphotometer data obtained by the ground-based Aerosol Robotic Network (AERONET), with sites scattered over much of the Earth's surface (Holben et al., 1998). A comparison of results from other satellite instruments has also been informative, helping to uncover potential biases in the retrieved aerosol properties which may be present and propagated to higher-level products. Field campaigns, during which coincident aircraft, surface and sa-

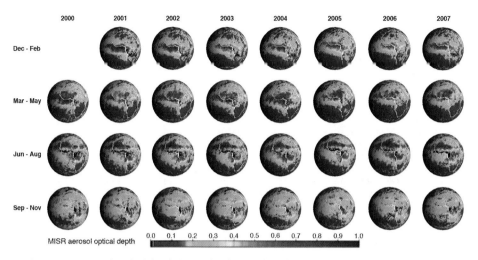

Fig. 9.2. MISR aerosol optical depth (green band) over the African continent as a function of year and season.

tellite observations were made, are also proving especially valuable for testing and refining MISR aerosol microphysical property retrievals. In this section the validation results are reviewed and selected study results are shown, illustrating how the MISR aerosol product over land can be used to investigate a variety of important aerosol phenomena.

A number of studies have been carried out to understand how well the MISR retrieval process has been performing over land. The first studies concentrated on aerosol optical depth (AOD) only, comparing MISR results to those from AERONET [Diner et al., 2001; Christopher and Wang, 2004; Liu et al., 2004a; Martonchik et al., 2004; Kahn et al., 2005; Jiang et al., 2007]. A similar study by Abdou et al. [2005] also included results from the Terra MODIS instrument. This MODIS–MISR comparison showed that MODIS AOD values over land were biased high compared to those from MISR and AERONET especially for AOD smaller than about 0.2. Overall, the results can be summarized by the statement that most of the MISR AOD values fall within 0.05 or 0.2 × AOD of the corresponding AERONET AOD values and more than a third are within 0.03 or 0.10 × AOD (see, e.g., Kahn et al. [2005]). Also, the studies have shown that when the AOD is larger than about 0.4 and when the aerosols are likely to be moderately to strongly absorbing, there is a tendency for the MISR values to be biased low (see, e.g., Kahn et al. [2005] and Jiang et al. [2007]). A global summary of the MISR-AERONET comparison extending from March 2001 through 2006 is shown in Fig. 9.3. The label above each plot denotes the aerosol type that typically (though not always) dominates at the sites included.

The Ångström exponent a is another retrieved aerosol parameter that can be validated with AERONET data. It is crudely related to particle size, with small particles having large values of a (Rayleigh scatterers have $a = 4$). The value of a can be determined from the spectral trend of the aerosol optical depth τ_λ as modeled by the expression

$$\tau_\lambda = \tau_{\lambda_0} \left[\frac{\lambda}{\lambda_0} \right]^{-a} \tag{23}$$

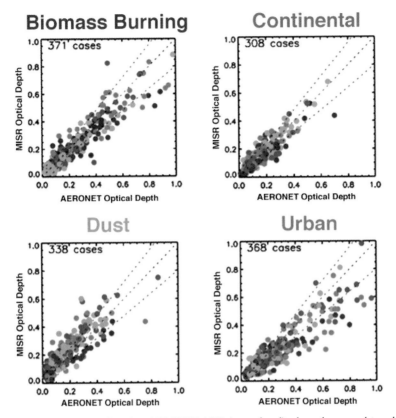

Fig. 9.3. MISR AOD regressed against AERONET AOD (green band) where the aerosol type has been segregated into four distinct classes. The circle color indicates season (brown = DJF, blue = MAM, green = JJA, orange = SON; MISR aerosol retrieval algorithm versions 16–20 are included in this comparison).

where λ_0 is a reference wavelength usually situated within the spectral range to be covered by a. Fig. 9.4 shows the MISR Ångström exponents associated with the AOD values plotted in Fig. 9.3, compared with the AERONET values. The smaller colored circles in Fig. 9.4 indicate associated AOD values less than 0.15. These cases have Ångström exponent values that are less strongly correlated with AERONET exponents than those cases with larger AOD values. This result is not unexpected since smaller aerosol amounts will have lower detection sensitivities to aerosol properties.

6.1 Particulate air quality study results

Air pollution can damage the health of human populations, primarily through respiratory and cardiovascular diseases. Fine airborne particulate matter with a diameter $< 2.5\,\mu m$ ($PM_{2.5}$) is a 'criteria pollutant' identified for monitoring under the US National Ambient Air Quality Standards (NAAQS). NAAQS implementation for $PM_{2.5}$, however, has encountered delays resulting from slow deployment of a nationwide *in situ* monitoring network, issues pertaining to how spatial averaging among monitors should be performed, and

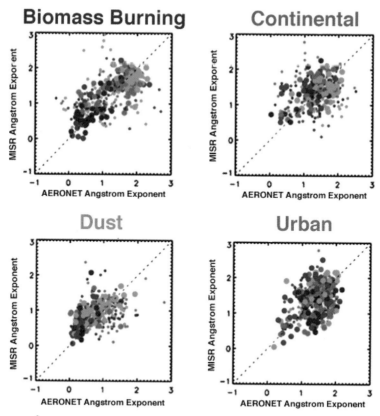

Fig. 9.4. MISR Ångström exponents regressed against AERONET values for the cases displayed in Fig. 9.3. The circle color indicates season (brown = DJF, blue = MAM, green = JJA, orange = SON). MISR aerosol retrieval algorithm versions 16–20 are included in this comparison.

contention over the contribution of non-local pollution sources [Esworthy, 2004]. Development of a satellite-based $PM_{2.5}$ measurement system could alleviate many of these issues and would also extend monitoring to many parts of the world where surface sensors do not exist.

MISR efforts are aimed at addressing how well retrievals of column AOD can be used to estimate surface $PM_{2.5}$ and to determine how much particulate pollution is imported from remote sources. Liu et al. [2004b, 2005, 2007a] have shown that an empirical regression model using MISR AOD and a select few geographical and meteorological parameters is capable of estimating the 24-hour average $PM_{2.5}$ concentrations at the surface to better than 50 %. A predictive relationship between surface $PM_{2.5}$ concentration and AOD was developed which relies on a global chemistry and transport model to provide a better physical basis for relating MISR total column AOD to the spatial and temporal variability of the $PM_{2.5}$ concentration. This technique was significantly improved [Liu et al., 2007b, 2007c] by using the fractional AOD contributed by each component of the MISR aerosol models instead of the total (i.e., combined) AOD. This information, coupled with simulated aerosol vertical profiles and detailed aerosol chemical speciation from a chemistry and transport model (GEOS-CHEM), allowed an estimate of the fractional anthropogenic

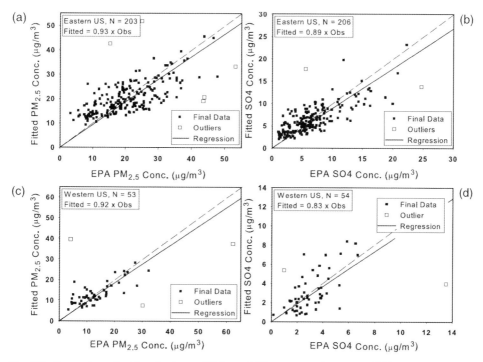

Fig. 9.5. Comparison of retrieved values of $PM_{2.5}$ and SO_4 concentrations versus EPA measurements in the eastern and western United States. From Liu et al. [2007c].

AOD in the lower atmosphere and were used to predict ground-level concentrations of total $PM_{2.5}$ mass and major particle species. Fig. 9.5 shows a comparison of the retrieved total $PM_{2.5}$ mass to the EPA-measured mass and the retrieved SO_4 concentration component to the EPA-measured SO_4 concentration for a number of sites across the United States in 2005. The data are separated into eastern and western regions because significant corrections for dust concentrations were made for EPA sites in the western region. The results show that $PM_{2.5}$ mass and SO_4 concentrations can be predicted reasonably well in both the eastern and the western regions compared with available EPA ground-truth. It is expected that further improvements can be made by the use of more mature MISR aerosol particle property and retrieval products and the analysis of larger EPA datasets, stratified by season and location.

6.2 Plume studies

Some of the difficulty in implementing pollution attainment standards is that not all pollution is local. Transported smoke can affect the surface $PM_{2.5}$ concentrations resulting in a violation of air-quality standards even if local air-quality measures are being followed. Also, long-range transport of smoke can affect the AOD-$PM_{2.5}$ relationship when the transported smoke is above the boundary layer, if the smoke vertical distribution is not properly taken into account. MISR can contribute in several ways, making use of stereoscopic parallax between the multi-angle views [Moroney et al., 2002; Muller et al., 2002]

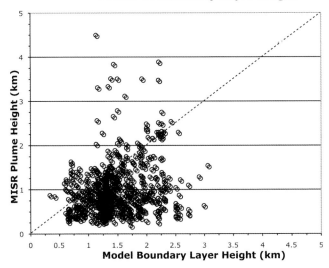

Fig. 9.6. MISR plume versus model-derived boundary layer heights above the terrain for 664 plumes in the Alaska-Yukon region, summer 2004. From Kahn et al. [2008].

to retrieve aerosol plume heights. Injection heights can be established using multi-angle stereo data when the aerosol plume has well defined spatial contrasts, providing initialization conditions for the modeling of wildfire, volcanic and dust aerosols near their sources and constraints on the physics of plume formation [Kahn et al., 2007]. Moreover, stereoscopic heights, AOD and physical and optical properties of transported aerosols downwind of sources provide validation information for transport models (e.g., Chen et al. [2007]). A study of smoke plumes in Alaska and the Yukon during the summer of 2004 has been performed to determine the frequency of occurrence of aerosol plume injection above the boundary layer [Kahn et al., 2008], using the MISR Interactive eXplorer (MINX) analysis tool to retrieve plume heights with greater detail than possible with the MISR standard stereo product. Fig. 9.6 shows the results for more than 600 identified plumes, indicating that at least 10 % of the plumes injected smoke above the boundary layer. Extension to other seasons and locations is under way to establish the robustness of this result.

7. Discussion and conclusion

The MISR instrument has been very successful in its ability to retrieve aerosol properties over a wide variety of surface types, including desert. The algorithms used to accomplish this task rely on a couple of relatively simple but important properties of surface reflectance that allow the separation of atmospheric path radiance from surface-reflected radiance in the TOA radiance measurements of MISR. These reflectance properties, namely the similarity of the angular shape of the surface BRF at different wavelengths (spectral invariance) and the limited variability of the surface BRF angular shape generally found within region-sized (10–20 km) areas (thus, the need for only a couple of eigenvectors in the MISR PCA description), usually provide sufficient constraints to retrieve aerosol properties over land. It is not clear, however, that the MISR implementation of these constraints, as described in this chapter, is the optimum way to proceed. Since the future trend of aerosol information from satellites will be in the direction of higher spatial resolution and more and better quality aerosol properties, it would be desirable to retire the PCA part of the MISR algorithm. The PCA requires a multi-angle and multi-pixel dataset, with the multi-pixel surface area having sufficient contrast to extract useful EOFs. Therefore, this need for a multi-pixel area implies a necessary reduction in the spatial resolution of the aerosol properties being retrieved, compared to the intrinsic spatial resolution of an individual pixel. The algorithm employing invariance of the surface spectral reflectance angular shape, however, has no multi-pixel restrictions and theoretically can be used at the highest spatial resolution of the multi-angle dataset. In fact, single pixel usage would seemingly be the best option since the assumptions inherent to the invariance principle should be best satisfied when the surface area being investigated is reasonably homogeneous. There is a need for further validation of the basic assumptions defining the invariance principle. Its general validity for all surface types encountered in remote sensing, and the degree to which its validity depends on spatial resolution, need to be confirmed before an algorithm of this type can play a stand-alone role in a future update to the MISR operational aerosol retrieval process.

In an effort to resolve these issues the MISR aerosol science team has initiated a study that uses both MISR and AERONET data to directly investigate AERONET site surface directional reflectance properties. A large number of different AERONET sites were chosen, each spanning the full range of seasonal change, to allow a large variety of surface types to be investigated in the study. Surface spectral HDRFs and BRFs are retrieved with spatial resolution ranging from 1.1 km to 17.6 km, employing the MISR operational surface retrieval algorithm [Martonchik et al., 1998b] and MISR TOA data. Instead of using the MISR retrieved aerosol properties as is normally done, however, the AERONET retrieved aerosol properties obtained during the time of MISR overpass are used, under the assumption that this aerosol information will be more accurate than that obtained from MISR. In addition only those overpass times where the AERONET AOD (green band) was 0.2 or less are being studied, since the surface reflectance retrieval is inherently more accurate for these smaller aerosol amounts. Therefore, it is reasonable to assume that the retrieved HDRFs and BRFs obtained using this procedure can be considered as a fairly accurate representation of the true surface reflectance properties. An example of retrieved BRFs are shown in Fig. 9.7 for the Avignon (France), AERONET site (the site picture indicates agricultural field surrounded by scattered trees) in Fall of 2000 at 1.1-km resolution. Also shown are three BRF model curves fitted to the retrieved values. One curve, labeled 'common-shape average', is for the model described by Eq. (2). The curve labeled 'common-shape mRPV' is described by the three-parameter modified Rahman–Pinty–Verstraete BRF model [Rahman et al., 1993; Martonchik et al., 1998b] whereby the two parameters defining the directional shape are the same for all four spectral bands. The curve labeled 'H-function' is described by the Hapke BRF model [Hapke, 1981] and is the only model of the three in which single and multiple scattering within the surface structure are treated separately. In this latter model the surface reflectance has an angular shape that depends on the single scattering albedo of the surface structure, resulting in spectral BRFs that are not precisely directionally invariant.

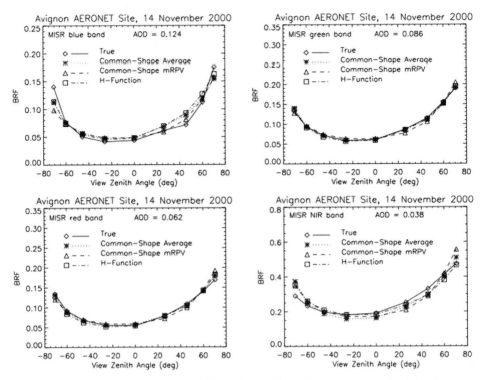

Fig. 9.7. Retrieved MISR spectral BRFs (labeled True) at the Avignon AERONET site. The other three curves are from semi-empirical BRF models described in the text. Note that the AERONET-derived AOD (green band) is only 0.086.

For this particular example the invariance of the directional shape of the reflectance is readily apparent with all three models providing reasonably good fits. Whether this type of behavior can be expected in general will be determined as the study progresses. Nevertheless, initial results suggest that surface spectral reflectance angular shape invariance appears to be a robust property of a wide variety of surfaces. There remains a need for improvements relative to how this constraint is implemented within the MISR standard retrieval but it seems certain that this particular surface reflectance phenomenon will play a major role in future space-based aerosol and surface remote sensing missions involving multi-angle observations.

Acknowledgments. We wish to thank Catherine Moroney and Larry Di Girolamo for helpful comments on various parts of the manuscript. This research is being carried out, in part, at the Jet Propulsion Laboratory, California Institute of Technology, under contract with NASA. Support of the NASA Radiation Sciences Program is also gratefully acknowledged.

References

Abdou, W.A., J.V. Martonchik, R.A. Kahn, R.A. West and D.J. Diner, 1997: A modified linear mixing method for calculating atmospheric path radiances of aerosol mixtures. *J. Geophys. Res.* **102**, D14, 16883–16888.

Abdou, W.A., D.J. Diner, J.V. Martonchik, C.J. Bruegge, R.A. Kahn, B.J. Gaitley and K.A.S. Crean, 2005: Comparison of coincident Multiangle Imaging Spectroradiometer and Moderate Resolution Imaging Spectroradiometer aerosol optical depths over land and ocean scenes containing Aerosol Robotic Network sites. *J. Geophys. Res.* **110**, D10S07, doi:10.1029/2004JD004693.

Andreae, M.O., 1995: Climate effects of changing atmospheric aerosol levels, in *Future Climate of the World: A Modeling Perspective. World Survey and Climatology*, A. Henderson-Sellers Ed.: Amsterdam Elsevier.

Charlson, R.J., S.E. Schwartz, J.M. Hales, R.D. Cess, J.A. Coakley, Jr, J.E. Hansen and D.J. Hoffman, 1992: Climate forcing by anthropogenic aerosols. *Science* **255**, 423–430.

Chen, Y., Q. Li, J. Randerson, E. Lyons, D. Nelson, D. Diner and R. Kahn, 2007: Improved temporal constraints on and vertical injections of biomass burning emissions: implications on global aerosol simulation. Eos Trans. AGU 88(52), Fall Meet. Suppl., Abstract A53C–1337.

Chen, W.-T., R. A. Kahn, D. Nelson, K. Yau and J.H. Seinfeld, 2008: Sensitivity to multi-angle imaging to the optical and microphysical propertiesof biomass burning aerosols. *J. Geophys. Res.*, in press.

Christopher, S., and J. Wang, 2004: Intercomparison between multi-angle imaging spectroradiometer (MISR) and sunphotometer aerosol optical thickness in dust source regions of China, implications for satellite aerosol retrievals and radiative forcing calculations. *Tellus*, Ser. B **56**, 451–456.

Chu, D.A., Y.J. Kaufman, C. Ichoku, L.A. Remer, D. Tanre and B. Holben, 2002: Validation of MODIS aerosol optical depth retrieval over land. *Geophys. Res. Lett.* **29**, doi:10.1029/2002GL013205.

Di Girolamo, L., and R. Davies, 1994: A band-differenced angular signature technique for cirrus cloud detection. *IEEE Trans. Geosci. Remote Sens.* **32**, 890–896.

Di Girolamo, L., and R. Davies, 1995:The image navigation cloud mask for the Multi-angle Imaging SpectroRadiometer (MISR). *J. Atmos. Oceanic Tech.* **12**, 1215–1228.

Di Girolamo, L., and M.J. Wilson, 2003: A first look at band-differenced angular signatures for cloud detection from MISR. *IEEE Trans. Geosci. Remote Sens.* **41**, 1730–1734.

Diner, D.J., J.C. Beckert, T.H. Reilly, C.J. Bruegge, J.E. Conel, R.A. Kahn, J.V. Martonchik, T.P. Ackerman, R. Davies, S.A.W. Gerstl, H.R. Gordon, J.-P. Muller, R.B. Myneni, P.J. Sellers, B. Pinty and M.M. Verstraete, 1998: Multi-angle Imaging SpectroRadiometer (MISR) instrument description and experiment overview. *IEEE Trans. Geosci. Remote Sens.* **36**, 1072–1087.

Diner, D.J., W.A. Abdou, J.E. Conel, K.A. Crean, B.J. Gaitley, M. Helmlinger, R.A. Kahn, J.V. Martonchik and S.H. Pilorz, 2001: MISR aerosol retrievals over southern Africa during the Safari-2000 dry season campaign. *Geophys. Res. Lett.* **28**, 3127–3130..

Diner, D.J., J.V. Martonchik, R.A. Kahn, B. Pinty, N. Gobron, D.L. Nelson and B.N. Holben, 2005: Using angular and spectral shape similarity constraints to improve MISR aerosol and surface retrievals over land. *Remote Sens. Environ.* **94**, 155–171.

Esworthy, R., 2004: Particulate matter (PM2.5): National Ambient Air Quality Standards (NAAQS) Implementation. Congressional Research Service, Order Code RL32431, 14 pp.

Flowerdew, R.J., and J.D. Haigh, 1995: An approximation to improve accuracy in the derivation of surface reflectances from multi-look satellite radiometers. *Geophys. Res. Lett.* **22**, 1693–1696.

Flowerdew, R.J., and J.D. Haigh, 1996: Retrieval of aerosol optical thickness over land using the ATSR-2 dual-look satellite radiometer. *Geophys. Res. Lett.* **23**, 351–354

Goodman, A.H., and A. Henderson-Sellers, 1988: Cloud detection analysis: a review of recent progress. *Atm. Res.* **21**, 203–228.

Grant, I.P., and G.E. Hunt, 1968: Solution of radiative transfer problems using the invariant Sn method. *Mon. Not. R. Astron. Soc.* **141** (1), 27–41.

Hapke, B., 1981: Bidirectional reflectance spectroscopy 1. Theory. *J. Geophys. Res.* **86**, 3039–3054.

Holben, B.N., T.F. Eck, I. Slutsker, D. Tanre, J.P. Buis, A. Setzer, E. Vermote, J.A. Reagan, Y.J. Kaufman, T. Nakajima, F. Lavenu, I. Jankowiak and A. Smirnov, 1998: AERONET – a federated instrument network and data archive for aerosol characterization. *Rem. Sens. Environ.* **66**, 1–16.

Hsu, N.C., S.-C. Tsay, M.D. King and J.R. Herman, 2006: Deep Blue retrievals of Asian aerosol properties during ACE-Asia. *IEEE Trans. Geosci. Remote Sens.* **44**, 3180–3195.

Jiang, X., Y. Liu, B. Yu and M. Jiang, 2007: Comparison of MISR aerosol optical thickness with AERO-NET measurements in Beijing metropolitan area. *Rem. Sens. Environ.* **107**, 45–53.

Kahn, R.A., P. Banerjee and D.McDonald, 2001: Sensitivity of multi-angle imaging to natural mixtures of aerosols over ocean, *J. Geophys. Res.* **106**, 18219–18238.

Kahn, R.A., B.J. Gaitley, J.V. Martonchik, D.J. Diner, K.A. Crean and B. Holben, 2005: Multiangle Imaging Spectroradiometer (MISR) global aerosol optical depth validation based on 2 years of coincident Aerosol Robotic Network (AERONET) observations. *J. Geophys. Res.* **110**, D10S04, doi:10.1029/2004JD004706.

Kahn, R.A., W.-H. Li, C. Moroney, D.J. Diner, J.V. Martonchik and E. Fishbein, 2007: Aerosol source plume physical characteristics from space-based multi-angle imaging. *J. Geophys. Res.* **112**, D11205,doi:10.1029/2006JD007647.

Kahn, R.A., Y. Chen, D.L. Nelson, F.-Y. Leung, Q. Li, D.J. Diner and J.A. Logan, 2008: Wildfire smoke injection heights – two perspectives from space. *Geophys. Res. Lett.* (in press).

Kalashnikova, O.V., R. Kahn, I.N. Sokolik and W.-H. Li, 2005: Ability of multi-angle remote sensing observations to identify and distinguish mineral dust types: optical models and retrievals of optically thick plumes, *J. Geophys. Res.* **110**, D18S14, doi:10.1029/2004JD004550.

Kalashnikova, O.V., and R.A. Kahn, 2006: Ability of multi-angle remote sensing observations to identiofy and distinguish mineral dust types: sensitivity over dark water. *J. Geophys. Res.* **111**, D11201, doi:10. 1029/2005JD006756.

Kaufman, Y.J., N. Gabron, B. Pinty, J.-L. Widlowski, and M. Verstraete, 2002: Relationship between surface reflectance in the visible and near IR used in MODIS aerosol algorithm-theory. *Geophys. Res. Lett.* **29**, doi:10.1029/2001GL04492.

Kiehl, J.T., and B.P. Briegleb, 1993: The relative role of sulfate aerosols and greenhouse gases in climate forcing. *Science* **260**, 311–314.

Liu, Y., J.A. Sarnat, B.A. Coull, P. Koutrakis and D.J. Jacob, 2004a: Validation of Multiangle Imaging Spectroradiometer (MISR) aerosol optical thickness measurements using Aerosol Robotic Network (AERONET) observations over the contiguous United States. *J. Geophys. Res.* **109**, D06205, doi:10.1029/2003JD003981.

Liu, Y., R.J. Park, D.J. Jacob, Q. Li, V. Kilaru and J.A. Sarnat, 2004b: Mapping annual mean ground-level PM$_{2.5}$ concentrations using Multiangle Imaging Spectroradiometer aerosol optical thickness over the contiguous United States. *J. Geophys Res.* **109**, D22206, doi:10.1029/2004JD005025.

Liu, Y., J.A. Sarnat, V. Kilaru, D.J. Jacob and P. Koutrakis, 2005: Estimating ground-level PM2.5 in the eastern United States using satellite remote sensing. *Environ. Sci. Technol.* **39**, 3269–3278.

Liu, Y., M. Franklin, R. Kahn and P. Koutrakis, 2007a: Using aerosol optical thickness to predict ground-level PM$_2$.5 concentrations in the St. Louis area: a comparison between MISR and MODIS. *Remote Sens. Environ.* **107**, 33–44.

Liu, Y., P. Koutrakis and R. Kahn, 2007b: Estimating fine particulate matter component concentrations and size distributions using satellite-retrieved fractional aerosol optical depth: Part 1 – Method development. *J. Air and Waste Manage. Assoc.* **57**, 1351–1359.

Liu, Y., P. Koutrakis, R. Kahn, S. Turquety and R.M. Yantosca, 2007c: Estimating fine particulate matter component concentrations and size distributions using satellite-retrieved fractional aerosol optical depth: Part 2 – A case study. *J. Air and Waste Manage. Assoc.* **57**, 1360–1369.

Martonchik, J.V., D.J. Diner, R.A. Kahn, T.P. Ackerman, M.M. Verstraete, B. Pinty and H.R. Gordon, 1998a: Techniques fot the retrieval of aerosol properties over land and ocean using multi-angle imagery. *IEEE Trans. Geosci. Remote Sens.* **36**, 1212–1227.

Martonchik, J.V. D.J. Diner, B. Pinty, M.M. Verstraete, R.B. Myneni, Y. Knyazikhin and H.R. Gordon, 1998b: Determination of land and ocean reflective, radiative, and biophysical properties using multi-angle imaging. *IEEE Trans. Geosci. Remote Sens.* **36**, 1268–12281.

Martonchik, J.V., C.J. Bruegge and A. Strahler, 2000: A review of reflectance nomenclature used in remote sensing. *Remote Sensing Reviews* **19**, 9–20.

Martonchik, J.V., D.J. Diner, K.A. Crean and M.A. Bull, (2002), Regional aerosol retrieval results from MISR. *IEEE Trans. Geosci. Remote Sens.* **40**, 1520–1531.

Martonchik, J.V., D.J. Diner, R. Kahn, B. Gaitley and B.N. Holben, 2004: Comparison of MISR and AERONET aerosol optical depths over desert sites. *Geophys. Res. Lett.* **31**, L16102, doi:10.1029/2004GL019807.

Moroney, C., R. Davies and J.-P. Muller, 2002: MISR stereoscopic image matchers: Techniques and results. *IEEE Trans. Geosci. Remote Sens.* **40**, 1547–1559.

Muller, J.-P., A. Mandanayake, C. Moroney, R. Davies, D.J. Diner and S. Paradise, 2002: Operational retrieval of cloud-top heights using MISR data. *IEEE Trans. Geosci. Remote Sens.* **40**, 1532–1546.

NOAA-EPA Global Ecosystems Database Project, 1992: Global Ecosystems Database Version 1.0 User's Guide, Documentation, Reprints and Digital Data on CD-ROM. U.S. DOC/NOAA National Geophysical Data Center, Boulder, CO, 36 pp.

North, P.R.J., 2002: Estimation of aerosol opacity and land surface bidirectional reflectance from ATSR-2 dual angle imagery: operational method and validation. *J. Geophys. Res.* **107**, D12, 4149, doi:10.1029/2000JD000207.

North, P.R.J., S.A. Briggs, S.E. Plummer and J.J. Settle, 1999: Retrieval of land surface bidirectional reflectance and aerosol opacity from ATSR-2 multi-angle imagery. *IEEE Trans. Geosci. Remote Sens.* **37**, 526–537.

Preisendorfer, R.W., 1988: *Principal Component Analysis in Meteorology and Oceanography.* Elsevier Science, New York.

Rahman, H., B. Pinty and M.M. Verstraete, 1993: Coupled surface–atmosphere reflectance (CSAR) model 2. Semiempirical surface model usable with NOAA Advanced Very High Resolution data. *J. Geophys. Res.* **98**, D11, 20791–20801.

Ramon, D., and R. Santer, 2001: Operational remote sensing of aerosols over land to account for directional effects. *Appl. Opt.* **40**, 3060–3075.

Rossow, W.B., F. Mosher, E. Kinsella, A. Arking, M. Desbois, E. Harrison, P. Minnis, E. Ruprecht, G. Seze, C. Simmer and E. Smith, 1985: ISCCP cloud algorithm intercomparison. *J. Clim. Appl. Met.* **24** (9), 887–903.

Rossow, W.B., 1989: Measuring cloud properties from space: a review. *J. Clim.* **2**, 201–213

Santer, R., V. Carrere, P. Dubuisson and J.C. Roger, 1999: Atmospheric correction over land for MERIS. *Int. J. Remote Sens.* **20**, 1819–1840.

Shaepman-Strub, G., M.E. Shaepman, T.H. Painter, S. Dangel and J.V. Martonchik, 2006: Reflectance quantities in optical remote sensing – definitions and case studies. *Remote Sens. Environ.* **103**, 27–42.

Veeffkind, J.P., G. de Leeuw and P. Durkee, 1998: Retrieval of aerosol optical depth over land using two-angle view satellite radiometry during TARFOX. *Geophys. Res. Lett.* **25**, 3135–3138.

Wang, M., and H.R. Gordon (1994), Radiance reflected from the ocean atmosphere system: synthesis from individual components of the aerosol size distribution. *Appl. Opt.* **33**, 7088–7095.

Yang, Y., L. Di Girolamo and D. Mazzoni, 2007: Selection of the automated thresholding algorithm for the Multi-angle Imaging SpectroRadiometer Camera-by-Camera Cloud Mask over land. *Remote Sens. Environ.* **107**, 159–171.

10　Polarimetric remote sensing of aerosols over land surfaces

Brian Cairns, Fabien Waquet, Kirk Knobelspiesse, Jacek Chowdhary, Jean-Luc Deuzé

1.　Introduction

Writing from the *S.S. Narkunda*, near Aden, C. V Raman noted [1921] that using a Nicol prism 'serves to cut off a great deal of the blue atmospheric "haze" which usually envelops a distant view, and mostly consists of polarized light.' Although the reason for the color and polarization of the sky had been explained some time before by J. W. Strutt [1871], later Lord Rayleigh, and the neutral points, where the polarization of the sky becomes zero, had already been named after their discoverers Arago [Barral, 1858], Babinet [1840] and Brewster [1842], this simple observation of Raman's was still considered noteworthy, because of the difference between the behavior of the object being observed and the haze. The reason for this difference is that light scattered by molecules and small aerosol is strongly polarized in a plane perpendicular to the scattering plane (the plane defined by the sun, the object being viewed and the observer) while light scattered by surfaces is only weakly polarized. Thus, when Raman oriented the polarizer to transmit light in the plane parallel to the scattering plane the contributions from light scattered by aerosols and molecules were suppressed while the lighthouse was made more visible (had more contrast). This difference between the polarizing properties of aerosols and molecules as compared to surfaces is used by modern polarimetric remote sensing instruments to determine the amount, size and type of aerosols that are present above the surface.

There are therefore three facets to the use of downward-looking polarimetric measurements for the remote sensing of aerosols over land surfaces. A method for the measurement of the polarization of the scene being observed, a quantitative model of the polarized reflectance behavior of the underlying surface and an understanding of what aspects of the aerosol loading, microphysics and vertical distribution are revealed in the polarization signal. This chapter is therefore organized as follows. In Section 2 we provide a brief review of the different instrumental approaches that have been taken to the Earth viewing measurement of polarization. This review focuses on the benefits and issues related to these measurement techniques with particular regard to their use over land. In Section 3 we summarize the conclusions that have been drawn in the existing literature related to the polarized reflectance of land surfaces and discuss how various modeling approaches, that can be used in remote sensing applications, take advantage of the polarized behavior of the surface. Although single scattering models of the polarized light reflected by the atmosphere–surface system of the Earth have been used with considerable success [Deschamps et al., 1994, Deuzé et al., 2001], when measurements are of sufficient accuracy, or extend into the blue, or ultraviolet spectral domain it is not sufficient to use such an approach for the modeling of the observed polarized radiances. In Section 4 we therefore describe how the land surface–atmosphere system can be modeled with different levels of fidelity depending on the accuracy and spectral domain of the measurements that are available. In Section 5 we review existing retrieval methods [Deuzé et al., 2001] and discuss how optimal methods [Rodgers, 2000; Dubovik and King, 2000] can be applied in the case of polarimetric remote sensing [Hasekamp and Landgraf, 2007, Lebsock et al., 2007]. In particular we emphasize the capability provided by the extended spectral range and more accurate measurements that are currently available in airborne polarimeters and are expected to be available in future spaceborne instruments.

2. Measuring polarization

In order to discuss how different instrument concepts implement the measurement of polarization we will define here the Stokes vector of light and how it is related to a simple measurement system [Hansen and Travis, 1974]. The intensity and polarization of light can be described by the Stokes vector $\boldsymbol{I}^T = (I;\ Q;\ U;\ V)$ where I is a measure of the intensity of the light, Q and U define the magnitude and orientation of the linearly polarized fraction of the light and V is a measure of the magnitude and helicity of the circular polarization. All four Stokes vector elements have the dimensions of intensity (e.g., W m^{-2} and have a simple relation to the time-averaged electric field of a superposition of transverse electromagnetic waves propagating in the same direction.

All four of the Stokes parameters describing a beam of light can be measured using a detector that is sensitive only to the intensity by first transforming the incident beam of light with a retarder and then using a polarizer to analyze the polarization state. The intensity that is observed by such a detector is given by the expression,

$$I(\theta,\delta) = \frac{1}{2}[I + Q\ \cos 2\theta + (U\ \cos\delta + V\sin\delta)\ \sin 2\theta], \tag{1}$$

in which δ is the relative phase delay between the electric vectors in orthogonal planes (E_x and E_y) and θ is the angle of rotation of the polarizer with respect to the x-plane. It is clear that by carefully choosing four measurement pairs δ and θ, the four Stokes parameters can be calculated. For the sake of discussing the instrumental issues involved in making polarization measurements it is worth examining a simple pair of measurements made by a detector with a polarizer in front of it that is oriented at $0°$ and then $90°$. Using Eq. (1) the relevant expressions are seen to be

$$\left.\begin{array}{l} I(90,0) = (I+Q)/2 \\ I(90,0) = (I-Q)/2 \end{array}\right\} \quad \Rightarrow \quad \left\{\begin{array}{l} I = [I(0°,0°) + I(90°,0°)]\ /2 \\ Q = [I(0°,0°) - I(90°,0°)]/2 \end{array}\right. . \tag{2}$$

The main difficulty in ensuring that the measurement of the Stokes vector component, Q, is accurate is that, for most Earth scenes, it is at least an order of magnitude smaller than the intensity, I. Operationally, in this case, it is therefore determined as the small difference between two large numbers. The reason for this is that the contribution to the Stokes vector element I by reflectance over land, is much larger than the contribution to the Stokes vector elements Q and U. This means that if there is a small error in either, or both, measurements there will be a large fractional error in the estimate of Q, even though the fractional error in the estimate of the intensity, I, might be quite small. Clearly if the $0°$ and $90°$ observations are made of slightly different scenes either through spatial mismatch, or a time delay between measurements there is the potential to create significant errors in the estimated Stokes vector elements and it is these errors that are usually called false polarization errors.

We have presented a very specific example here, but examining Eq. (1) it is apparent that any method for estimating Q, U and V will require the use of differences, or some form of modulation of a polarizer or a retarder and that in all of these schemes the intensity is a large static or invariant term that is the most readily estimated. The determination of Q, U and V is therefore a test of the capability of a polarimeter to accurately determine a small modulation, or difference, in the presence of a large background signal.

In the following subsections we describe polarimeter designs that use oriented polarizers [Deschamps et al., 1994; Cairns et al., 1999], temporal modulation of retarders [Diner et al., 2007] and spectral modulation induced by crystal elements [Jones et al., 2004] in order to measure polarization. In order to fully exploit the information content available from polarimetric observations it also necessary to observe the same scene from multiple angles and we therefore also briefly note how the various sensors using these polarimetric analysis methods achieve this multi-angle sampling. Although we make no claims for the completeness of this set of polarimeters they do represent the three basic methods that have been used, or are planned, for use in measurements of polarization from aircraft, or satellites. We also note that since the amount of sunlight that is circularly polarized by reflection off natural objects, aerosols, or molecules is negligibly small the focus of the Earth observing instruments developed to measure polarization has been on the first three Stokes parameters I, Q and U [Kawata, 1978]. The interactions of incident sunlight with the surface–atmosphere system can therefore be modeled using the upper left three by three block of the atmosphere–surface Mueller reflection matrix [Hansen and Travis, 1974].

2.1 Oriented polarizers

The two instruments that have provided most of the downward-looking measurements of the polarization of the Earth and its atmosphere and that have been used in the remote sensing retrieval of aerosol properties are the Polarization and Directionality of the Earth Reflectance (POLDER) sensor [Deschamps et al., 1994] and the Research Scanning Polarimeter (RSP) [Cairns et al., 1999]. The POLDER sensor has flown on three satellite missions, ADEOS I, ADEOS II and PARASOL while the RSP has flown on four different types of aircraft in a number of large field experiments. The NASA Glory mission [Mishchenko et al., 2007], scheduled for launch in June 2009, will also carry a polarimeter, the Aerosol Polarimetry Sensor (APS), which uses the same conceptual design as the RSP. Both the POLDER and RSP use oriented polarizers to analyze the polarization state of incident radiation. However the way in which they do this is substantially different with the POLDER instrument using sequential rotation of polarizers while the RSP measures the intensity in four polarization orientations in multiple spectral bands simultaneously. There are benefits and costs to each approach and we therefore describe them separately.

2.1.1 Sequential measurement

The simplest way to estimate the first three Stokes vector elements of a scene is to rotate a polarizer in front of a detector. This method is the basis of the POLDER instrument that measures the linear polarization state of light in three spectral bands at 443, 673 and 865 nm using polarizers that are sequentially oriented at $0°$, $60°$ and $120°$. The POLDER instrument also makes intensity only measurements in a number of other bands and provides images of the scene being viewed on a two-dimensional (2D) charge-couple device (CCD) array. The Stokes parameters I, Q and U can be determined from the three sequential polarizer measurements using the formulae

$$I = \frac{2}{3}[I(0°,0°) + I(60°,0°) + I(120°,0°)],$$

$$Q = \frac{2}{3}\{[I(0°,0°) - I(60°,0°)] + [I(0°,0°) - I(120°,0°)]\}, \tag{3}$$

$$U = \frac{2}{\sqrt{3}}[I(60°,0°) - I(120°,0°)].$$

The frame rate of POLDER is sufficiently high that multiple images are captured of every pixel providing multi-angle views of the same scene. However, as the satellite flies over the Earth the scene that is being viewed moves across the focal plane array as the polarizers with different orientations sequentially analyze the scene being viewed. In order to minimize the errors that are caused by this motion, the POLDER sensor uses a motion-compensation plate to provide the best possible spatial matching of the different scenes that are viewed with the polarizers in their different orientations. This is difficult to achieve over the full field of a wide-angle camera such as POLDER and not even possible if there are actual time variations in the scene being observed, such as when there are moving clouds within a pixel. As we note above, false polarization is a concern for sequential, or spatially mismatched measurements, and the consequent uncertainty in the estimates of Q and U that POLDER provides are expected to be of the order of 1–2 % larger than for I over heterogeneous surfaces [Hagolle et al., 1999], with better accuracy possible over homogeneous surfaces such as the open ocean. Nonetheless the POLDER instrument has demonstrated that an instrument of this kind has a valuable capability to retrieve aerosol properties over land and one that is not as sensitive to the underlying surface as sensors that only use reflectance measurements [Deuzé et al., 2001; Fan et al., 2008], as we will discuss in Section 4.

2.1.2 Simultaneous measurements

Wollaston prisms are optical elements that when illuminated by a collimated incident beam provide angular separation, and consequently spatial separation, of the beam into orthogonal polarization states. They can therefore be used, when combined with an optical system such as a relay telescope, to measure two orthogonal polarization states of the same scene simultaneously, eliminating false polarization from the estimate of the Stokes vector parameters. The RSP uses a refractive relay telescope to define the field of view and provide collimated illumination of a Wollaston prism that spatially separates beams with orthogonal polarization states that then illuminate paired detectors. By making measurements with one telescope in which the Wollaston prism is oriented to analyze orthogonal polarization states of $0°$ and $90°$ (see Eq. (2)) and with a second telescope in which the Wollaston prism is oriented to analyze orthogonal polarization states of $45°$ and $135°$ the RSP allows all three Stokes parameters, I, Q and U, to be determined simultaneously. Wollaston prisms are broadband optical elements and this allows the RSP design to use each telescope to make measurements in three spectral bands that are separated by dichroic beam splitters. There are a total of six telescopes in the RSP. Three of the telescopes measure the Stokes parameters I and Q in nine spectral bands, while the other three telescopes measure the Stokes parameters I and U in the same nine spectral bands providing simultaneous measurements of I, Q and U in all nine spectral bands. The nine spectral bands are at 410, 470 (443), 555, 670, 865, 960 (910), 1590, 1880 (1378) and 2250 nm. The three spectral band centers in parentheses are those that will be used for the Aerosol Polarimetry Sensor that are different from the RSP. The band at 443 nm will be used for APS because it is better for ocean color estimation, is darker over land and is less affected by trace gas absorption than the RSP 470 nm band. The band at 910 nm will be used for APS because it provides a broader dynamic range for water vapor estimates than the RSP band at 96 nm and allows for better detector performance when using silicon detectors. Both the 1378 and 1880 nm bands are extremely effective for screening for thin cirrus clouds, but in the event of a volcanic eruption the shorter wavelength 1378 nm band would allow for better detection and characterization of stratospheric aerosols and so this band was chosen for the APS. The instantaneous field of view (14 mrad) of each telescope is scanned continuously, with data being taken over a view-angle range of $120°$ ($\pm\,60°$ from nadir), using a polarization-insensitive scan mirror system. This system consists of two mirrors each operating with a $45°$ angle of incidence with their planes of incidence oriented orthogonally. This ensures that the polarization orientation that is perpendicular to the plane of reflection at the first mirror is parallel to the plane of reflection at the second mirror so that all polarization states are transmitted equally. The scan is oriented along the aircraft, or satellite, ground track in order to provide multiple views of the same scene from multiple angles. The RSP also incorporates a calibration system that allows the relative responsivity of the detectors measuring orthogonal polarization states to be tracked continuously allowing a polarimetric accuracy of better than 0.2 % [Cairns et al., 1999] to be achieved independent of the scene that is being viewed.

2.2 Temporal modulation

As can be seen from Eq. (1) varying the retardance of an element that is then followed by a polarizer at a fixed angle also allows the incident polarization state of a beam of light to be analyzed. A practical implementation of this concept, that is somewhat more complicated than the simple system modeled by Eq. (1), is to use a photo-elastic modulator (PEM) to provide a time-varying retardance that is placed between a pair of quarter-wave plates such that the combination acts as a time-varying circular retarder [Chipman, 1994]. The circular retarder modulates Q and U such that if it is followed by an appropriately oriented polarization analyzer and detector the temporal modulation of the detector signal is well represented by the formula,

$$I(0°,\delta) = \frac{1}{2}(I + Q\ \cos\delta + U\ \sin\delta). \tag{4}$$

In astronomical applications it is feasible to use high-speed demodulation at the resonant frequency (typically between 10 and 100 kHz) and harmonics of the PEM in order to determine Q and U using the different phase and harmonic content of their modulation [Keller, 2002; Povel et al., 1990]. For Earth remote sensing applications this is an extremely high speed at which to operate a focal plane, given that the frame rate of a push broom imager with a resolution of hundreds of meters in low Earth orbit is expected to be on the order of tens of milliseconds. An ingenious approach to imaging polarimetry for Earth-viewing satellite applications has therefore been proposed in which a pair of PEMs are used as the source of retardance modulation with the analysis of the signal being performed at the beat frequency [Diner et al., 2007]. In this method analyzers oriented at $0°$ and $45°$ are used to provide estimates of I, Q and U as indicated in the following equations:

$$I(0°,\delta_b) = 1/2[I + QJ_0(\delta_b)], \tag{5}$$

$$I(45°,\delta_b) = 1/2[I + UJ_0(\delta_b)],$$

in which J_0 is the zero-order Bessel function and the effective magnitude of the beat frequency retardance modulation, δ_b, is

$$\delta_b = 2\delta_0\ \cos(\omega_b t - \eta), \tag{6}$$

with ω_b being the beat frequency, η the phase difference between the modulating waveforms of the two PEMs, δ_0 is the magnitude of the retardance modulation of a single PEM and t is time. Although 'false'' polarization can contaminate the estimate of Stokes vector elements using this measurement approach it is expected to be a weak effect since only temporal variations in the scene that are similar to the terms modulating Q and U will alias into those elements. This polarimetric concept is being implemented as an imaging polarimeter in which a push broom imaging mode provides cross-track coverage while multi-angle views are obtained by the brute force approach of having individual cameras for each view angle. The polarimetric analysis concept underlying this measurement approach has been demonstrated by a laboratory prototype [Diner et al., 2007] but has not, thus far, been used for remote sensing measurements.

2.3 Spectral modulation

The final approach to remote sensing measurements of polarization that we review here has only been developed recently. It uses spectral modulation to encode the Stokes vector into an intensity measurement [Jones et al., 2004; Oka and Kato, 1999]. In this method a system of polarization analysis optics is inserted between the scene being viewed and an imaging spectrometer. These polarization analysis systems have the advantage, for implementation in a remote sensing system, that they have no moving parts. The way that they work is by imposing a variation on the incident spectrum that is rapid (hyperspectral) compared with the spectral variations of atmospheric aerosol and molecular scattering. This rapid variation depends on the Stokes vector, with one particular implementation being described by Eq. (7),

$$I(v) = 1/2[I + Q \ \cos(\phi)] + 1/4U[\cos(\Delta\phi) - \cos(\Sigma\phi)] + V[\sin(\Sigma\phi) - \sin(\Delta\phi)], \quad (7)$$

where the modulating terms in the sinusoids are

$$\begin{aligned}
\phi &= 2\pi v(n_e - n_o)l_2/c, \\
\Delta\phi &= 2\pi v(n_e - n_o)(l_1 - l_2)/c, \\
\Sigma\phi &= 2\pi v(n_e - n_o)(l_1 + l_2)/c,
\end{aligned} \qquad (8)$$

with l_1 and l_2 being the optical path lengths through two birefringent crystal elements that compose the polarization analysis optics and n_e and n_o being the refractive indices for the extraordinary and the ordinary rays in those crystals. The frequency and speed of light are represented by v and c respectively. Since the modulating terms are 'fast' compared to variations in atmospheric scattering the Stokes vector can be estimated within broader windows (typically of 10–20 nm) over the entire spectrum that is measured by the imaging spectrometer. Multi-angle views can be provided along the ground track of a plane, or satellite, by orienting the image direction of the imaging spectrometer in the direction of motion. In principle this is a robust and accurate way to estimate the Stokes vector since there are no moving parts and the Stokes vector elements are all encoded onto the spectrum simultaneously. These measurements are therefore inherently insensitive to 'false' polarization caused by temporal or spatial variations in the scene intensity that is observed being aliased into the estimated polarization state. Of course since the polarization state is modulated onto the spectrum, care must be taken to avoid, or correct for atmospheric absorption or scattering features that could themselves cause 'false' polarization [Jones et al., 2004]. In practice, as with all accurate measurement systems, great care needs to be taken to ensure the thermal and mechanical stability of the system so that it can be characterized and calibrated on the ground in a way that is applicable to the remote sensing measurements themselves. A remote sensing instrument that uses hyperspectral modulation (Hyperspectral Polarimeter for Aerosol Retrievals HySPAR) has been developed and successfully flown [Jones et al., 2004], and although the scientific analysis of the data obtained is still in its early phases the comparisons of the degree of linear polarization $(DoLP = \sqrt{(Q^2 + U^2)}/I)$ with simultaneous RSP measurements showed reasonable (within a few percent) agreement, given that this was one of the first flights of this sensor.

3.   Surface polarization

In the following discussion we will refer frequently to the polarized reflectance because it is a particularly useful quantity to use in discussing the polarizing properties of surfaces. This is a result of its simplicity, in terms of representing polarization properties, and also in understanding why those properties behave the way they do. The solutions to multiple scattering problems and the effects of surface reflection on arbitrary radiation fields can be conveniently expressed as reflection matrices, $\mathbf{R}$, each composed of four rows and four columns such that

$$\mathbf{I}_v(\mu_v, \varphi_v) = \frac{1}{\pi} \int_0^1 \mu_s \ d\mu_s \int_0^{2\pi} d\varphi_s \mathbf{R}(\mu_v, \mu_s; \varphi_s - \varphi_v) \mathbf{I}_s(\mu_s, \varphi_s), \qquad (9)$$

where μ and φ are the cosine of the zenith angle and the azimuth angle, with the subscripts s and v indicating that the angles referred to are the direction of the source radiation and the viewing direction respectively. It is assumed here that there is no preferred direction so that the only azimuthal dependence of the reflection matrices is on the difference between the solar and view azimuths, $\Delta\varphi = \varphi_s - \varphi_v$. The Stokes vector $\mathbf{I}_\eta$ describes the source of radiation incident from the top on the surface, or surface-atmosphere system, represented by its reflection matrix, $\mathbf{R}$, and v is the Stokes vector of the reflected radiation field that is being observed. For the purposes of this discussion the source of radiation is the sun which is essentially unpolarized, at least in terms of the average value across the solar disk, and for which the Stokes vector is well approximated by the expression $\mathbf{I}_s^T = [F_s\delta(\mu - \mu_s)\delta(\phi - \phi_s), 0, 0, 0]$ where the Greek delta symbol is used to represent the Dirac delta function and F_s is the solar irradiance at the top of the atmosphere. Substituting the particular behavior of the Stokes vector of the solar radiation into Eq. (9) yields the formula,

$$\mathbf{I}(\mu_v, \varphi_v) = \frac{\mu_s F_s}{\pi} \begin{pmatrix} R_{11}(\mu_v, \mu_s; \Delta\varphi) \\ R_{21}(\mu_v, \mu_s; \Delta\varphi) \\ R_{31}(\mu_v, \mu_s; \Delta\varphi) \\ R_{41}(\mu_v, \mu_s; \Delta\varphi) \end{pmatrix} \tag{10a}$$

and the complementary relations that define the reflectances in terms of the Stokes parameters

$$\begin{pmatrix} R_{11} \\ R_{21} \\ R_{31} \\ R_{41} \end{pmatrix} = \begin{pmatrix} \frac{\pi I}{\mu_s F_s} \\ \frac{\pi Q}{\mu_s F_s} \\ \frac{\pi U}{\mu_s F_s} \\ \frac{\pi V}{\mu_s F_s} \end{pmatrix}. \tag{10b}$$

Since the circular polarization is negligibly small [18] the magnitude of the polarized reflectance, R_p, can therefore be defined to be

$$R_p = \pm\sqrt{R_{21}^2 + R_{31}^2}, \tag{11}$$

which, like the usual definition of reflection, has the benefit that it is independent of the brightness of the illuminating source. This expression does not capture the orientation of the polarization, but for single scattering the polarization is either parallel to or perpendicular to the plane of scattering (the plane that contains the solar and viewing directions). Therefore, if the polarized radiance is defined to be positive when the polarization direction is perpendicular to the plane of scattering and negative when it is parallel to the plane of scattering, all of the information in the polarized reflectance is captured by the polarized reflectance defined in Eq. (9). It has also been found that, even for multiple scattering, rotating the reference frame for Q and U into the scattering plane causes U to become small, with the predominant information content of the observations being captured by Q. Thus, calculating the polarized reflectance using Eq. (11) with a sign that is assigned based on whether Q in the scattering plane is positive, or negative, captures the majority of the information in polarization measurements without recourse to multi-dimensional vectors. We do note, however, that the best accuracy can be retained, when actually performing calculations or retrievals, by using the Stokes vector elements themselves.

One of the main beauties of the Earth when viewed in remote sensing measurements from space, or aircraft, is its bright, many-hued underlying surface. This also presents one of the main difficulties in retrieving aerosol amounts and types using passive remote sensing measurements of the intensity over land surfaces, as noted elsewhere in this volume, since the background is brighter than our object of interest, the aerosols in the atmosphere. Although the polarized reflectance of the land surface at visible wavelengths is typically smaller than the signal from aerosols, it is clear that in order to use polarization measurements to provide an accurate determination of the type and amount of aerosols present in the atmosphere we need a quantitative understanding of the polarization properties of the land surface. Such a quantitative understanding, though not the subject of this discussion, would also be of use in remote sensing of the surface, being indicative of the texture of soils, or the leaf inclination distribution of vegetation. Fig. 10.1 shows an image of the different behavior of reflectance and polarized reflectance. The strong contrast between vegetated and bare fields shown by the reflectance image (Fig. 10.1, left panel) is absent in the polarized reflectance image (Fig. 10.1, right panel) which shows the dependence of

Fig. 10.1. False-color images created using RSP observations obtained on an aircraft flying at 3000 m above Oxnard and Ventura, California, USA. The red, green and blue colors are the 2250, 865 and 410 nm reflectances (left) and polarized reflectances (right) respectively.

molecular scattering on viewing geometry, decreasing from left to right with scan angle as the scattering angle increases from near 90° towards the backscattering direction.

The polarizing properties of natural surfaces as understood from ground- and aircraft-based measurements and observations from several space shuttle flights have been summarized by Coulson [1988]. This summary is still an excellent reference to the historical measurements that have been made of the polarizing properties of mineral and vegetated surfaces, although the tendency to present linear polarization measurements in terms of the *DoLP* makes some of this information difficult to use. This is because the *DoLP*, which is a ratio, mixes the effects of the polarization properties of the surface (numerator) and the reflection properties of the surface (denominator) that are to a large extent caused by the different mechanisms of surface and volume scattering respectively. There have also been a significant number of satellite measurements provided by the POLDER sensor flown on the ADEOS-I, -II and PARASOL missions, and airborne measurements made by the RSP sensor since Coulson's review was published. In this Section we will therefore present the understanding of the polarizing properties of the surface that has been developed based on these more recent measurements, together with the historical measurements, and how the surface polarized reflectance can therefore be modeled.

Our current understanding is that the polarization of surfaces is primarily generated by external reflections off the facets of soil grains, or the cuticles of leaves. A consequence of this surface polarization being generated by external reflections is that its magnitude will tend to be spectrally neutral as long as the real refractive index of the surface varies little. This is generally true of both minerals [Pollack et al., 1973] and the surface cuticles of

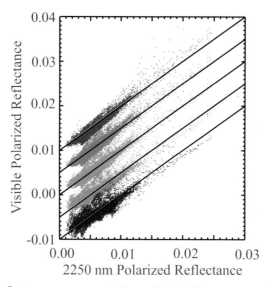

Fig. 10.2. Polarized reflectance measurements taken with the RSP sensor mounted on a small survey plane over agricultural land near Oxnard, California, USA, at an altitude of 3000 m. The visible polarized reflectances at 410, 470, 555, 670 and 865 nm are plotted against that at 2250 nm in blue, mauve, turquoise, green and red respectively. The bands are offset by 0.005 from one another to allow any spectral differences in behavior to be identified. The solid (1:1) lines show what is expected if the surface has a gray polarized reflectance.

vegetation [Vanderbilt et al., 1985; Rondeaux and Herman, 1991] and has been experimentally verified to be true for forests, bare soils, agricultural fields and urban landscapes over the spectral range from the deep blue to the infrared. Thus, although the surface reflectance is both colorful and spatially variable, the surface polarized reflectance is spatially variable but spectrally gray. It is this feature of the surface polarized reflectance that makes polarization measurements such a useful tool for aerosol retrievals over land. The key remaining questions regarding the surface polarized reflectance, if we are to use polarization for remote sensing of aerosols, are the predictability of its absolute value, for a particular surface, and its angular distribution.

Measurements taken with the RSP sensor over agricultural land containing bare soil and a range of different crops that are grown near Oxnard, CA (broccoli, peppers etc.) are shown in Fig. 10.2. This provides an example of the absence of spectral variability that is seen in the observed polarized reflectance for a wide range of viewing geometries and surface types. These measurements, which were obtained with the RSP installed in a small survey plane flying at an altitude of 3000 m, have been atmospherically corrected using simultaneous, collocated sunphotometer measurements. In Fig. 10.3 we show the angular variation of the polarized reflectance separately for soil (ploughed fields) and vegetation (soy bean, or winter wheat fields) in the Southern Great Plains of the United States (during the ALIVE field experiment) and over the Dismal Swamp in Virginia (during the CLAMS field experiment). Although the two surface types have different magnitudes of polarized reflectance they both have spectrally neutral polarized reflectance as evidenced in the residual differences between all the shorter wavelength channels and the longest RSP channel at 2250 nm shown in the lower parts of Figs 10.3(a) and (b). They also show a similar functional variation with view angle. These surface reflectance estimates

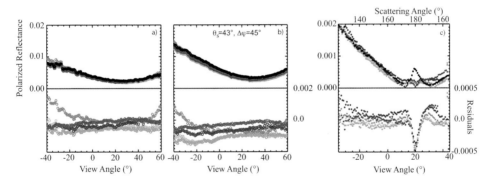

Fig. 10.3. Polarized reflectance measurements taken with the RSP sensor (mounted on the Sky Research Inc. BAE J-31 research plane) at an altitude of 200 m above the Southern Great Plains in Oklahoma (a) and (b) during the ALIVE field experiment. For (a) and (b) the solar zenith angle (θ_s) is 43° and the relative solar azimuth angle ($\Delta\varphi$) is 45°. The data in (c) was obtained with the RSP sensor on a Cessna 310 at 200 m over the Dismal Swamp in Virginia during the CLAMS field experiment; the solar zenith angle was 20° and the relative solar azimuth angle is less than 1°. The measurements were atmospherically corrected and have been separated into vegetated (a), bare soil (b) and (c) forested swamp surface types with the bands at 410, 470, 555, 670, 865, 1590 and 2250 shown in blue, mauve, turquoise, green, red, purple and black respectively. The residuals are the differences between the polarized reflectance at 2250 nm and the polarized reflectance in each of the other bands, using the same color scheme as used for the polarized reflectances figures.

are based on measurements that were taken at 200 m above the surface and that were corrected to the surface level using data from a nearby ground-based sunphotometer [Holben et al., 1998] and a sunphotometer on top of the aircraft [Redemann et al., 2006] to define the aerosol scattering and a microwave radiometer to define the amount of water vapor present. Other surfaces such as swamps, cityscapes and forests (not shown) show similar behavior with little spectral variation in the surface polarization.

These measurements show that if a long-wavelength measurement (e.g., 2200 nm) that is only weakly affected by the aerosol and molecular scattering is available then this can be used to characterize the polarized reflectance across the entire solar spectrum. A similar concept has been put forth for the retrieval of aerosols over land using radiance measurements in a more limited set of spectral bands [Kaufman et al., 1997]. The only limiting factor in using a long-wavelength measurement to characterize the surface is whether the observational viewing geometry is sufficient to predict the behavior of the surface polarization for other viewing geometries. This is primarily of concern for short wavelengths where the molecular and aerosol scattering is sufficiently strong that diffuse surface–atmosphere interactions do have to be modeled. When diffuse interactions are significant a substantial fraction of the radiation incident at the surface is not in the direction of the direct solar beam and a substantial fraction of the observed radiation was not reflected by the surface into the viewing angle. This is why a model of the surface is necessary for accurate forward modeling of the observed polarized reflectance at short wavelengths. In Fig. 10.4 we show an observation in the solar principal plane (the plane that contains the local vertical and the sun) that is used to estimate the parameters in a simple Fresnel model of the surface polarized reflectance, viz.

$$R_p^{Surf}(\mu_s, \mu_v, \Delta\varphi) = \zeta R_p^F(\gamma). \tag{12}$$

R_p^F is the Fresnel coefficient for polarized light calculated for a surface refractive index of $1.5.\gamma$ is the reflection angle that can be expressed as a function of the scattering angle $\Theta = \mathrm{acos}[-\mu_v\mu_s + \sqrt{(1 - \mu_v)^2(1 - \mu_s)^2}\cos(\phi_v - \phi_s)]$ by $\gamma = \pi - \Theta/2$ with ζ being a coefficient that provides the best fit to the measurements at 2250 nm in the solar principal plane. This model assumes that the surface does not have a preferred azimuth, as a ploughed field would have for example, and is therefore applicable to natural surfaces, or urban landscapes over a sufficiently large area that this assumption is valid. The model is then used to predict the polarized reflectance of this surface in a different scan plane from the original observations. It can be seen that the model prediction of the polarized reflectance for a different viewing geometry to that used to estimate the model is in good agreement with the observations in this other scan plane. The conclusion we draw is that a simple Fresnel model, fitted to observations of the surface (or long-wavelength observations), is sufficient to predict the angular variation of the surface polarized reflectance at all view angles that are not close to the backscatter direction.

When a long-wavelength measurement is not available it is necessary to predict the magnitude of the surface polarized reflectance using other observations, or prior information in order to use polarization measurements for remote sensing of aerosols. If the polarized reflectance of vegetation showed the same magnitude of variability as the polarized reflectance at Brewster's angle that is seen in single leaves [Grant et al., 1993], it unlikely that such an approach would work. However, the variability in the polarized reflectance

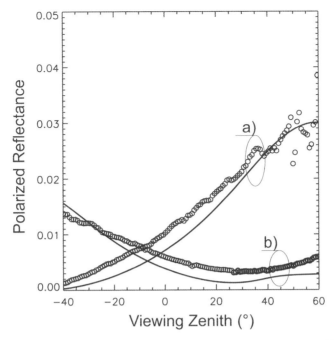

Fig. 10.4. Atmospherically corrected polarized reflectance measurements at 2250 nm with the data taken in meridional planes (a) close to the principal plane (b) at $45°$ to the principal plane.

between different plant types is not that great, which indicates that the macroscopic behavior of plant canopies, or ensembles is not well represented by measurements of individual leaves at a single angle. A more important predictor of the polarized reflectance of surfaces is the fractional coverage of vegetation and soil, since the differences between the polarized reflectance of soils and vegetation are larger than the variability within either class (see Fig. 10.2). This property of the surface polarized reflectance was used in an analysis of POLDER data [Nadal and Bréon, 1999] to generate a model that is only dependent on the surface type and the normalized difference vegetation index (NDVI) [Tucker, 1979]. The NDVI serves as a proxy for the fractional coverage of soil and vegetation, while the partitioning of the POLDER data into different surface types allows the model, viz.

$$R_p^{Surf}(\mu_s,\mu_v,\Delta\varphi) = a_0\left[1 - \exp\left(-\frac{\beta_0 R_p^F(\gamma)}{(\mu_s+\mu_v)}\right)\right],\tag{13}$$

to account for the fact that NDVI depends both on fractional coverage of vegetation and vegetation type by allowing the parameters a_0 and β_0 to be different for the various surface types of the International Geosphere–Biosphere Program classification [Belward et al., 1999]. This surface model has been successfully used in the operational processing of POLDER data for the retrieval of aerosol properties over land [Deuzé et al., 2001], although it has been found that the retrievals are affected by errors in the model prediction of surface polarized reflectance [Waquet et al., 2007]. This model limits the magnitude of polarized reflectance, at high view and solar zenith angles when the argument of the exponent in Eq. (13) is large to a_0. At lower view and zenith angles the exponent in Eq. (13) is small and the surface polarization can be approximated, by the expression

$$R_p^{Surf}(\mu_s, \mu_v, \Delta\varphi) \approx \frac{a_0 \beta_0 R_p^F(\gamma)}{(\mu_s + \mu_v)}, \tag{14}$$

which is similar to the vegetation model proposed by Breon et al. [1995] with a scaling factor of $a_0\beta_0$ that plays the same role as ζ in Eq. (12), although the dependence on view and zenith angles of Eqs (12) and (14) is different.

4.   Modelling atmosphere–surface interactions

In this Section we describe how the surface models introduced above can be incorporated into a radiative model of how the atmosphere and surface interact. It is forward models of this kind that are then used either in optimal estimation, or look-up-table (LUT) approaches, to perform the aerosol retrievals that are described in Section 5. In order to describe the main processes that generate the up-welling polarized reflectance, we provide in Eq. (15) an approximate expression for the surface–atmosphere Mueller reflection matrix, viz.

$$\mathbf{R}^{Atm+Surf} = \mathbf{R}^{Atm} + t_+\mathbf{R}^{Surf}t_- + \left[\mathbf{T}_+^d\mathbf{R}^{Surf}t_- + t_+\mathbf{R}^{Surf}\mathbf{T}_-^d + \mathbf{T}_+^d\mathbf{R}^{Surf}\mathbf{T}_-^d\right], \tag{15}$$

in which we have suppressed the dependence of the matrices on the viewing geometry $(\mu_s, \mu_v\varphi_s - \varphi_v)$ and wavelength for the sake of clarity. The plus (minus) subscripts refer to a downward (upward) direction. $\mathbf{R}^{Atm}$ and $\mathbf{R}^{Surf}$ are respectively the atmospheric and surface reflection matrices. $\mathbf{T}^d$ corresponds to the diffuse transmission matrix of the atmosphere and t is a direct transmission term given by

$$t_{-,+} = \exp\left[-\left(\frac{\tau_{a,\lambda} + \tau_{r,\lambda}}{\mu_{s,v}}\right)\right], \tag{16}$$

where $\tau_{a,\lambda}$ and $\tau_{r,\lambda}$ are respectively the aerosol and molecular total optical thicknesses at a particular wavelength λ.

Eq. (15) separates the terms that contribute to the observed reflectance into three distinct components. The first term $\mathbf{R}^{Atm}$ describes the contribution of the upwelling light scattered from the atmosphere without interactions with the surface. The second term describes the surface contribution transmitted directly through the atmosphere while the third term, between brackets, models the diffuse interactions between the surface and the atmosphere. An exact expression for $\mathbf{R}^{Atm+Surf}$ would require the inclusion of multiple surface reflections (i.e. multiple scattering interactions between surface and atmosphere).

Fig. 10.5 shows an example of polarized reflectances calculated at the top of the atmosphere (TOA) levels (solid lines) and the same quantities calculated when suppressing the process of multiple surface reflections (dots). The effective radius (r_{eff}) and effective variance (v_{eff}) characterize the aerosol size distribution [Hansen and Travis, 1974] and m_r is the refractive index of the aerosols with the calculations being performed for a bimodal

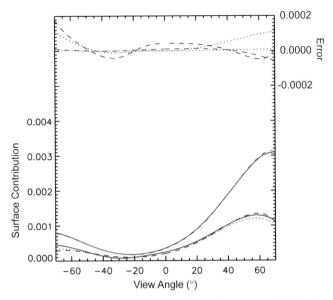

Fig. 10.5. Contribution of the surface to the TOA polarized reflectance at 410 (blue lines) and 865 (red lines) nm: Exact calculations (solid lines), calculations without multiple scattering interactions between surface and atmosphere (dotted lines) and calculations performed using direct transmission term and scaling optical thicknesses (dashed lines). The errors (absolute magnitudes of the difference between exact and approximate calculations) on the surface and TOA polarized reflectances are shown in more detail at the top of the figure. Aerosol layer: $r_{eff=0.14\ \mu m}$, $v_{eff}^f = 0.15$, $r_{eff}^c = 2.5\ \mu m$, $v_{eff}^c = 0.5$, $m_r^f = m_r^c = 1.4$ and $\tau_a = 0.1$ at 550 nm with coarse mode optical depth being 10% of the total. Calculations made for a solar zenith angle of $30°$, $\Delta\varphi = 45°$. The surface albedo was 0.1.

size distribution composed of fine (superscript f) and coarse (superscript c) mode particles. It is apparent that the contribution of multiple surface–atmosphere interactions to the observed polarization at the top of the atmosphere is small. Note that this process is generally accounted for in multiple scattering calculations [Hansen and Travis, 1974] but can, if necessary, be neglected and the surface–atmosphere interaction for polarization can be treated using Eq. (15) rather than an exact calculation.

The polarized reflectance measured over land is usually modeled by considering only the upwelling polarized light scattering from the atmosphere and a single reflection off the surface [Deuzé et al., 2001; Waquet et al., 2007]. The polarized reflectance can then be written in the following form,

$$R_{p,\lambda}^{Calc}(\mu_s, \mu_v, \Delta\varphi) = R_p^{Atm}(\mu_s, \mu_v; \Delta\varphi) + t_- {}^* R_p^{Surf}(\mu_s, \mu_v, \Delta\varphi) t_+ {}^*. \qquad (17)$$

In this equation R_p^{Atm} is the atmospheric polarized reflectance (i.e. calculation made with a black surface) and R_p^{Surf} is the surface polarized reflectance. $t_{-,+}{}^*$ is a direct transmission term where the aerosol and molecular optical thicknesses are scaled respectively by factors κ and ϕ. These factors are empirically derived and account for the neglect of the diffuse surface–atmosphere interactions [Lafrance, 1997]. In Fig. 10.5, we show the results of using these different approaches to modeling the polarized reflectance at the TOA.

The coefficients κ and ϕ are respectively equal to 0.63 and 0.44. In this example these factors were adjusted to minimize model errors for this particular aerosol model and load over the full range of solar zenith and azimuth angles. Although this approach is optimized, it still introduces some significant errors in the modeling of the surface contribution, resulting in errors in the TOA polarized reflectance that are as large as 0.0001 (2.5 % of the signal). These errors increase at shorter wavelengths for which the diffuse transmission of the atmosphere increases relative to the direct beam transmission.

At these shorter wavelengths, typically less than 670 nm, where the simple model of Eq. (17) is not sufficiently accurate the surface model introduced in Eq. (12) can be used to calculate the surface reflection matrix so that the diffuse interactions between the surface and the atmosphere can be accurately calculated. An approach to modeling the polarized reflectances that has been found to have certain advantages for modeling real data, particular data with high spatial resolution is presented in Eq. (18). For a given viewing geometry, the surface contribution can be calculated by considering only the Fresnel reflectance that is multiplied by a coefficient in order to scale the surface to that which is observed. In practice, this can be implemented model by the formula,

$$R_{p,\lambda}^{Calc} = R_{p,\lambda}^{Atm} + \left[R_{p,\lambda,Fresnel\_surface}^{Atm+Surf} - R_{p,\lambda}^{Atm} \right] \xi. \tag{18}$$

The dependence of the quantities on viewing geometry (μ_s, μ_v, $\varphi_s - \varphi_v$) is once again suppressed for clarity. $R_{p,\lambda,Fresnel\_surface}^{Atm+Surf}$ is the polarized reflectance calculated when the elements of the reflection matrix are calculated according to the Fresnel law using a surface refractive index equal to 1.5. The term in brackets corresponds to the contribution from surface interactions (see Eq. (15)), which, since multiple surface interactions are negligible (see Fig. 10.5) can simply be multiplied by a scale factor ξ in order to properly include diffuse reflections. The factor ξ that provides an appropriate scaling of the surface reflectance to match the observations is estimated using the formula

$$\xi(\mu_s, \mu_v, \Delta\varphi) = R_{p,2250nm}^{Meas}(\mu_s, \mu_v, \Delta\varphi) \Big/ R_p^F(\gamma), \tag{19}$$

where R_p^{Meas} is the polarized reflectance measured at 2250 nm. Clearly, if the polarized reflectance at 2250 nm is not affected by aerosols and Fresnel reflectance for a refractive index of 1.5 is a perfect model of the surface then the parameter ξ will be unity and the calculated polarized reflectance will be $R_{p,\lambda,Fresnel\_surface}^{Atm+Surf}$ as it should be. Similarly if the actual surface polarized reflectance is negligibly small then ξ will be equal to zero and the calculated polarized reflectance will be simply $R_{p\lambda}^{Atm}$. Eq. (19) is therefore being used to extrapolate from the simple Fresnel model using actual observations of the 2250 nm polarized reflectance. This approach also has the advantage that the scaling factor, ξ, is derived for each view angle and allows for the fact that not all view angles may see exactly the same surface as a result of aircraft attitude variations. For instruments in which the spectral polarization measurements are simultaneously acquired in each view, this modeling allows the surface contribution to be eliminated from the measurements even when different angular samples view different surfaces. This is apparent in Fig. 10.10 (a) where the total polarized reflectance observed with the RSP during ALIVE at an altitude of 4000 m is shown together with the residual errors in the model fit to the data. The large fluctuations in the polarized reflectance (a) are caused by the RSP observing different surface types at

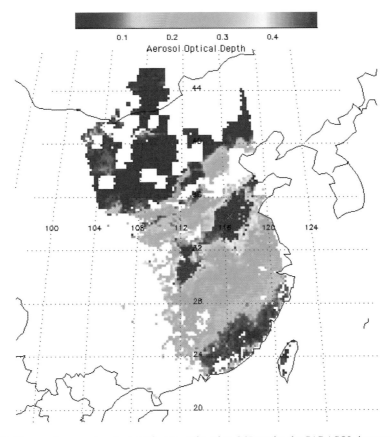

Fig. 10.6. Fine mode aerosol optical thickness retrieved at 865 nm by the PARASOL instrument over China (monthly average for February 2007).

different view angles. The fact that these fluctuations are not apparent in the residual errors demonstrates the effectiveness of Eqs. (18) and (19) for modeling the effects of surface polarized reflectance. The polarization generated by Fresnel reflection (i.e. R_p^f) progressively decreases with the scattering angle and becomes null in the backscattering direction ($\Theta = 180°$). Near backscattering the polarized reflectance is therefore determined by effects (multiple and internal scattering) that are not captured by a Fresnel model and that may also have some spectral dependence. It is therefore necessary to restrict the use of these simple Fresnel models to scattering angles smaller than $160°$ and indeed aerosol retrievals using polarized reflectances are only feasible with this limit on scattering geometries.

5.   Aerosol retrievals using polarimetric observations

The use of polarized reflectance measurements for aerosol retrievals over land surfaces is substantially simpler and more robust than that using reflectance measurements because of the simpler lower boundary condition provided by the gray nature of the polarized reflectance of natural surfaces. The aerosol retrievals over land using polarized reflectance that have thus far been implemented for the POLDER and RSP instruments [Deuzé et al., 2001; Waquet et al., 2008] have therefore restricted themselves to using only polarized reflectance measurements, even though multi-angle reflectance measurements are also available, because reflectance measurements are more susceptible to cloud contamination and erroneous assumptions regarding the properties of surface reflectance. In this Section we describe two approaches to using polarized reflectances to retrieve aerosols over land.

5.1 The POLDER experience

The POLDER team pioneered the use of polarimetric measurements for aerosol retrievals over land surfaces. The principle of the algorithm currently used for the analysis of the PARASOL observations is based on the one developed for the POLDER instrument [Deuzé et al., 2001]. Improvements concerning the multiple scattering computations have been implemented in the present algorithm. The approach is to use a search of look-up tables (LUTs) to find the size and optical depth of aerosol that best fits the observations in two bands at 670 and 865 nm. The observations are simulated using a model based on Eqs (13) and (17) where the surface contribution is estimated based on its IGBP [Belward et al., 1999] classification and the current NDVI [Tucker, 1979]. The atmospheric contribution is calculated by interpolation within LUTs computed using a multiple scattering code. In order to account for the altitude of the pixel, the calculations are made for two values of molecular optical thickness corresponding to surface levels of 0 and 2 km. Calculations are made for several aerosol optical thicknesses varying between 0.0 and 2.0 at 865 nm and for 10 aerosol models. The aerosol polarization mainly comes from the small spherical particles [Vermeulen et al., 2000] whereas the coarse mode particles (e.g., mineral dust) do not polarize much [Herman et al., 2005] and therefore cannot be accurately detected with polarized measurements, at least in the spectral range between 670 and 865 nm. The aerosol models used in the algorithm consist of single lognormal size distributions of small spherical particles with effective radius varying between 0.075 and 0.225 μm and an effective variance of 0.175 (corresponding to the Ångström exponent varying between 3 and 1.8). Polarization measurements at 670 nm and 870 nm do not contain sufficient information to constrain the aerosol refractive index. As the method is mainly sensitive to fine mode particles, a refractive index of $1.47 - 0.01i$ is assumed for all retrievals, which is the mean value for urban-industrial and biomass burning aerosols [Dubovik et al., 2002]. As the majority of the anthropogenic particles have sizes that are within the fine mode, the POLDER missions allow one to locate the main sources of anthropogenic aerosols at a global scale. Fig. 10.6 shows the optical thickness due to small particles retrieved by PARASOL over China for February 2007.

The highest aerosol optical thicknesses appear in red on the images. We observe high optical thicknesses in the vicinity of Beijing with strong gradients across the China. Re-

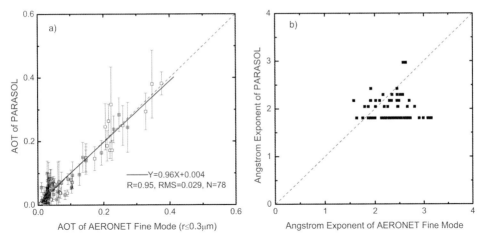

Fig. 10.7. PARASOL AOT and Ångström exponent versus those of AERONET for the fine mode fraction ($r_{eff} < 0.3$ μm) over Beijing.

gional validation performed over the northeast part of China has shown that the AOTs retrieved by PARASOL are consistent with the ones given by the federated Aerosol Robotic Network (AERONET) sunphotometers for the fine mode. The operational inversion algorithm developed for the analysis of the AERONET sunphotometer measurements provides the spectral aerosol optical thickness, the aerosol complex refractive index and the particle size distribution between 0.05 and 15 μm [Dubovik and King, 2000]. Fig. 10.7(a) shows a comparison between the PARASOL AOT and the ones given by AERONET for the fine mode ($r_{eff} < 0.3$ μm) for the city of Beijing [Fan et al., 2008]. A similar comparison of the Ångström exponent is shown in Fig. 10.7(b).

As shown in different studies [Chowdhary et al., 2005; Waquet et al., 2008] and discussed here after, polarization measurements at short visible wavelengths contain important information on the aerosol microphysical properties and the aerosol vertical extent. The PARASOL observations at 490 nm are well calibrated [Fougnie et al., 2007] and are also sensitive to the presence of mineral dust particles in the air [Fan et al., 2008] and to the altitude of the layer. Future efforts will therefore focus on the inclusion of the PARASOL band at 490 nm in the algorithm.

The use of a semi-empirical surface model and restricted spectral range (490–865 nm) limits the capacity of PARASOL to retrieve detailed aerosol microphysical properties over land [Deuze et al., 2001]. A multi-spectral airborne polarimeter has been recently developed by the Laboratoire d'Optique Atmosphérique and the French Centre National d'Etudes Spatiales. This instrument, called OSIRIS (Observing System Including PolaRisation in the Solar Infrared Spectrum), is based on the POLDER concept, but with the spectral range extended into the solar near-infrared spectrum with spectral bands at 1600 and 2100 nm. This instrument constitutes a crucial step in the development of a new European imaging polarimeter dedicated to the monitoring of the climate and air quality.

5.2   The RSP experience

The approach that is used for the analysis of the RSP observations [Waquet et al., 2008] and which is planned for use with the APS is an optimal estimation method. This approach is preferred to the use of a LUT because of the substantially higher information content of the seven RSP spectral bands that are located in atmospheric windows compared with the two spectral bands currently used in the POLDER approach. We therefore break this Section into two subsections: the theory of optimal estimation and our experimental implementation.

5.2.1 Theory

The principle of an optimal estimate is to determine the most probable atmospheric state conditional on the value of the measurements and some *a priori* knowledge of this medium and errors in measurement [Rodgers et al., 2000]. The determination of the most probable atmospheric state is identical to the minimization of a cost function Φ that accounts for these different quantities:

$$\Phi = (\mathbf{Y} - \mathbf{F})^T \mathbf{C}_T^{-1} (\mathbf{Y} - \mathbf{F}) + (\mathbf{X} - \mathbf{X}_a)^T \mathbf{C}_a^{-1} (\mathbf{X} - \mathbf{X}_a), \tag{20}$$

where $\mathbf{Y}$ is the measurement vector, $\mathbf{F}$ is the simulation vector, $\mathbf{C}_T^{-1}$ is the total error covariance matrix, $\mathbf{X}$ is the atmospheric state vector, $\mathbf{X}_a$ is the *a priori* atmospheric state vector and $\mathbf{C}_A^{-1}$ is the *a priori* error covariance matrix.

The first term in Eq. (20) corresponds to a weighted least squares error term that measures the distance between the observed polarized reflectances and the modeled polarized reflectances. We use the first six spectral bands of the RSP instrument that are in atmospheric windows (410, 470, 555, 670, 873 and 1590 nm) to constrain the aerosol properties. The observations are simulated using Eqs (18) and (19) and the polarized reflectance measured at 2250 nm is used to accurately model the surface. The residual atmospheric effect at 2250 nm is also accounted for in our retrieval process. We use the aerosol model retrieved with the shorter bands and a rearranged form of Eq. (18) to perform an atmospheric correction the polarized reflectance measured at 2250 nm.

The total error covariance matrix accounts for the measurement errors and some potential modeling errors. We assume that the different sources of errors are independent. Then, the total covariance matrix is given by the sum of the different error covariance matrices:

$$C_T = C_\varepsilon + C_{cal} + C_{pol} + \mathbf{C}_F. \tag{21}$$

The first term accounts for the effects of the instrumental noise, the second for uncertainty in the absolute calibration and the third one for the polarimetric accuracy. In the analysis presented here these error covariance matrices are filled out according to the signal, noise and accuracy obtained with the RSP instrument which is similar to that expected for the APS instrument (i.e. signal to noise ratio greater than 300 over dark oceans with typical aerosol loads, polarimetric accuracy for most scenes of 0.2 % or better and absolute radiometric accuracy of 5 %). $\mathbf{C}_F$ accounts for the effects of a change of the surface refractive index with wavelength and for the residual errors introduced by our modeling of the polarized reflectances (see Fig. 10.5).

Table 10.1. A priori knowledge of the aerosol parameters and associated uncertainties

	r_g, μm	σ	m_r	m_i
Fine mode	0.15 (0.1)	0.4 (0.1)	1.47 (0.14)	0.01 (0.015)
Coarse mode	0.8 (0.2)	0.6 (0.3)	1.53 (0.05)	0.005 (0.005)

The second term in Eq. (20) is a penalty function that constrains the solution to lie near the *a priori* state where the 'near' is quantified by the *a priori* error covariance matrix. The state vector **X** contains the aerosol parameters that allow characterizing each mode separately: N^f, r_g^f, σ^f, m_r^f, m_i^f, N^c, r_g^c, σ^c, m_r^c, m_i^c. The parameter N is the number density of aerosol particles (1/cm), r_g and σ are the parameters that define a lognormal size distribution (see Eq. (9) in Chapter 2 of this book), m_r and m_i are the real and imaginary refractive indices and the superscripts f and c denote the fine and coarse modes. We assume that aerosols are uniformly from the surface to a pressure level P, which is a reasonable assumption for a well-mixed boundary layer. The pressure level P should therefore correspond to the mixed layer depth. The *a priori* knowledge of the aerosol parameters is based on the aerosol climatology [Dubovik et al., 2002]. In Table 10.1, we provide the values of the *a priori* aerosol parameters (i.e. vector $\mathbf{X}_a$) and the associated uncertainties (standard deviation).

The covariance matrix $\mathbf{C}_a$ is assumed diagonal where the diagonal elements correspond to the square of the standard deviation values given in Table 10.1. The *a priori* values for N^f and N^c are derived using a LUT approach as explained in the following.

The determination of the best solution **X** that minimizes the cost function requires the resolution of a nonlinear equation. Nonlinear systems are usually solved using the Newton–Gauss iteration procedure. In practice, the Newton–Gauss procedure may not converge and needs to be modified. The most widely used modification is known as the Levenberg–Marquardt method, which is implemented by the following equation:

$$\mathbf{X}_{i+1} = \mathbf{X}_i - [\mathbf{H}(\mathbf{X}_i) + \gamma \cdot \mathbf{I}]^{-1} \cdot \nabla_x \Phi(\mathbf{X}_i), \tag{22}$$

where **I** is the identity matrix with the dimensionality of the state vector, i indicates the number of the iteration and γ is a positive coefficient that aids in the convergence of the iteration. **H** is known as the Hessian matrix,

$$\mathbf{H}(\mathbf{X}_i) = \nabla_X^2 \Phi(\mathbf{X}_i) \approx \mathbf{C}_a^{-1} + \mathbf{K}_i^T \cdot \mathbf{C}_T^{-1} \cdot \mathbf{K}_i \tag{23}$$

where

$$\mathbf{K}_i = \frac{\partial \mathbf{F}(\mathbf{X})}{\partial \mathbf{X}_i}. \tag{24}$$

K is the Jacobian matrix, which represents the sensitivity of the forward model to the retrieved quantity (i.e. sensitivity of the polarized reflectances to the aerosol parameters). The criteria for changing the γ value is dependent on the convergence behavior. If $\Phi(\mathbf{X}_{i+1}) > \Phi(\mathbf{X}_i)$ then we reject the solution $\mathbf{X}_{i+1}$ and we increase γ whereas if $\Phi(\mathbf{X}_{i+1}) < \Phi(\mathbf{X}_i)$ then we accept the solution $\mathbf{X}_{i+1}$ and we decrease γ. For larger γ values the steepest descent dominates and the convergence is slow (i.e. small step size) but robust whereas for smaller γ values, the search turns to the faster Newtonian descent. The iteration process is stopped when there is no change of the cost function between two successive iteration steps.

The optimal estimate method also provides an error diagnostic of the retrieved parameters. The Hessian matrix obtained at the final step of iteration can be used to calculate the retrieval error covariance matrix $\mathbf{C}_x$:

$$\mathbf{C}_x = \left(\mathbf{C}_a^{-1} + \mathbf{K}_i^T \cdot \mathbf{C}_T^{-1} \cdot \mathbf{K}_i\right)^{-1}. \tag{25}$$

The square roots of the diagonal elements of $\mathbf{C}_x$ give the standard deviation associated with each retrieved parameter. The aerosol microphysical parameters contained in the vector $\mathbf{X}$ and the error retrieval covariance matrix $\mathbf{C}_x$ can also be used to calculate the standard deviation, or retrieval uncertainty, associated with the optical aerosol parameters. For the aerosol optical thickness, the standard deviation is given by:

$$\sigma_\tau = \sqrt{\left(\sum_{i=1}^{N}\sum_{j=1}^{N} C_{x,i,j} \frac{\partial \tau}{\partial X_i} \frac{\partial \tau}{\partial X_j}\right)}. \tag{26}$$

A similar formula applies for the single scattering albedo ϖ_0.

A first guess of the aerosol parameters $(\mathbf{X}_0)$ is required to start the iterative process. A good first guess allows the number of iterations to be reduced and alleviates the problem of finding a solution $\mathbf{X}$ that is only a local minimum of the cost function Φ. A LUT provides a simple and effective approach to derive a first estimate of the aerosol optical thickness and aerosol model. In the LUT used for aerosols retrievals over land the polarized reflectances are calculated for various aerosol optical thicknesses and aerosol models. Twelve fine mode models are considered ($r_g^f = 0.05$, 0.1, 0.15, 0.2 μm, $\sigma_f = 0.4$, $m_r = 1.4$, 1.47, 1.54, $m_i = 0.01$). We also include the coarse mode particle model described in Table 10.1. The polarized reflectances are calculated for aerosol optical thicknesses at a reference wavelength of 550 nm equal to 0, 0.05, 0.1, 0.2, 0.4, 0.8, 1 and 2 and an interpolation process is used to create a fine step. The first guess corresponds to the aerosol model and aerosol optical thickness that minimize the least squares error term (see Eq. (20)) calculated between the measurements and the simulations. The properties of the coarse mode particles are not given the same level of detail in the LUT because the sensitivity of polarized reflectance measurements to coarse mode particles is limited by their weak spectral variation (similar to the surface) and weak polarization signal. As a first guess, the aerosol optical thickness of the coarse mode (at 550 nm) is assumed to be a tenth of the total aerosol optical thickness. The properties of the aerosol models considered in the LUT allow the number density of particles associated with each mode to be derived. In retrievals the LUT is assumed to reduce the uncertainties in r_g^f and m_r^f given in Table 10.1 to 0.05 and 0.07, respectively, and we assume that the relative uncertainty for both N^f and N^c is 100 %.

The residual atmospheric effect is accounted for in the retrieval process. This effect is usually small and mainly depends on the aerosol size and load [Waquet et al., 2007]. The aerosol model retrieved with the shorter bands is used in a rearranged form of Eq. (17) to perform an atmospheric correction on the polarized reflectance measured at 2250 nm. This correction is performed before the first iteration using the retrieved parameters obtained with the LUT approach and is refined after each iteration step. The final results from the retrieval are the estimate of the state vector **X** that describes the aerosol model and the uncertainty in that state vector calculated using Eq. (25).

The information contained in polarization measurements is mainly dominated by the single scattering properties of the aerosol. The angular dependence of the single scattering of polarized light is given by the polarized phase function (P_{12} element of the aerosol scattering phase matrix), which is a particular function of the scattering angle. The variations of the polarized phase function with the scattering angle and wavelength are strongly dependent on the aerosol microphysical properties [Mishchenko et al., 2006]. Fig. 10.8 illustrates the sensitivity of the observed polarized reflectance to various aerosol parameters as a function of the wavelength and scattering angle.

We note the large angular and spectral dependences of the polarized reflectance on the particle size and on the real part of the refractive index. Fig. 10.8 also shows that polarization measurements in the shortest visible bands are sensitive to the aerosol absorption (i.e. imaginary part of the complex refractive index) and to the height that they are mixed to (i.e. the mixed layer depth, P). The sensitivity to these latter parameters mainly appears in the UV and blue part of the spectrum where the molecular and multiple scattering contributions are significant.

The retrieval error covariance matrix defined in Eq. (25) can be used to simulate the retrieval errors obtained for any instrument type and surface–atmosphere system with synthetic measurements. Fig. 10.9 shows an example of retrieval errors obtained for $\tau(\lambda)$, $v_0(\lambda)$, $m_r\lambda$, r_{eff}, v_{eff} of each mode and pressure as a function of the aerosol optical thickness at 550 nm. Calculations are made for an aerosol model representative of pollutant particles following Dubovik et al. [2002]. The error covariance matrix is modeled accord-

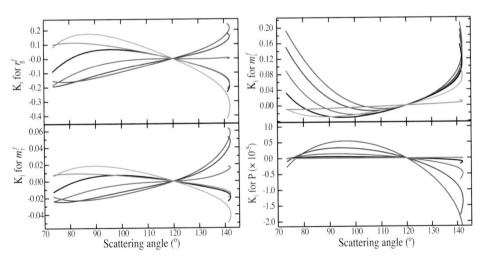

Fig. 10.8. Elements of the Jacobian matrix calculated for m_r^f, m_i^f, r_g^f and P as a function of the scattering angle and wavelength. Polarized reflectance calculated for an aerosol model described by a single log-normal size distribution ($r_g = 0.1$, $\sigma = 0.403$) and a complex refractive index of $1.44-0.01i$. Calculations performed for $\theta_s = 60°$ and $\Delta\varphi = 45°$ and an aerosol optical thickness of 0.5 at 550 nm and including a surface polarized reflectance. The wavelengths are: 0.41, 0.47, 0.55, 0.67, 0.865 and 1.6 μm, respectively in red, blue, magenta, black, green and dark green.

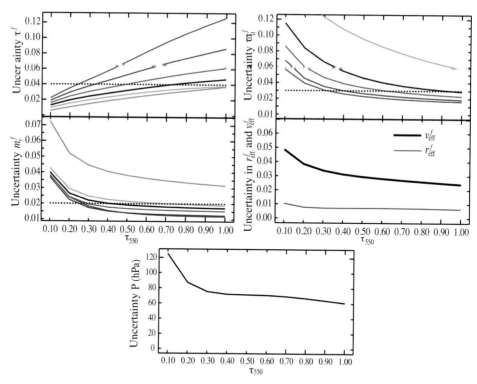

Fig. 10.9. Retrieval errors (see Eq. (26)) of the spectral aerosol optical thickness (a), single scattering albedo (b), real part of the refractive index (c), effective radius and effective variance (d) and pressure of the top of the aerosol layer as a function of the aerosol optical thickness at 550 nm. The dashed lines correspond to the absolute accuracy requirements suggested by Mishchenko et al. [2004]. Computations are performed at TOA for $\theta_s = 45°$ and $\Delta\varphi = 45°$. The wavelengths are 410, 470, 555, 670, 865 and 1600 nm, respectively in red, blue, magenta, black, green and dark green.

ing to the instrumental characteristics of the RSP instrument and the *a priori* information given in Table 10.1.

Mishchenko et al. [2004] suggested the following retrieval requirements for climate research over land: 0.04 or (10 %) for the aerosol optical thickness, 0.03 for the single scattering albedo, 0.1 micron (or 10 %) for the effective radius, 0.3 (or 50 %) for the effective variance, and 0.02 for the real part of the refractive index. This sensitivity analysis shows that the uncertainties in aerosol optical thickness (AOT) increase with AOT while the uncertainties in the microphysical model decrease. The uncertainty in the single scattering albedo (SSA) is notably less than 0.05 by the time the AOT is greater than 0.2. The increase in the errors associated with the AOT is explained by the fact that the errors in the column number density of particles N increase with N (not shown) and because the AOT is closely connected to this parameter. The relative retrieval errors in the AOT, however, decrease for increasing AOT and remain under the required value for all the wavelengths considered here $(\sigma_{-\tau}(\lambda)/\tau(\lambda) < 10\,\%)$. The requirements for r_{eff} and v_{eff} are reached for any AOT values larger than 0.1 at 550 nm, whereas for m_r and v_0, the requirements are only reached in three spectral bands (410, 470 and 550 nm) and for AOT respectively larger than

0.3 and 0.6. Figure 10.9 shows that there is also useful information about the height of the top of the aerosol layer. The error in the pressure level of the top of the aerosol layer is less than 70 hPa for AOT greater than 0.6 at 550 nm when the true mixed layer depth is $P = 700$ hPa. This corresponds to an error of 0.8 km when the aerosol altitude is 3 km. It is important to note that we are primarily interested in retrieving P to ensure that large retrieval and/or fitting errors do not occur in the presence of thick aerosol layers such as smoke plumes and that erroneous assumptions about P do not therefore cause a bias in our retrievals. This aerosol layer pressure top estimate is based on the fact that aerosols reduce the polarized reflectance at 410 and 470 nm. This happens because the aerosols attenuate molecular scattering below the aerosol layer, while their own polarized reflectance contribution is less than that from molecular scattering, even for very small particles.

5.2.2 Experiment

We applied the method described above to a set of field experiment observations performed with the RSP that provide a test of the realism of the analysis for various aerosol optical depth regimes. Figure 10.10 shows three examples of scans of polarized reflectance acquired during the Aerosol Lidar Validation Experiment (ALIVE) and the Megacity Initiative: Local And Global Research Observations (MILAGRO) field campaign. The ALIVE experiment took place over the Department of Energy Atmospheric Radiation Measurements [Ackerman and Stokes, 2003] program facility in the Southern Great Plains (SGP). The RSP instrument participated in this campaign onboard the Sky Research Inc. Jetstream-31 (J31) research aircraft and acquired data throughout all twelve flights performed between September 12 and 22, 2005. The MILAGRO experiment was an international, multi-agency campaign involving numerous academic and research institutions from the USA, Mexico and other countries [Fast et al., 2007]. This campaign was designed to study pollution from Mexico City and regional biomass burning (sources, transport, transformations and effects) and involved intensive aircraft and ground-based measurements. The RSP instrument flew onboard the J31 between March 3 and 20, 2006, and participated in 13 flights. A variety of aerosol, cloud, water vapor, and surface conditions were sampled over Mexico City and the Gulf of Mexico. During these two campaigns the NASA Ames Airborne Tracking 14-Channel Sun-photometer (AATS-14) was also integrated onto the J31. These measurements allowed the aerosol optical thickness of the column above the aircraft to be derived in 13 bands between 353 and 2105 nm [Schmid et al., 1997; Russell et al., 1999]. A cloud-screening algorithm has been developed for the analysis of the AATS measurements and allowed us to select scenes for which the direct beam of the sun was cloud-free above the aircraft that eliminated concerns regarding cirrus clouds.

In Table 10.2 we summarize the observational and illumination geometry of the three example scenes presented here. Panels (a) and (b) in Fig. 10.10 show scans performed on September 16 and 19, 2005, during the ALIVE experiment, close to the AERONET station denoted to be CART (N36.6069°, W97.4858°). The spikes in these scans, that are correlated across all spectral bands and that are present particularly in Fig. 10.10(a), are the result of different view angles seeing different surface types, since the data have not been reorganized to view the same point on the ground. During the ALIVE experiment,

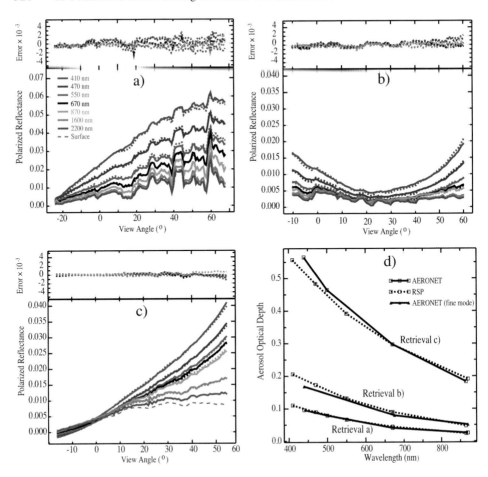

Fig. 10.10. (a), (b), (c) Polarized reflectance measured at the aircraft level (solid lines) and simulated polarized reflectance (dashed lines) as a function of the viewing angle. The wavelengths are 410, 470, 555, 670, 865, 1600 and 2250 nm, respectively in red, blue, magenta, black, green, dark green and brown. The dashed brown line corresponds to the direct surface contribution (measurements at 2250 nm corrected from the atmospheric effects). The error is the difference between the simulated and measured polarized reflectances and is shown at the top of the figures. Scans (a) and b) were obtained on 09/16/05 and 09/19/05 during the ALIVE experiment (Southern Great Plains, USA). Scan (c) was obtained over Mexico City on 03/15/06 during the MILAGRO experiment. (d) Aerosol optical thicknesses retrieved by the RSP instrument from the measurements shown in (a), (b) and (c) and coincident AERO-NET measurements (or retrievals when the fine mode AOT is reported) as a function of the wavelength.

the flights were performed for low aerosol loading ($\tau < 0.165$ at 500 nm) and we therefore did not try to retrieve the absorption and the altitude of the aerosol since the sensitivity to these parameters is weak at such low AOT. The two scans shown here were performed in different planes of observation, which provided measurements with different ranges of scattering angle. We observe that the use of the 2250 nm measurement in Eqs. (18) and (19) allows the effects of the various surface types to be compensated extremely effectively. The coarse-mode AOT is negligible for the scan shown in Fig. 10.10(a) and

Table 10.2. Illumination and viewing geometry of the three datasets presented in Fig. 10.10.

Scan	Viewing geometry		
	Scattering angle, deg	Solar zenith angle, deg	The difference of azimuths, deg
a	70–160	56.0	– 186
b	135–160	36.5	37
c	85–170	24.1	– 175

reaches 0.05 at 670 nm for Fig. 10.10(b). Figure 10.10(d) shows a comparison between the AOT retrieved by the RSP and those measured by AERONET. These results show that our approach allows the AOT of the fine mode particles to be retrieved with a maximal error of 0.01 for an AOT of 0.08 at 670 nm. The spectral dependence of the AOT estimate agrees well with the sunphotometer measurements, which suggests that the microphysical properties of the observed aerosol (i.e. the particle size and the real part of the refractive index), are also correctly estimated. This conclusion is born out by the agreement of the RSP inversion of the data shown in Fig. 10.10(b) with an inversion of AERONET sun-photometer measurements that was made close to the time of the RSP observations. The difference between the effective radii and refractive index retrievals from these two data sets is 0.025 μm and 0.003 respectively. This comparison also indicated that there is little sensitivity of the polarized reflectance observations to the coarse particle mode at least for coarse-mode AOTs less than 0.05 with a very large coarse particle mode.

Figure 10.10(c) shows a scan acquired on March 15, 2006, around 1800 UT during the MILAGRO experiment over the ground-based site called T0. This site was located in the urban area of Mexico City and was equipped with an AERONET sunphotometer (N19.4900°, W99.1478°). For this scan, the data are reorganized so that each part of the scan sees the same target at the ground, as is the case for observations taken from satellites. It is also interesting to notice that the residual atmospheric effect at 2250 nm is not negligible for this case and must be taken into account (see the differences between the solid and dashed brown lines in Fig. 10.10(c)). The contribution of the coarse-mode particles to the total AOT is quite stable during the day and has an average value of 0.02 at 550 nm.

The retrieved AOT is equal to 0.3 (0.005) at 670 nm, the effective radius and the effective variance are respectively equal to 0.15 (0.005) μm and 0.485 (0.01), the real and imaginary refractive indices are equal to 1.54 (0.01) and 0.027 (0.005), respectively and P is equal to 627 hPa (10), corresponding to a height of about 3.85 km. The parenthetic values are the uncertainties in the retrievals calculated using Eq. (26). We retrieve a small effective radius and a high real refractive index, which is characteristic of the properties of the biomass burning particles [Dubovik et al., 2002]. The retrieved imaginary refractive index leads to a single scattering albedo of 0.865 (0.005) at 550 nm. This value is similar to estimates made for African savanna biomass burning particles ($0.84 < \omega_0 < 0.88$) and for fresh biomass burning particles observed in the vicinity of the source (0.86) [Schmid et al., 2003]. Figure 10.10(d) shows that the AOT retrieved by the RSP is in good agreement with the coincident AOT measured by AERONET over a large spectral range, which indicates that our retrieved aerosol model is valid. An inversion of the AERONET sunphotometer measurements was made at the T0 site earlier in the afternoon (15:25 UT). However the

Ångström exponent measured by AERONET is equal to 1.65 at 15:25 UT when this inversion was performed but changes to 1.9 by 18:00 UT close to the time that the RSP observations were made. This indicates that the fine mode particles observed at 15:25 UT were different to those observed at 18:00 UT. A comparison of the RSP and AERONET microphysical retrievals is therefore not useful in this case. Analysis of the chemical composition of the aerosols observed at the T0 site was made for the period March 15–27, 2006 [Moffett et al., 2008]. It was found that biomass burning and aged organic carbon particle types respectively constituted 40 % and 31 % of the submicrometer mode on average and that the fresh biomass fraction dominated in the afternoons (around 65 % at the time of the RSP observations). This composition is the result of many fires to the south and southeast of the city on the surrounding mountains and the local meteorological conditions and is coherent with the RSP microphysical retrieval showing quite strongly absorbing aerosols. The NASA Langley Research Center High Spectral Resolution Lidar (HSRL) [Hair et al., 2008] flew over the T0 site an hour prior the RSP overpass. This instrument measures, among other things, the aerosol backscattering coefficient at 532 nm. The HSRL measurements show that the aerosols were located between the ground and 3.4 km, which indicates that our estimate of the aerosol top layer height is also realistic.

6.   Conclusions

It has become apparent over the last few years, based on the measurements presented here as well as previous work that the surface polarized reflectance is indeed almost perfectly gray (less than 0.001 variation in the polarized reflectance) over the entire solar and shortwave infrared spectrum from 400 to 2,300 nm. This is of particular importance given the apparently similar spectral invariance of the angular shape of the surface reflectance and the consequent correlation between reflectance shape properties and the surface polarized reflectance [Elias et al., 2004]. We have hitherto not used the reflectance measurements over land for aerosol retrievals from POLDER, or RSP, because of a tendency to degrade the quality of the retrievals obtained using only polarized reflectances. However, based on the results presented in this book, it would appear that careful use of the spectral invariance of the angular shape of the surface reflectance in conjunction with the spectral invariance of the polarized reflectance would provide even stronger constraints on the aerosol retrievals than the current approaches used for analysis of POLDER and RSP data. This extension of the retrieval approach using polarized reflectances presented here to incorporate the techniques used for multi-angle reflectance measurements presented elsewhere in this book is an area of ongoing research.

The POLDER retrieval algorithms have proven themselves to be robust and globally applicable with particular sensitivity to the fine mode particles that are primarily of anthropogenic origin in many areas and excellent coherence between retrievals over ocean and land. The addition of the shorter wavelength band at 490 nm is expected to improve detection of dust aerosols that are currently difficult to identify using just two bands at 670 and 865 nm. The RSP retrieval algorithms are more computationally intensive than those used for POLDER, but do have the advantage that the addition of other measurements sources (such as the multi-angle reflectance measurements) are fairly straightforward and assumptions regarding sizes and composition (refractive indices) are minimized. A retrieval approach of this kind is in fact necessary for the analysis of RSP data because of the larger number of spectral bands (seven as compared to three) and viewing angles (152 as compared to 12) and the need to appropriately weight the information in each measurement. In addition a major advantage of optimal estimation methods is the provision of uncertainties with the retrieval products [Rodgers, 2000] so that appropriate statistical weighting can be used when estimating mean values and their uncertainties. Having retrieval uncertainties available is particularly helpful when one is interested in something that is derived from the retrieval products. One example of this is the absorption AOT which is the product of the AOT and the single scattering co-albedo (i.e. $(1 - \nu)\tau$). The uncertainty in this quantity and not the uncertainty in the single scattering albedo is what is really important in estimating the effect of absorption by aerosols. If one examines the uncertainty in this quantity as a function of AOT using Figure 10.9 it is apparent that the absorption AOT can be retrieved to better than for an AOT of 0.2 and that the uncertainty in its retrieval increases to almost 0.03 at an AOT of 0.8. The RSP retrievals of absorption AOT are therefore useful at all optical depths, even though the single scattering albedo is highly uncertain at low optical depths. In summary, the sensitivity analyses and results presented here show that passive polarimetric remote sensing has an excellent capability to determine the size, composition, amount and useful vertical distribution information for aerosols over land.

Acknowledgments. Figure 10.7 was prepared by X. Fan from the Institute of Atmospheric Physics of the Chinese Academy of Sciences and we would like to express our thanks to him for providing this to us. We would also like to thank the personnel from NASA Ames Research Center led by P.B. Russell, B. Schmid (now at PNNL, Richmond, Washington, USA) and J. Redemann for facilitating the participation of the RSP instrument in the ALIVE and MILAGRO field experiments. The work described here has been carried out at NASA Goddard Institute for Space Studies and the Laboratoire d'Optique Atmosphérique of the University of Lille. Support of the NASA Radiation Sciences Program managed by H. Maring and the NASA Glory project is gratefully acknowledged. Finally we would like to thank M. Mishchenko, L. Travis, R.A. Chandos and E. Russell for assistance in the acquisition of the data used here and in the preparation of this paper.

References

Ackerman, T.P., and G.M. Stokes, 2003: The atmospheric radiation measurement program, *Physics Today*, **56**, DOI:10.1063/1.1554135.

Babinet, J., 1840: Sur un nouveau point neutre dans l'atmosphère, *C. R. Acad. Sci. Paris*, **11**, 618–620.

Barral, M.J.A., 1858: *Oeuvres de Francois Arago I–V*, Gide, Paris.

Belward, A.S., J.E. Estes, K.D. Kline, 1999: The IGBP-DIS global 1-km land-cover data set discover: A project overview, *Photogramm. Eng. Rem. Sensing*, **65**, 1013–1020.

Bréon, F.M., D. Tanré, P. Leconte, and M. Herman, 1995: Polarized reflectance of bare soils and vegetation: measurements and models, *IEEE Trans. Geosci. Remote Sens.*, **33**, 487–499.

Brewster, D., 1842: On the existence of a new neutral point and two secondary neutral points, *Rep. Brit.-Assoc. Adv. Sci.*, **2**, 13–25.

Cairns, B., L.D. Travis, and E.E. Russell, 1999: The Research Scanning Polarimeter: calibration and ground-based measurements, *Proc. SPIE*, **3754**, 186–197.

Chipman, R.A., 1994: Polarimetry, in *Handbook of Optics: Devices, Measurements and Properties*, 2nd edn. M. Bass (Ed.), McGraw-Hill, New York, Vol. II, pp. 1–37.

Chowdhary, J. et al., 2005: Retrieval of aerosol scattering and absorption properties from photo-polarimetric observations over the ocean during the CLAMS experiment, *J. Atmos. Sci.*, **62**, 1093–1117, doi:10.1175/JAS3389.1.

Coulson, K.L., 1988: *Polarization and Intensity of Light in the Atmosphere*, A Deepak, Hampton, VA.

d'Almeida, G.A., P. Koepke, and E.P. Shettle, 1991: *Atmospheric Aerosols, Global Climatology and Radiative Characteristics*, A. Deepak, Hampton, VA.

Deschamps et al., 1994: The POLDER mission: instrument characteristics and scientific objectives, *IEEE Trans. Geosci. Remote Sens.*, **32**, 598–615.

Deuzé, J.-L. et al., 2001: Remote sensing of aerosols over land surfaces from POLDER-ADEOS-1 polarized measurements, *J. Geophys. Res.*, **106**, 4913–4926.

Diner, D. J. et al., 2007: Dual-photoelastic-modulator-based polarimetric imaging concept for aerosol remote sensing, *Appl. Opt.*, **46**, 8428–8440.

Dubovik, O., and M. D. King, 2000: A flexible inversion algorithm for retrieval of aerosol optical properties from Sun and sky radiance measurements, *J. Geophys. Res.*, **105**, 20673–20696.

Dubovik, O. et al., 2002: Variability of absorption and optical properties of key aerosol types observed in worldwide locations, *J. Atmos. Sci.*, **59**, 590–608.

Elias, T., B. Cairns, and J. Chowdhary J., 2004: Surface optical properties measured by the airborne research scanning polarimeter during the CLAMS experiment, in *Remote Sensing of Clouds and the Atmosphere VIII*, K.P. Schäfer, A. Cameron, M.R. Carleer, R.H. Picard (Eds), *Proc. SPIE, 5235*(#79), SPIE, Bellingham, WA, pp. 595-606.

Fan, X., P. Goloub, J.-L. Deuzé, H. Chen, W. Zhang, D. Tanré, and Z. Li, 2008: Evaluation of PARASOL aerosol retrieval over North East Asia, *Rem. Sens. Env.*, **112**, 697–707.

Fast, J.D. et al., 2007: A meteorological overview of the MILAGRO field campaigns, *Atmos. Chem. Phys.*, **7**, 2233–2257.

Hair, J.W. et al., 2008: Airbonc High Spectral Resolution Lidar for profiling aerosol optical properties, *Appl. Opt.*, **47**, 6734–6752, doi:10.1364/AO.47.006734.

Fougnie, B., G. Bracco, B. Lafrance, C. Ruffel, O. Hagolle, and C. Tinel, 2007: PARASOL in-flight calibration and performance, *Appl. Opt.*, **46**, 5435–5451.

Grant, L., C.S.T. Daughtry, and V.C. Vanderbilt, 1993: Polarized and specular reflectance variation with leaf surface features, *Physiologia Plantarum*, **88**, 1–9.

Hagolle, O. et al., 1999: Results of POLDER in-flight calibration, *IEEE Trans. Geosci. Remote Sens.*, **37**, 1550–1566.

Hansen, J.E., and L.D. Travis, 1974: Light scattering in planetary atmospheres, *Space Sci. Rev.*, **16**, 527–610.

Hasekamp, O.P., and J. Landgraf, 2007: Retrieval of aerosol properties over land surfaces: capabilities of multiple-viewing-angle intensity and polarization measurements, *Appl. Opt.*, **46**, 3332–3344, doi:10.1364/AO.46.003332.

Herman, M., et al., 2005: Aerosol remote sensing from POLDER/ADEOS over the ocean: improved retrieval using a nonspherical particle model, *J. Geophys. Res.*, **110**, D10S02, doi:10.1029/2004JD004798.

Holben, B., et al., 1998: AERONET – a federated instrument network and data archive for aerosol characterization. *Rem. Sens. Environ.*, **66**, 1–16.

Jones, S., F. Iannarilli, and P. Kebabian, 2004: Realization of quantitative-grade fieldable snapshot imaging spectropolarimeter, *Opt. Express*, **12**, 6559–6573.

Kaufman, Y.J. et al., 1997: Remote sensing of tropospheric aerosol from EOS-MODIS over the land using dark targets and dynamic aerosol models. *J. Geophys. Res.*, **102**, 17051–17067.

Kawata, Y., 1978: Circular polarization of sunlight reflected by planetary atmospheres, *Icarus*, *33*, 217–232.

Keller, C.U., 2002: Instrumentation for astrophysical spectropolarimetry, in *Astrophysical Spectropolarimetry*, J. Trujillo-Bueno, F. Moreno-Insertis, and F. Sánchez (Eds), Cambridge University Press, Cambridge, UK, pp. 303–354.

Lafrance, B., 1997: Modélisation simplifiée de la lumière polarisée émergeant de l'atmosphère. Correction de l'impact des aérosols stratosphériques sur les mesures de POLDER. Thèse, Université des Sciences et Techniques de Lille. Lebsock, M.D., T.S. L'Ecuyer, and G.L. Stephens, 2007: Information content of near-infrared spaceborne multiangular polarization measurements for aerosol retrieval, *J. Geophys. Res.*, **112**, doi:10.1029/2007JD008535.

Mishchenko, M.I., B. Cairns, J.E. Hansen, L.D. Travis, R. Burg, Y.J. Kaufman, J.V. Martins, and E.P. Shettle 2004. Monitoring of aerosol forcing of climate from space: analysis of measurement requirements. *J. Quant. Spectrosc. Radiat. Transfer* **88**, 149–161, doi:10.1016/j.jqsrt.2004.03.030.

Mishchenko, M.I., L.D. Travis, and A.A. Lacis, 2006: *Multiple Scattering of Light by Particles: Radiative Transfer and Coherent Backscattering*, Cambridge University Press, Cambridge, UK.

Mishchenko, M.I., B. Cairns, G. Kopp, C.F. Schueler, B.A. Fafaul, J.E. Hansen, R.J. Hooker, T. Itchkawich, H.B. Maring, and L.D. Travis, 2007: Precise and accurate monitoring of terrestrial aerosols and total solar irradiance: introducing the Glory mission. *Bull. Amer. Meteorol. Soc.*, **88**, 677–691, doi:10.1175/BAMS-88-5-677.

Moffet, R.C., B. de Foy, L.T. Molina, M.J. Molina, and K.A. Prather, 2008: Measurement of ambient aerosols in northern Mexico City by single particle mass spectrometry, *Atmos. Chem. Phys.*, **8**, 4499–4516.

Nadal, F., and F.M. Bréon, 1999: Parameterization of surface polarized reflectances derived from POLDER spaceborne measurements, *IEEE Trans. Geosci. Remote Sens.*, **37**, 1709–1719.

Oka, K., and T. Kato, 1999: Spectroscopic polarimetry with a channeled spectrum, *Opt. Lett.*, **24**, 1475–1477.

Pollack, J.B., O.B. Toon, and B.N. Khare, 1973: Surface refractive index, *Icarus*, **19**, 372–389.

Povel, H., H. Aebersold, and J.O. Stenflo, 1990: Charge-coupled device image sensor as a demodulator in a 2-D polarimeter with a piezoelastic modulator, *Appl. Opt.*, **29**, 1186–1190.

Raman, C.V., 1921: A method of improving the visibility of distant objects, *Nature* (London) **108**, 242.

Redemann, J., Q. Zhang, B. Schmid, P.B. Russell, J.M. Livingston, H. Jonsson, and L.A. Remer, 2006: Assessment of MODIS-derived visible and near-IR aerosol optical properties and their spatial variability in the presence of mineral dust, *Geophys. Res. Lett.*, **33**, L18814, doi:10.1029/2006GL026626.

Rodgers C.D., 2000: *Inverse Methods for Atmospheric Sounding, World Scientific.*

Rondeaux, G., and M. Herman, 1991: Polarization of light reflected by crop canopies. *Rem. Sens. Environ.*, **38**, 63–75.

Russell, P.B., J.M. Livingston, P. Hignett, S. Kinne, J. Wong, A. Chien, R. Bergstrom, P. Durkee, and P.V. Hobbs (1999), Aerosol-induced radiative flux changes off the United States mid-Atlantic coast: comparison of values calculated from sunphotometer and in situ data with those measured by airborne pyranometer, *J. Geophys. Res.*, **104**(D2), 2289–2307.

Schmid, B., et al., 1997: Retrieval of optical depth and particle size distributions of tropospheric and stratospheric aerosols by means of sun photometry, *IEEE Trans. Geosci. Remote Sens.*, **35**, 172–181.

Schmid, B. et al., 2003: Column closure studies of lower tropospheric aerosol and water vapor during ACE-Asia using airborne Sun photometer and airborne in situ and ship-based lidar measurements, *J. Geophys. Res.*, **108**(D23), 8656, doi:10.1029/2002JD003361.

Shettle, E.P., and R.W. Fenn, 1979: Models for the aerosols of the lower atmosphere and the effect of humidity variations on their optical properties, Tech. Rep., AFGL-TR-79-0214, Air Force Geophys. Lab., Hanscomb AFB, MA.

Strutt, J.W. (Lord Rayleigh), 1871: On the light from the sky it polarization and color, *Phil. Mag.* **XLI**, 107–120. Reprinted in *Scientific Papers*, Vol. I, Cambridge University Press, Cambridge, UK, 1899.

Torres, O., P.K. Bhartia, J.R. Herman, Z. Ahmad, and J. Gleason (1998), Derivation of aerosol properties from satellite measurements of backscattered ultraviolet radiation: theoretical basis, *J. Geophys. Res.*, **103**(D14), 17099–17110.

Tucker, C.J., 1979: Red and photographic infrared linear combination for monitoring vegetation. *Rem. Sens. Environ.* **8**, 127–150.

Vanderbilt V.C. et al., 1985: Specular, diffuse, and polarized light scattered by two wheat canopies, *Appl. Opt.*, **24**, 2408–2418.

Vermeulen, A., C. Devaux, and M. Herman, 2000: Retrieval of the scattering and microphysical properties of aerosols from ground-based optical measurements including polarization. I. Method, *Appl. Opt.*, **39**, 6207–6220.

Waquet F., P. Goloub, J.-L. Deuzé, J.-F. Léon, F. Auriol, C. Verwaerde, J.-Y. Balois, and P. Fran\c{cois, 2007: Aerosol retrieval over land using a multi-band polarimeter and comparison with path radiance method, *J. Geophys. Res.*, **112**, D111214, doi:10.1029/2006JD008029.

Waquet, F., B. Cairns, K. Knobespiesse, J. Chowdhary, L.D. Travis, B. Schmid, and M.I. Mishchenko, 2009: Polarimetric remote sensing of aerosols over land, *J. Geophys. Res.*, **114**, DOI 1206, doi:1029/2008JD010619.

11 Optimal estimation applied to the joint retrieval of aerosol optical depth and surface BRF using MSG/SEVIRI observations

Yves M. Govaerts, Sébastien Wagner, Alessio Lattanzio, Philip Watts

1. Introduction

Passive spaceborne imagers observe radiation that has interacted both with the atmosphere and the surface. Interactions with the atmosphere include gaseous absorption and scattering by molecules and particulate matter or aerosols. Their characterization from remote sensing observations relies essentially on their capacity to modify differently the amount of radiation observed as a function of the wavelength, the viewing directions or the polarization. One of the major issues when retrieving tropospheric aerosol properties using spaceborne imager observations is to discriminate the contribution of the observed signal reflected by the surface from the one scattered and absorbed by the aerosols. In particular, it is crucial when the retrieval occurs above land surfaces which might be responsible for a non-negligible part of the total signal. Conceptually, this is equivalent to solving a radiative system composed of minimum two layers, where the upper layers includes aerosols and the bottom ones represents the soil/vegetation strata. This problem is further complicated by the intrinsic anisotropic radiative behavior of natural surfaces and its coupling with atmospheric radiative processes. An increase in aerosol optical thickness is responsible for an increase of the fraction of diffuse sky radiation which, in turn, smooths the effects of surface anisotropy.

The accuracy with which it is possible to retrieve aerosol properties, i.e., to determine the radiative properties of the upper layer(s), is therefore intimately controlled by the knowledge of the underlying surface characteristics. As the number of independent observations is not large enough to fully characterize the radiative properties of all these layers, i.e., the problem is ill-posed or under-constrained, it is necessary to constrain it by providing some assumptions or *a priori* knowledge on the radiative properties of this medium. Since the primary objective is to determine the characteristics of the upper aerosol layer(s), this additional knowledge concerns the lower layer(s), i.e., the surface properties.

Different approaches have been proposed so far to define *a priori* information on surface properties according to the type of radiometers or the nature of the retrieval algorithms [14]. These approaches are briefly reviewed in Section 2 for passive remote sensing radiometers. An original method is proposed here where surface reflectance anisotropy and atmospheric scattering properties are retrieved simultaneously, explicitly accounting for the radiative coupling between these two systems. The algorithm exploits the frequent repeat cycle of the Spinning Enhanced Visible and InfraRed Imager (SEVIRI) observations onboard Meteosat Second Generation (MSG) to characterize the surface bi-directional reflectance factor (BRF). This instrument scans the Earth disk every 15 min in

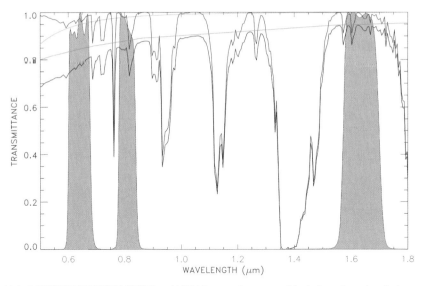

Fig. 11.1. MSG/SEVIRI VIS06, VIS08 and NIR16 spectral response (shaded gray) used to derive aerosol optical depth. The red line represents the total gaseous transmittance in a US standard atmosphere. The green line represents aerosol transmittance for a continental aerosol optical depth [34] of 0.2 at 0.55 μm. The light blue line represents Rayleigh transmittance. The blue line represents the total atmospheric transmittance.

11 spectral channels ranging from 0.6 to 13 μm with a sampling distance at the sub-satellite point of 3 km. The solar channels (see Fig. 11.1) are calibrated with the method proposed in [4] with an estimated accuracy of $\pm 5\%$.

A priori information on surface BRF assumes that surface albedo temporal variations are much slower than aerosol ones. The use of *a priori* information is mathematically rigorously combined with the information derived from SEVIRI observations in the framework of the Optimal Estimation (OE) method. The retrieval approach is briefly described in Section 3. The benefit of the OE approach on the characterization of the surface–atmosphere coupling is illustrated through the analysis of the error covariance matrix. Finally, in this chapter we analyse in detail the impact of improving the surface properties prior knowledge on the accuracy of the retrieved Aerosol Optical Depth (AOD).

2.   Characterization of surface *a priori* information

The determination of *a priori* information on surface reflectance is a critical issue when retrieving aerosol optical depth over land surfaces. Essentially three different approaches have been proposed according to the nature of the observations.

In case of mono-directional multi-spectral imagers, the common approach consists in assuming that the spectral shape of surface reflectance is known or that the ratio between surface reflectance in various wavelengths can be determined. Such an approach has been widely applied to spectral imager such as the MEdium Resolution Imaging Spectrometer (MERIS), the Moderate Resolution Imaging Spectroradiometer (MODIS), or Sea-viewing Wide Field-of-view Sensor (SeaWiFS), (e.g., [32, 38, 36, 28, 16]). The application of this method is often limited to the case of dark surfaces as the magnitude of atmospheric scattering prevails upon surface reflectance in such conditions. For instance, in [32] authors use the atmospherically resistant vegetation index [12] to detect dark dense vegetation pixels for land aerosol remote sensing. In [38] it is assumed that the land surface reflectance is estimated by a linear mixing model of vegetation and non-vegetation spectra, tuned by the normalized differential vegetation index. The method proposed by [28, 16] assumes a predefined ratio between the intensity of the reflected surface radiation in various wavelengths. This ratio might change according to the illumination and viewing conditions. In [36] authors applied an algorithm based on the OE theory where the prior information on the surface albedo relies on the MODIS surface BRF product [11]. The proposed method includes a rigorous mathematical use of *a priori* information, but relies on external datasets which are acquired at a different spatial and temporal resolution. The surface albedo is retrieved by first assuming an albedo spectral shape for the 0.55, 0.67, 0.87 and 1.6 μm channels. The retrieval procedure searches for the solution with the lowest cost by varying the albedo in the 0.55 μm channel and keeping constant the respective ratios of reflectances for all the other channels to that for this channel. Usually, the approaches listed in this paragraph also assume aerosol classes where the spectral shape of the single scattering albedo is prescribed.

A second type of approach imposes constraint on the spectral invariance through the angular shape of the surface directional reflectance. Such an approach is clearly dedicated to multi-angular observations. [19, 20, 8, 7] developed a simple physical model of light scattering that is pertinent to the dual-angle sampling of the Along-Track Scanning Radiometer (ATSR) instrument and can be used to separate the surface bi-directional reflectance from the atmospheric aerosol properties without recourse to prior information on the magnitude of surface reflectance. Studies of modeled and real bi-directional reflectance data have shown that the angular variation of bi-directional reflectance at the different optical bands of the ATSR-2 instrument are comparable [37]. A similar approach has also been proposed in [1] to derive aerosol properties from observations acquired by the Multiangle Imaging SpectroRadiometer (MISR). These authors assume a spectral invariance of the normalized shape of the surface BRF to improve the retrieval of aerosol properties.

Finally, it is worth mentioning an approach based on the temporal stability of surface reflectance and the fact that aerosols tend to increase the observed signal at short wavelengths, as surface is generally dark in this spectral region. Such a principle has been used by different authors. In the method proposed in [13, 24, 23], the surface contribution is determined from temporal compositing of visible imagery, where darker pixels correspond to less atmospheric scattering. Surface reflectance is deduced from the composite using radiative transfer calculations. The proposed approach suffers from a lack of rigorous treatment of the radiative transfer as the shape and intensity of surface reflectance depends on the illumination conditions (e.g., the amount of aerosol) over anisotropic surfaces. Additionally, these maps are contaminated by remaining aerosols or cloud shadows, and the surface reflectance has to be assumed constant over a 28-day period. In [10] authors also proposed a similar approach for MODIS observations, assuming Lambertian surface reflectance. A reference surface reflectance map is generated from the lowest observed values over a month. This method relies on the hypothesis that, in the blue spectral region, surface reflectance is low and that any departure of the observed top-of-atmosphere (TOA) BRF from the reference map is caused by the presence of aerosols in the atmosphere.

3.   Overview of the optimal estimation retrieval method

The main objective of the Land Daily Aerosol (LDA) algorithm described in this chapter is to derive the mean daily tropospheric AOD for various aerosol classes over land surfaces. The aerosol properties are inferred from the inversion of a forward radiative transfer model against daily accumulated observations in the 0.6, 0.8 and 1.6 μm SEVIRI bands.

The proposed algorithm capitalizes on the capability of SEVIRI to acquire data every 15 minutes to perform an angular sampling of the same radiance field under various solar geometries (Fig. 11.3). The temporal accumulation of data acquired under different illumination conditions are used to form a virtual multi-angular measurement system. SEVIRI data are daily accumulated in the 0.6, 0.8 and 1.6 μm bands (noted $\tilde{\lambda}_1$, $\tilde{\lambda}_2$, $\tilde{\lambda}_3$, see Fig. 11.1) to form the measurement vector. The inversion procedure takes place at the end of this daily accumulation period to retrieve the surface reflectance in each band and optical depth for different aerosol classes. The optical thickness is delivered at 0.55 μm. Retrievals at all channels are processed simultaneously.

The retrieval approach is based on the Optimal Estimation (OE) theory (e.g., [31], [35]). The goal of such 1D variational retrieval is to seek an optimal balance between information that can be derived from the observations, and the one that is derived from *a priori* knowledge on the system. These techniques are traditionally used for the retrieval of vertical temperature, pressure or humidity profiles within the atmosphere from sounding observations (e.g., [29]). The basic principle of Optimal Estimation is thus to maximize the probability of the retrieved atmospheric and surface state $\mathbf{x}$ conditional on the value of the measurements and any prior information. A forward model y_m represents the conversion from the state $\mathbf{x}$ to the observation vector $\mathbf{y}_o$. The departure of a potential solution $y_m(\mathbf{x})$ to a given observation vector, $\mathbf{y}_o$, and to an *a priori*, $\mathbf{x}_b$, is expressed by a scalar function, namely the cost function (Section 5.2). It can be shown that maximising probability is equivalent to minimizing a cost function which combines these two pieces of information.

The measurement system in an OE estimation framework is composed of the forward model y_m (Section 4.2), the observations $\mathbf{y}_o$ (Section 5.1) and the *a priori* information $\mathbf{x}_b$ (Section 5.10). For each of these elements, the characterization of the corresponding error covariance matrix is essential and represents thus one of the most difficult aspect of this method (e.g., [30, 17]). Additionally, the minimization of the cost function requires knowledge of the Jacobian of the forward model.

4.   Forward modelling

4.1   State vector

The state parameters defining the radiative properties of the observed medium are divided into two categories. The first one, noted $\mathbf{x}$, represents the state variables that are retrieved from the observations and is referred to as the *state vector*. The second one, noted $\mathbf{b}$, represents those parameters that also have non-negligible radiative effects on the observed medium but may not be reliably estimated from those measurements. They are referred to as the *model parameters* and are obtained from external sources of information. In the present case, the model parameters are composed of the ozone and water vapor total column content $\mathbf{b} = \{U_{O_3}, U_{H_2O}\}$ and are taken from European Centre for Medium-Range Weather Forecasts (ECMWF) data.

The state vector $\mathbf{x}$ is composed of the surface reflectance $\rho_s(\tilde{\lambda})$ in the three SEVIRI bands, and the aerosol optical depth τ delivered at $0.55\,\mu m$. The surface reflectance is represented by the Rahman–Pinty–Verstraete (RPV) model [26] which has four parameters that are all wavelength-dependent:

ρ_0 controls the mean amplitude of the BRF. This parameter strongly varies with the wavelength and mainly controls the surface albedo. It varies between 0 and 1.

k is the modified Minnaert contribution that determines the bowl shape of the BRF. It varies between -1 and 1.

Θ is the asymmetry parameter of the Henyey–Greenstein phase function and also varies between -1 and 1.

h controls the amplitude of the hot-spot, i.e., the 'porosity' of the medium. This parameter takes only positive values and typically varies between 0 and 5. In the present case h is set to a fixed value.

In [3] authors analysed in detail the performance and limits of applicability of this parametric model. This model formally writes

$$\rho_s(\rho_0(\lambda), k(\lambda), \Theta(\lambda), h(\lambda); \mu_0, -\mu_v, \phi_r) = \rho_0(\lambda)\breve{\rho}_s(k(\lambda), \Theta(\lambda), h(\lambda); \mu_0, -\mu_v, \phi_r) \quad (1)$$

where ρ_0 and $\breve{\rho}_s(k(\lambda), \Theta(\lambda), h(\lambda); \mu_0, -\mu_v, \phi_r)$ describe the amplitude and the angular field of the surface BRF, respectively. This latter quantity is expressed by:

$$\breve{\rho}_s(k(\lambda), \Theta(\lambda), h(\lambda); \mu_0, -\mu_v, \phi_r) = M_I(\mu_0, -\mu_v; k(\lambda))F(g; \Theta(\lambda))H(h(\lambda); G) \quad (2)$$

where:

$$M_I(\mu_0, -\mu_v; k(\lambda)) = \frac{\mu_0^{k-1}\mu_v^{k-1}}{(\mu_0 + \mu_v)^{1-k}} \quad (3)$$

$$F(g; \Theta(\lambda)) = \frac{1 - \Theta^2(\lambda)}{[1 + 2\Theta(\lambda)\cos g + \Theta^2(\lambda)]^{3/2}} \quad (4)$$

$$H(h(\lambda); G) = 1 + \frac{1 - h(\lambda)}{1 + G} \quad (5)$$

$$\cos g = \cos\theta_v\cos\theta_0 + \sin\theta_v\sin\theta_0\cos\phi_r \quad (6)$$

$$G = [\tan^2\theta_0 + \tan^2\theta_v - 2\tan\theta_0\tan\theta_v\cos\phi_r]^{1/2} \quad (7)$$

The relative azimuth angle, ϕ_r, is zero when the source of illumination is behind the satellite. Here μ_v is the cosine of the satellite zenith angle θ_v and μ_v is the cosine of the solar zenith angle.

Aerosol optical depth is delivered at $0.55\,\mu m$ for three standard pre-defined aerosol classes (e.g., large particles [33], small non-absorbing particles [27] and continental aerosol model [34]) shown in Fig. 11.2. All these aerosol models assume spherical particles.

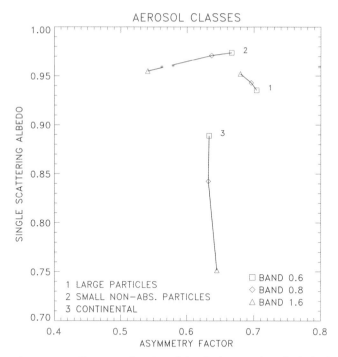

Fig. 11.2. Aerosol asymmetry factor as a function of the single scattering albedo in the three SEVIRI solar bands for the various classes used in LDA.

4.2   Model formulation

The forward model $y_m(\mathbf{x}, \mathbf{b})$ represents the TOA BRF in the SEVIRI solar channels. This model depends on the retrieved state variables $\mathbf{x}$ and the model parameters $\mathbf{b}$. It also formally depends on the observation conditions $\mathbf{m}$, i.e., the illumination and viewing geometry, the acquisition time and location, and finally the wavelength. However, this index is skipped in the notation for the sake of clearness. This model expressed the TOA BRF as a sum of the atmospheric contribution and the contribution specifically due to surface scattering effects. In addition, gaseous absorption is treated separately from the molecular and aerosol scattering-absorbing effects as represented in Fig. 11.3. However, the SEVIRI 0.8 μm band includes a water vapor absorption band close to 0.83 μm so that this band cannot be considered as totally transparent as can be seen on Fig. 11.1. As both water vapor absorption and aerosol scattering occur in the lower part of the atmosphere, it is no longer possible to assume that gaseous absorption is decoupled from the molecular and aerosol scattering–absorbing effects. In order to account for the coupling between aerosol scattering and water vapor absorption, a correction factor η, applied on the water vapor transmittance, has been introduced in the original formulation proposed in [21]. In simplified notation, the forward model writes

$$y_m(\mathbf{x}, \mathbf{b}) = T_g(\mathbf{b})(\eta(\mathbf{x}, \mathbf{b})\rho_a(\mathbf{x}) + \rho_0 \breve{\rho}_s(\mathbf{x})) \qquad (8)$$

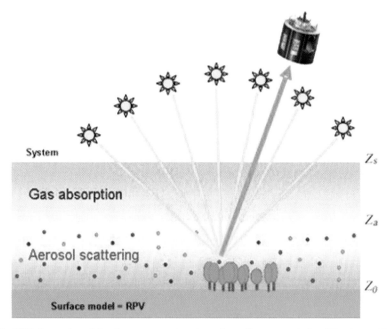

Fig. 11.3. LDA forward model and measurement vector schematic representation. The observation vector $\mathbf{y}_o$ is composed of daily accumulated SEVIRI data acquired under different illumination geometries. The forward model assumes the atmosphere is composed of two layers: the scattering layer ranging from level z_0 (surface) to z_a. The absorbing gaseous layer ranges from level z_a to z_s. These two layers are coupled. The underlying surface at level z_0 is anisotropic.

where

$T_g(\mathbf{b})$ is the total gaseous transmittance;
$\eta(\mathbf{x})$ is the coupling factor between the scattering and gaseous layers;
$\rho_a(\mathbf{x})$ is the scattering layer intrinsic reflectance;
ρ_0 is the amplitude of the surface reflectance;
$\breve{\rho}_s(\mathbf{x})$ is the shape of the surface reflectance at level z_s.

The derivatives $\mathbf{K}_x$ (Jacobian matrix) with respect to the forward model $y_m(\mathbf{x}, \mathbf{b})$ are calculated as forward finite differences with the exception of the derivatives with respect to $\mathbf{b}$.

This model explicitly accounts for the surface anisotropy and its coupling with atmospheric scattering [21]. The costs associated with the computation of the forward model are reduced by pre-computing the values of these functions and storing them in look-up tables. This, however, implies that a limited number of pre-defined solutions to the inverse problem will be considered, especially for the parameters k and Θ controlling the shape of the surface BRF and the optical depths where only the following values are considered : 0.05, 0.1, 0.2, 0.3, 0.4, 0.6, 0.8, 1.0, 1.5. To reduce further the computing time, no interpolation is performed between these values. These discretized values are noted $k_\Delta(\tilde{\lambda})$, $\Theta_\Delta(\tilde{\lambda})$, $h_\Delta(\tilde{\lambda})$ and τ_Δ.

5.   Inverse problem

5.1   Measurement vector

The proposed approach exploits SEVIRI capability to acquire data every 15 minutes over a same pixel to perform an angular sampling of the reflectance changes due to changing illumination conditions. Using the principle of reciprocity applied to the three SEVIRI solar channels [15], this daily temporal accumulation is thus used to form a virtual multi-angular and multispectral measurement vector $\mathbf{y}_o$ that can be used for retrieving simultaneously information on the atmospheric aerosol load-, and the Earth surface reflectance. Formally, this measurement vector contains the TOA Bi-directional Reflectance Factor in

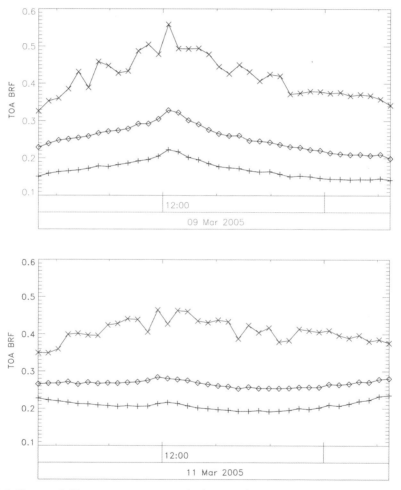

Fig. 11.4. Top panel: Measurement vectors acquired on March 9, 2005, over the Dakar AERONET station (14.38N, 16.95W) in the three SEVIRI solar channels (VIS06 + symbol, VIS08: ◇ symbol and NIR16: × symbol). The daily average AOD estimated at 0.55 μm during that day in the range of 0.06 ± 0.015. **Bottom panel**: The same but for March 11, 2005, where daily average AOD is in the range 1.18 ± 0.17.

the three SEVIRI solar channels. This approach assumes that the aerosol load does not vary during the course of the day. As explained in Section 5.3, the error induced by this assumption is converted into an equivalent aerosol autocorrelation noise which is added to the estimated measurement uncertainties [5]. This vector contains only clear-sky observations, cloudy observations are disregarded with the high-frequency filtering method proposed by [22].

Fig. 11.4 illustrates two examples of measurement vectors accumulated during respectively March 9 and 11, 2005, over the Dakar AERONET station (14.38 N, 16.95 W) in the three SEVIRI solar channels. March 9, 2005, corresponds to a very clear day where the average AOD measured during that day is in the range of 0.06 ± 0.015. On March 11, 2005, the mean observed daily AOD was as large as 1.18 ± 0.17. The impact of this AOD difference on the measurement vector magnitude and shape is clearly visible. An important increase is observed in the VIS06 band, from 0.15 to more than 0.2, whereas a decrease is observed in the NIR16 band. As concern the shape of $\mathbf{y}_o$, the effects of the AOD magnitude are also well pronounced. When the AOD is very low (March 9, 2005),the influence of the surface BRF on the TOA is large. As can be seen on Fig. 11.5, during that day, the satellite and the sun zenith and viewing angles are similar, a configuration responsible for the so-called 'hot-spot' effect. In this particular geometry, both the sun and the radiometer are aligned with respect to the observed pixel so that no surface shadow is seen by SEVIRI. This particular condition translates into a sharp increase of the BRF around

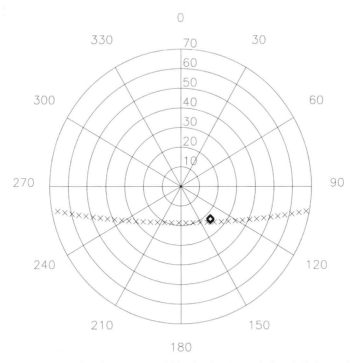

Fig. 11.5. Polar plot representing the geometry of illumination (× symbol) and of observation (◇ symbol) over the Dakar AERONET station in March 2005. Circles represent zenith angles and polar angles represent azimuth angles with zero azimuth pointing to the north.

12:00 UTC. Conversely, large AOD values tend to increase the scattering of incoming solar radiation in the atmosphere, increasing thereby the contribution of the diffuse sky radiation reaching the surface. Under such conditions, shadows are much smoother, leading to a general increase of the surface BRF and a decrease of the 'hot-spot' phenomenon. These figures also illustrate the clear need to retrieve jointly surface reflectance and AOD, as these two parameters are radiatively coupled.

5.2 Definition of the cost function

The basic principle of optimal estimation is to maximize the probability of the retrieved state conditional on the value of the measurements and any *a priori* information. Formally, it is required to maximize the conditional probability $P = P(\mathbf{x}|\mathbf{y}_o, \mathbf{x_b}, \mathbf{b})$ with respect to the values of the state vector $\mathbf{b}$ (Section 4.1), where $\mathbf{x_b}$ is the *a priori* value of the state vector (Section 5.10) and $\mathbf{b}$ are all the other elements of the radiative transfer, called forward model parameters. The assumption is made that errors in the measurements ($\mathbf{S}_y$), and *a priori* ($\mathbf{S}_x$) are normally distributed with zero mean. Then, the conditional probability takes the form:

$$P(\mathbf{x}) \propto \exp\left[-(y_m(\mathbf{x}, \mathbf{b}) - \mathbf{y}_o)\mathbf{S}_y^{-1}(y_m(\mathbf{x}, \mathbf{b}) - \mathbf{y}_o)^T\right] \times \exp\left[-(\mathbf{x} - \mathbf{x}_b)\mathbf{S}_x^{-1}(\mathbf{x} - \mathbf{x}_b)^T\right] \quad (9)$$

where the first term represents weighted deviations from measurements and the second one represents deviations from the *a priori* state parameters. Maximizing probability $P(\mathbf{x})$ is equivalent to minimizing the so-called cost function J

$$J(\mathbf{x}) = (y_m(\mathbf{x}, \mathbf{b}) - \mathbf{y}_o)\mathbf{S}_y^{-1}(y_m(\mathbf{x}, \mathbf{b}) - \mathbf{y}_o)^T + (\mathbf{x} - \mathbf{x}_b)\mathbf{S}_x^{-1}(\mathbf{x} - \mathbf{x}_b)^T \quad (10)$$

Notice that J is minimized with respect to the state variable $\mathbf{x}$, so that the derivative of J is independent of the model parameters $\mathbf{b}$ which therefore cannot be part of the solution. The first term of the right-hand side of Eq. (10) expresses the contribution due to the observation and is noted J_y whereas the second term represents the cost due to the *a priori* information and is noted J_x so that $J = J_y + J_x$.

The diagonal elements of the $\mathbf{S}_y$ matrix represent the variance of the mismatch between the forward model and the observations, which is the sum of the variances associated with all error components of both the observations and the forward model. The estimation of these various contributions is one of the most important aspects of the OE approach. Their estimation is described in Section 5.3. These covariance errors also determine the posterior covariance or retrieval uncertainty as explained in Section 5.8. Consequently, their correct definition is of paramount importance for a reliable estimation of the solution uncertainty. Large values of $\mathbf{S}_y$ tend to reduce the importance of the observations in the retrieval system, increasing thereby the contribution of the *a priori* information.

The diagonal elements of the matrix $\mathbf{S}_x$, on the other hand, represent the error variance of the prior state vector estimate and specify the extent to which the actual solution is expected to deviate from this prior estimate. Any non-diagonal terms of $\mathbf{S}_x$ represent how the state parameters are correlated within a SEVIRI band and between the bands. The associated magnitude is inversely proportional to the strength of this correlation.

When the diagonal elements of $\mathbf{S}_x$ are taking large values, the cost function J is poorly constraint by the *a priori* information and will essentially rely on the observations.

5.3   Measurement system error

A reliable estimation of the measurement system error, i.e., the $\mathbf{S}_y$ matrix, is one of the most critical aspects of the OE approach as it strongly determines the likelihood of the solution. It should therefore be carefully estimated. In the present case, the measurement system total error covariance matrix is defined as

$$\mathbf{S}_y = \mathbf{S}_N + \mathbf{S}_C + \mathbf{S}_A + \mathbf{S}_B + \mathbf{S}_F \tag{11}$$

with

S_N the radiometric noise error matrix;
S_C the calibration uncertainty error matrix;
S_A the aerosol autocorrelation error matrix;
S_B the equivalent model parameter error matrix;
S_F the forward model error matrix.

The first two terms on the right side characterize instrumental noise whereas the last three terms characterize the forward model and associated assumption uncertainty. In the current algorithm, all errors are assumed to have a Gaussian distribution but this assumption has not been rigorously verified.

Radiometric noise $\mathbf{S_N}$

The radiometric noise is composed of (*i*) the instrument noise due to the dark current; (*ii*) the difference between the detector gains; (*iii*) the number of digitalization levels; and finally (*iv*) the geo-location/coregistration accuracy. As SEVIRI images are accumulated during the course of the day to form a virtual multi-angular observation, any inaccuracies in the rectification are converted into an equivalent radiometric error.

Calibration uncertainty $\mathbf{S_C}$

Calibration uncertainties are responsible for systematic error that are present in all measurements. These errors might be responsible for biases that translate into underestimation or overestimation of optical depth. It is, however, not possible to optimize the cost function against such type of bias so that, strictly speaking, it should not appear in the $\mathbf{S}_y$ matrix when it is assembled for the cost function estimation. Conversely, when $\mathbf{S}_y$ is used for the estimation of retrieval total error (see Section 5.8), this contribution should be accounted for.

Equivalent aerosol autocorrelation noise $\mathbf{S_A}$

One of the main assumptions of this algorithm is the relative stability of the aerosol load during the day. The error induced by this assumption is modeled by the equivalent aerosol autocorrelation noise $\mathbf{S}_A$ that accounts for the variation of the AOD around the value taken at a reference time. It increases as the duration of the accumulation period increases.

Equivalent model parameter noise $\mathbf{S_B}$

This noise is due to the errors in the state vector $\mathbf{b}(\bar{t})$. It is converted in Equivalent Model Parameter Noise (EOMPN).

Forward model error $\mathbf{S_F}$

This term represents the intrinsic errors of the fast forward model y_m, i.e., the errors resulting from the forward model assumptions and approximations. This error is estimated comparing the fast model with a very accurate radiative transfer model [6].

Minimization of J

This section addresses the problem of finding the solution $\hat{\mathbf{x}}$ that minimizes J. This is a topic where many techniques and methods can be deployed and where tuning of the adopted scheme can turn out to be as important as the scheme itself. Essentially any method of finding the minimum is acceptable in a sense, with the caveat that in an operational context it must be robust and fast. The particular characteristics of this problem are that:

- *First and second derivatives of J (with respect to* $\mathbf{x}$*) are available and continuous.* This condition implies descent algorithms that make use of the local gradient are possible and these are generally faster than methods that do not.
- *Multiple minima are unlikely.* This condition is, however, not met in the present case, and the solution is estimated for different first guess values (see Section 5.5).
- *J is likely to be approximately quadratic in the region of the solution, far from quadratic elsewhere.* This characteristics is a result of the reasonably strongly nonlinear nature of the forward (radiative transfer) problem. It means that quick convergence from a poor starting position is unlikely.

The cost function $J(\mathbf{x})$ is minimized, for example using a steepest descend method, or the Marquardt–Levenberg method (combined steepest-descent method/Newton method, as described in Section 5.6). This minimization is performed individually for each defined aerosol class. The first and second derivatives of J with respect to $\mathbf{x}$ are given by:

$$\mathbf{J}' = \frac{\partial J}{\partial \mathbf{x}} = \mathbf{K}_x^T \mathbf{S}_y^{-1}(\mu_m(\mathbf{x}, \mathbf{b}) - \mathbf{y}_o) + \mathbf{S}_x^{-1}(\mathbf{x} - \mathbf{x}_b) \tag{12}$$

and

$$\mathbf{J}'' = \frac{\partial^2 J}{\partial \mathbf{x}^2} = \mathbf{K}_x^T \mathbf{S}_y^{-1} \mathbf{K}_x + \mathbf{S}_x^{-1}. \tag{13}$$

The expression for $\mathbf{J}''$ is a commonly used approximation in that $\mathbf{K}_x$ is assumed to be independent of $\mathbf{x}$, i.e., the radiative transfer is linear in $\hat{\mathbf{x}}$. This is only strictly true near the solution (in the region where J is quadratic) but (see next section) since $\mathbf{J}''$ is only employed near the solution the approximation is acceptable. Fig. 11.6 shows an example of the joint retrieval of AOD $\hat{\tau}$ and surface parameters over the Dakar region for a clear (March 9, 2005) and heavy aerosol load (March 11, 2005) day. In this Figure, the BRF surface parameters are used to estimate the surface albedo or more precisely the BiHemispherical Reflectance ($BHR_\bullet$) [18] corresponding to perfectly isotropic illumination (or white sky albedo in the MODIS terminology).

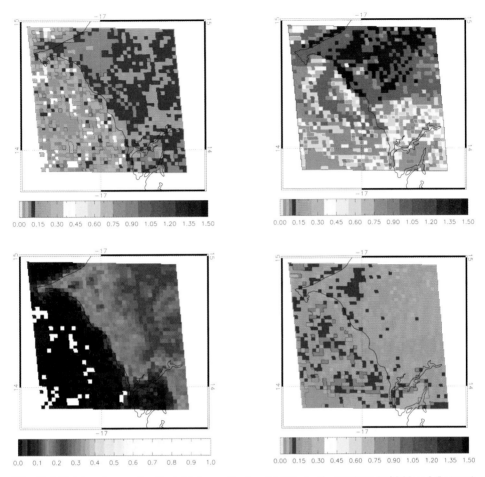

Fig. 11.6. Retrieved aerosol optical thickness (top) and BHR, in the SEVIRI VIS06 band (bottom) over the Dakar area for March 9, 2005 (left) and March 11, 2005 (right).

5.4   First guesses

The presence of multiple local minima are quite likely due to the multi-layer nature of the problem. Hence, when a minimum value is found, an exploration process should be made around that minimum in order to determine whether or not it is a local minimum. Such exploration phase could be computationally expensive. In order to avoid explorations, several independent first guesses are used in order to start the minimization from four different places in the state variable space. For the surface parameters, $k_\Delta(\tilde{\lambda})$, $\Theta_\Delta(\tilde{\lambda})$ and $h_\Delta(\tilde{\lambda})$ are randomly chosen in each wavelength to best sample the state parameter domain, i.e., corresponding to low and high surface reflectance values. For each of these cases, an high and low aerosol load τ_Δ is also randomly selected for each aerosol class. A first guess $\mathbf{x}_0$ is defined as a set of values for $k_\Delta(\tilde{\lambda})$, $\Theta_\Delta(\tilde{\lambda})$, $h_\Delta(\tilde{\lambda})$, and τ_Δ and is chosen among the following four possible combinations of state parameter intervals:

1. Low surface reflectance, low AOD value.
2. Low surface reflectance, high AOD value.
3. High surface reflectance, high AOD value.
4. High surface reflectance, low AOD value.

Once a set of values for $k_\Delta(\tilde\lambda)$, $\Theta_\Delta(\tilde\lambda)$, $h_\Delta(\tilde\lambda)$, and τ_Δ has been randomly selected within each of the four prescribed intervals, the corresponding ρ_0 parameter is estimated in each channel $\tilde\lambda$, minimizing the following expression:

$$\rho_0(\tilde\lambda) = \frac{\sum_\Omega W_{inv}(\Omega, \tilde\lambda) \left[\mathbf{y}_o(\Omega)/T_g(\mathbf{b}) - \eta(\mathbf{x}, \mathbf{b})\rho_a(\mathbf{x}) \right]}{\sum_\Omega W_{inv}(\Omega, \tilde\lambda)\breve\rho_s(\mathbf{x})} \tag{14}$$

where $W_{inv}(\Omega, \tilde\lambda)$ is the weight $(1/\sigma_y^2)$ given to the BRF value observed with the angle Ω. The selected solution $\hat{\mathbf{x}}$ among the four first guesses is the one which provides the smallest cost function J.

5.5 Marquardt descent algorithm

The descent algorithm aims at finding the value $\hat{\mathbf{x}}$ that minimizes J. To find the minimum a 'first guess' state vector $\mathbf{x}_0$ (Section 5.5) is randomly selected and takes steps, $\delta\mathbf{x}_n$, based on some algorithmic theory. Assuming the value of J decreases at each step then the updated $\mathbf{x}$ vector is taking the process towards the cost function's minimum. The algorithm uses an implementation of the Marquardt algorithm [25] to define the value of $\delta\mathbf{x}_n$. The use of this algorithm is consistent with the three points made above. The rationale of the Marquardt algorithm is to use a weighted combination of steepest descent and Newtonian descent according to the characteristics of the cost function. Thus, when the cost function is near quadratic (generally near the solution) the efficiency of the Newtonian scheme is employed, and when the cost function is far from quadratic (generally when far from the solution) the robustness of the Steepest Descent algorithm is favoured.

The Steepest Descent algorithm is intuitively the simplest. The vector $-\mathbf{J}'$ defines the 'downward' direction of the local steepest gradient. A move $\delta\mathbf{x} = \mathbf{J}'$ is almost certainly at least approximately in the direction of the minimum although it may be too far or barely far enough. The step is therefore usually scaled, $\delta\mathbf{x} = -a\mathbf{J}'$ where a is variable. If J is found to be decreasing a can be increased to move faster; if J increases then a is reduced until J decreases. J must eventually decrease with this method otherwise something is wrong with the calculation of $\partial J/\partial\mathbf{x}$. The problem with steepest descent is that it can be very slow to converge, especially near the solution where the gradient necessarily becomes small. It is, however, very robust.

Newtonian descent on the other hand is very fast near the solution because it will find it in one iteration if J is quadratic. Newton's method finds the root of an equation and is therefore applied here in the form to find the root of $\mathbf{J}' = 0$. The Newton step is therefore defined as $\delta x = -\mathbf{J}'/\mathbf{J}''$. The problem with Newtonian descent is that, away from the solution, J can be very non-quadratic; $\mathbf{J}''$ can easily have the 'wrong' sign and the step is taken away from the solution. Scaling can not solve this problem.

The combined use of Steepest Descent and Newtonian methods constitutes the method of Marquardt. Before each step, the method checks whether the resulting cost is reduced. If so the step is taken and an adjustment is made to make the next step 'more Newtonian'. If an increase in J is detected, then the step is not taken and an adjustment is made towards more steepest descent. In this way, the Marquardt method adopts Steepest Descent away from the solution and makes use of Newtonian descent near the solution. Formally, the increment in the Marquardt method is:

$$\delta\mathbf{x} = -(\mathbf{J}'' + a\mathbf{I})^{-1}\mathbf{J}' \qquad (15)$$

where $\mathbf{I}$ is the unit matrix (size $n_x \times n_x$) and a is the variable that controls the contribution of the Steepest Descent or the exploration. When a is large (compared to the 'average' size of $\mathbf{J}''$) the step tends to that of the Steepest Descent; when a is small the step is close to the Newtonian one. To initialize the descent, a is set proportional to the average of the diagonals of $\mathbf{J}''$ to obtain a reasonable value

$$a_o = MQ_{start} \times \mathrm{trace}(\mathbf{J}''). \qquad (16)$$

With a successful (decreasing J) step, then the control parameter is decreased

$$a_{n+1} = a_n \div MQ_{step} \qquad (16)$$

and with an unsuccessful step the control parameter is increased

$$a_{n+1} = a_n \times MQ_{step}. \qquad (18)$$

The parameters MQ_{start} and MQ_{step} determine respectively the initial value of a and the contribution of the steepest descend method. The Marquardt a parameter is set and the iteration loop starts. This involves calculation of δx via the quantities $\mathbf{J}'$ and $\mathbf{J}''$. S_y is updated at each iteration step. The step $\delta\mathbf{x}$ is checked to see if any parameter will go out of bounds and limits an offending parameter to the bound value if this is the case. Following this, the cost J at the prospective new state is calculated and a check is made to ensure that it is less than the previous value. If it is not, then the step is not made; the weight given to the Steepest Descent is *increased* and the step recalculated. If the cost is decreasing, the step is made and the Steepest Descent weight is *decreased*. Following this, the step is checked for convergence and the results output if so, a further iteration taken if not.

The iteration process is stopped when the decrease in J between iterations, δJ_n, is so small as to be negligible, determined as smaller than a preset value. A parameter that is not well constrained in a particular situation can oscillate or be unstable (because the cost function is 'flat' in that direction). This instability makes it look like the solution is not yet found whereas, in fact, the cost is minimal and cannot decrease further. The use of a convergence criteria on δJ_n avoids this problem.

5.6 Quality control

In the case of $\mathbf{S}_x$ and $\mathbf{S}_y$ being both diagonal matrices, the sum of the squared residuals scaled by their uncertainties represents a χ^2, distribution. In such condition, the magnitude of $J(\hat{\mathbf{x}})$ is a good indicator of the reliability of the solution, i.e., the goodness-of-fit between the observations to the model on one hand and prior and posterior estimate on the other hand. This indicator might, however, be difficult to spatially or temporally interpret as the number of degrees of freedom v_y can change from pixel to pixel or from day to day due to changing number of available observations caused by cloud or illumination conditions. The number of degrees of freedom to be used in this case is equal to the total number of independent pieces of information introduced into the retrieval system, i.e., the number of observational data n_y, plus the number of prior estimates minus the size n_x of the state vector $\mathbf{x}$. The normalization of $J(\hat{\mathbf{x}})$ by the number of degrees of freedom, should be close to 1, which means that the mismatch between the observations and the modeled data (or between the prior and posterior estimate) has the same magnitude as $\mathbf{S}_y$ ($\mathbf{S}_x$) [17].

This method might, however, be misleading in the present case as the same criteria would be applied whatever the number of degrees of freedom. In case only a very limited number of observations is available, the requirement on the quality of the fit should be much higher as in the case where a large number of observations are available. Hence, a probabilistic approach is proposed here based only on J_y, i.e., the model–observation mismatch. Assuming that the differences between y_m and $\mathbf{y}_o$ are normally distributed, the probability Q that a value of J_y as poor as the values given by Eq. (10) of the solution $\hat{\mathbf{x}}$ should occur by chance writes

$$P_{\hat{\mathbf{x}}} = Q(J_y(\hat{\mathbf{x}})|v_y) = 1 - \frac{1}{2^{v_y/2}\Gamma(v_y/2)} \int_0^{\chi_t^2} e^{-t/2} t^{v_y/2-1} \, dt, \tag{19}$$

where $v_y = n_y - n_x$ is the number of degrees of freedom, and Γ is the Gamma function. n_y and n_x represent respectively the size of the $\mathbf{y}_o$ and $\mathbf{x}$ vectors. This statistics formulates the probability $P_{\hat{\mathbf{x}}}$ to find a solution $\hat{\mathbf{x}}$ with a cost $J_y(\hat{\mathbf{x}})$ greater than the cost that has been calculated. A high probability means that there is a high probability not to find a solution at a lower cost. So the calculated cost is possibly a good solution to the problem. On the other hand, a small probability means that the possibility of finding a greater cost is low. So, it means that it would be highly probable to find a better solution, and therefore such a probability should be interpreted as a poor agreement between the observations and the

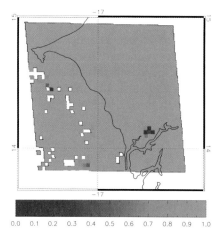
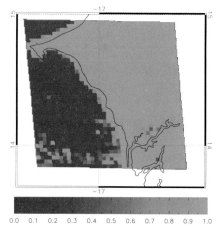

Fig. 11.7. Estimated probability as defined by Eq. (19) over the Dakar area for March 9, 2005 (left) and March 11, 2005 (right).

direct model. An example of estimated probability corresponding to a clear and an aerosol loaded day is shown on Fig. 11.7. Probabilities are taking very low values over sea on March 11, 2005, when the AOD is larger than one. Retrieved AOD and $BHR_\bullet$ over sea for that day are therefore not reliable. The retrieved $BHR_\bullet$, shown on the lower right panel of Fig. 11.6, is taking suspiciously high values over sea that should be rejected. The comparison between the retrieved $BHR_\bullet$ for these two dates (Fig. 11.6) overland surfaces also shows discrepancies. The LDA algorithm still requires some improvements in the definition of the aerosol classes (see Section 7),which affect particularly retrievals with high optical depth.

5.7 Linear error analysis

The shape of $J(\mathbf{x})$ in the vicinity of the solution $\hat{\mathbf{x}}$ is partially determined by the magnitude of the elements of the matrices $\mathbf{S}_y$ and $\mathbf{S}_x$. In the case of the observations, large values of $\mathbf{S}_y$ tend to reduce the $J(\mathbf{x})$ value but also to flatten its shape. In this condition, any minor perturbation in $\mathbf{y}_o$, resulting from measurement noise or forward model uncertainties, can lead to large displacement of $\hat{\mathbf{x}}$, increasing thereby the solution uncertainty. Large values of $\mathbf{S}_y$ decrease the importance of the *a priori* information and therefore its contribution to the estimated *a posteriori* error.

The retrieval error is based on the OE theory, assuming a linear behavior of $y_m(\mathbf{x}, \mathbf{b})$ in the vicinity of the solution $\hat{\mathbf{x}}$. In this condition, the *a posteriori* error covariance matrix $\mathbf{S}_\varepsilon$ writes

$$\sigma_{\hat{\mathbf{x}}}^2 = \left(\frac{\partial^2 J(\mathbf{x})}{\partial \mathbf{x}^2}\right)^{-1} = \left(\mathbf{K}_x^T \mathbf{S}_y^{-1} \mathbf{K}_x + \mathbf{S}_x^{-1}\right)^{-1}, \tag{20}$$

where $\mathbf{K}_x$ is the gradient of $y_m(\mathbf{x}, \mathbf{b})$ calculated at the point $\hat{\mathbf{x}}$.

The error matrix $\mathbf{S}_\varepsilon$ provides information on the correlation between the retrieved model state variables of the solution state vector $\hat{\mathbf{x}}$. Comparison of the values in $\mathbf{S}_x$ and $\mathbf{S}_\varepsilon$, re-

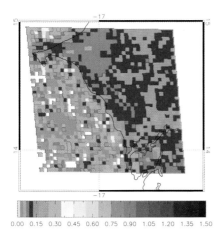 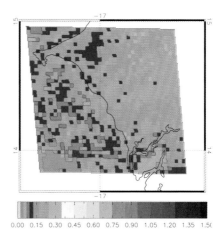

Fig. 11.8. Estimated AOD error over the Dakar area for the March 9, 2005 (left) and March 11, 2005 (right).

spectively the prior and posterior covariance matrices, expresses the knowledge gain from the measurement system and its associated uncertainties. Examples of error covariance matrices $\mathbf{S}_\varepsilon$ are given in Section 6. These estimated errors should be interpreted under the assumption that the forward model is actually representing the observed medium, i.e., reasonably plane-parallel in nature, and that the observed aerosols are consistent with the aerosol classes. Fig. 11.8 shows an example of the estimated AOD error derived from Eq. (20) over the Dakar region for a clear day (March 9, 2005) and a day with heavy aerosol load (March 11, 2005). On March 9, 2005, the estimated AOD error are larger over land than sea surfaces. It is not possible to drawn a similar conclusion for March 11, 2005, as retrievals over sea surface are not reliable according to results shown on Fig. 11.7.

5.8   Aerosol class selection

For each processed SEVIRI pixel, the solution $\hat{\mathbf{x}}$ is calculated independently for each pre-defined aerosol class. As an aerosol class is not expected to change erratically from one pixel to another, a spatial analysis is performed to determine the most probable aerosol class within a given region. For each pixel (p_x, p_y), a restricted geographical area of $n_p \times n_p$ pixels centered in (p_x, p_y) is defined. The mean probability $\bar{P}_{\hat{\mathbf{x}}}$ over this $n_p \times n_p$ area is estimated for each aerosol class. The class with the highest mean probability $\bar{P}_{\hat{\mathbf{x}}}$ is taken as the best one for the pixel (p_x, p_y).

This computation is performed only for the aerosol class with a successful retrieval in (p_x, p_y), i.e., with a probability higher than a pre-defined minimum threshold value. When two aerosol classes have the same mean probability $\bar{P}_{\hat{\mathbf{x}}}$, the one with the smallest relative error $\sigma_{\hat{\tau}}/\hat{\tau}$ is chosen.

5.9   *A priori* information

As already mentioned, one of the major issues in retrieving the aerosol load over land surfaces is to separate the aerosol contribution from the surface one. In order to improve the retrieval of the aerosol load, it is therefore essential to characterize the surface as accurately as possible. To do so, a mechanism has been implemented to update the prior knowledge $\mathbf{x}_b$ on the surface reflectance is implemented. This prior update procedure relies on the assumption that the surface BRF temporal stability is stronger than the aerosol load one. Hence, a time-series analysis can be performed on the previous day retrievals, seven in the present case, to provide *a priori* information on the expected amplitude and shape of the surface BRF. The estimated retrieval error can thus subsequently be used to estimate the *a priori* information error on the state of the surface.

The most obvious way to perform this temporal analysis and to select the 'best day' d_b is to take the one with the highest probability. However, as clouds tend to increase the signal received at the satellite level, the selection of the solution with the smallest $BHR_\bullet$ value in the VIS06 band will tend to minimize the impact of the clouds. $BHR_\bullet$ is the surface BiHemispherical Reflectance corresponding to purely isotropic incident flux. Hence, the most representative solution $\bar{\bar{x}}(d)$ within a compositing period $[1 \dots n_a]$ is selected according to the following mechanism:

1. Solutions with too large $BHR_\bullet$ relative errors or a probabilities $P_{\hat{\mathbf{x}}}$ smaller than 0.1 are disregarded from the time series. The purpose of this test is to discard unlikely solutions

with large uncertainties. An error is too large when the corresponding relative error exceeds a threshold value, in at least one of the processed spectral bands.

2. Within the remaining solutions of the time-series, solutions with too high $BHR_\bullet$ values in the VIS06 band are disregarded. This test aims at removing solutions that might still be contaminated by clouds. It also privileges snow free solution. To identify these too high values, the weighted mean $\bar{BHR}_\bullet$ and standard deviations

$$\sigma_d(\bar{BHR}_\bullet) = \sqrt{\frac{n_a \sum_{d=1}^{n_a} \hat{\kappa}(d)(BHR_\bullet(d) - \bar{BHR}_\bullet)^2}{(n_a - 1) \sum_{d=1}^{n_a} \hat{\kappa}(d)}} \tag{21}$$

values are computed in the VIS06 band, where the weight $\hat{\kappa}(d)$ is equal to $1/\sigma_{BHR_\bullet}^2(d)$. A $BHR_\bullet(d)$ is rejected from the time-series when

$$BHR_\bullet(d) > \bar{BHR}_\bullet + \sigma_d(\bar{BHR}_\bullet). \tag{22}$$

3. The selected solution corresponding to day d_b is the one which fullfils the following criteria within the remaining solutions of the time-series:
   (a) the highest probability;
   (b) the largest number of degrees of freedom;
   (c) the smallest AOD value;
   (d) the smallest AOD error.

In case there is more than one valid solution in the time-series, the error of most representative solution $\bar{\hat{x}}(d_b)$ is recalculated with

$$\sigma_{\bar{\hat{x}}}(d_b) = \text{MAX}\left\{ \sqrt{\sum_{d=1}^{n_a} \hat{\kappa}(d)(\hat{x}(d) - \bar{\hat{x}})^2}, \sigma_{\hat{x}}(d_b) \right\} \tag{23}$$

where the sum is performed only on all valid solutions. It is expected that the parameter error decreases as the number of days increases. In case only one likely solution has been found during the temporal compositing period, the error is taken from Eq. (20). This temporal analysis only starts after the processing of at least n_a days. Hence, during this 'training' period, a constant value is assigned to $\mathbf{x}_b$ with a very large error $\sigma_{\bar{\mathbf{x}}}$.

6.   Interpretation of the error and autocorrelation matrices

The *a posteriori* error covariance matrix $\mathbf{S}_\varepsilon$ provides useful information on the correlation between the state variables. In order to interpret the non-diagonal terms of $\mathbf{S}_\varepsilon$ a re-scaling is performed, leading to the autocorrelation matrix Σ whose components write

$$\Sigma_{ij} = \frac{\mathbf{S}_{\varepsilon,ij}}{\sqrt{\mathbf{S}_{\varepsilon,ii} \cdot \mathbf{S}_{\varepsilon,jj}}} \tag{24}$$

such that $-1 \leq \Sigma_{ij} \leq 1$. Three different cases could be differentiated for the interpretation of the terms of Σ_{ij}:

1. When a term $\Sigma_{ij} \to 0$, uncertainties on the state variables i and j are uncorrelated. In other words, a retrieval error on the state variable i does not affect the retrieval accuracy of variable j.
2. When $\Sigma_{ij} \to 1$, then uncertainties on the state variables i and j are correlated, i.e., any error on i affects j, leading to an overestimation or underestimation of both variables.
3. When $\Sigma_{ij} \to -1$ then uncertainties on the state variables i and j are anti-orrelated. Any overestimation of i translates into an underestimation of j.

In order to illustrate the interpretation of $\mathbf{S}_\varepsilon$ and Σ, the algorithm was applied with SEVIRI data acquired over AERONET [9] stations for a time period extending from February 15, 2005, to April 15, 2005. 56 AERONET stations with data available during that period over the Meteosat Second Generation (MSG) disk are shown in Fig. 11.9. Unfortunately, these data are not uniformly distributed within the disk but are essentially concentrated over Europe.

Fig. 11.9. Location of the AERONET stations used in this study. The * symbol indicates the restricted list of stations for which the time-series analysis was performed shown on Fig. 11.16.

The effect of the *a priori* information **x** on $\mathbf{S}_\varepsilon$ is assessed by comparing matrices derived for each station with and without update of the prior information. The efficiency of the *a priori* information can be measured by the decrease in the correlation between the AOD τ_{550} and the surface parameters. This analysis essentially reveals four different types of impact:

1. Only the correlations between the amplitude (ρ) of the surface contribution and the AOD τ_{550} are reduced.
2. Only the correlations between the anisotropy surface (Θ) and the AOD τ_{550} are reduced.
3. Both correlations between τ_{550} and the amplitude (ρ), and τ_{550} and the surface anisotropy (Θ) are reduced.
4. The update of the prior has little impact on the retrieval. Fig. 11.10 represents the Σ matrices respectively without (top panel) and with (bottom panel) update of prior information on the surface. Note that all these matrices are symmetric and should be interpreted the following way: the first column indicates how the AOD parameter is correlated with the surface parameters in each SEVIRI band. The diagonal blocks show the correlation between the surface parameters within a SEVIRI band. The non-diagonal blocks provides information on the spectral correlation of the surface parameters between the SEVIRI bands.

The comparison between the two panels of Fig 11.10 illustrates thus the impact of updating the surface *a priori* knowledge on the autocorrelation matrix. These figures correspond to the case where both the crossed terms τ/amplitude (ρ), and τ- surface anisotropy (Θ) are reduced. The presented matrices give an averaged value of the results obtained for all stations where such a behavior was observed. In all cases, only retrievals with a probability $P_{(\mathbf{x})}$ larger than 0.9 are considered for analysis.

When the surface prior information is updated with the results of the previous days (Fig. 11.10 bottom panel), it is clear that the overall correlations are decreased. The correlation between the errors on τ and the surface variables is reduced by at least 37 % (with a maximum of 94 % for the crossed terms τ/k) in the VIS08 and NIR16 channels. In the VIS06 channel, the decrease ranges between 4 % and 31 %. In general, the lowest gain is observed for the crossed terms τ/ρ, or τ/θ, showing that the coupling between the aerosol contribution and the surface contribution to the TOA BRF is made through both the amplitude of the surface contribution and the anisotropic properties of the surface. The anticorrelation between τ and the surface parameters ρ and Θ is an evidence of the compensation mechanism that radiatively couples the surface contribution to the aerosol one. As natural surface reflectance tends to exhibit predominantly back scattering and aerosol forward scattering, any under (over)-estimation Θ translates into an over (under)-estimation of the aerosol contribution. It demonstrates the importance of the angular sampling of the observation vector $\mathbf{y}_o$ to characterize both the surface anisotropy and aerosol phase function.

For the diagonal blocks, the gain obtained by improving the *a priori* knowledge ranges between 10 % and 46 %, with the lowest values for the crossed terms ρ/θ. Within each SEVIRI band, the ρ_0, k and Θ parameters are strongly coupled, which indicates the difficulty to discriminate the individual contribution of each of these parameters in determining the shape and amplitude of the surface BRF. This behavior supports the concept of introducing a spectral invariance of the BRF shape as proposed by [1]. In the OE framework, this *a priori* information is equivalent to having the non-diagonal terms corresponding to the couple of parameters (k, Θ) in the various SEVIRI bands different from zero. The

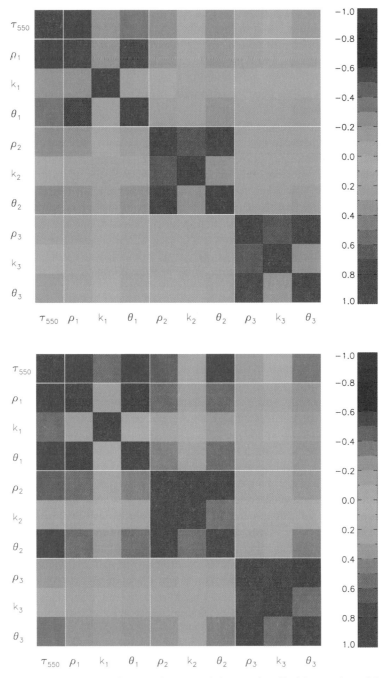

Fig. 11.10. Top panel: Example of averaged autocorrelation matrices Σ without update of the prior information. ρ is the amplitude of the surface contribution. k is the modified Minnaert contribution. θ is the Henyey–Greenstein contribution (surface anisotropy). The indices 1, 2, and 3 stand respectively for the channels VIS06, VIS08, and NIR16. Bottom panel: The same but with updated prior information.

determination of the strength of this correlation is clearly outside the scope of this chapter. These terms are thus kept equal to zero for the time being.

The non-diagonal blocks provides information on the spectral correlation between the SEVIRI bands. As no *a priori* information is imposed as concerns the surface spectral shape, the only spectral correlation between the surface parameters results from the pre-defined aerosol classes, where the spectral shape of the single scattering albedo and asymmetry parameter are imposed as shown on Fig. 11.2. This translates into a positive correlation between the ρ_0 values derived in the SEVIRI bands. This conclusion also holds for the Θ parameter. The ρ_0 and Θ parameters are also coupled between bands, e.g., ρ_0 in the VIS06 band with the Θ in the VIS08 band. The k parameters do not exhibit inter-band correlation. There is, however, a weak anti-correlation between k derived in the VIS06 band and ρ_0 and Θ derived in the VIS08 band. For some extra-diagonal elements of Σ, some negative gains have been locally observed (i.e., an increasing correlation despite an improved *a priori* information), but they correspond to negligible absolute values in any case when compared with the order of most of the matrix values. This example illustrates the necessity to determine correctly both the surface anisotropy and the aerosol phase function for the AOD retrieval over land surfaces.

Fig. 11.11 presents the corresponding relative errors, respectively, for the simulations without update of the *a priori* information, and the simulations with improved *a priori* knowledge. The significant decrease of the error terms when the *a priori* information on the surface is improved is clear, in particular concerning the estimation of the Θ parameters.

7.   Temporal analysis of prior information update

Section 6 illustrates the role of the surface *a priori* information mechanism described in Section 5.10 in terms of its impact on the error covariance and autocorrelation matrices. Both matrices show an overall positive impact. The performance of this mechanism is examined now against AOD ground measurements. A series of experiments are defined to evaluate the algorithm performance, comparing the retrieved aerosol optical depths with those from the AERONET dataset. The reference experiment is made without updating the prior information on the surface. In this case, the default prior information is fixed once and for all, and is given with a very large error covariance $\mathbf{S}_x$ so that $\mathbf{x}_b$ has no significant impact on $J(\mathbf{x})$. Sensitivity analysis is next performed using, from one day to another, updated prior information on the surface, as described in Section 5.10. The results of these retrievals are compared with the reference cases, that is to say the cases without prior update of the surface reflectance.

As explained in Section 5.10, the surface prior update mechanism consists in updating the surface components of $\mathbf{x}_b$ and $\mathbf{S}_x$ with reliable retrieved values from previous days. The impact of this mechanism on the retrieved solution $\hat{\mathbf{x}}$ is first examined for a time-series derived over the Dakar AERONET station (Fig. 11.12). As can be seen on the bottom panel of this figure, a major dust event has been recorded by the AERONET station on March 11, 2005, with a mean daily ground estimated AOD values as large as 1.18 at 0.55 μm.

When no *a priori* is used on the surface reflectance (dashed line),high $BHR_\bullet$ values are retrieved in all channels by the LDA algorithm. This situation illustrates the difficulty to

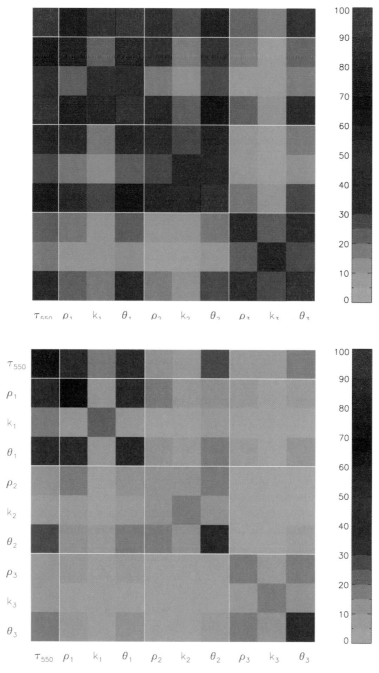

Fig. 11.11. Top panel: Example of averaged relative error matrices without update of the prior information. Values range from 0 % (in green) up to a predefined threshold (here 100 %) (in red). p is the amplitude of the surface contribution. k is the modified Minnaert contribution. θ is the Henyey–Greenstein contribution (surface anisotropy). The indices 1, 2, and 3 stand respectively for the channels VIS06, VIS08, and NIR16. Bottom panel: The same but with updated prior information.

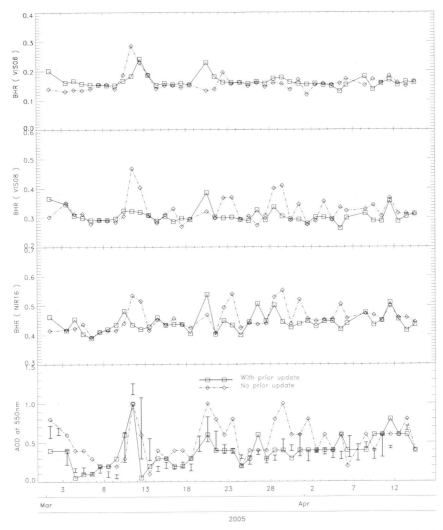

Fig. 11.12. Impact of updating the prior information on the surface BRF and the AOD over the Dakar AERONET station. Dashed line (symbol): no update of the prior information. Plain line (symbol): update of the prior information. Vertical bars in the lower panel indicate the daily variability of AERONET retrieved AOD.

discriminate atmospheric from surface contribution for large AOD values. This event is further illustrated on Figs 11.13 to 11.15. As can be seen on these figures, the retrieval of all surface parameters ρ_0, k and Θ are affected by the March, 11 event which translates by a decrease of k and an increase of ρ_0 and Θ. In this specific event, the use of the prior information has a major impact on the AOD error estimation. The retrieved AOD on March 11, 2005 is equal to 1.0, with or without providing surface prior information (see Fig. 11.7). However, as already seen in Section 6, providing surface prior information has a major impact on the estimated AOD error which decreases from 0.693 without prior down to 0.188 with prior surface information.

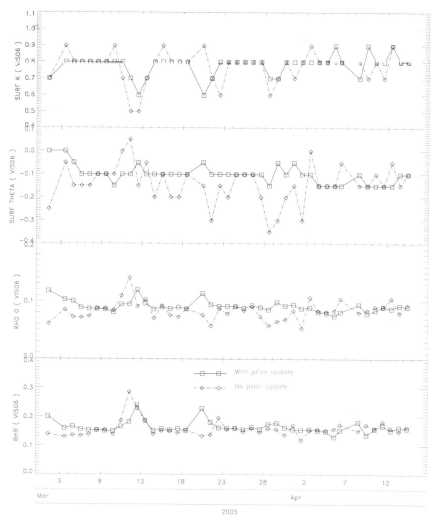

Fig. 11.13. Impact of updating the prior surface information on retrieved surface parameters over the Dakar AERONET station. Dashed line (symbol): no update of the prior information. Plain line (symbol): update of the prior surface information. Results are shown for the SEVIRI VIS06 band. The first panel is for k, the second for Θ, the third for ρ_0 and the last one for the $BHR_\bullet$.

When *a priori* information is provided on the surface reflectance, surface parameter time-series, shown with the solid line on Figs 11.13 to 11.15 are more stable than without update (dashed line). Consequently, the retrieved surface $BHR_\bullet$ remains closer to the mean value which corresponds to a decrease of the $BHR_\bullet$ standard deviation (Table 11.1). This is essentially true in the VIS06 and VIS08 bands. Additionally, the root-mean-square error (RMSE) between AERONET and LDA AOD retrieval decreases from 0.27 (without *a priori* update) to 0.21 (with *a priori* update) during that period. These results clearly illustrate the stabilization of the surface parameters and improvement of AOD retrieval resulting from the *a priori* information.

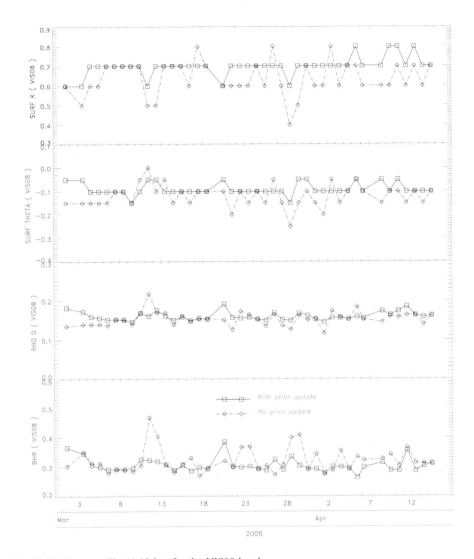

Fig. 11.14. Same as Fig. 11.13 but for the VIS08 band.

Table 11.1. Impact of updating the surface prior information on the surface $BHR_\bullet$ and the AOD over the Dakar AERONET station. The $BHR_\bullet$ columns represent the weight mean retrieved value during the processed period. The σ columns represent the weighted standard deviation estimated with Eq. 21. The Bias column is the bias between AERONET and LDA AOD values at 0.55 μm. The RMSE column represents the root-mean-square error between LDA and AERONET AOD retrieval

	VIS06		VIS08		NIR16		AOD	
	$BHR_\bullet$	σ	$BHR_\bullet$	σ	$BHR_\bullet$	σ	Bias	RMSE
No prior	0.156	0.021	0.311	0.033	0.453	0.037	− 0.085	0.272
Prior	0.164	0.020	0.304	0.022	0.440	0.030	0.007	0.209

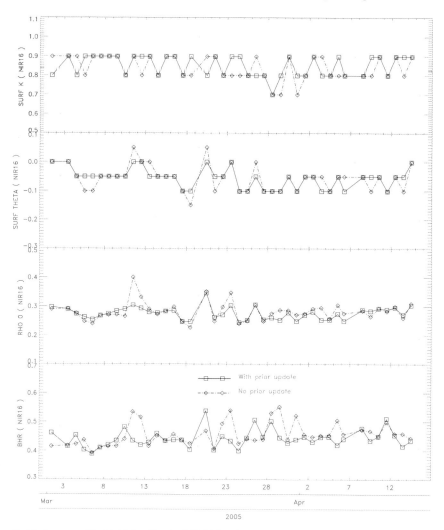

Fig. 11.15. Same as Fig. 11.13 but for the NIR16 band.

When observed daily AOD variations are large, as is the case on March 10 and 12, the LDA algorithm performs poorly. These situations violate the algorithm assumptions and lead to unexpected behavior. On March 10ß, the retrieved *BHR.* value in the NIR16 is too large, whereas large *BHR.* value is found in the VIS06 band on March 12.

This result indicates that the performance of the algorithm should globally increase in time, as the accuracy of the surface characterization increases, leading thereby to a better separation between the surface and the atmospheric contributions. This analysis is limited to stations that have a complete record during the 2-month time interval, eliminating stations for which too scarce observations are available due to the presence of clouds or other limiting factors. These stations are indicated with the * symbol on Fig. 11.9. When no update of the *a priori* takes place, the overall RMSE between the retrieved

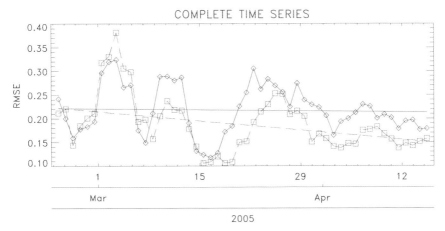

Fig. 11.16. RMSE time-series analysis between retrieved aerosol optical depth from SEVIRI observations and AERONET data shown with a * symbol in Fig. 11.9. The solid line (symbol) represents the reference experiment, i.e., without update of the *a priori* information on surface reflectance. The dashed line (□ symbol) represents the RMSE with update of the *a priori* information.

AODs and AERONET (shown with the solid line on Fig. 11.16) does not show any particular trend during the period of interest, despite some sharp variations from one day to another. These variations are essentially due to the limited number of samples used to derive these statistics. Conversely, when *a priori* information on the surface is provided, the RMSE decreases in time (dashed line). This result shows the positive impact of the 'memory' mechanism put in place for the surface reflectance. The RMSE exhibits a slow decrease in time, with an average reduction of about 35 % at the end of the processed period.

8. Quantitative effects of prior updating

The overall impact of the prior update mechanism is analysed now considering all AERONET observations available during the investigation period (Section 6). As can be seen on Fig. 11.17, the improvement of the surface BRF *a priori* information tends to reduce the RMSE between the AERONET and our retrieval from 0.30 to 0.28. However, our retrieval tends to underestimate AERONET values. Part of this difference might be explained by the difference in the measurement spatial resolution.

There is a major difference as function of the type of aerosol. Fewer large particles are found when *a priori* information on the surface is provided. This mechanism has, however, no positive impact in this case (Fig. 11.18) and the retrieved values are totally underestimated. Conversely, retrievals are much better when LDA identify the small non-absorbing aerosol as the best class (Fig. 11.19) where the RMSE decreases from 0.32 to 0.27. These results demonstrate the importance of the aerosol class definition in the accuracy of the retrieval. Large particles defined in the aerosol model are assumed spherical, which is clearly unappropriated in the current situation. It is necessary to consider correctly the shape of large aerosol particles when estimating the phase function (e.g., [39, 2]).

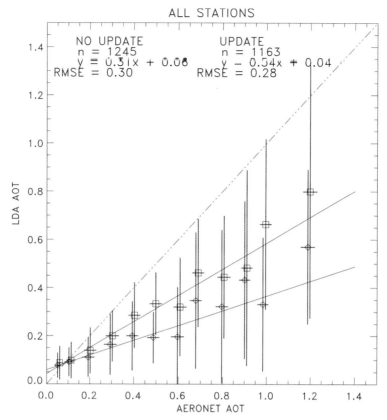

Fig. 11.17. AERONET versus LDA AOD for 0.1 bins using all stations shown in Fig. 11.9 from February 15, 2005 to April 15, 2005. The ◇ symbol represents the reference experiment. The □ symbol represents the experiment with *a priori* update for all aerosol classes.

9.   Discussion and conclusion

An original method has been proposed to retrieve simultaneously aerosol load and surface reflectance using the SEVIRI observations. The algorithm has proved to be capable of separating the aerosol contribution from the surface one in the overall TOA observed signal. The method that has been put in place, based on Optimal Estimation, insures a rigorous control of the system errors, and allows a quality check on the retrieved information. The *a priori* information on surface reflectance is based on its temporal stability, an assumption which is rigorously handled in the framework of the OE theory.

The analysis of the autocorrelation matrices has shown the strong coupling between surface and aerosol contributions, and provides a detailed understanding of, the compensation mechanism that occurs between the various state variables. This demonstrates the necessity of retrieving simultaneously the surface reflectance and the aerosol optical depth, and the importance of the angular sampling in this joint retrieval.

Comparisons with AERONET time-series over a period of 2 months, and at 56 locations around the Earth disk covered by the SEVIRI instrument have demonstrated the robustness

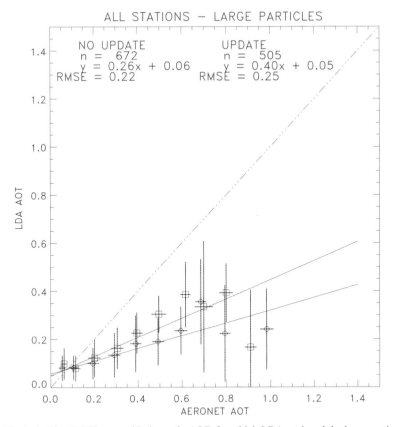

Fig. 11.18. As in Fig. 11.17 but considering only AOD for which LDA retrieved the large particle class.

of the algorithm. They have also shown the stabilization effect on the surface properties of improving the *a priori* knowledge on the surface state variables, and the resulting benefit on the AOD retrieved values when compared to AERONET. The retrieval error has shown to decrease significantly when the *a priori* information on the surface variables is improved and better constrained.

The benefits in time of updating the prior can be summarized by the decreasing trend of the RMSE. These comparisons have also shown the limitations of the aerosol class definitions currently used in LDA, in particular in the case of large particles. The use of specific phase functions for non-spherical aerosols will be part of future work on the LDA algorithm.

More generally, the use of aerosol classes does not fit very well in the context of OE theory. Indeed, an aerosol class represents an ensemble of stable variables which also represents some *a priori* knowledge on the observed medium. This *a priori* information is however not handled correctly, as no error covariance matrix is associated with it. It is clear that the this limitation represents an area which will require improvement before the proposed method becomes fully mature.

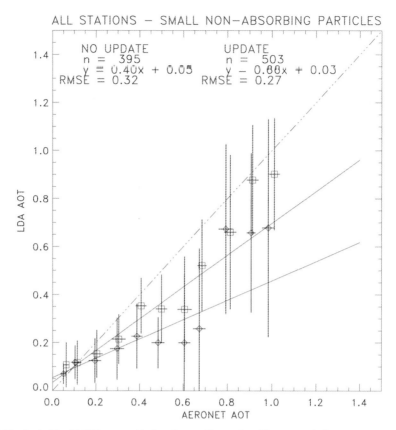

Fig. 11.19. As in Fig. 11.17 but considering the small non-absorbing aerosol class.

References

D.J. Diner, J.V. Martonchik, R.A. Kahn, B. Pinty, N. Gobron, D.L. Nelson, and B.N. Holben. Using angular and spectral shape similarity constraints to improve MISR aerosol and surface retrievals over land. *Remote Sensing of Environment*, **94**:155–171, 2005.

O. Dubovik, A. Sinyuk, T. Lapyonok, B. N. Holben, M. Mishchenko, P. Yang, T. F. Eck, H. Volten, O. Munoz, B. Veihelmann, W. J. van der Zande, J. F. Leon, M. Sorokin, and I. Slutsker. Application of spheroid models to account for aerosol particle nonsphericity in remote sensing of desert dust. *Journal of Geophysical Research–Atmospheres*, **111**(D11):11208–11208, 2006.

O. Engelsen, B. Pinty, M.M. Verstraete, and J.V. Martonchik. Parametric bidirectional reflectance factor models: evaluation, improvements and applications. Technical Report EUR 16426 EN, Space Applications Institute, JRC, 1996.

M.Y. Govaerts and M. Clerici. Operational vicarious calibration of the msg/seviri solar channels. In EU-METSAT, editor, *The 2003 EUMETSAT Meteorological Satellite Data User's Conference*, pages 147–154,Weimar, Germany, 2003. EUMETSAT.

Y. Govaerts and A. Lattanzio. Retrieval error estimation of surface albedo derived from geostationary large band satellite observations: Application to Meteosat-2 and -7 data. *Journal of Geophysical Research*, **112**(D05102):doi:10.1029/2006JD007313, 2007.

Y. Govaerts, A. Lattanzio, and J. Schmetz. Global surface albedo derived from geostationary satellites. *GEWEX News*, **16**(1):15, 2006.

W.M.F. Grey, P.R.J. North, and S.O. Los. Computationally efficient method for retrieving aerosol optical depth from ATSR-2 and AATSR data. *Applied Optics*, **45**(12):2786–2795, 2006.

W.M.F. Grey, P.R.J. North, S.O. Los, and R.M. Mitchell. Aerosol optical depth and land surface reflectance from multiangle AATSR measurements: global validation and intersensor comparisons. *IEEE Transactions on Geoscience and Remote Sensing*, **44**(8):2184–2197, 2006.

B.N. Holben, T.F. Eck, I. Slutsker, D. Tanre, J.P. Buis, A. Setzer, E. Vermote, J.A. Reagan, Y.J. Kaufman, T. Nakajima, F. Lavenu, I. Jankowiak, and A. Smirnov. AERONET – a federated instrument network and data archive for aerosol characterization. *Remote Sensing of Environment*, **66**:1–16, 1998.

N.C. Hsu, S.C. Tsay, M.D. King, and J.R. Herman. Aerosol properties over bright-reflecting source regions. *Ieee Transactions on Geoscience and Remote Sensing*, **42**(3):557–569, 2004.

Y. Jin, C.B. Schaaf, F. Gao, X. Li, A.H. Strahler, W. Lucht, and S. Liang. Consistency of modis surface bidirectional reflectance distribution function and albedo retrievals: 1. algorithm performance. *J. Geophys. Res.*, **108**(D5):4158, doi:10.1029/2002JD002803, 2003.

Y.J. Kaufman and D. Tanré. Atmospherically resistant vegetation index (ARVI) for EOS-MODIS. *IEEE Transcations on Geoscience and Remote Sensing*, **20**:261–270, 1992.

K.R. Knapp, R. Frouin, S. Kondragunta, and A. Prados. Toward aerosol optical depth retrievals over land from goes visible radiances: determining surface reflectance. *International Journal of Remote Sensing*, **26**(18):4097–4116, 2005.

A.A. Kokhanovsky, F.M. Breon, A. Cacciari, E. Carboni, D. Diner, W. Di Nicolantonio, R.G. Grainger, W.M.F. Grey, R. Holler, K.H. Lee, Z. Li, P.R.J. North, A.M. Sayer, G.E. Thomas, and W. von Hoyningen-Huene. Aerosol remote sensing over land: a comparison of satellite retrievals using different algorithms and instruments. *Atmospheric Research*, **85**(3–4):372–394, 2007.

A. Lattanzio, Y. Govaerts, and B. Pinty. Consistency of surface anisotropy characterization with Meteosat observations. *Advanced Space Research*, doi:10.1016/j.asr.2006.02.049, 2006.

R. Levy, L. Remer, S. Mattoo, E. Vermote, and Y. Kaufman. Second-generation operational algorithm: retrieval of aerosol properties over land from inversion of moderate resolution imaging spectroradiometer spectral reflectance. *Journal of Geophysical Research*, **112**(D13211):doi:10.1029/2006JD007811, 2007.

A.M. Michalak, A. Hirsch, L. Bruhwiler, K.R. Gurney, W. Peters, and P.P. Tans. Maximum likelihood estimation of covariance parameters for bayesian atmospheric trace gas surface flux inversions. *Journal of Geophysical Research–Atmospheres*, **110**:24107, 2005.

F. E. Nicodemus, J.C. Richmond, J.J. Hsia, I.W. Ginsberg, and T. Limperis. Geometrical considerations and nomenclature for reflectance. Technical report, National Bureau of Standards, 1977.

P.R.J. North, S.A. Briggs, S.E. Plummer, and J.J. Settle. Retrieval of land surface bidirectional reflectance and aerosol opacity from ATSR-2 multiangle imagery. *IEEE Transactions on Geoscience and Remote Sensing*, **37**(1):526–537, 1999. Part 2.

P.R.J. North. Estimation of FAPAR, LAI, and vegetation fractional cover from ATSR-2 imagery. *Remote Sensing Environment*, **80**:114–121, 2002.

B. Pinty, F. Roveda, M.M. Verstraete, N. Gobron, Y. Govaerts, J.V. Martonchik, D.J. Diner, and R.A. Kahn. Surface albedo retrieval from Meteosat: Part 1: Theory. *Journal of Geophysical Research*, **105**:18099–18112, 2000.

B. Pinty, F. Roveda, M.M. Verstraete, N. Gobron, Y. Govaerts, J.V. Martonchik, D.J. Diner, and R.A. Kahn. Surface albedo retrieval from Meteosat: Part 2: Applications. *Journal of Geophysical Research*, **105**:18113–18134, 2000.

C. Popp, A. Hauser, N. Foppa, and S. Wunderle. Remote sensing of aerosol optical depth over central Europe from MSG-SEVIRI data and accuracy assessment with ground-based AERONET measurements. *Journal of Geophysical Research*, **112**:doi:10.1029/2007JD008423, 2006.

A.I. Prados, S. Kondragunta, P. Ciren, and K.R. Knapp. GOES aerosol/smoke product (GASP) over north america: Comparisons to AERONET and MODIS observations. *Journal of Geophysical Research-Atmospheres*, **112**(D15):15201–15201, 2007.

W.H. Press, S.A. Teukolsky, W.T. Vetterling, and B.P. Flannery. *Numerical Recipes in FORTRAN: The Art of Scientific Computing*. Cambridge University Press, Cambridge, 2nd edition, 1992.

H. Rahman, B. Pinty, and M.M. Verstraete. Coupled surface-atmosphere reflectance (CSAR) model. 2. Semiempirical surface model usable with NOAA Advanced Very High Resolution Radiometer data. *Journal of Geophysical Research*, **98**(D11):20,791–20,801, 1993.

L.A. Remer, Y. Kaufman, and B. Holben. The size distribution of ambient aerosol particles: smoke versus urban/industrial aerosol. In *Global Biomass Burning*, pages 519–530. MIT Press, Cambridge, MA, 1996.

L.A. Remer, Y.J. Kaufman, D. Tanré, S. Mattoo, D.A. Chu, J.V. Martins, R.R. Li, C. Ichoku, R.C. Levy, R.G. Kleidman, T.F. Eck, E. Vermote, and B.N. Holben. The MODIS aerosol algorithm, products and validation. *Journal of the Atmospheric Sciences*, **62**:947–973, 2005.

C.D. Rodgers. Retrieval of atmospheric temperature and composition from remote measurements of thermal radiation. *Rev. Geophys. and Space Phys*, **14**:609, 1976.

C.D. Rodgers. The characterization and error analysis of proŸles retrieved from remote sounding measurements. *Journal of Geophysical Research*, **95**:5587, 1990.

C.D. Rodgers. *Inverse Methods for Atmospheric Sounding*, volume 2 of Series on Atmospheric Oceanic and Planetary Physics. World Scientfic, 2000.

R. Santer, V. Carrere, P. Dubuisson, and J.C. Roger. Atmospheric correction over land for MERIS. *International Journal of Remote Sensing*, **20**(9):1819–1840, 1999.

E.P. Shettle. Optical and radiative properties of a desert aerosol model. In A. Deepak, editor, *Radiation in the Atmosphere*, pages 74–77, Perugia, Italy, 1984.

E.P. Shettle and R.W. Fenn. Models for the aerosols of the lower atmosphere and the effects of humidity variations on their optical properties. Technical Report AFGL-TR-79-0214, U.S. Air Force Geophysics Laboratory, 1979.

A. Tarantola. *Inverse Problem Theory: Methods for Data Fitting and Model Parameter Estimation*. Elsevier, Amsterdam, 3rd edition, 1998.

G.E. Thomas, C.A. Poulsen, R.L. Curier, G. De Leeuw, S.H. Marsh, E. Carboni, R.G. Grainger, and R Siddans. Comparison of AATSR and SEVIRI aerosol retrievals over the Northern Adriatic. *Quarterly Journal of the Royal Meteorological Society*, **133**(S1):85–95, 2007.

J. P. Veefkind, G. de Leeuw, and Ph. A. Durkee. Retrieval of aerosol optical depth over land using two-angle view satellite radiometry during TARFOX. *Geophysical Research Letters*, **25**:3135–3138, 1998.

W. von Hoyningen-Huene, M. Freitag, and J. B. Burrows. Retrieval of aerosol optical thickness over land surfaces from top-of-atmosphere radiance. *Journal of Geophysical Research Atmospheres*, **108**(D9):4260–4260, 2003.

J. Wang, X. Liu, S.A. Christopher, J.S. Reid, E. Reid, and H. Maring. The effects of non-sphericity on geostationary satellite retrievals of dust aerosols. *Geophysical Research Letters*, **30**(24):2293–2293, 2003.

12 Remote sensing data combinations: superior global maps for aerosol optical depth

Stefan Kinne

1. Introduction

Aerosol remote sensing from space is predominantly based on sensor data of reflected sunlight in solar spectral regions, where the attenuation by trace-gases can be neglected or easily accounted for. But even at these spectral regions retrievals of aerosol properties are by no means a simple task, as explained in the previous chapters of this book. This is mostly due to the following major reasons:

1. *Cloud contamination.* The solar light reflection attributed to aerosol is small compared to that of clouds and identifying cloud-free and cloud-influence-free regions is a challenge, especially with sensor limitations to spatial resolution. Also at low sun-elevations retrievals near clouds are complicated by cloud shadow scene darkening or sidescatter scene brightening.

2. *Surface contributions.* The solar reflection attributed to aerosol can be smaller than surface signals. Thus, surface albedo (also as function of the sun-elevation) needs to be known to high accuracy. To minimize the surface albedo problem innovative methods are applied. They rely on spectral dependencies (Kaufman et al., 1997), multi-angular views (Martonchik et al., 1998), polarization (Deuzé et al., 1999) or retrievals in the UV (Torres et al., 2002). Higher and variable surface albedos still remain the major reason that most aerosol satellite products have no or only limited coverage over land.

3. *A-priori assumptions.* The relationship that associates changes of solar reflection to aerosol amount in cloud-free conditions is modulated by aerosol composition and even atmospheric environment. Even when combining different sensor data sources, any potential solution is under-determined in the context of dependencies to aerosol amount, particle size, shape and composition. Thus, a-priori assumptions are required. Some of these assumptions, usually to absorption, size and shape, have been locally and/or seasonally validated, but their regional (or even global) and annual application in the context of aerosol temporal and spatial variability is rarely justifiable.

The availability of remote sensing data with respect to aerosol properties is uneven. Table 12.1 illustrates that the most commonly retrieved aerosol property is the (mid-visible) aerosol optical depth (AOD, representing aerosol amount) over oceans. The different approaches listed in Table 12.1 can provide – next to AOD – important constraints on other aerosol and environmental properties. Thus, among the different available AOD maps from satellite remote sensing quality differences can be expected. However, since more capable sensors are not always matched with better retrieval assumptions, there is ambiguity about which satellite AOD products to believe. Thus, there is a strong

Table 12.1. Aerosol properties, availability and associated techniques in satellite remote sensing

Property	Ocean		Land	
	Availability	Technique	Availability	Technique
amount	good	solar reflection	limited	multi-directional
absorption	poor	glint	poor	UV-spectral
size	limited	multi-spectral	poor	polarization
shape	poor	polarization	poor	multi-directional
altitude	poor	multi-directional	limited	lidar

community interest in comparing and assessing available AOD datasets and in providing needed recommendations for AOD measurement use (e.g., for model input or evaluation).

2. Satellite AOD datasets

The use of different and often complementary techniques is desirable, but at the same time complicates data comparisons, as different sensors, retrievals methods and assumptions are applied. To minimize additional complications by differences in data-sampling,

Table 12.2. Multi-annual available AOD datasets from remote sensing, temporal coverage, literature references, major data limitations and recognized (pos (+) or neg (−)) biases

	Sensor	Year	References	Limitation	Bias
Aer	AERONET	96–06	Holben et al., 1998	local, land only	
Mc4	MODIS, T + A	00–05	Tanré et al., 1997, Kaufman et al., 1997	no deserts	+ over land
Mc5	MODIS, T + A	00–05	Tanré et al., 1997, Remer et al., 2005	no deserts	
MIS	MISR	00–05	Kahn et al., 1998, Martonchik et al., 1998 Diner et al., 2005	6-day repeat	+ over ocean
AVn	AVHRR, NOAA	81–90	Stowe et al., 1997	no land, *a priori*	
AVg	AVHRR, GACP	84–00	Geogdzhyev et al., 2002	no land	+ cloud contamination
TOo	TOMS	79–01	Torres et al., 2002	50 km pixel, old r_s	++ cloud contamination
TOn	TOMS	79–01	Torres (private comm.)	50 km pixel	−
POL	POLDER	97, 03	Deuzé et al., 1999, Deuzé et al., 2001	Small size detection only over land	+ at high elevation

* Aer: quality AOD reference data are provided by ground based monitoring networks. Providers include passive remote sensing by sunphotometry, as AERONET http://aeronet.gsfc.nasa.gov/, SKYNET http://atmos.cr.chiba-u.ac.jp/ or GAW http://wdca.jrc.it/ and active remote sensing by lidar as http://www.earlinet.org/ for active remote sensing by lidar.
* AVn: AOD estimates benefit from cancellation of errors as impacts of overestimates to size are often partially offset by underestimates to aerosol absorption
* TOn/TOo: the newer processing assumes stronger surface reflectance contributions. This is the major cause for the significantly reduced AOD values compared to the older processing AOD estimates.

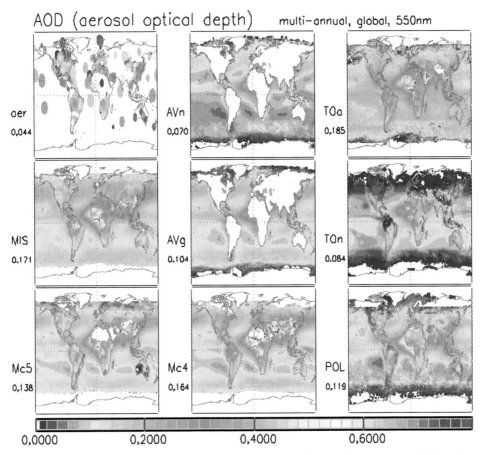

Fig. 12.1. Annual global mid-visible AOD maps from remote sensing (at 0.55 μm wavelength). Sunphotometer data (aer – enlarged for visual purposes) of ground-based monitoring networks by AERONET, SKYNET and GAW are compared to satellite sensor retrievals of MISR (MIS), of MODIS collections 5 (Mc5) and 4 (Mc4) – (of both Terra and Aqua platforms), of AVHRR by NOAA (AVn) and by GACP (AVg) of TOMS older (TOo) and newer (TOn) data processing and of POLDER (POL). Numbers at labels display annual global averages of locations with (nonzero) data.

only monthly mean properties of multi-annual AOD datasets are considered. The dataset comparison includes suggestions by MODIS collections 5 and 4 (2000–2005), MISR (2000–2005), TOMS (1979–2001) new and old processing, POLDER (1987, 2002), and AVHRR NOAA (1981–1990) and AVHRR GACP (1984–2001). Time periods with enhanced stratospheric aerosol loading (e.g., after the El Chichon 1982–1985 and after the Mt. Pinatubo 1991–1994 volcanic eruptions) are excluded from these averages. Annual global AOD maps of these eight remote sensing data products are compared to data from ground-based monitoring in Fig. 12.1. Background information is provided in Table 12.2.

Despite some similarity in major AOD annual patterns, differences among the available AOD data at a particular time and location can be large (Liu and Mishchenko, 2008). Already for annual AOD retrievals, locally the range among different AOD retrievals

seasonal aer

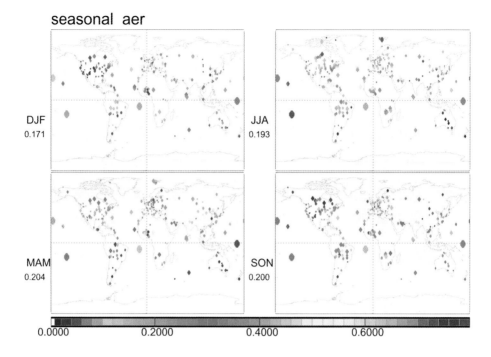

seasonal aex

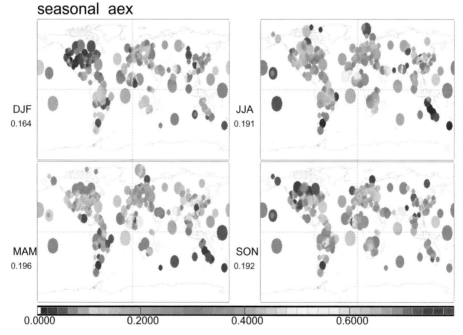

Fig. 12.2. Seasonal sunphotometer AOD maps of AERONET (1996–2006), SKYNET (2003–2004) and GAW (2003–2004) ground-based monitoring networks. The upper set of panels shows the actual data-points for all networks, while the lower set of panels displays local site statistics spatially stretched for better visibility. Seasonal global averages of non-zero data are listed below the labels.

can often exceed the AOD average. In need for a reference, local statistics from ground-based monitoring is applied. Ground-based monitoring of direct solar attenuation (a transmission) by sunphotometry has the advantage that AOD can be directly measured without a need to prescribe aerosol composition and (unless aerosols are large) aerosol size. In addition, the background is well defined as its contributions are negligible.

3.   AOD data reference

Monthly multi-annual data from ground-based sunphotometry are selected as reference for an assessment and ranking of the different satellite AOD regional maps. The sunphotometer monthly statistics is based primarily on 1996–2006 AERONET data (Holben et al., 1998) further enhanced by 2003–2004 data of SKYNET (Aoki et al., 2006) and GAW. Seasonal AOD maps from sunphotometry are displayed in Fig. 12.2.

To simplify the assessments, individual station statistics of sunphotometer data were combined onto the common $1° \times 1°$ lat/lon grid of the satellite remote sensing datasets. This gridding procedure considered known differences in quality and regional representation of the local data (T. Eck, personal communication). With the overall goal to combine identified regional retrievals strengths for a superior satellite AOD composite, assessments were conducted on a regional basis.

4.   Regional stratification

It is expected that a dataset that combines all regionally best-performing satellite retrievals will be superior in coverage and accuracy over any individual satellite retrieval. For the necessary regional stratification six oceanic and six continental zonal bands were chosen. The zonal bands separate Arctic, northern mid-latitudes, dust-belt, biomass belt, southern oceans and Antarctica, as illustrated in Fig. 12.3. Fig. 12.3 also compares zonal annual

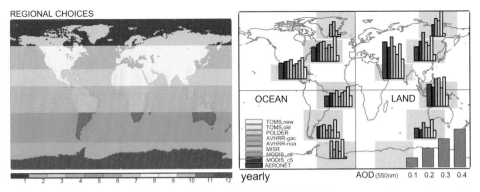

Fig. 12.3. Regional choices and a comparison of mid-visible AOD of all remote sensing annual maps in Fig.12.1. Averages of zonal bands over oceans are to the left and those over land are to the right. Regional averages follow the scale in the lower right and are only displayed if satellite data had at least 25 % or gridded sunphotometer data (AERONET) had at least 2.5 % coverage.

averages of the remote sensing datasets, already introduced in Fig. 12.1 and Table 12.2, separately over oceanic (left comparisons) and continental regions (right comparisons).

In Fig. 12.3 regional AOD averages are displayed only if spatial coverage in that region exceeded 25 % for satellite data or 2.5 % for sunphotometer (1 × 1 gridded) data. In light of the different spatial sub-samples (see Fig. 12.1) for satellite data over land and at high latitudes and certainly for sunphotometer statistics due to their local nature, these comparisons are more general in nature. More meaningful assessments of satellite AOD data (to the sunphotometry reference) requires that all satellite data are sub-sampled in any region only at grid locations where sunphotometer data exist.

5.   Regional comparisons

Comparisons of matching data are presented in Fig. 12.4 for regions where nonzero (gridded 1° × 1° lat/lon) sunphotometer data (AERONET) cover at least 5 % of that region. This limits satellite AOD dataset assessments to low and mid-latitudes over land and only to northern hemispheric low- and mid-latitudes over oceans.

This more visible evaluation indicates that in reference to ground-based monitoring by sunphotometry almost all satellite retrievals overestimate AOD over oceans. Tendencies over land are more diverse and also a function of the dominant aerosol type (e.g., industrial

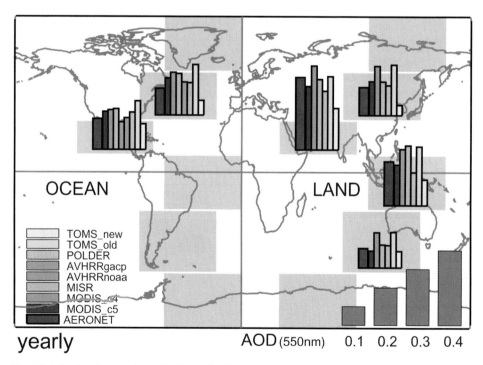

Fig. 12.4. Regional comparison of subsets of mid-visible AOD remote sensing data of Fig. 12.1. All datasets were sub-sampled only at grid-points with nonzero sunphotometer data and are only displayed, if the regional coverage of sunphotometer nonzero gridded data exceeded 5 % in that region. Averages of zonal bands over oceans are to the left and those over land to the right.

pollution, desert dust or biomass burning). In comparison to sunphotometry, there is no single best retrieval for all regions. For the determination of a region's superior satellite AOD retrieval a more objective rank scoring scheme was applied to the satellite-retrieval versus sunphotometry data pairs. The overall idea is to rank satellite datasets as they compare locally to AERONET sunphotometer data and then use the results to create a merged dataset from the highest scoring satellite retrievals.

6. Scoring concept

In order to score a data-set performance with respect to a quality reference, aspects of bias, spatial and seasonal variability should be addressed. Such scoring should start on small temporal and spatial scales for performance analysis detail. Still, in order to get a better overview, there is also the need to combine sub-scores into summarizing overall scores. Thus a total score $\mathbf{S_T}$ is defined ($\mathbf{S_T} = \text{sign}(\mathbf{E_B}) \times [1 - |\mathbf{E_B}|] \times [1 - \mathbf{E_V}] \times [1 - \mathbf{E_S}]$) with '0' for poor and '1' for perfect. This total score $\mathbf{S_T}$ combines error scores of bias ($\mathbf{E_B}$), spatial variability ($\mathbf{E_V}$) and seasonality ($\mathbf{E_S}$). Hereby, the sign of $\mathbf{S_T}$ represents the direction of the average bias with respect to the reference data. Error-scores ($\mathbf{E_B}$, $\mathbf{E_V}$ and $\mathbf{E_S}$) are defined for a range from 0 for 'perfect' and 1 for 'poor'. All error-scores are based on ranks (not values) to minimize the impact of data outliers. For the bias score $\mathbf{E_B}$ involving N data-pairs from a dataset D and a reference dataset R, all $2 \times N$ elements are placed in one single array C and ranked by value (rank = 1 for the lowest value, rank = $2 \times N$ for the largest value). Then the ranks of C associated with D and R are summed and compared. If the two sums of rank-values for D (D_{SUM}) and R (R_{SUM}) differ, then a bias and also its sign are identified by $\mathbf{E_B}$:

$$\mathbf{E_B} = w \times [(D_{\text{SUM}} - R_{\text{SUM}})/(D_{\text{SUM}} + R_{\text{SUM}})],$$
$$\text{with } w = [_{\text{RANGE}}D + _{\text{RANGE}} R]/[_{\text{MEDIAN}}D + _{\text{MEDIAN}} R].$$

The weight w is applied to avoid an overemphasis of an error in the overall score, in case most elements are similar in magnitude (note than the RANGE in the weight formula represents the difference between the 90th and 10th percentile of the cumulative distribution function). The same weight is also applied for the (regional) variability score $\mathbf{E_V}$ and the (temporal) seasonality score $\mathbf{E_S}$:

$$\mathbf{E_V} = w \times [1 - R_C]/2 \qquad \mathbf{E_S} = w \times [1 - R_C]/2,$$
$$\text{with } w = [_{\text{RANGE}}D + _{\text{RANGE}} R]/[_{\text{MEDIAN}}D + _{\text{MEDIAN}} R].$$

$\mathbf{E_V}$ examines the spatial distribution at a given time-interval, whereas $\mathbf{E_S}$ examines temporal development of the median value during an annual cycle. Both variability scores are based on the (Spearman) rank correlation coefficient R_C, which ranges from 1 for 'correlated' to $- 1$ for 'anti-correlated'. Note that a lack of correlation [$R_C = 0$] still leads to a positive score for $\mathbf{E_V}$ and $\mathbf{E_S}$.

Scores of any dataset D with respect to reference dataset R are first determined on a monthly and regional basis – scoring regional variability and bias. Then these regional variability and bias scores are annually averaged and an additional seasonality score is added, which is based on monthly median data pairs for D and R. Finally all three regional annual scores are combined according to their surface area fraction into annual global scores.

7. Global scores

Annual global scores(S_T) for remote sensing AOD data of Fig. 12.1 are presented in Table 12.3 based on regional and monthly sub-scores to AERONET sunphotometer statistics. Table 12.3 also provides contributing sub-scores for seasonality, bias and regional variability. In addition, scores for AERONET sky-radiance data, a subset of sunphotometry, data, are presented. Note, however, that due to different regional samples and coverage, the scores below do not represent a uniform test of satellite retrieval accuracy, which is beyond the scope of the current chapter, but is subject of continuing work.

7.1 Sky versus sun data

The AERONET sky-photometer (radiance) data are a subset of the (AERONET) sunphotometer (direct attenuation) data reference. Thus, better scores compared to satellite remote sensing are expected. The overall 'sky' score of -0.82 indicates by its negative sign that sky-data are biased low compared to sun-data and its absolute value (0.82) that the match of monthly statistics between sky- and sun-data is not perfect (1.0). Sub-scores of 0.94, -0.93 and 0.93 reveal that these deductions are almost evenly from differences in variability, seasonality and bias. As sky-data are a subset of the sun-data reference, this suggest any total score larger than (absolute) 0.8 seems excellent.

The negative bias of the sky-data (with respect to sun-data) was expected, as its radiance symmetry requirement seemed more clear-sky case conservative than the temporal varia-

Table 12.3. Global annual scores of remote sensing AOD datasets versus the AERONET sunphotometer data reference (the higher the absolute value of the score, the better)

Rank	Label	Overall score S_T	Season $1 - E_S$	Bias $1 - E_B$	Variability $1 - E_V$	Dataset
1	sky	-0.82	0.94	-0.93	0.93	AERONET-sky
2	AVn	0.69	0.92	0.90	0.83	AVHRR-NOAA[a]
3	Mc5	0.64	0.95	0.83	0.82	MODIS-coll.5
4	MIS	0.64	0.93	0.83	0.82	MISR-vers.22
5	AVg	0.61	0.93	0.82	0.79	AVHRR-GACP[a]
6	Mc4	0.59	0.94	0.78	0.81	MODIS-coll.4
7	POL	0.58	0.88	0.81	0.80	POLDER
8	TOn	0.48	0.89	0.73	0.74	TOMS, old
9	TOo	-0.47	0.84	-0.81	0.68	TOMS, new

[a] Note: AVHRR scores are based on ocean data only. AVHRR-NOAA data have a slight advantage as they were calibrated against the AERONET reference at selected sites. MISR data (with a narrow swath) and TOMS data (with a 50 km × 50 km footprint) have fewer overall samples. And MISR and TOMS scores also include more difficult retrievals over bright desert surfaces, which are avoided in MODIS retrievals. Scores limited to either ocean or land regions are given in Tables 12.4 and 12.5.

bility method for the sun-data. However, on a regional monthly basis the sky-data bias is often positive, especially for land regions in the northern hemisphere. In those instances the largest (above the 90th percentile) AOD values are usually significantly larger. This suggests that, against expectations, cloud-contamination is an issue for the sky-data.

7.2 Satellite versus sun data

The overall scores for the satellite AOD retrievals range from 0.6 (better) to 0.47 (poorer). Almost all satellite remote sensing data display positive overall scores. Thus, if AERONET sunphotometer AOD data are trusted, this indicates that (on average) almost all satellite retrievals overestimate AOD. Sub-scores indicate that the largest discrepancies are usually associated with spatial variability and bias. Regional and temporal variability are poorer over land than over oceans when comparing scores in Tables 12.4 (ocean) and 12.3 (land). Updated retrievals of the same sensor data (e.g., MODIS, MISR usually) show improved overall scores. This indicates the untouched potential of sensor data and the need for continued retrieval improvements.

In order to create a satellite composite, the strongest performing retrievals on a regional basis must be identified. Scores for all (except polar) regions are listed for the continental

Table 12.4. Ocean annual scores for remote sensing AOD data versus the AERONET sun data reference

Rank	Label	Overall score S_T	Season $1 - E_S$	Bias $1 - E_B$	Variability $1 - E_V$	Dataset
1	sky	− 0.78	0.93	− 0.92	0.92	AERONET-sky
2	AVn	0.69	0.92	0.87	0.83	AVHRR-NOAA
3	Mc5	0.66	0.97	0.81	0.85	MODIS-coll.5
4	Mc4	0.64	0.95	0.80	0.85	MODIS-coll.4
5	AVg	0.61	0.93	0.81	0.81	AVHRR-GACP
6	MIS	0.61	0.92	0.79	0.84	MISR-vers.22
7	POL	0.60	0.91	0.80	0.83	POLDER
8	TOn	− 0.54	0.87	− .86	0.72	TOMS, new
9	TOo	0.51	0.90	0.73	0.78	TOMS, old

[a] See footnote to Table 12.3.

Table 12.5. Land annual scores for remote sensing AOD data versus the AERONET sun data reference

Rank	Label	Overall score S_T	Season $1 - E_S$	Bias $1 - E_B$	Variability $1 - E_V$	Dataset
1	sky	0.90	0.98	0.97	0.96	AERONET-sky
2	MIS	0.70	0.96	0.92	0.80	MISR-vers.22
3	Mc5	0.58	0.90	0.87	0.74	MODIS-coll.5
4	POL	0.52	0.80	0.86	0.77	POLDER
5	Mc4	0.48	0.90	0.74	0.72	MODIS-coll.4
6	TOo	0.41	0.86	0.73	0.65	TOMS, old
7	TOn	− 0.32	0.78	− 0.71	0.57	TOMS, new

zonal latitude bands in Table 12.6 and for the oceanic zonal latitude bands in Table 12.7. As a reference, also the scores of the model median AOD fields of global model simulations by AeroCom (Kinne et al., 2006) are provided. When comparing scores, however, it should be kept in mind that the sample volume differs among datasets. It is possible that poorer sampling for MISR and TOMS contributed to their lower scores. Thus, unless the sampling impact is better understood, scores should not be interpreted as retrieval error. Also the scoring is only based on AERONET locations, where satellite data are available (e.g., MODIS does not provide data over bright land surfaces), which also explains differences for the reference median (R_{MEDIAN}), especially over land.

No individual satellite retrieval displays the highest score. Thus, a satellite composite was created by combining the regionally highest scoring retrievals.

8.   Satellite composite

Over continents the highest scores among satellite retrievals are achieved by MISR, especially for northern hemispheric urban pollution. MODIS (collection 5) is usually a close second and even scores better than MISR over southern mid-latitude land regions. Over oceans the AVHRR-NOAA retrieval yields the highest score except for northern hemispheric higher latitudes, where MISR performs best. MODIS (collection 5) is always the second best choice. (In fairness to more advanced recent retrievals over oceans (e.g. MISR, MODIS) it should be admitted that the AVHRR-NOAA retrieval is tied to sunphotometry for sensor calibration.) In order to create a satellite remote sensing composite separate for the non-polar 4 ocean and 4 land regions the two best retrievals are combined, such as giving the primary choice a 0.67 weight and the secondary choice a 0.33 weight. Then the regional monthly AOD maps are globally combined. To reduce the potential for abrupt changes when switching between different retrievals at zonal band boundaries and land/ocean transitions, a 6_o latitudinal transition zone and smoothing in coastal regions was permitted. The annual global AOD maps of the composite and its three contributing retrievals are presented in Fig. 12.5.

Seasonal AOD maps of the composite are presented in Fig. 12.6. They illustrate that AOD maxima due to biomass burning and dust have a strong seasonal character, which gets lost in the presentation of annual global maps.

Although retrievals with the regionally highest scores with respect to sunphotometer statistics are applied, none of these retrievals scores close to perfect. Thus, differences to the reference data can be expected and are illustrated on a seasonal basis in Fig. 12.7.

The AOD satellite composite still displays strong differences to the reference from sunphotometry. Higher AOD values in northern mid-latitudes during winter and spring suggest snow contamination. Lower AOD values near fast-growing urban pollution areas and dust outflow regions seem to suggest that some of the larger AOD events are missed due to sampling limitations or removed in conservative cloud-screening algorithms. The same reason may also be responsible for the most severe deviations as the timing of the tropical biomass maxima occurs too early (in late summer) compared to the sunphotometer reference (early autumn). To reduce these deviations in a measurement-based AOD composite, a modified version of the satellite composite has been developed, where the monthly AOD fields of the composite are drawn to the AERONET data.

Table 12.6. Oceanic regional annual scoring and sub-scoring of remote sensing AOD datasets with respect to AERONET sunphotometer data. Scores of median fields by global modeling are added

	S_T	1-E_S	1-E_B	1-E_V	D_{MED}	R_{MED}	ERR$_{REL}$	BIAS$_{REL}$	NH3058_Ocean
1	− 0.88	0.95	− 0.97	0.95	0.141	0.142	0.16	− 0.051	AERONET-sky
2	0.65	0.96	0.81	0.85	0.171	0.120	0.37	0.27	MISR-vers.22
3	0.64	0.89	0.81	0.87	0.179	0.120	0.37	0.27	MODIS-coll.5
4	0.62	0.91	0.84	0.81	0.161	0.120	0.36	0.21	POLDER
5	0.61	0.92	0.84	0.79	0.167	0.120	0.36	0.19	AVHRR-GACP
6	0.60	0.88	0.84	0.81	0.149	0.109	0.31	0.14	AVHRR-NOAA
7	0.60	0.92	0.75	0.87	0.209	0.120	0.45	0.40	MODIS-coll.4
8	0.43	0.92	0.69	0.68	0.271	0.129	0.62	0.55	TOMS, old
9	− 0.37	0.92	− 0.68	0.60	0.092	0.120	0.63	− 0.58	TOMS, new
m	*0.67*	*0.97*	*0.83*	*0.83*	*0.182*	*0.133*	*0.37*	*0.27*	*median model*

	S_T	1-E_S	1-E_B	1-E_V	D_{MED}	R_{MED}	ERR$_{REL}$	BIAS$_{REL}$	NH1030_Ocean
1	− 0.83	0.91	− 0.94	0.97	0.107	0.131	0.15	− 0.062	AERONET-sky
2	0.78	0.95	0.94	0.87	0.109	0.103	0.26	− 0.013	AVHRR-NOAA
3	0.75	0.99	0.83	0.91	0.154	0.103	0.36	0.27	MODIS-coll.5
4	0.73	0.99	0.82	0.90	0.162	0.103	0.37	0.30	MODIS-coll.4
5	0.73	0.97	0.82	0.92	0.157	0.103	0.35	0.28	MISR-vers.22
6	0.70	0.96	0.83	0.88	0.153	0.103	0.39	0.28	POLDER
7	0.65	0.86	0.87	0.86	0.137	0.103	0.35	0.15	AVHRR-GACP
8	− 0.62	0.97	− 0.92	0.69	0.106	0.103	0.40	− 0.11	TOMS, new
9	0.57	0.91	0.75	0.85	0.208	0.103	0.61	0.53	TOMS, old
m	*− 0.78*	*0.93*	*− 0.96*	*0.88*	*0.117*	*0.114*	*0.28*	*− 0.016*	*median model*

	S_T	1-E_S	1-E_B	1-E_V	D_{MED}	R_{MED}	ERR$_{REL}$	BIAS$_{REL}$	EQ2210_Ocean
1	− 0.80	0.98	− 0.93	0.88	0.083	0.071	0.24	− 0.051	AVHRR-NOAA
2	− 0.77	0.95	− 0.88	0.92	0.061	0.070	0.21	− 0.14	AERONET-sky
3	0.72	0.98	0.82	0.89	0.112	0.071	0.29	0.24	MODIS-coll.5
4	0.72	0.95	0.85	0.89	0.104	0.071	0.26	0.20	MODIS-coll.4
5	0.69	0.98	0.83	0.85	0.124	0.071	0.37	0.26	AVHRR-GACP
6	0.67	0.96	0.79	0.89	0.124	0.071	0.37	0.33	MISR-vers.22
7	0.57	0.96	0.73	0.82	0.165	0.071	0.51	0.48	POLDER
8	0.56	0.85	0.89	0.74	0.117	0.071	0.39	0.13	TOMS, new
9	0.51	0.89	0.71	0.80	0.198	0.071	0.61	0.59	TOMS, old
m	*− 0.60*	*0.86*	*− 0.79*	*0.87*	*0.057*	*0.070*	*0.40*	*− 0.37*	*median model*

	S_T	1-E_S	1-E_B	1-E_V	D_{MED}	R_{MED}	ERR$_{REL}$	BIAS$_{REL}$	SH5822_Ocean
1	− 0.75	0.92	− 0.93	0.88	0.046	0.054	0.21	− 0.020	AERONET-sky
2	0.63	0.90	0.87	0.80	0.084	0.073	0.33	0.20	AVHRR-NOAA
3	0.59	0.97	0.78	0.76	0.119	0.073	0.57	0.51	MODIS-coll.5
4	0.56	0.96	0.75	0.78	0.130	0.073	0.66	0.63	AVHRR-GACP
5	0.56	0.83	0.83	0.81	0.084	0.073	0.35	0.24	POLDER
6	0.55	0.95	0.76	0.77	0.126	0.073	0.65	0.63	MODIS-coll.4
7	0.54	0.80	0.88	0.77	0.089	0.073	0.41	0.24	TOMS, new
8	0.52	0.90	0.76	0.76	0.204	0.068	0.98	0.98	TOMS, old
9	0.48	0.85	0.77	0.73	0.127	0.073	0.64	0.57	MISR-vers.22
m	*0.49*	*0.88*	*0.86*	*0.65*	*0.086*	*0.071*	*0.45*	*0.25*	*median model*

Table 12.7. Land regional annual scoring and sub-scoring of remote sensing AOD datasets with respect to AERONET sunphotometer data. Scores of median fields by global modeling are added.

	S_T	$1\text{-}E_S$	$1\text{-}E_B$	$1\text{-}E_V$	D_{MED}	R_{MED}	ERR_{REL}	$BIAS_{REL}$	NH3058_Land
1	0.92	0.97	0.98	0.96	0.132	0.129	0.16	0.034	AERONET-sky
2	0.77	0.99	0.96	0.81	0.131	0.121	0.34	0.019	MISR-vers.22
3	0.51	0.87	0.84	0.70	0.171	0.121	0.46	0.24	MODIS-coll.5
4	0.51	0.80	0.85	0.75	0.155	0.140	0.45	0.14	POLDER
5	0.45	0.96	0.68	0.70	0.268	0.120	0.76	0.64	TOMS, old
6	0.42	0.91	0.66	0.69	0.255	0.122	0.67	0.63	MODIS-coll.4
7	− 0.29	0.86	− 0.65	0.52	0.054	0.121	0.92	− 0.82	TOMS, new
m	*0.73*	*0.98*	*0.89*	*0.84*	*0.162*	*0.128*	*0.41*	*0.21*	*median model*

	S_T	$1\text{-}E_S$	$1\text{-}E_B$	$1\text{-}E_V$	D_{MED}	R_{MED}	ERR_{REL}	$BIAS_{REL}$	NH1030_Land
1	0.88	0.95	0.97	0.96	0.402	0.384	0.12	0.014	AERONET-sky
2	− 0.69	0.93	− 0.90	0.82	0.321	0.361	0.31	− 0.15	MISR-vers.22
3	− 0.66	0.93	− 0.89	0.80	0.321	0.366	0.35	− 0.15	MODIS-coll.5
4	0.64	0.96	0.90	0.74	0.458	0.383	0.38	0.16	MODIS-coll.4
5	− 0.51	0.75	− 0.86	0.80	0.296	0.301	0.40	− 0.16	POLDER
6	0.50	0.86	0.89	0.65	0.376	0.361	0.41	0.039	TOMS, old
7	− 0.47	0.93	− 0.74	0.69	0.242	0.366	0.67	− 0.60	TOMS, new
m	*− 0.62*	*0.93*	*− 0.89*	*0.75*	*0.332*	*0.374*	*0.36*	*− 0.18*	*median model*

	S_T	$1\text{-}E_S$	$1\text{-}E_B$	$1\text{-}E_V$	D_{MED}	R_{MED}	ERR_{REL}	$BIAS_{REL}$	EQ2210_Land
1	− 0.89	0.99	− 0.96	0.93	0.125	0.135	0.17	− 0.047	AERONET-sky
2	− 0.66	0.96	− 0.88	0.78	0.159	0.138	0.35	− 0.051	MISR-vers.22
3	− 0.65	0.96	− 0.89	0.77	0.156	0.140	0.40	− 0.049	MODIS-coll.5
4	0.64	0.92	0.89	0.78	0.141	0.110	0.38	0.19	POLDER
5	0.46	0.89	0.74	0.71	0.254	0.138	0.54	0.39	MODIS -coll.4
6	0.37	0.75	0.73	0.67	0.301	0.138	0.63	0.47	TOMS, old
7	− 0.25	0.51	− 0.79	0.63	0.081	0.138	0.75	− 0.54	TOMS, new
m	*− 0.51*	*0.89*	*− 0.78*	*0.74*	*0.098*	*0.137*	*0.51*	*− 0.31*	*median model*

	S_T	$1\text{-}E_S$	$1\text{-}E_B$	$1\text{-}E_V$	D_{MED}	R_{MED}	ERR_{REL}	$BIAS_{REL}$	SH5822_Land
1	0.93	0.98	0.97	0.98	0.076	0.079	0.13	0.034	AERONET-sky
2	0.64	0.91	0.95	0.74	0.073	0.068	0.47	− 0.020	MODIS-coll.5
3	0.62	0.89	0.87	0.80	0.095	0.068	0.40	0.17	MISR-vers.22
4	0.45	0.96	0.66	0.72	0.186	0.068	0.74	0.70	MODIS-coll.4
5	0.42	0.89	0.88	0.54	0.089	0.068	0.53	0.13	TOMS, new
6	0.35	0.84	0.64	0.66	0.181	0.068	0.82	0.76	TOMS, old
7	0.32	0.57	0.78	0.72	0.120	0.075	0.48	0.33	POLDER
m	*0.48*	*0.78*	*0.88*	*0.71*	*0.072*	*0.079*	*0.45*	*0.17*	*median model*

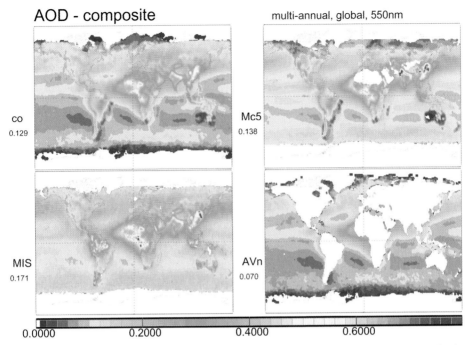

Fig. 12.5. Comparison of the annual mid-visible AOD maps of the satellite composite (co) to contributing multi-annual data of MISR (MIS), MODIS collection 5 (Mc5) and AVHRR-NOAA (AVn).

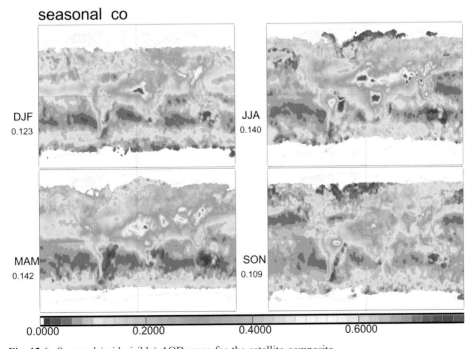

Fig. 12.6. Seasonal (mid-visible) AOD maps for the satellite composite.

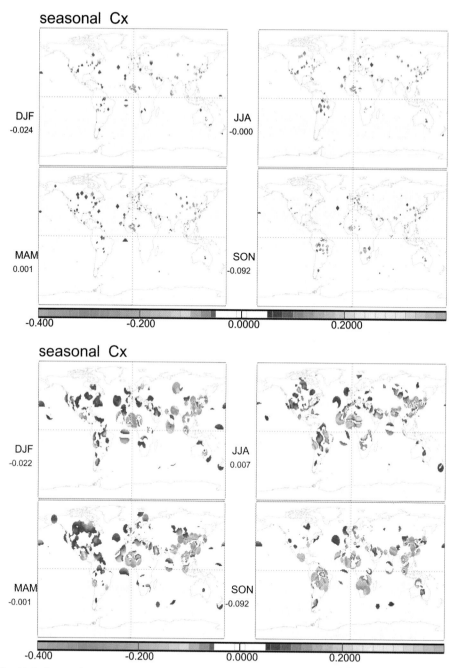

Fig. 12.7. Seasonal AOD differences between the satellite composite and (sparse and visually enlarged) AERONET site statistics. Positive differences indicate overestimates and negative values indicate under-estimates. Only AOD differences exceeding +/− 0.05 are displayed. Annual global averages for deviations at non-zero points are listed below the labels. Note, the site domain magnification causes minor distortions with respect to the differences.

9.   Enhanced composite

Information of sunphotometer monthly statistics was added to the AOD satellite composite. This merging of data gives priority to sunphotometer statistics when locally available. First background ratio fields are defined by globally spreading available sunphotometer to satellite composite ratios with weights that decay with distance. Then these ratios are applied in spatial domains surrounding each surface site. The domain size was defined by regional representation score attributed to each site (T. Eck and the AERONET staff, personal communication). Global annual (mid-visible) AOD maps of this new enhanced satellite composite in comparison to contributing fields of the original satellite composite and of sunphotometry (AERONET) samples are presented in Fig. 12.8. For comparison the annual AOD map from global modeling median is given as well.

The new enhanced AOD composite had modified the initial satellite composite mainly over the western part of North America and over the tropical biomass regions of South America. Seasonal maps of the (new) enhanced AOD composite are presented in Fig. 12.9.

The comparison of seasonal AOD maps between the new (Fig. 12.9) and the initial (Fig. 12.6) satellite composites, demonstrates for the enhanced composite reduced AOD values during continental winters of the northern mid-latitudes and increased AOD values near tropical biomass-burning regions and urban pollution in Asia during autumn. Remaining differences of the enhanced composite to the sunphotometer reference are illustrated on a seasonal basis in Fig. 12.10.

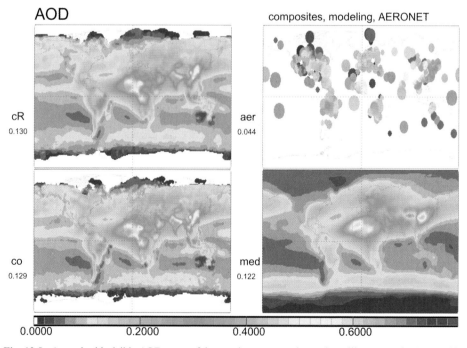

Fig. 12.8. Annual mid-visible AOD maps of the sunphotometry enhanced satellite composite (cR) and its contributing maps of the satellite composite (co) and AERONET (aer). For comparison the annual global AOD map based on local monthly median values from global modeling is displayed (med) as well.

seasonal cR

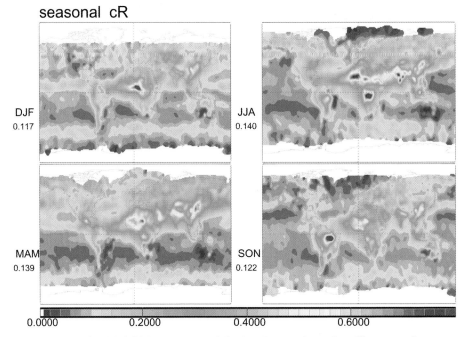

Fig. 12.9. Seasonal (mid-visible) AOD maps of the AERONET enhanced satellite composite.

The differences of the (new) enhanced composite and the AERONET reference (Fig. 12.10) have been reduced compared to those of the initial composite (Fig. 12.7). Some larger deviations near urban centers (e.g., Mexico City, East Asia) or biomass sources (e.g., central South America) remain, as the local influence on surrounding regions was apparently over-extended in the applied data-merging. With improved merging procedures a closer fit to AERONET (and smaller deviations) can be expected. As the (new) enhanced composite displays smaller deviations from the reference overall, quantitative scores should be significantly improved. And they do, as illustrated in Tables 12.8, 12.9 and 12.10 for global, oceanic and continental scores. The sub-scores demonstrate that the biggest improvements are to the temporal and the spatial variability.

10.   Global modeling

An alternative source for global AOD maps is the use of simulated distributions from global modeling. For characteristic AOD maps from global modeling, a composite of monthly local median values of simulations with twenty different models of AeroCom exercises was chosen (Kinne et al., 2006). The scores of the model median dataset rank with the best datasets from remote sensing and are comparable to those of the composite.

Median data of an ensemble reduce the impact of outliers by individual models. Thus, relatively good scores can be expected which are at least superior to the ensemble average. Nonetheless, there are apparent biases of this model median dataset compared to regional and seasonal distributions suggested by remote sensing. Based on the regional (bias) scores

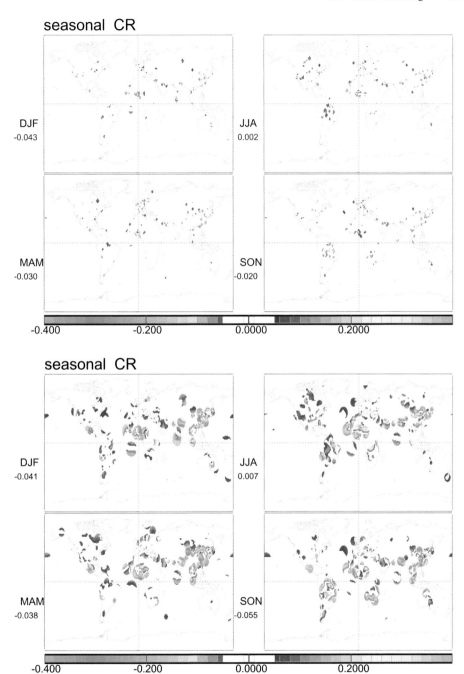

Fig. 12.10. Seasonal AOD differences between the AERONET enhanced satellite composite and (sparse and visually enlarged) AERONET site statistics. Positive deviations show overestimates and negative values show underestimates. Only AOD differences exceeding +/− 0.05 are shown. Annual global averages for deviations are listed below the labels. Note, the site domain magnification causes minor distortions with respect to the differences.

Table 12.8. Global annual scoring of AOD data composites and median fields of global modeling

Rank	Label	Overall score S_T	Season $1 - E_S$	Bias $1 - E_B$	Variability $1 - E_V$	Dataset
1	CR	0.75	0.96	0.91	0.86	enhanced composite
2	Cx	0.68	0.94	0.88	0.83	composite
3	med	− 0.59	0.88	− 0.85	0.78	median model

Table 12.9. Ocean annual scoring of AOD data composites and median fields of global modeling

Rank	Label	Overall score S_T	Season $1 - E_S$	Bias $1 - E_B$	Variability	Dataset
1	CR	0.74	0.96	0.90	0.86	enhanced composite
2	Cx	0.68	0.94	0.86	0.85	composite
3	med	− 0.60	0.89	− 0.85	0.79	median model

Table 12.10. Land annual scoring of AOD data composites and median fields of global modeling

Rank	Label	Overall score S_T	Season $1 - E_S$	Bias $1 - E_B$	Variability $1 - E_V$	Dataset
1	CR	0.77	0.97	0.93	0.85	enhanced composite
2	Cx	0.68	0.96	0.91	0.78	composite
3	med	0.57	0.88	0.86	0.76	median model

in Tables 12.6 and 12.7, global modeling tends to overestimate AOD over both land and ocean at northern mid-latitudes (outdated emission inventories?) and tends to underestimate AOD over land regions with major dust and biomass-burning sources and over remote oceanic regions of the southern hemisphere. These general modeling biases are illustrated in differences to the (data-based) enhanced composite on a seasonal basis in Fig. 12.11.

The deviations are significant, as these are not individual events but seasonal averages. There seems to be a general tendency in global modeling to overestimate contributions from urban pollution and dust and to underestimate contributions from biomass burning. Models also tend to underestimate aerosol amounts in remote regions with the lowest AOD values. The permitted uncertainty of $+/- 0.05$ in Fig.12.11, however, is too large to show this low (but – in a relative sense – significant) bias.

11.   Conclusion

When climatological aerosol data are needed, there is a temptation to adopt data from global modeling. The advantage of global modeling is that the data are complete (e.g., no data gaps) and consistent (e.g., all aerosol properties). However, there should be awareness that data, and in particular aerosol data, from global modeling are based on often poorly constrained interactions and input data. More specifically, despite the ability to distinguish between different aerosol types in advanced aerosol modules, the required ae-

seasonal co

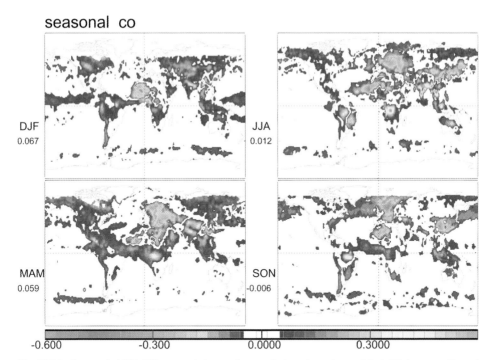

Fig. 12.11. Seasonal AOD differences between the sunphotometry enhanced (satellite) composite and median fields of global modeling. Positive differences suggest model underestimates, while negative deviations indicate model overestimates. Only AOD differences exceeding +/− 0.05 in absolute value are displayed. Annual global averages for deviations are listed below the labels.

rosol input data (e.g., emission source strength and location) and the aerosol process parameterizations (e.g., transport and removal) are highly uncertain. In fact, the resulting AOD maps in the end are usually tuned towards available AOD data from observations. Thus, to take advantage of the extra detail by global modeling (e.g., information on properties that cannot be measured) there is a demand for quality constrains by measurement-based data. In the case of aerosol, reliable global monthly maps for (mid-visible) AOD (and AOD spectral dependence) would be extremely useful. Unfortunately, individual data sources usually lack either spatial coverage (e.g., sunphotometry) or accuracy (e.g., satellite remote sensing). Thus, methods need to be explored that combine the strength of individual data sources in order to create a superior data product that can really help modeling. The enhanced AOD composite developed in this contribution is far from perfect. But it demonstrates that there are ways to more useful data-products. Thus, as an incentive to continue on that path, Fig. 12.12 displays monthly global maps of the enhanced AOD composite from remote sensing.

Acknowlegements

This study was supported by the EU-projects GEMS and EUCAARI. The help and assistance of the many remote sensing data-teams, including the AERONET staff, is gratefully acknowledged.

monthly cR

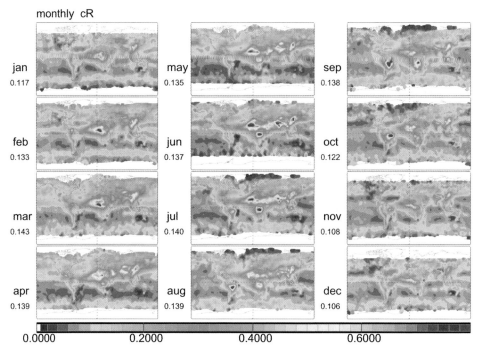

Fig. 12.12. Monthly global maps for the mid-visible aerosol optical depth based on an AERONET enhanced satellite retrievals composite.

References

Aoki, K., 2006: Aerosol optical characteristics in Asia from measurements of SKYNET sky radiometers, in *IRS 2004: Current Problems in Atmospheric Radiation*, H. Fischer and B-J. Sohn (Eds), A. Deepak, Hampton, VA, 311–313.

Deuzé, J.L., M. Herman and P. Goloub, 1999: Characterization of aerosols over ocean from POLDER/ADEOS-1, *Geophys. Res. Lett.*, **26**, 1421–1424.

Deuzé, J.L, F.M. Breon, C. Devaux, P. Goloub, M. Herman, B. Lafrance, F. Maignan, A. Marchand, F. Nadal, G. Perry and D.Tanré, 2001: Remote sensing of aerosol over land surfaces from POLDER/ADEOS-1 polarized measurements, *J. Geophys. Res.*, **106**, 4912–4926.

Diner, D.J., J.V. Martonchik, R.A. Kahn, B. Pinty, N. Gobron, D.L. Nelson, and B.N. Holben: Using angular and spectral shape similarity constraints to improve MISR aerosol and surface retrievals over land. *Rem. Sens. Environ* **94**, 155-171, 2005.

Geogdzhyev, I., M. Mishchenko, W. Rossow, B. Cairns and A. Lacis, 2002: Global 2-channel AVHRR retrieval of aerosol properties over the ocean for the period of NOAA-9 observations and preliminary retrievals using NOAA-7 and NOAA-11 data, *J. Atmos. Sci.*, **59**, 262–278.

Holben, B., T. Eck, I. Slutsker, D. Tanré, J. Buis, E. Vermote, J. Reagan, Y. Kaufman, T. Nakajima, F. Lavenau, I. Jankowiak and A. Smirnov, 1998: AERONET, a federated instrument network and data-archive for aerosol characterization, *Rem. Sens. Environ.*, **66**, 1–66.

Kahn, R., P. Banerjee, D. McDonald and D.J. Diner, 1998: Sensitivity of multi-angle imaging to aerosol optical depth and to pure particle size distribution and composition over ocean, *J. Geophys. Res.*, **103**, 32195–32213.

Kaufman, Y., D. Tanré, L. Remer, E. Vermote, D. Chu and B. Holben, 1997: Operational remote sensing of tropospheric aerosol over the land from EOS-MODIS, *J. Geophys. Res.*, **102**, 17051–17061.

Kinne, S., M. Schulz, C. Textor, S. Guibert, Y. Balkanski, S. Bauer, T. Berntsen, T. Berglen, O. Boucher, M. Chin, F. Dentener, T. Diehl, H. Feichter, D. Fillmore, S. Ghan, P. Ginoux, S. Gong, A. Grini, J. Hendricks, L. Horowitz, I. Isaksen, T. Iversen, A. Kirkevåg, S. Kloster, D. Koch, J.E. Kristjansson, M. Krol, A. Lauer, J.F. Lamarque, X. Liu, V. Montanaro, G. Myhre, J. Penner, G. Pitari, S. Reddy, Ø. Seland, P. Stier, T. Takemura and X. Tie, 2006: An AeroCom initial assessment – optical properties in aerosol component modules of global models. *Atmos. Chem. Phys.*, **6**, 1815–1834.

Liu, L. and M.I. Mishchenko, 2008: Toward unified satellite climatology of aerosol properties: direct comparisons of advanced level 2 aerosol products, *J. Quant. Spectr. Rad. Transfer*, **109**, 2376–2385, doi:10.1016/j.jqsrt.2008.05.003.

Martonchik J.V., D.J. Diner, R.A. Kahn, T.P. Ackerman, M.E. Verstraete, B. Pinty and H.R. Gordon, 1998: Techniques for the retrieval of aerosol properties over land and ocean using multi-angle imaging, *IEEE Trans. Geosci. Remote Sens.*, **36**, 1212–1227.

Remer, L.A., Y.J. Kaufman, D. Tanré, S. Mattoo, D.A. Chu, J.V. Martins, R.-R. Li, C. Ichoku, R.C. Levy, R.G. Kleidman, T.F. Eck, E. Vermote and B.N. Holben, 2005: The MODIS aerosol algorithm, products and validation, *J. Atmos. Sci., 62, 947–973.*

Schulz, M., C. Textor, S. Kinne, S. Guibert, Y. Balkanski, S. Bauer, T. Berntsen, T. Berglen, O. Boucher, M. Chin, F. Dentener, T. Diehl, H. Feichter, D. Fillmore, S. Ghan, P. Ginoux, S. Gong, A. Grini, J. Hendricks, L. Horowitz, I. Isaksen, T. Iversen, A. Kirkevåg, S. Kloster, D. Koch, J.E. Kristjansson, M. Krol, A. Lauer, J.F. Lamarque, X. Liu, V. Montanaro, G. Myhre, J. Penner, G. Pitari, S. Reddy, Ø. Seland, P. Stier, T. Takemura and X. Tie, 2006: Radiative forcing by aerosols as derived from the AeroCom present-day and pre-industrial simulations. *Atmos. Chem. Phys. Disc.*, **6**, 5095–5136.

Stowe L., A. Ignatov and R. Singh, 1997: Development, validation and potential enhancements to the second generation operational aerosol product at NOAA/NESDIS, *J. Geophys. Res.* **102**, (D14), 16923–16934.

Tanré, D., Y.J. Kaufman, M. Herman and S. Mattoo, 1991: Remote Sensing of aerosol properties over ocean using the MODIS/EOS spectral radiances, *J. Geophys. Res.*, **102**, 16971–16988.

Torres, O., P.K. Barthia, J.R. Herman, A. Sinyuk, P. Ginoux and B. Holben, 2002: A long-term record of aerosol optical depth from TOMS observations and comparisons to AERONET measurements. *J. Atmos. Sci.* **59**, 398–413.

Index

Printing: Mercedes-Druck, Berlin
Binding: Stein+Lehmann, Berlin